Philosophy and Model Theory

Philosophy and Model Theory

Tim Button and Sean Walsh
with a historical appendix by Wilfrid Hodges

OXFORD
UNIVERSITY PRESS

Great Clarendon Street, Oxford, OX2 6DP,
United Kingdom

Oxford University Press is a department of the University of Oxford.
It furthers the University's objective of excellence in research, scholarship,
and education by publishing worldwide. Oxford is a registered trade mark of
Oxford University Press in the UK and in certain other countries

First Edition published in 2018
Impression: 1

Published in the United States of America by Oxford University Press
198 Madison Avenue, New York, NY 10016, United States of America

British Library Cataloguing in Publication Data
Data available

Library of Congress Control Number: 2017959066

ISBN: 978–0–19–879039–6 (hbk.)
978–0–19–879040–2 (pbk.)

Printed and bound by
CPI Group (UK) Ltd, Croydon, CR0 4YY

Preface

Philosophy and model theory frequently meet one another. This book aims to understand their interactions.

Model theory is used in every 'theoretical' branch of analytic philosophy: in philosophy of mathematics; in philosophy of science; in philosophy of language; in philosophical logic; and in metaphysics. But these wide-ranging appeals to model theory have created a highly fragmented literature. On the one hand, many philosophically significant results are found only in mathematics textbooks: these are aimed squarely at mathematicians; they typically presuppose that the reader has a serious background in mathematics; and little clue is given as to their philosophical significance. On the other hand, the philosophical applications of these results are scattered across disconnected pockets of papers.

The first aim of our book, then, is to consider the *philosophical uses of model theory*. We state and prove the best versions of results for philosophical purposes. We then probe their philosophical significance. And we show how similar dialectical situations arise repeatedly across fragmented debates in different areas.

The second aim of our book, though, is to consider the *philosophy of model theory*. Model theory itself is rarely taken as the subject matter of philosophising (contrast this with the philosophy of biology, or the philosophy of set theory). But model theory is a beautiful part of pure mathematics, and worthy of philosophical study in its own right.

Both aims give rise to challenges. On the one hand: the philosophical uses of model theory are scattered across a disunified literature. And on the other hand: there is scarcely any literature on the philosophy of model theory.

All of which is to say: *philosophy and model theory isn't really 'a thing' yet.* This book aims to start carving out such a thing. We want to chart the rock-face and trace its dialectical contours. But we present this book, not as a final word on what philosophically inclined model theorists and model-theoretically inclined philosophers should do, but as an invitation to join in.

So. This is not a book in which a single axe is ground, page by page, into an increasingly sharp blade. No fundamental line of argument—arching from Chapter 1 through to Chapter 17—serves as the spine of the book. What knits the chapters together into a single book is not a single thesis, but a sequence of overlapping, criss-crossing themes.

Topic selection

Precisely because philosophy and model theory isn't yet 'a thing', we have had to make some difficult decisions about what topics to discuss.

On the one hand, we aimed to pick topics which should be of fairly mainstream philosophical concern. So, when it comes to the philosophical *uses* of model theory, we have largely considered topics concerning *reference*, *realism*, and *doxology* (a term we introduce in Chapter 6). But, even when we have considered questions which fall squarely within in the philosophy *of* model theory, the questions that we have focussed on are clearly instances of 'big questions'. We look at questions of *sameness* of theories/structure (Chapter 5); of taking *diverse perspectives* on the same concept (in Chapter 14); of how to draw *boundaries* of logic (in Chapter 16); and of *classification* of mathematical objects (Chapter 17).

On the other hand, we also wanted to give you a decent bang for your buck. We figured that if you were going to have to wrestle with some new philosophical idea or model-theoretic result, then you should get to see it put to decent use. (This explains why, for example, the Push-Through Construction, the just-more theory manoeuvre, supervaluational semantics, and the ideas of moderation and modelism, occur so often in this book.) Conversely, we have had to set aside debates—no matter how interesting—which would have taken too long to set up.

All of which is to say: this book is not comprehensive. Not even close.

Although we consider models of set theory in Chapters 8 and 11, we scarcely scratch the surface. Although we discuss infinitary logics in Chapters 15–16, we only use them in fairly limited ways.[1] Whilst we mention Tennenbaum's Theorem in Chapter 7, that is as close as we get to computable model theory. We devote only one brief section to o-minimality, namely §4.10. And although we consider quantifiers in Chapter 16 and frequently touch on issues concerning logical consequence, we never address the latter topic head on.[2]

The grave enormity, though, is that we have barely scratched the surface of model theory itself. As the table of contents reveals, the vast majority of the book considers model theory as it existed before Morley's Categoricity Theorem.

Partially correcting for this, Wilfrid Hodges' wonderful historical essay appears as Part D of this book. Wilfrid's essay treats Morley's Theorem as a pivot-point for the subject of model theory. He looks back critically to the history, to uncover the notions at work in Morley's Theorem and its proof, and he looks forward to the riches that have followed from it. We have both learned so much from Wilfrid's *Model Theory*,[3] and we are delighted to include his 'short history' here.

---

[1] We do not, for example, consider the connection between infinitary logics and supervenience, as Glanzberg (2001) and Bader (2011) do.

[2] For that, we would point the reader to Blanchette (2001) and Shapiro (2005a).

[3] Hodges (1993).

Still, concerning all those topics which we have omitted: we intend no slight against them. But there is only so much one book can do, and this book is already much (much) longer than we originally planned. We are sincere in our earlier claim, that this book is not offered as a final word, but as an invitation to take part.

Structuring the book

Having selected our topics, we needed to arrange them into a book. At this point, we realised that these topics have no natural linear ordering.

As such, we have tried to strike a balance between three aims that did not always point in the same direction: to order by *philosophical theme*, to order by increasing *philosophical sophistication*, and to order by increasing *mathematical sophistication*. The book's final structure of represents our best compromise between these three aims. It is divided into three main parts: *Reference and realism*, *Categoricity*, and *Indiscernibility and classification*.

Each part has an introduction, and those who want to dip in and out of particular topics, rather than reading cover-to-cover, should read the three part-introductions after they have finished reading this preface. The part-introductions provide thematic overviews of each chapter, and they also contain diagrams which depict the dependencies between each section of the book. In combination with the table of contents, these diagrams will allow readers to take shortcuts to their favourite destinations, without having to stop to smell every rose along the way.

Presuppositions and proofs

So far as possible, the book assumes only that you have completed a 101-level logic course, and so have some familiarity with first-order logic.

Inevitably, there are some exceptions to this: we were forced to assume some familiarity with analysis when discussing infinitesimals in Chapter 4, and equally some familiarity with topology when discussing Stone spaces in Chapter 14. We do not prove Gödel's incompleteness results, although we do state versions of them in §5.A. A book can only be so self-contained.

By and large, though, this book *is* self-contained. When we invoke a model-theoretic notion, we almost always define the notion formally in the text. When it comes to proofs, we follow these rules of thumb.

The *main text* includes both brief proofs, and also those proofs which we wanted to discuss directly.

The *appendices* include proofs which we wanted to include in the book, but which were too long to feature in the main text. These include: proofs concerning elementary topics which our readers should come to understand (at least one day); proofs which are difficult to access in the existing literature; proofs of certain folk-lore results; and proofs of new results.

But the *book omits* all proofs which are both readily accessed and too long to be self-contained. In such cases, we simply provide readers with citations.

The quick moral for readers to extract is this. If you encounter a proof in the main text of a chapter, you should follow it through. But we would add a note for readers whose primary background is in philosophy. If you really want to understand a mathematical concept, you need to see it in action. Read the appendices!

Acknowledgements

The book arose from a seminar series on philosophy and model theory that we ran in Birkbeck in Autumn 2011. We turned the seminar into a paper, but it was vastly too long. An anonymous referee for *Philosophia Mathematica* suggested the paper might form the basis for a book. So it did.

We have presented topics from this book several times. It would not be the book it is, without the feedback, questions and comments we have received. So we owe thanks to: an anonymous referee for *Philosophia Mathematica*, and James Studd for OUP; and to Sarah Acton, George Anegg, Andrew Arana, Bahram Assadian, John Baldwin, Kyle Banick, Neil Barton, Timothy Bays, Anna Bellomo, Liam Bright, Chloé de Canson, Adam Caulton, Catrin Campbell-Moore, John Corcoran, Radin Dardashti, Walter Dean, Natalja Deng, William Demopoulos, Michael Detlefsen, Fiona Doherty, Cian Dorr, Stephen Duxbury, Sean Ebels-Duggan, Sam Eklund, Hartry Field, Branden Fitelson, Vera Flocke, Salvatore Florio, Peter Fritz, Michael Gabbay, Haim Gaifman, J. Ethan Galebach, Marcus Giaquinto, Peter Gibson, Tamara von Glehn, Owen Griffiths, Emmylou Haffner, Bob Hale, Jeremy Heis, Will Hendy, Simon Hewitt, Kate Hodesdon, Wilfrid Hodges, Luca Incurvati, Douglas Jesseph, Nicholas Jones, Peter Koellner, Brian King, Eleanor Knox, Johannes Korbmacher, Arnold Koslow, Hans-Christoph Kotzsch, Greg Lauro, Sarah Lawsky, Øystein Linnebo, Yang Liu, Pen Maddy, Kate Manion, Tony Martin, Guillaume Massas, Vann McGee, Toby Meadows, Richard Mendelsohn, Christopher Mitsch, Stella Moon, Adrian Moore, J. Brian Pitts, Jonathan Nassim, Fredrik Nyseth, Sara Parhizgari, Charles Parsons, Jonathan Payne, Graham Priest, Michael Potter, Hilary Putnam, Paula Quinon, David Rabouin, Erich Reck, Sam Roberts, Marcus Rossberg, J. Schatz, Gil Sagi, Bernhard Salow, Chris Scambler, Thomas Schindler, Dana Scott, Stewart Shapiro, Gila Sher, Lukas Skiba, Jönne Speck, Sebastian Speitel, Will Stafford, Trevor Teitel, Robert Trueman, Jouko Väänänen, Kai Wehmeier, J. Robert G. Williams, John Wigglesworth, Hugh Woodin, Jack Woods, Crispin Wright, Wesley Wrigley, and Kino Zhao.

We owe some special debts to people involved in the original Birkbeck seminar.

First, the seminar was held under the auspices of the Department of Philosophy at Birkbeck and Øystein Linnebo's European Research Council-funded project 'Plurals, Predicates, and Paradox', and we are very grateful to all the people from the project and the department for participating and helping to make the seminar possible. Second, we were lucky to have several great external speakers visit the seminar, whom we would especially like to thank. The speakers were: Timothy Bays, Walter Dean, Volker Halbach, Leon Horsten, Richard Kaye, Jeff Ketland, Angus Macintyre, Paula Quinon, Peter Smith, and J. Robert G. Williams. Third, many of the external talks were hosted by the Institute of Philosophy, and we wish to thank Barry C. Smith and Shahrar Ali for all their support and help in this connection.

A more distant yet important debt is owed to Denis Bonnay, Brice Halimi, and Jean-Michel Salanskis, who organised a lovely event in Paris in June 2010 called 'Philosophy and Model Theory.' That event got some of us first thinking about 'Philosophy and Model Theory' as a unified topic.

We are also grateful to various editors and publishers for allowing us to reuse previously published material. Chapter 5 draws heavily on Walsh 2014, and the copyright is held by the Association for Symbolic Logic and is being used with their permission. Chapters 7–11 draw heavily upon on Button and Walsh 2016, published by *Philosophia Mathematica*. Finally, §13.7 draws from Button 2016b, published by *Analysis*, and §15.1 draws from Button 2017, published by the *Notre Dame Journal of Formal Logic*.

Finally, though, a word from us, as individuals.

From Tim. I want to offer my deep thanks to the Leverhulme Trust: their funding, in the form of a Philip Leverhulme Prize (PLP–2014–140), enabled me to take the research leave necessary for this book. But I mostly want to thank two very special people. Without Sean, this book could not be. And without my Ben, I could not be.

From Sean. I want to thank the Kurt Gödel Society, whose funding, in the form of a Kurt Gödel Research Prize Fellowship, helped us put on the original Birkbeck seminar. I also want to thank Tim for being a model co-author and a model friend. Finally, I want to thank Kari for her complete love and support.

Contents

A

Reference and realism

Introduction to Part A

The two central themes of Part A are *reference* and *realism*.

Here is an old philosophical chestnut: *How do we (even manage to) represent the world?* Our most sophisticated representations of the world are perhaps linguistic. So a specialised—but still enormously broad—version of this question is: *How do words (even manage to) represent things?*

Enter model theory. One of the most basic ideas in model theory is that a structure assigns interpretations to bits of vocabulary, and in such a way that we can make excellent sense of the idea that the structure makes each sentence (in that vocabulary) either true or false. Squint slightly, and model theory seems to be providing us with a perfectly precise, formal way to understand certain aspects of linguistic representation. It is no surprise at all, then, that almost any philosophical discussion of linguistic representation, or reference, or truth, ends up invoking notions which are recognisably model-theoretic.

In Chapter 1, we introduce the building blocks of model theory: the notions of signature, structure, and satisfaction. Whilst the bare technical bones should be familiar to anyone who has covered a 101-level course in mathematical logic, we also discuss the philosophical question: *How should we best understand quantifiers and variables?* Here we see that philosophical issues arise at the very outset of our model-theoretic investigations. We also introduce second-order logic and its various semantics. While second-order logic is less commonly employed in contemporary model theory, it is employed frequently in philosophy of model theory, and understanding the differences between its various semantics will be important in many subsequent chapters.

In Chapter 2, we examine various concerns about the determinacy of reference and so, perhaps, the determinacy of our representations. Here we encounter famous arguments from Benacerraf and Putnam, which we explicate using the formal Push-Through Construction. Since isomorphic structures are elementarily equivalent—that is, they make exactly the same sentences true and false—this threatens the conclusion that it is radically indeterminate, which of many isomorphic structures accurately captures how language represents the world.

Now, one might think that the reference of our word 'cat' is constrained by the causal links between cats and our uses of that word. Fair enough. But there are no causal links between mathematical objects and mathematical words. So, on certain conceptions of what humans are like, we will be unable to answer the question: *How do we (even manage to) refer to any particular mathematical entity?* That is, we will have to accept that we *do not* refer to particular mathematical entities.

Whilst discussing these issues, we introduce Putnam's famous *just-more-theory manoeuvre*. It is important to do this both clearly and early, since many versions of this dialectical move occur in the philosophical literature on model theory. Indeed, they occur especially frequently in Part B of this book.

Now, philosophers have often linked the topic of reference to the topic of realism. One way to draw the connection is as follows: If reference is radically indeterminate, then my word 'cabbage' and my word 'cat' fail to pick out anything determinately. So when I say something like 'there is a cabbage and there is a cat', I have *at best* managed to say that there are at least two distinct objects. That seems to fall far short of expressing any real commitment to *cats* and *cabbages* themselves.[1] In short, radical referential indeterminacy threatens to undercut certain kinds of realism altogether. But only certain kinds: we close Chapter 2 by suggesting that some versions of mathematical platonism can live with the fact that mathematical language is radically referentially indeterminate by embracing a supervaluational semantics.

Concerns about referential indeterminacy also feature in discussions about realism within the philosophy of science. In Chapter 3, we examine a particular version of scientific realism that arises by considering Ramsey sentences. Roughly, these are sentences where all the 'theoretical vocabulary' has been 'existentially quantified away'. Ramsey sentences seem promising, since they seem to incur a kind of existential commitment to theoretical entities, which is characteristic of realism, whilst making room for a certain level referential indeterminacy. We look at the relation between Newman's objection and the Push-Through Construction of Chapter 2, and between Ramsey sentences and various model-theoretic notions of conservation. Ultimately, by combining the Push-Through Construction with these notions of conservation, we argue that the dialectic surrounding Newman's objection should track the dialectic of Chapter 2, surrounding Putnam's permutation argument in the philosophy of mathematics.

The notions of conservation we introduce in Chapter 3 are crucial to Abraham Robinson's attempt to use model theory to salvage Leibniz's notion of an 'infinitesimal'. Infinitesimals are quantities whose absolute value is smaller than that of any given positive real number. They were an important part of the historical calculus; they fell from grace with the rise of ε–δ notation; but they were given a new lease of life within model theory via Robinson's non-standard analysis. This is the topic of Chapter 4. Here we introduce the idea of *compactness* to prove that the use of infinitesimals is consistent.

Robinson believed that this vindicated the viability of the Leibnizian approach to the calculus. Against this, Bos has questioned whether Robinson's non-standard analysis is genuinely faithful to Leibniz's mathematical practice. In Chapter 4, we

[1] Cf. Putnam (1977: 491) and Button (2013: 59–60).

offer a novel defence of Robinson. By building valuations into Robinson's model theory, we prove new results which allow us to approximate more closely what we know about the Leibnizian conception of the structure of the infinitesimals. Indeed, we show that Robinson's non-standard analysis can rehabilitate various historical methods for reasoning with and about infinitesimals that have fallen far from fashion.

The question remains, of whether we should *believe* in infinitesimals. Leibniz himself was tempted to treat his infinitesimals as 'convenient fictions'; Robinson explicitly regarded his infinitesimals in the same way; and their method of introduction in model theory allows for perhaps the cleanest possible version of a fictionalist-cum-instrumentalist attitude towards 'troublesome' entities. Indeed, we can prove that reasoning *as if* there are infinitesimals will only generate results that one could have obtained *without* that assumption. One can have anti-realism, then, with a clear conscience.

In Chapter 5, we take a step back from these specific applications of model theory, to discuss a more methodological question about the philosophical application of model theory: *under what circumstances should we call two structures 'the same'?* This question can be posed within mathematics, where its answer will depend upon the similarities and differences that matter for the mathematical purposes at hand. But the question can also be given a metaphysical gloss. In particular, consider a philosopher who thinks (for example) that: (a) there is a *single*, abstract, entity which is 'the natural number structure', and that (b) there is a *single*, abstract entity which is 'the structure of the integers'; but that (c) these two entities are distinct. Then this philosopher must provide an account of identity and distinctness between 'structures', so construed; and we show just how hard this is.

Notions of sameness of structure also induce notions of sameness of theory. After surveying a wide variety of formal notions of sameness of structure and theory, we discuss three ambitious claims concerning what sameness of theory preserves, namely: truth; arithmetical provability; and proof. We conclude that more philosophically ambitious versions of these preservation-theses generally fail.

This meta-issue of sameness of structure and theory is a good place to end Part A, though, both because (a) the discussion is enhanced by the specific examples of structures and theories discussed earlier in the text, and because (b) questions about sameness of structure and theory inform a number of the discussions and debates which we treat in later Parts of the book.

Readers who only want to dip into particular topics of Part A can consult the following Hasse diagram of dependencies between the sections of Part A, whilst referring to the table of contents. A section y depends upon a section x iff there is a path leading downwards from x to y. So, a reader who wants to get straight to the discussion of fictionalism about infinitesimals will want to leap straight to §4.7; but they should know that this section assumes a prior understanding of §§2.1, 4.1, 4.2, and much (but not all) of Chapter 1. (We omit purely technical appendices from this diagram.)

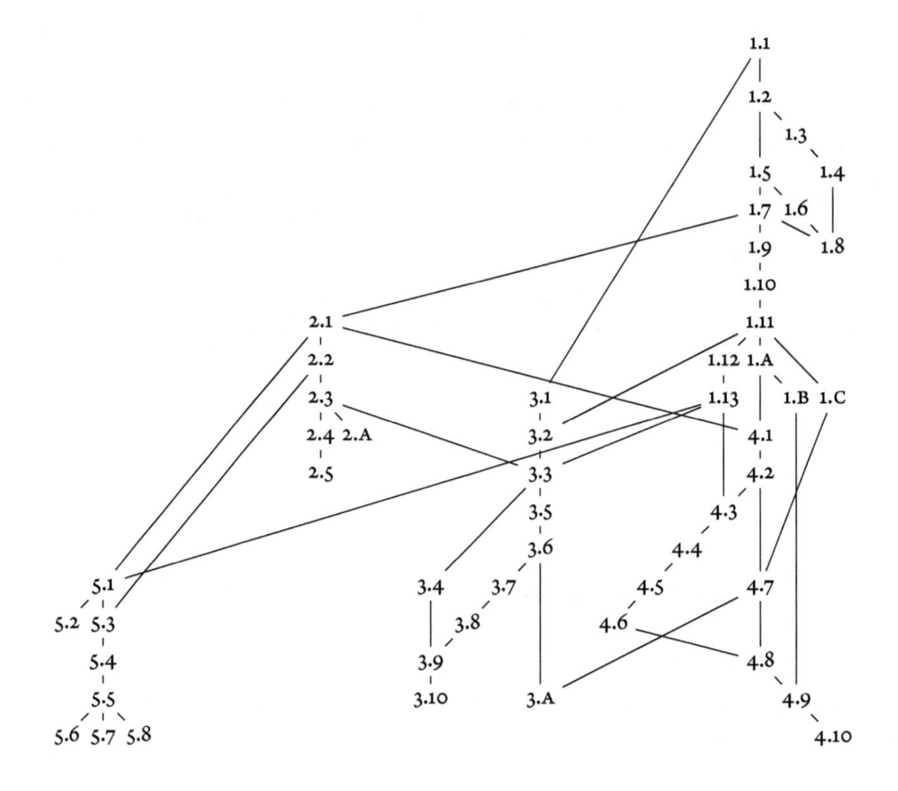

1

Logics and languages

Model theory begins by considering the relationship between languages and structures. This chapter outlines the most basic aspects of that relationship.

One purpose of the chapter will therefore be immediately clear: we want to lay down some fairly dry, technical preliminaries. Readers with some familiarity with mathematical logic should feel free to skim through these technicalities, as there are no great surprises in store.

Before the skimming commences, though, we should flag a second purpose of this chapter. There are at least three rather different approaches to the semantics for formal languages. In a straightforward sense, these approaches are technically equivalent. Most books simply choose one of them without comment. We, however, lay down all three approaches and discuss their comparative strengths and weaknesses. Doing this highlights that there are philosophical discussions to be had from the get-go. Moreover, by considering what is invariant between the different approaches, we can better distinguish between the merely idiosyncratic features of a particular approach, and the things which really matter.

One last point, before we get going: tradition demands that we issue a caveat. Since Tarski and Quine, philosophers have been careful to emphasise the important distinction between *using* and *mentioning* words. In philosophical texts, that distinction is typically flagged with various kinds of quotation marks. But within model theory, context almost always disambiguates between use and mention. Moreover, including too much punctuation makes for ugly text. With this in mind, we follow model-theoretic practice and avoid using quotation marks except when they will be especially helpful.

1.1 Signatures and structures

We start with the idea that formal languages can have primitive vocabularies:

Definition 1.1: A signature, $\mathscr{L}$, *is a set of symbols, of three basic kinds: constant symbols, relation symbols, and function symbols. Each relation symbol and function symbol has an associated number of* places *(a natural number), so that one may speak of an n-place relation or function symbol.*

Throughout this book, we use script fonts for signatures. Constant symbols should be thought of as *names* for entities, and we tend to use c_1, c_2, etc. Relation symbols, which are also known as predicates, should be thought of as picking out *properties* or *relations*. A two-place relation, such as *x is smaller than y*, must be associated with a two-place relation symbol. We tend to use R_1, R_2, etc. for relation symbols. Function symbols should be thought of as picking out functions and, again, they need an associated number of places: the function of *multiplication on the natural numbers* takes two natural numbers as inputs and outputs a single natural number, so we must associate that function with a two-place function symbol. We tend to use f_1, f_2, etc. for function symbols.

The examples just given—*being smaller than*, and *multiplication on the natural numbers*—suggest that we will use our formal vocabulary to make determinate claims about certain objects, such as people or numbers. To make this precise, we introduce the notion of an $\mathscr{L}$*-structure*; that is, a structure whose signature is $\mathscr{L}$. An $\mathscr{L}$-structure, $\mathcal{M}$, is an underlying domain, M, together with an assignment of $\mathscr{L}$'s constant symbols to elements of M, of $\mathscr{L}$'s relation symbols to relations on M, and of $\mathscr{L}$'s function symbols to functions over M. We always use calligraphic fonts $\mathcal{M}, \mathcal{N}, \ldots$ for structures, and $M, N, \ldots$ for their underlying domains. Where s is any $\mathscr{L}$-symbol, we say that $s^{\mathcal{M}}$ is the object, relation or function (as appropriate) assigned to s in the structure $\mathcal{M}$. This informal explanation of an $\mathscr{L}$-structure is always given a set-theoretic implementation, leading to the following definition:

Definition 1.2: *An $\mathscr{L}$-*structure*, $\mathcal{M}$, consists of:*

- *a non-empty set, M, which is the underlying domain of $\mathcal{M}$,*
- *an object $c^{\mathcal{M}} \in M$ for each constant symbol c from $\mathscr{L}$,*
- *a relation $R^{\mathcal{M}} \subseteq M^n$ for each n-place relation symbol R from $\mathscr{L}$, and*
- *a function $f^{\mathcal{M}} : M^n \longrightarrow M$ for each n-place function symbol f from $\mathscr{L}$.*

As is usual in set theory, M^n is just the set of n-tuples over M, i.e.:[1]

$$M^n = \{(a_1, \ldots, a_n) : a_1 \in M \text{ and } \ldots \text{ and } a_n \in M\}$$

Likewise, we implement a function $g : M^n \longrightarrow M$ in terms of its set-theoretic graph. That is, g will be a subset of M^{n+1} such that if $(x_1, \ldots, x_n, y)$ and $(x_1, \ldots, x_n, z)$ are elements of g then $y = z$ and such that for every $(x_1, \ldots, x_n)$ in M^n there is y in M such that $(x_1, \ldots, x_n, y)$ is in g. But we continue to think about functions in the normal way, as maps sending n-tuples of the domain, M^n, to elements of the co-domain, M, so tend to write $(x_1, \ldots, x_n, y) \in g$ just as $g(x_1, \ldots, x_n) = y$.

[1] The full definition of X^n is by recursion: $X^1 = X$ and $X^{n+1} = X^n \times X$, where $A \times B = \{(a, b) : a \in A \text{ and } b \in B\}$. Likewise, we recursively define ordered n-tuples in terms of ordered pairs by setting e.g. $(a, b, c) = ((a, b), c)$.

Given the set-theoretic background, $\mathscr{L}$-structures are individuated *extensionally*: they are identical iff they have exactly the same underlying domain and make exactly the same assignments. So, where $\mathcal{M}, \mathcal{N}$ are $\mathscr{L}$-structures, $\mathcal{M} = \mathcal{N}$ iff both $M = N$ and $s^{\mathcal{M}} = s^{\mathcal{N}}$ for all s from $\mathscr{L}$. To obtain different structures, then, we can either change the domain, change the interpretation of some symbol(s), or both. Structures are, then, individuated rather finely, and indeed we will see in Chapters 2 and 5 that this individuation is too fine for many purposes. But for now, we can simply observe that there are many, *many* different structures, in the sense of Definition 1.2.

1.2 First-order logic: a first look

We know what ($\mathscr{L}$-)structures are. To move to the idea of a *model*, we need to think of a structure as making certain sentences true or false. So we must build up to the notion of a sentence. We start with their syntax.

Syntax for first-order logic

Initially, we restrict our attention to *first-order sentences*. These are the sentences we obtain by adding a basic starter-pack of logical symbols to a signature (in the sense of Definition 1.1). These logical symbols are:

- variables: u, v, w, x, y, z, with numerical subscripts as necessary
- the identity sign: $=$
- a one-place sentential connective: $\neg$
- two-place sentential connectives: $\wedge$, $\vee$
- quantifiers: $\exists$, $\forall$
- brackets: (,)

We now offer a recursive definition of the syntax of our language:[2]

Definition 1.3: *The following, and nothing else, are first-order* $\mathscr{L}$-terms*:*

- *any variable, and any constant symbol c from $\mathscr{L}$*

[2] A pedantic comment is in order. The symbols 't_1' and 't_2' are not being used here as expressions in the object language (i.e. first-order logic with signature $\mathscr{L}$). Rather, they are being used as expressions of the metalanguage, within which we describe the syntax of first-order $\mathscr{L}$-terms and $\mathscr{L}$-formulas. Similarly, the symbol 'x', as it occurs in the last clause of Definition 1.3, is not being used as an expression of the object language, but in the metalanguage. So the final clause in this definition should be read as saying something like this. *For any variable and any formula φ which does not already contain a concatenation of a quantifier followed by that variable, the following concatenation is a formula: a quantifier, followed by that variable, followed by φ.* (The reason for this clause is to guarantee that e.g. $\exists v \forall v F(v)$ is not a formula.) We could flag this more explicitly, by using a different font for metalinguistic variables (for example). However, as with flagging quotation, we think the additional precision is not worth the ugliness.

- $f(t_1, \ldots, t_n)$, *for any* $\mathscr{L}$*-terms* $t_1, \ldots, t_n$ *and any n-place function symbol* f *from* $\mathscr{L}$

The following, and nothing else, are first-order $\mathscr{L}$-formulas*:*

- $t_1 = t_2$, *for any* $\mathscr{L}$*-terms* t_1 *and* t_2
- $R(t_1, \ldots, t_n)$, *for any* $\mathscr{L}$*-terms* $t_1, \ldots, t_n$ *and any n-place relation symbol* R *from* $\mathscr{L}$
- $\neg\varphi$, *for any* $\mathscr{L}$*-formula* φ
- $(\varphi \wedge \psi)$ *and* $(\varphi \vee \psi)$, *for any* $\mathscr{L}$*-formulas* φ *and* ψ
- $\exists x\varphi$ *and* $\forall x\varphi$, *for any variable* x *and any* $\mathscr{L}$*-formula* φ *which contains neither of the expressions* $\exists x$ *nor* $\forall x$.

Formulas of the first two sorts—i.e. terms appropriately concatenated either with the identity sign or an $\mathscr{L}$*-predicate—are called* atomic $\mathscr{L}$-formulas.

As is usual, for convenience we add two more sentential connectives, $\rightarrow$ and $\leftrightarrow$, with their usual abbreviations. So, $(\varphi \rightarrow \psi)$ abbreviates $(\neg\varphi \vee \psi)$, and $(\varphi \leftrightarrow \psi)$ abbreviates $((\varphi \rightarrow \psi) \wedge (\psi \rightarrow \varphi))$. We will also use some extremely common bracketing conventions to aid readability, so we sometimes use square brackets rather than rounded brackets, and we sometimes omit brackets where no ambiguity can arise.

We say that a variable is *bound* if it occurs within the scope of a quantifier, i.e. we have something like $\exists x(\ldots x \ldots)$. A variable is *free* if it is not bound. We now say that an $\mathscr{L}$*-sentence* is an $\mathscr{L}$-formula containing no free variables. When we want to draw attention to the fact that some formula φ has certain free variables, say x and y, we tend to do this by writing the formula as $\varphi(x, y)$. We say that $\varphi(x, y)$ is a formula *with free variables displayed* iff x and y are the *only* free variables in φ. When we consider a sequence of n-variables, such as $v_1, \ldots, v_n$, we usually use overlining to write this more compactly, as $\overline{v}$, leaving it to context to determine the number of variables in the sequence. So if we say '$\varphi(\overline{x})$ is a formula with free variables displayed', we mean that all and only its free variables are in the sequence $\overline{x}$. We also use similar overlining for other expressions. For example, we could have phrased part of Definition 1.3 as follows: $f(\overline{t})$ is a term whenever each entry in $\overline{t}$ is an $\mathscr{L}$-term and f is a function symbol from $\mathscr{L}$.

Semantics: the trouble with quantifiers

We now understand the syntax of first-order sentences. Later, we will consider logics with a more permissive syntax. But first-order logic is something like the *default*, for both philosophers and model theorists. And our next task is to understand its *semantics*. Roughly, our aim is to define a relation, $\vDash$, which obtains between a structure and a sentence just in case (intuitively) the sentence is true in the structure. In fact, there are many different but extensionally equivalent approaches to defining this relation, and we will consider three in this chapter.

To understand why there are several different approaches to the semantics for first-order logic, we must see why the most obvious approach fails. Our sentences have a nice, recursive syntax, so we will want to provide them with a nice, recursive semantics. The most obvious starting point is to supply semantic clauses for the two kinds of atomic sentence, as follows:

$$\mathcal{M} \vDash t_1 = t_2 \text{ iff } t_1^{\mathcal{M}} = t_2^{\mathcal{M}}$$
$$\mathcal{M} \vDash R(t_1, \ldots, t_n) \text{ iff } (t_1^{\mathcal{M}}, \ldots, t_n^{\mathcal{M}}) \in R^{\mathcal{M}}$$

Next, we would need recursion clauses for the quantifier-free sentences. So, writing $\mathcal{M} \nvDash \varphi$ for *it is not the case that* $\mathcal{M} \vDash \varphi$, we would offer:

$$\mathcal{M} \vDash \neg\varphi \text{ iff } \mathcal{M} \nvDash \varphi$$
$$\mathcal{M} \vDash (\varphi \wedge \psi) \text{ iff } \mathcal{M} \vDash \varphi \text{ and } \mathcal{M} \vDash \psi$$

So far, so good. But the problem arises with the quantifiers. Where the notation $\varphi(c/x)$ indicates the formula obtained by replacing every instance of the free variable x in $\varphi(x)$ with the constant symbol c, an obvious thought would be to try:

$$\mathcal{M} \vDash \forall x\varphi(x) \text{ iff } \mathcal{M} \vDash \varphi(c/x) \text{ for every constant symbol } c \text{ from } \mathcal{L}$$

Unfortunately, *this recursion clause is inadequate.* To see why, suppose we had a very simple signature containing a single one-place predicate R and *no* constant symbols. Then, for any structure $\mathcal{M}$ in that signature, we would *vacuously* have that $\mathcal{M} \vDash \forall vR(v)$. But this would be the case even if $R^{\mathcal{M}} = \varnothing$, that is, even if *nothing* had the property picked out by R. Intuitively, that is the wrong verdict.

The essential difficulty in defining the semantics for first-order logic therefore arises when we confront quantifiers. The three approaches to semantics which we consider present three ways to overcome this difficulty.

Why it is worth considering different approaches

In a straightforward sense, the three approaches are technically equivalent. So most books simply adopt one of these approaches, without comment, and get on with other things. In deciding to present all three approaches here, we seem to be trebling our reader's workload. So we should pause to explain our decision.

First: the three approaches to semantics are so intimately related, at a technical level, that the workload is probably only *doubled*, rather than trebled.

Second: readers who are happy ploughing through technical definitions will find nothing very tricky here. And such readers should find that the additional technical investment gives a decent philosophical pay-off. For, as we move through the chapter, we will see that these (quite dry) technicalities can both generate and resolve philosophical controversies.

Third: we expect that even novice philosophers reading this book will have at least a rough and ready idea of what is coming next. And such readers will be better served by reading (and perhaps only partially absorbing) multiple *different* approaches to the semantics for first-order logic, than by trying to rote-learn one *specific* definition. They will thereby get a sense of what is important to supplying a semantics, and what is merely an idiosyncratic feature of a particular approach.

1.3 The Tarskian approach to semantics

We begin with the Tarskian approach.[3] Recall that the 'obvious' semantic clauses fail because $\mathcal{L}$ may not contain enough constant symbols. The Tarskian approach handles this problem by assigning interpretations to the *variables* of the language. In particular, where $\mathcal{M}$ is any $\mathcal{L}$-structure, a *variable-assignment* is any function σ from the set of variables to the underlying domain M. We then define satisfaction with respect to pairs of structures with variable-assignments.

To do this, we must first specify how the structure / variable-assignment pair determines the behaviour of the $\mathcal{L}$-terms. We do this by recursively defining an element $t^{\mathcal{M},\sigma}$ of M for a term t with free variables among $x_1, \ldots, x_n$ as follows:

$$t^{\mathcal{M},\sigma} = \sigma(x_i), \text{ if } t \text{ is the variable } x_i$$
$$t^{\mathcal{M},\sigma} = f^{\mathcal{M}}(s_1^{\mathcal{M},\sigma}, \ldots, s_k^{\mathcal{M},\sigma}), \text{ if } t \text{ is the term } f(s_1, \ldots, s_k)$$

To illustrate this definition, suppose that $\mathcal{M}$ is the natural numbers in the signature $\{0, 1, +, \times\}$, with each symbol interpreted as normal. (This licenses us in dropping the '$\mathcal{M}$'-superscript when writing the symbols.) Suppose that σ and τ are variable-assignments such that $\sigma(x_1) = 5$, $\sigma(x_2) = 7$, $\tau(x_1) = 3$, $\tau(x_2) = 7$, and consider the term $t(x_1, x_2) = (1 + x_1) \times (x_1 + x_2)$. Then we can compute the interpretation of the term relative to the variable-assignments as follows:

$$t^{\mathcal{M},\sigma} = (1 + x_1^{\mathcal{M},\sigma}) \times (x_1^{\mathcal{M},\sigma} + x_2^{\mathcal{M},\sigma}) = (1 + 5) \times (5 + 7) = 72$$
$$t^{\mathcal{M},\tau} = (1 + x_1^{\mathcal{M},\tau}) \times (x_1^{\mathcal{M},\tau} + x_2^{\mathcal{M},\tau}) = (1 + 3) \times (3 + 7) = 40$$

We next define the notion of satisfaction relative to a variable-assignment:

---

[3] See Tarski (1933) and Tarski and Vaught (1958), but also §12.A.

$$\mathcal{M}, \sigma \vDash t_1 = t_2 \text{ iff } t_1^{\mathcal{M},\sigma} = t_2^{\mathcal{M},\sigma}, \text{for any } \mathcal{L}\text{-terms } t_1, t_2$$
$$\mathcal{M}, \sigma \vDash R(t_1, \ldots, t_n) \text{ iff } (t_1^{\mathcal{M},\sigma}, \ldots, t_n^{\mathcal{M},\sigma}) \in R^{\mathcal{M}}, \text{for any } \mathcal{L}\text{-terms } t_1, \ldots, t_n$$
$$\text{and any } n\text{-place relation symbol } R \text{ from } \mathcal{L}$$
$$\mathcal{M}, \sigma \vDash \neg\varphi \text{ iff } \mathcal{M}, \sigma \nvDash \varphi$$
$$\mathcal{M}, \sigma \vDash (\varphi \wedge \psi) \text{ iff } \mathcal{M}, \sigma \vDash \varphi \text{ and } \mathcal{M}, \sigma \vDash \psi$$
$$\mathcal{M}, \sigma \vDash \forall x\varphi(x) \text{ iff } \mathcal{M}, \tau \vDash \varphi(x) \text{ for every variable-assignment } \tau$$
$$\text{which agrees with } \sigma \text{ except perhaps on the value of } x$$

We leave it to the reader to formulate clauses for disjunction and existential quantification. Finally, where φ is any first-order $\mathcal{L}$*-sentence,* we say that $\mathcal{M} \vDash \varphi$ iff $\mathcal{M}, \sigma \vDash \varphi$ for all variable-assignments σ.

1.4 Semantics for variables

The Tarskian approach is technically flawless. However, the apparatus of variable-assignments raises certain philosophical issues.

A variable-assignment effectively gives variables a particular interpretation. In that sense, variables are treated rather like names (or constant symbols). However, when we encounter the clause for a quantifier binding a variable, we allow ourselves to consider all of the *other* ways that the bound variable might have been interpreted. In short, the Tarskian approach treats variables as something like *varying names.*

This gives rise to a philosophical question: *should* we regard variables as varying names? With Quine, our answer is *No*: 'the "variation" connoted [by the word "variable"] belongs to a vague metaphor which is best forgotten.'[4]

To explain why we say this, we begin with a simple observation. A Tarskian variable-assignment may assign different semantic values to the formulas $x > 0$ and $y > 0$. But, on the face of it, that seems mistaken. As Fine puts the point, using one variable rather than the other 'would appear to be as clear a case as any of a mere "conventional" or "notational" difference; the difference is merely in the choice of the symbol and not in its linguistic function.'[5] And this leads Fine to say:

(a) 'Any two variables (ranging over a given domain of objects) have the same semantic role.'

---

[4] Quine (1981: §12). For ease of reference, we cite the 1981-edition. However, the relevant sections are entirely unchanged from the (first) 1940-edition. We owe several people thanks for discussion of material in this section. Michael Potter alerted us to Bourbaki's notation; Kai Wehmeier alerted us to Quine's (cf. Wehmeier forthcoming); and Robert Trueman suggested that we should connect all of this to Fine's antinomy of the variable.

[5] Fine (2003: 606, 2007: 7), for this and all subsequent quotes from Fine.

But, as Fine notes, this cannot be right either. For, 'when we consider the semantic role of the variables in the same expression—such as "$x > y$"—then it seems equally clear that their semantic role is different.' So Fine says:

(b) 'Any two variables (ranging over a given domain of objects) have a different semantic role.'

And now we have arrived at Fine's *antinomy of the variable.*

We think that this whole antinomy gets going from the mistaken assumption that we can assign a 'semantic role' to a variable in isolation from the quantifier which binds it.[6] As Quine put the point more than six decades before Fine: 'The variables [...] serve merely to indicate cross-references to various positions of quantification.'[7] Quine's point is that $\exists x \forall y \varphi(x, y)$ and $\exists y \forall x \varphi(y, x)$ are indeed just typographical variants, but that both are importantly different from $\forall x \exists y \varphi(x, y)$. And to illustrate this graphically, Quine notes that we could use a notation which abandons typographically distinct variables altogether. For example, instead of writing:

$$\exists x \forall y((\varphi(x, y) \wedge \exists z \varphi(x, z)) \rightarrow \varphi(y, x))$$

we might have written:[8]

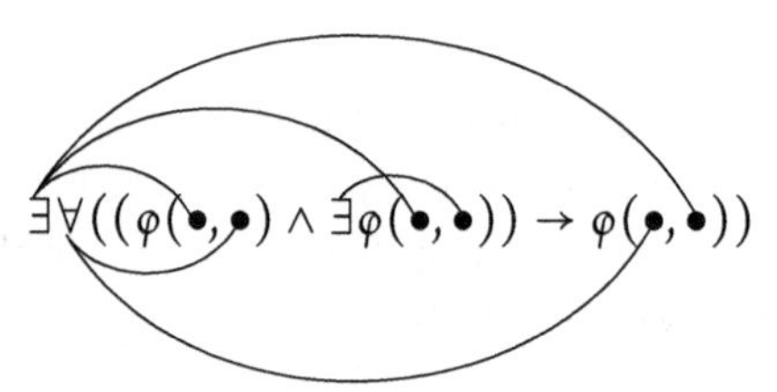

Bourbaki rigorously developed Quine's brief notational suggestion.[9] And the resulting *Quine–Bourbaki notation* is evidently just as expressively powerful as our ordinary notation. However, if we adopt the Quine–Bourbaki notation, then we will not even be able to *ask* whether typographically distinct variables like 'x' and 'y' have different 'semantic roles', and Fine's antinomy will dissolve away.[10]

[6] Fine (2003: 610–14, 2007: 12–16) considers this thought, but does not consider the present point.

[7] Quine (1981: 69–70). See also Curry (1933: 389–90), Quine (1981: iv, 5, 71), Dummett (1981: ch.1), Kaplan (1986: 244), Lavine (2000: 5–6), and Potter (2000: 64).

[8] Quine (1981: §12).

[9] Bourbaki (1954: ch.1), apparently independently. The slight difference is that Bourbaki uses Hilbert's epsilon operator instead of quantifiers.

[10] Pickel and Rabern (2017: 148–52) consider and criticise the Quine–Bourbaki approach to Fine's antinomy. Pickel and Rabern assume that the Quine–Bourbaki approach will be coupled with Frege's idea that one obtains the predicate '( ) ≤ ( )' by taking a sentence like '7 ≤ 7' and deleting the names. They then insist that Frege must distinguish between the case when '( ) ≤ ( )' is regarded as a one-place predicate, and the case where it is regarded as a two-place predicate. And they then maintain: 'if Frege were to introduce marks capable of typographically distinguishing between these predicates, then that mark would need its own semantic significance, which in this context means designation.' We disagree with the last part of this claim. *Brackets* are semantically significant, in that $\neg(\varphi \wedge \psi)$ is importantly different from $(\neg\varphi \wedge \psi)$; but brackets do not denote. Fregeans should simply insist that any 'marks' on predicate-positions have a similarly *non-denotational* semantic significance. After all, their ultimate purpose is just to account for the different 'cross-referencing' in $\forall x \exists y \varphi(x, y)$ and $\forall x \exists y \varphi(y, x)$.

To be clear, no one is recommending that we *should adopt* the Quine–Bourbaki notation in practice: it would be hard to read and a pain to typeset. To dissolve the antimony of the variable, it is enough to know that we *could in principle* have adopted this notation.

But there is a catch. Just as this notation leaves us unable to formulate Fine's antinomy of the variable, it leaves us unable to define the notion of a variable-assignment. So, until we can provide a non-Tarskian approach to semantics, which does *not* essentially rely upon variable-assignments, we have no guarantee that we *could* have adopted the Quine–Bourbaki notation, even in principle. Now, we can of course use the Tarskian approach to supply a semantics for Quine–Bourbaki sentences derivatively.[11] But if we were to do that that, we would lose the right to say that we could, in principle, have done away with typographically distinct variables altogether, for we would still be relying upon them in our semantic machinery.

In sum, we want an approach to semantics which (unlike Tarski's) accords variables with no more apparent significance than is suggested by the Quine–Bourbaki notation. Fortunately, such approaches are available.

1.5 The Robinsonian approach to semantics

To recall: difficulties concerning the semantics for quantifiers arise because $\mathscr{L}$ may not contain names for every object in the domain. One solution to this problem is obvious: just *add* new constants. This was essentially Robinson's approach.[12]

To define how to *add* new symbols, it is easiest to define how to *remove* them. Given a structure $\mathcal{M}$, its $\mathscr{L}$-reduct is the $\mathscr{L}$-structure we obtain by *ignoring* the interpretation of the symbols in $\mathcal{M}$'s signature which are not in $\mathscr{L}$. More precisely:[13]

Definition 1.4: *Let $\mathscr{L}^+$ and $\mathscr{L}$ be signatures with $\mathscr{L}^+ \supseteq \mathscr{L}$. Let $\mathcal{M}$ be an $\mathscr{L}^+$-structure. Then $\mathcal{M}$'s $\mathscr{L}$-*reduct*, $\mathcal{N}$, is the unique $\mathscr{L}$-structure with domain M such that $s^{\mathcal{N}} = s^{\mathcal{M}}$ for all s from $\mathscr{L}$. We also say that $\mathcal{M}$ is a* signature-expansion *of $\mathcal{N}$, and that $\mathcal{N}$ is a* signature-reduct *of $\mathcal{M}$.*

In Quinean terms, the difference between a model and its reduct is not *ontological* but *ideological*.[14] We do not add or remove any entities from the domain; we just add or remove some (interpretations of) symbols.

---

[11] Where φ is any Quine–Bourbaki sentence, let φ^{fo} be the sentence of first-order logic which results by: (a) inserting the variable v_n after the nth quantifier in φ, counting quantifiers from left-to-right; (b) replacing each blob connected to the nth-quantifier with the variable v_n and (c) deleting all the connecting wires. Then say $\mathcal{M} \vDash \varphi$ iff $\mathcal{M} \vDash \varphi^{\text{fo}}$, with $\mathcal{M} \vDash \varphi^{\text{fo}}$ defined via the Tarskian approach.

[12] A. Robinson (1951: 19–21), with a tweak that one finds in, e.g., Sacks (1972: ch.4).

[13] Cf. Hodges (1993: 9ff) and Marker (2002: 31).

[14] Quine (1951: 14).

We can now define the idea of 'adding new constants for every member of the domain'. The following definition explains how to add, for each element $a \in M$, a new constant symbol, c_a, which is taken to name a:

Definition 1.5: *Let $\mathscr{L}$ be any signature. For any set M, $\mathscr{L}(M)$ is the signature obtained by adding to $\mathscr{L}$ a new constant symbol c_a for each $a \in M$. For any $\mathscr{L}$-structure $\mathcal{M}$ with domain M, we say that $\mathcal{M}^\circ$ is the $\mathscr{L}(M)$-structure whose $\mathscr{L}$-reduct is $\mathcal{M}$ and such that $c_a^{\mathcal{M}^\circ} = a$ for all $a \in M$.*

Since $\mathcal{M}^\circ$ is flooded with constants, it is very easy to set up its semantics. We start by defining the interpretation of the $\mathscr{L}(M)$-terms which contain no variables:

$$t^{\mathcal{M}^\circ} = f^{\mathcal{M}^\circ}(s_1^{\mathcal{M}^\circ}, \ldots, s_k^{\mathcal{M}^\circ}), \text{ if } t \text{ is the variable-free } \mathscr{L}(M)\text{-term } f(s_1, \ldots, s_k)$$

For each atomic first-order $\mathscr{L}(M)$-sentence, we then define:

$$\mathcal{M}^\circ \vDash t_1 = t_2 \text{ iff } t_1^{\mathcal{M}^\circ} = t_2^{\mathcal{M}^\circ}, \text{for any variable-free } \mathscr{L}(M)\text{-terms } t_1, t_2$$

$$\mathcal{M}^\circ \vDash R(t_1, \ldots, t_n) \text{ iff } (t_1^{\mathcal{M}^\circ}, \ldots, t_n^{\mathcal{M}^\circ}) \in R^{\mathcal{M}^\circ}, \text{for}$$

any variable-free $\mathscr{L}(M)$-terms $t_1, \ldots, t_n$ and
any n-place relation symbol R from $\mathscr{L}(M)$

And finally we offer:

$$\mathcal{M}^\circ \vDash \neg\varphi \text{ iff } \mathcal{M}^\circ \nvDash \varphi$$
$$\mathcal{M}^\circ \vDash (\varphi \wedge \psi) \text{ iff } \mathcal{M}^\circ \vDash \varphi \text{ and } \mathcal{M}^\circ \vDash \psi$$
$$\mathcal{M}^\circ \vDash \forall x\varphi(x) \text{ iff } \mathcal{M}^\circ \vDash \varphi(c_a/x) \text{ for every } a \in M$$

We now have what we want, in terms of $\mathcal{M}^\circ$. And, since $\mathcal{M}^\circ$ is uniquely determined by $\mathcal{M}$, we can now extract what we really wanted: definitions concerning $\mathcal{M}$ itself. Where $\varphi(\bar{v})$ is a first-order $\mathscr{L}$-formula with free variables displayed, and $\bar{a}$ are from M, we define a *three*-place relation which, intuitively, says that $\varphi(\bar{v})$ is *true of* the entities $\bar{a}$ according to $\mathcal{M}$. Here is the definition:

$$\mathcal{M} \vDash \varphi(\bar{a}) \text{ iff } \mathcal{M}^\circ \vDash \varphi(\overline{c_a}/\bar{v})$$

The notation $\varphi(\bar{c}/\bar{v})$ indicates the $\mathscr{L}(M)$-formula obtained by substituting the kth constant in the sequence $\bar{c}$ for the kth variable in the sequence $\bar{v}$. So we have defined a *three*-place relation between an $\mathscr{L}$-formula, entities $\bar{a}$, and a structure $\mathcal{M}$, in terms of a *two*-place relation between a structure $\mathcal{M}^\circ$ and an $\mathscr{L}(M)$-formula. For readability, we will write $\varphi(\bar{c})$ instead of $\varphi(\bar{c}/\bar{v})$, where no confusion arises.

As a limiting case, a sentence is a formula with no free variables. So for each $\mathscr{L}$-sentence φ, our definition states that $\mathcal{M} \vDash \varphi$ iff $\mathcal{M}^\circ \vDash \varphi$. And, intuitively, we can read this as saying that φ is *true* in $\mathcal{M}$.

To complete the Robinsonian semantics, we will define something similar for *terms*. So, where $\bar{a}$ are entities from M and $t(\bar{v})$ is an $\mathscr{L}$-term with free variables displayed, we define a function $t^{\mathcal{M}} : M^n \longrightarrow M$, by:

$$t^{\mathcal{M}}(\bar{a}) = (t(\overline{c_a}/\bar{v}))^{\mathcal{M}^\circ}$$

This completes the Robinsonian approach. And the approach carries no taint of the antinomy of the variable, since it clearly accords variables with no more semantic significance than is suggested by the Quine–Bourbaki notation. Indeed, it is easy to give a Robinsonian semantics directly for Quine–Bourbaki sentences, via: $\mathcal{M}^\circ$ satisfies a Quine–Bourbaki sentence beginning with '$\forall$' *iff* for every $a \in M$ the model $\mathcal{M}^\circ$ satisfies the Quine–Bourbaki sentence which results from replacing all blobs connected to the quantifier with 'c_a' and then deleting the quantifier and the connecting wires.

1.6 Straining the notion of 'language'

For all its virtues, the Robinsonian approach has some eyebrow-raising features of its own. To define satisfaction for the sentences of the first-order $\mathscr{L}$-sentences, we have considered the sentences in some *other* formal languages, namely, those with signature $\mathscr{L}(M)$ for any $\mathscr{L}$-structure $\mathcal{M}$. These languages can be *enormous*. Let $\mathcal{M}$ be an infinite $\mathscr{L}$-structure, whose domain M has size κ for some very big cardinal κ.[15] Then $\mathscr{L}(M)$ contains at least κ symbols. Can such a beast really count as a *language*, in any intuitive sense?

Of course, there is no technical impediment to defining these enormous languages. So, if model theory is just regarded as a branch of *pure* mathematics, then there is no real reason to worry about any of this. But we might, instead, want model theory to be regarded as a branch of *applied* mathematics, whose (idealised) subject matter is the languages and theories that mathematicians *actually* use. And if we regard model theory that way, then we will not want our technical notion of a 'language' to diverge too far from the kinds of things which we would ordinarily count as languages.

There is a second issue with the Robinsonian approach. In Definition 1.5, we introduced a new constant symbol, c_a, for each $a \in M$. But we did not say what, exactly, the constant symbol c_a *is*. Robinson himself suggested that the constant c_a should just be the object *a itself*.[16] In that case, every object in $\mathcal{M}^\circ$ would name itself. But this is both philosophically strange and also technically awkward.

On the philosophical front: we might want to consider a structure, $\mathcal{W}$, whose domain is the set of all living wombats. In order to work out which sentences are

[15] As is standard, we use κ to denote a cardinal; see the end of §1.B for a brief review of cardinals.

[16] A. Robinson (1951: 21).

true in $\mathcal{W}$ using Robinson's own proposal, we would have to treat each wombat as a name for itself, and so imagine a language whose syntactic parts are live wombats.[17] This stretches the ordinary notion of a language to breaking point.

There is also a technical hitch with Robinson's own proposal. Suppose that c is a constant symbol of $\mathcal{L}$. Suppose that $\mathcal{M}$ is an $\mathcal{L}$-structure where the *symbol* c is itself an *element* of $\mathcal{M}$'s underlying domain. Finally, suppose that $\mathcal{M}$ interprets c as naming some element other than c itself, i.e. $c^{\mathcal{M}} \neq c$. Now Robinson's proposal requires that $c^{\mathcal{M}^\circ} = c$. But since $\mathcal{M}^\circ$ is a signature expansion of $\mathcal{M}$, we require that $c^{\mathcal{M}^\circ} = c^{\mathcal{M}}$, which is a contradiction.

To fix this bug whilst retaining Robinson's idea that $c_a = a$, we would have to tweak the definition of an $\mathcal{L}$-structure to ensure that the envisaged situation cannot arise.[18] A better alternative—which also spares the wombats—is to abandon Robinson's suggestion that $c_a = a$, and instead define the symbol c_a so that it is guaranteed *not* to be an element of $\mathcal{M}$'s underlying domain.[19] So this is our official Robinson*ian* semantics (even if it was not exactly Robinson's).

1.7 The Hybrid approach to semantics

Tarskian and Robinsonian semantics are technically equivalent, in the following sense: they use the same notion of an $\mathcal{L}$-structure, they use the same notion of an $\mathcal{L}$-sentence, and they end up defining exactly the same relation, $\vDash$, between structures and sentences. But, as we have seen, neither approach is exactly ideal. So we turn to a third approach: a *hybrid* approach.

In the Robinsonian semantics, we used $\mathcal{M}^\circ$ to define the expression $\mathcal{M} \vDash \varphi(\bar{a})$. Intuitively, this states that φ is true of $\bar{a}$ in $\mathcal{M}$. If we *start* by defining this notation—which we can do quite easily—then we can use it to present a semantics with the following recursion clauses:

$$\mathcal{M} \vDash t_1 = t_2 \text{ iff } t_1^{\mathcal{M}} = t_2^{\mathcal{M}}, \text{for any variable-free } \mathcal{L}\text{-terms } t_1, t_2$$

$$\mathcal{M} \vDash R(t_1, \ldots, t_n) \text{ iff } (t_1^{\mathcal{M}}, \ldots, t_n^{\mathcal{M}}) \in R^{\mathcal{M}}, \text{for any variable-free } \mathcal{L}\text{-terms } t_1, \ldots, t_n \text{ and any } n\text{-place relation symbol } R \text{ from } \mathcal{L}$$

$$\mathcal{M} \vDash \neg\varphi \text{ iff } \mathcal{M} \nvDash \varphi$$

$$\mathcal{M} \vDash (\varphi \wedge \psi) \text{ iff } \mathcal{M} \vDash \varphi \text{ and } \mathcal{M} \vDash \psi$$

$$\mathcal{M} \vDash \forall v\varphi(v) \text{ iff } \mathcal{M} \vDash \varphi(a) \text{ for all } a \in M$$

[17] Cf. Lewis (1986: 145) on 'Lagadonian languages'.

[18] We would have to add a clause: if $\mathcal{M}$ is an $\mathcal{L}$-structure and $s \in \mathcal{L} \cap M$, then $s^{\mathcal{M}} = s$.

[19] A simple way to do this is as follows: let c_a be the ordered pair (a, M). By Foundation in the background set theory within which we implement our model theory, $(a, M) \notin M$.

All that remains is to define $\mathcal{M} \vDash \varphi(\bar{a})$ without going all-out Robinsonian. And the idea here is quite simple: we just add new constant symbols when we need them, but not before. Here is the idea, rigorously developed. Let $\mathcal{M}$ be an $\mathcal{L}$-structure with $\bar{a}$ from M. For each a_i among $\bar{a}$, let c_{a_i} be a constant symbol not occurring in $\mathcal{L}$. Intuitively, we interpret each c_{a_i} as a name for a_i. More formally, we define $\mathcal{M}[\bar{a}]$ to be a structure whose signature is $\mathcal{L}$ together with the new constant symbols among $\overline{c_a}$, whose $\mathcal{L}$-reduct is $\mathcal{M}$, and such that $c_{a_i}^{\mathcal{M}[\bar{a}]} = a_i$ for each i. Where $\varphi(\bar{v})$ is an $\mathcal{L}$-formula with free variables displayed, the Hybrid approach defines:

$$\mathcal{M} \vDash \varphi(\bar{a}) \text{ iff } \mathcal{M}[\bar{a}] \vDash \varphi(\overline{c_a}/\bar{v})$$

When we combine our new definition of $\mathcal{M} \vDash \varphi(a)$ with the clause for universal quantification, we see that universal quantification effectively amounts to considering all the different ways of expanding the signature of $\mathcal{M}$ with a *new* constant symbol which could be interpreted to name *any* element of M. (So the Hybrid approach offers a semantics by simultaneous recursion over structures and languages.) Finally, we offer a similar clause for terms:

$$t^{\mathcal{M}}(\bar{a}) = t^{\mathcal{M}[\bar{a}]}(\overline{c_a}/\bar{v})$$

thereby completing the Hybrid approach.[20]

1.8 Linguistic compositionality

Unsurprisingly, the Hybrid approach is technically equivalent to the Robinsonian and Tarskian approaches. However, its philosophical merits come out when we revisit some of the potential defects of the other approaches. The Tarskian approach does not distinguish sufficiently between names and variables; the Hybrid approach has no such issues. Indeed, just like the Robinsonian approach, the Hybrid approach accords variables with no greater semantic significance than is suggested by the Quine–Bourbaki notation. But the Robinsonian approach involved vast, peculiar 'languages'; the Hybrid approach has no such issues. And, following Lavine, we will pause on this last point.[21]

It is common to insist that languages should be *compositional*, in some sense. One of the most famous arguments to this effect is due to Davidson. Because natural languages are *learnable*, Davidson insists that 'the meaning of each sentence [must be] a function of a finite number of features of the sentence'. For, on the one hand,

[20] The hybrid approach is hinted at by Geach (1962: 160), and Mates (1965: 54–7) offers something similar. But the clearest examples we can find are Boolos and Jeffrey (1974: 104–5), Boolos (1975: 513–4), and Lavine (2000: 10–12).

[21] See Lavine's (2000: 12–13) comments on compositionality and learnability.

if a language has this feature, then we 'understand how an infinite aptitude can be encompassed by finite accomplishments.' Conversely, if 'a language lacks this feature then no matter how many sentences a would-be speaker learns to produce and understand, there will remain others whose meanings are not given by the rules already mastered.'[22]

Davidson's argument is too quick. After all, it is a *wild* idealisation to suggest that any actual human can indeed understand or learn the meanings of *infinitely* many sentences: some sentences are just too long for any actual human to parse. It is unclear, then, why we should worry about the 'learnability' of such sentences.

Still, something in the *vicinity* of Davidson's argument seems right. In §1.6, we floated the idea that model theory should be regarded as a branch of *applied* mathematics, whose (idealised) subject matter is the languages and theories that (pure) mathematicians *actually* use. But here is an apparent phenomenon concerning that subject matter: once we have a fixed interpretation in mind, we tend to act as if that interpretation fixes the truth value of *any* sentence of the appropriate language, no matter how long or complicated that sentence is.[23] All three of our approaches to formal semantics accommodate this point. For, given a signature $\mathscr{L}$ and an $\mathscr{L}$-structure $\mathcal{M}$—i.e. an interpretation of the range of quantification and an interpretation of each $\mathscr{L}$-symbol—the semantic value of every $\mathscr{L}$-sentence is completely determined within $\mathcal{M}$, in the sense that, for every $\mathscr{L}$-sentence φ, either $\mathcal{M} \vDash \varphi$, or $\mathcal{M} \vDash \neg\varphi$, but not both.

But the Hybrid approach, specifically, may allow us to go a little further. For, when $\mathscr{L}$ is finite,[24] and we offer the Hybrid approach to semantics, we may gain some insight into how a finite mind might *fully understand* the rules by which an interpretation fixes the truth-value of every sentence. That understanding seems to reduce to three rather tractable components:

(a) an understanding of the finitely many recursion clauses governing satisfaction for atomic sentences (finitely many, as we assumed that $\mathscr{L}$ is finite);
(b) an understanding of the handful of recursion clauses governing sentential connectives; and
(c) an understanding of the recursion clauses governing quantification

On the Hybrid approach, point (c) reduces to an understanding of two ideas: (i) the *general* idea that names can pick out objects,[25] and (ii) the intuitive idea that, for any object, we could expand our language with a new name for that object. In short:

[22] Davidson (1965: 9).

[23] A theme of Part B is whether, in certain circumstances, axioms can also fix truth values.

[24] We can make a similar point if $\mathscr{L}$ can be recursively specified.

[25] There are some deep philosophical issues concerning the question of how names pick out objects (see Chapters 2 and 15). However, the general notion seems to be required by *any* model-theoretic semantics, so that there is no *special* problem here for the Hybrid approach.

the Hybrid semantics seems to provide a truly *compositional* notion of meaning. But we should be clear on what this means.

First, we are not aiming to escape what Sheffer once called the 'logocentric predicament', that '*In order to give an account of logic, we must presuppose and employ logic.*'[26] Our semantic clause for object-language conjunction, $\land$, always involves conjunction in the metalanguage. On the Hybrid approach, our semantic clause for object-language universal quantification, $\forall$, involved (metalinguistic) quantification over all the ways in which a new name could be added to a signature. We do not, of course, claim that anyone could read these semantic clauses and come to *understand* the very idea of conjunction or quantification from scratch. We are making a much more mundane point: to understand the hybrid approach to semantics, one need only understand a tractable number of ideas.

Second, in describing our semantics as compositional, we are *not* aiming to supply a semantics according to which the meaning of $\forall x F(x)$ depends upon the separate meanings of the expressions $\forall$, x, F, and x.[27] Not only would that involve an oddly inflexible understanding of the word 'compositional'; the discussion of §1.4 should have convinced us that variables do not have semantic values in isolation.[28] Instead, on the hybrid approach, the meaning of $\forall x F(x)$ depends upon the meanings of the quantifier-expression $\forall x \ldots x$ and the predicate-expression $F(\ )$. The crucial point is this: the Hybrid approach delivers the truth-conditions of infinitely many sentences using only a small 'starter pack' of principles.

Having aired the virtues of the Hybrid approach, though, it is worth repeating that our three semantic approaches are technically equivalent. As such, we can in good faith use whichever approach we like, whilst claiming all of the pleasant philosophical features of the Hybrid approach. Indeed, in the rest of this book, we simply use whichever approach is easiest for the purpose at hand.

This concludes our discussion of first-order logic. It also concludes the 'philosophical' component of this chapter. The remainder of this chapter sets down the purely technical groundwork for several later philosophical discussions.

1.9 Second-order logic: syntax

Having covered first-order logic, we now consider *second*-order logic. This is much less popular than first-order logic among working model-theorists. However, it has

---

[26] Sheffer (1926: 228).

[27] Pickel and Rabern (2017: 155) call this 'structure intrinsicalism', and advocate it.

[28] Nor would it help to suggest that the meaning of $\forall x F(x)$ depends upon the separate meanings of the two composite expressions $\forall x$ and $F(x)$. For if we think that open formulas possess semantic values (in isolation), we will obtain an exactly parallel (and exactly as confused) 'antinomy of the open formula' as follows: clearly $F(x)$ and $F(y)$ are notational variants, and so should have the same semantic value; but they cannot have the same value, since $F(x) \land \neg F(y)$ is not a contradiction.

certain philosophically interesting dimensions. We explore these philosophical issues in later chapters; here, we simply outline its technicalities.

First-order logic can be thought of as allowing quantification into *name* position. For example, if $\varphi(c)$ is a formula containing a constant symbol c, then we also have a formula $\forall v\varphi(v/c)$, replacing c with a variable which is bound by the quantifier. To extend the language, we can allow quantification into *relation symbol* or *function symbol* position. For example, if $\varphi(R)$ is a formula containing a relation symbol R, we would want to have a formula $\forall X\varphi(X/R)$, replacing the relation symbol R with a relation-variable, X, which is bound by the quantifier. Equally, if $\varphi(f)$ is a formula containing a function symbol f, we would want to have a formula $\forall p\varphi(p/f)$.

Let us make this precise, starting with the syntax. In addition to all the symbols of first-order logic, our language adds some new symbols:

- relation-variables: U, V, W, X, Y, Z
- function-variables: p, q

both with numerical subscripts and superscripts as necessary. In more detail: just like relation symbols and functions symbols, these higher-order variables come equipped with a number of places, indicated (where helpful) with superscripts. So, together with the subscripts, this means we have countably many relation-variables and function-symbols for each number of places. We then expand the recursive definition of a term, to allow:

- $q^n(t_1, \ldots, t_n)$, for any $\mathcal{L}$-terms $t_1, \ldots, t_n$ and n-place function-variable q^n

and we expand the notion of a formula, to allow

- $X^n(t_1, \ldots, t_n)$, for any $\mathcal{L}$-terms $t_1, \ldots, t_n$ and n-place relation-variable X^n
- $\exists X^n\varphi$ and $\forall X^n\varphi$, for any n-place relation-variable X^n and any second-order $\mathcal{L}$-formula φ which contains neither of the expressions $\exists X^n$ nor $\forall X^n$
- $\exists q^n\varphi$ and $\forall q^n\varphi$, for any n-place function-variable q^n and any second-order $\mathcal{L}$-formula φ which contains neither of the expressions $\exists q^n$ nor $\forall q^n$

We will also introduce some abbreviations which are particularly helpful in a second-order context. Where Ξ is any one-place relation symbol or relation-variable, we write $(\forall x : \Xi)\varphi$ for $\forall x(\Xi(x) \rightarrow \varphi)$, and $(\exists x : \Xi)\varphi$ for $\exists x(\Xi(x) \land \varphi)$. We also allow ourselves to bind multiple quantifiers at once; so $(\forall x, y, z : \Xi)\varphi$ abbreviates $\forall x\forall y\forall z((\Xi(x) \land \Xi(y) \land \Xi(z)) \rightarrow \varphi)$.

1.10 Full semantics

The syntax of second-order logic is straightforward. The semantics is more subtle; for here there are some genuinely *non*-equivalent options.

We start with *full semantics* for second-order logic (also known as *standard* semantics). This uses $\mathcal{L}$-structures, exactly as we defined them in Definition 1.2.

The trick is to add new semantic clauses for our second-order quantifiers. In fact, we can adopt any of the Tarskian, Robinsonian, or Hybrid approaches here, and we sketch all three (leaving the reader to fill in some obvious details).

Tarskian. Variable-assignments are the key to the Tarskian approach to first-order logic. So the Tarskian approach to second-order logic must expand the notion of a variable-assignment, to cover both relation-variables and function-variables. In particular, we take it that σ is a function which assigns every variable to some entity $a \in M$, every n-place relation-variable to some subset of M^n, and every function-variable to some function $M^n \longrightarrow M$. We now add clauses:

$$\mathcal{M}, \sigma \vDash X^n(t_1, \ldots, t_n) \text{ iff } (t_1^{\mathcal{M},\sigma}, \ldots, t_n^{\mathcal{M},\sigma}) \in (X^n)^{\mathcal{M},\sigma} \text{ for any } \mathcal{L}\text{-terms } t_1, \ldots, t_n$$

$$\mathcal{M}, \sigma \vDash \forall X^n \varphi(X^n) \text{ iff } \mathcal{M}, \tau \vDash \varphi(X^n) \text{ for every variable-assignment } \tau \text{ which agrees with } \sigma \text{ except perhaps on } X^n$$

$$\mathcal{M}, \sigma \vDash \forall q^n \varphi(q^n) \text{ iff } \mathcal{M}, \tau \vDash \varphi(q^n) \text{ for every variable-assignment } \tau \text{ which agrees with } \sigma \text{ except perhaps on } q^n$$

Robinsonian. The key to the Robinsonian approach to first-order logic is to introduce a new constant symbol for every entity in the domain. So the Robinsonian approach to second-order logic must introduce a new relation symbol for every possible relation on M, and a new function symbol for every possible function. Let $\mathcal{M}^\blacksquare$ be the structure which expands $\mathcal{M}$ in just this way. So, for each n and each $S \subseteq M^n$, we add a new relation symbol R_S with $S = R_S^{\mathcal{M}^\blacksquare}$, and for each function $g : M^n \longrightarrow M$ we add a new function symbol f_g with $g = f_g^{\mathcal{M}^\blacksquare}$. We can now simply rewrite the first-order semantics, replacing $\mathcal{M}^\circ$ with $\mathcal{M}^\blacksquare$, and adding:

$$\mathcal{M}^\blacksquare \vDash \forall X^n \varphi(X^n) \text{ iff } \mathcal{M}^\blacksquare \vDash \varphi(R_S/X^n) \text{ for every } S \subseteq M^n$$

$$\mathcal{M}^\blacksquare \vDash \forall q^n \varphi(q^n) \text{ iff } \mathcal{M}^\blacksquare \vDash \varphi(f_g/q^n) \text{ for every function } g : M^n \longrightarrow M$$

Hybrid. The key to the Hybrid approach to second-order logic is to define, upfront, the three-place relation between $\mathcal{M}$, a formula φ, and a relation (or function) on $\mathcal{M}$.[29] We illustrate the idea for the case of relations (the case of functions is exactly similar). Let S be a relation on M^n. Let R_S be an n-place relation symbol not occurring in $\mathcal{L}$. We define $\mathcal{M}[S]$ to be a structure whose signature is $\mathcal{L}$ together with the new relation symbol R_S, such that $\mathcal{M}[S]$'s $\mathcal{L}$-reduct is $\mathcal{M}$ and $R_S^{\mathcal{M}[S]} = S$. Then where $\varphi(X)$ is an $\mathcal{L}$-formula with free relation-variable displayed, we define:

$$\mathcal{M} \vDash \varphi(S) \text{ iff } \mathcal{M}[S] \vDash \varphi(R_S/X) \text{ for any relation symbol } R_S \notin \mathcal{L}$$

$$\mathcal{M} \vDash \forall X^n \varphi(X^n) \text{ iff } \mathcal{M} \vDash \varphi(S) \text{ for every relation } S \subseteq M^n$$

[29] Trueman (2012) recommends a semantics like this as a means for overcoming philosophical resistance to the use of second-order logic.

The three approaches ultimately define the same semantic relation. And we call the ensuing semantics *full* second-order semantics.

The relative merits of these three approaches are much as before. So: the Tarskian approach unhelpfully treats relation-variables as if they were varying predicates; the Robinsonian approach forces us to stretch the idea of a language to breaking point; but the Hybrid approach avoids both problems and provides us with a reasonable notion of compositionality. (It is worth noting, though, that all three approaches effectively assume that we understand notions like 'all subsets of M^n'. We revisit this point in Part B.)

1.11 Henkin semantics

The Tarskian, Robinsonian, and Hybrid approaches all yielded the same relation, $\vDash$. However, there is a *genuinely alternative* semantics for second-order logic. Moreover, the availability of this alternative is an important theme in Part B of this book. So we outline that alternative here.

In *full* second-order logic, universal quantification into relation-position effectively involves considering *all possible* relations on the structure. Indeed, using $\wp(A)$ for A's powerset, i.e. $\{B : B \subseteq A\}$, we have the following: if X is a one-place relation-variable, then the relevant 'domain' of quantification in $\forall X\varphi$ is $\wp(M)$; and if X is an n-place relation-variable, then the relevant 'domain' of quantification in $\forall X\varphi$ is $\wp(M^n)$. An alternative semantics naturally arises, then, by considering more *restrictive* 'domains' of quantification, as follows:

Definition 1.6: *A* Henkin $\mathscr{L}$-structure, $\mathcal{M}$, *consists of:*

- *a non-empty set, M, which is the underlying domain of $\mathcal{M}$*
- *a set $M_n^{\text{rel}} \subseteq \wp(M^n)$ for each $n < \omega$*
- *a set $M_n^{\text{fun}} \subseteq \{g \in \wp(M^{n+1}) : g$ is a function $M^n \longrightarrow M\}$ for each $n < \omega$*
- *an object $c^{\mathcal{M}} \in M$ for each constant symbol c from $\mathscr{L}$*
- *a relation $R^{\mathcal{M}} \subseteq M^n$ for each n-place relation symbol R from $\mathscr{L}$*
- *a function $f^{\mathcal{M}} : M^n \longrightarrow M$ for each n-place function symbol f from $\mathscr{L}$.*

In essence, M_n^{rel} serves as the domain of quantification for the n-place relation-variables, and M_n^{fun} serves as the domain of quantification for the n-place function-variables. As before, though, we can make this idea precise using any of our three approaches to formal semantics. We sketch all three.

Tarskian. Where $\mathcal{M}$ is a Henkin structure, we take our variable-assignments σ to be restricted in the following way: σ assigns each variable to some entity $a \in M$, each n-place relation-variable to some element of M_n^{rel}, and each n-place function-variable to some element of M_n^{fun}. We then rewrite the clauses for the full semantics, exactly as before, but using this more restricted notion of a variable-assignment.

Robinsonian. Where $\mathcal{M}$ is a Henkin structure, we let $\mathcal{M}^{\square}$ be the structure which expands $\mathcal{M}$ by adding new relation symbols R_S such that $S = R_S^{\mathcal{M}^{\square}}$ for every relation $S \in M_n^{\text{rel}}$, and new function symbols f_g such that $g = f_g^{\mathcal{M}^{\square}}$ for every function $g \in M_n^{\text{fun}}$. We then offer these clauses:

$$\mathcal{M}^{\square} \vDash \forall X^n \varphi(X^n) \text{ iff } \mathcal{M}^{\square} \vDash \varphi(R_S/X^n) \text{ for every relation } S \in M_n^{\text{rel}}$$
$$\mathcal{M}^{\square} \vDash \forall q^n \varphi(q^n) \text{ iff } \mathcal{M}^{\square} \vDash \varphi(f_g/q^n) \text{ for every function } g \in M_n^{\text{fun}}$$

Hybrid. We need only tweak the recursion clauses, as follows:

$$\mathcal{M} \vDash \forall X^n \varphi(X^n) \text{ iff } \mathcal{M} \vDash \varphi(S) \text{ for every relation } S \subseteq M_n^{\text{rel}}$$
$$\mathcal{M} \vDash \forall q^n \varphi(q^n) \text{ iff } \mathcal{M} \vDash \varphi(g) \text{ for every function } g \in M_n^{\text{fun}}$$

We say that *Henkin semantics* is the semantics yielded by any of these three approaches, as applied to Henkin structures. Importantly, Henkin semantics generalises the *full* semantics of §1.10. To show this, let $\mathcal{M}$ be an $\mathcal{L}$-structure in the sense of Definition 1.2. From this, define a Henkin structure $\mathcal{N}$ by setting, for each $n < \omega$, $N_n^{\text{rel}} = \wp(N^n)$ and N_n^{fun} as the set of all functions $N^n \longrightarrow N$. Then *full* satisfaction, defined over $\mathcal{N}$, is exactly like *Henkin* satisfaction, defined over $\mathcal{N}$.

The notion of a Henkin structure may, though, be a bit *too* general. To see why, consider a Henkin $\mathcal{L}$-structure $\mathcal{M}$, and suppose that R is a one-place relation symbol of $\mathcal{L}$, so that $R^{\mathcal{M}} \subseteq M$. Presumably, we should want $\mathcal{M}$ to satisfy $\exists X \forall v(R(v) \leftrightarrow X(v))$, for $R^{\mathcal{M}}$ should *itself* provide a witness to the second-order existential quantifier. But this holds if and only if $R^{\mathcal{M}} \in M_1^{\text{rel}}$, and the definition of a Henkin structure does not guarantee this. For this reason, it is common to insist that the following axiom schema should hold in all structures:

Comprehension Schema. $\exists X^n \forall \bar{v}(\varphi(\bar{v}) \leftrightarrow X^n(\bar{v}))$, *for every formula* $\varphi(\bar{v})$ *which does not contain the relation-variable* X^n

We must block X^n from appearing in $\varphi(\bar{v})$, since otherwise an axiom would be $\exists X \forall v(\neg X(v) \leftrightarrow X(v))$, which will be inconsistent. However, we allow other free first-order and second-order variables, because this allows us to form new concepts from old concepts. For instance, given the two-place relation symbol R, we have as an axiom $\exists X^2 \forall v_1 \forall v_2(\neg R(v_1, v_2) \leftrightarrow X^2(v_1, v_2))$, i.e. M_2^{rel} must contain the set of all pairs not in $R^{\mathcal{M}}$, i.e. $M^2 \setminus R^{\mathcal{M}}$. So: if we insist that (all instances) of the Comprehension Schema must hold in all Henkin structures, then we are insisting on further properties concerning our various M_n^{rel}s. There is also a *predicative* version of Comprehension:

Predicative Comprehension Schema. $\exists X^n \forall \bar{v}(\varphi(\bar{v}) \leftrightarrow X^n(\bar{v}))$, *for every formula* $\varphi(\bar{v})$ *which neither contains the relation-variable* X^n *nor any second-order quantifiers*

When we want to draw the contrast, we call the (plain vanilla) Comprehension Schema the *Impredicative* Comprehension Schema. But this will happen only rarely; we only mention Predicative Comprehension in §§5.7, 10.2, 10.c, and 11.3.

We could provide a similar schema to govern functions. But it is usual to make the stronger claim, that the following should hold in all structures (for every n):[30]

Choice Schema. $\forall X^{n+1}\left(\forall \bar{v}\exists y\, X^{n+1}(\bar{v},y) \rightarrow \exists p^n \forall \bar{v}\, X^{n+1}(\bar{v}, p^n(\bar{v}))\right)$

To understand these axioms, let S be a two-place relation on the domain, and suppose that the antecedent is satisfied, i.e. that for any x there is some y such that $S(x,y)$. The relevant Choice instance then states that there is then a one-place function, p, which 'chooses', for each x, a *particular* entity $p(x)$ such that $S(x,p(x))$. For obvious reasons, this p is known as a *choice function*. Hence, just like the Comprehension Schema, the Choice Schema guarantees that the domains of the higher-order quantifiers are well populated.

This leads to a final definition: a *faithful Henkin structure* is a Henkin structure within which both (impredicative) Comprehension and Choice hold.[31]

1.12 Consequence

We have defined satisfaction for first-order logic and for both the full- and Henkin-semantics for second-order logic. However, any definition of satisfaction induces a notion of consequence, via the following:

Definition 1.7: *A* theory *is a set of sentences in the logic under consideration. Given a structure $\mathcal{M}$ and a theory T, we say that $\mathcal{M}$ is a* model *of T, or more simply $\mathcal{M} \vDash T$, iff $\mathcal{M} \vDash \varphi$ for all sentences φ from T. We say that T has φ as a* consequence, *or that T entails φ, or more simply just $T \vDash \varphi$, iff: if $\mathcal{M} \vDash T$ then $\mathcal{M} \vDash \varphi$ for all structures $\mathcal{M}$.*

Note that this definition is relative to a semantics. So there are as many notions of logical consequence as there are semantics.

Here are some examples to illustrate the notation. Consider the natural numbers $\mathcal{N}$ and the integers $\mathcal{Z}$ in the signature consisting just of the symbol $<$, where this is given its natural interpretation. It is easy to see that both structures satisfy the following axioms:

$$\forall x \forall y \forall z((x < y \wedge y < z) \rightarrow x < z)$$
$$\forall x(x \not< x)$$
$$\forall x \forall y(x < y \vee x = y \vee y < x)$$

[30] For more, see Shapiro (1991: 67).

[31] See e.g. Shapiro (1991: 98–9).

These are the axioms of a *linear order*. Let T_{LO} be the theory consisting of just these three axioms. Then we would write $\mathcal{N} \models T_{\text{LO}}$ and $\mathcal{Z} \models T_{\text{LO}}$. But if we drop the third axiom, we obtain the related notion of a *partial order*. For an example of a partial order which is not a linear order, consider any set X with more than two elements, and consider the structure $\mathcal{P}$ whose first-order domain is the powerset $\wp(X)$ of X, with $<$ interpreted in $\mathcal{P}$ as the subset relation. If a, b are distinct elements of X, then $\mathcal{P} \models \{a\} \nless \{b\} \wedge \{a\} \neq \{b\} \wedge \{b\} \nless \{a\}$. So $\mathcal{P} \not\models T_{\text{LO}}$.

1.13 Definability

In addition to a notion of consequence, a semantics will induce a notion of definability, as follows:

Definition 1.8: *Let $\mathcal{M}$ be any structure and $n \geq 1$. We say that a subset X of M^n is* definable *iff there is both a formula $\varphi(v_1, \ldots, v_n, x_1, \ldots, x_m)$ with all free variables displayed and also elements $b_1, \ldots, b_m \in M$ such that:*

$$X = \{(a_1, \ldots, a_n) \in M^n : \mathcal{M} \models \varphi(a_1, \ldots, a_n, b_1, \ldots, b_m)\}$$

Here, the elements $b_1, \ldots, b_m$ are called *parameters*. Many authors allow parameters to be tacitly suppressed, and so say that X is definable iff $X = \{(a_1, \ldots, a_n) \in M^n : \mathcal{M} \models \varphi(a_1, \ldots, a_n)\}$ for some $\varphi(v_1, \ldots, v_n)$ which is (tacitly) allowed to contain further unmentioned parameters. If parameters are not allowed, such authors typically say this explicitly. We will be similarly explicit. When parameters are not allowed, the resulting sets are called *parameter-free definable sets*. Clearly a set is $\mathcal{M}$-definable iff it is parameter-free definable in some signature-expansion of $\mathcal{M}$ (see Definition 1.4).

To illustrate the idea of definability, consider again the natural numbers $\mathcal{N}$ in the signature consisting just of $<$, again with its natural interpretation. Here is a simple definable set:

$$\{0\} = \{n \in N : \mathcal{N} \models \neg\exists x\, x < n\}$$

As a slightly more complicated example, the graph of the successor operation in $\mathcal{N}$ is definable, since intuitively $n = m + 1$ iff m is less than n and there is no natural number strictly between m and n. More precisely:

$$G = \{(n, m) \in N^2 : \mathcal{N} \models (m < n \wedge \neg\exists z(m < z < n))\}$$

Now, both of these sets are *parameter-free* definable. And so it follows that *all* definable sets over $\mathcal{N}$ are parameter-free definable. For, where S is the successor function on the natural numbers, each natural number n is equal to the term $S^n(0)$, which we define recursively as follows:

$$S^0(a) = a \qquad\qquad S^{n+1}(a) = S(S^n(a)) \qquad\qquad (numerals)$$

(We label this definition '(*numerals*)' for future reference.) Hence, to say that $2 = S^2(0)$ is just a fancy way of saying that two is the second successor of zero. The terms $S^n(0)$ are sometimes called the *numerals*, and clearly $\mathcal{N} \vDash n = S^n(0)$ for each natural number $n \geq 0$. So, we can explicitly define the numerals in terms of the less-than relation using G, any definable set on $\mathcal{N}$ is *parameter-free* definable, by the following:

$$\begin{aligned} &\{(a_1, ..., a_n) \in N^n : \mathcal{N} \vDash \varphi(a_1, ..., a_n, b_1, ..., b_m)\} \\ =&\{(a_1, ..., a_n) \in N^n : \mathcal{N} \vDash \varphi(a_1, ..., a_n, S^{b_1}(0), ..., S^{b_m}(0))\} \end{aligned}$$

For an example of a structure with definable sets which are not parameter-free definable, let $\mathcal{L}$ be a countable signature and let $\mathcal{M}$ be an uncountable $\mathcal{L}$-structure. Since there are only countably many $\mathcal{L}$-formulas, there are only countably many parameter-free definable sets. But trivially the singleton $\{a\}$ of any element a from M is definable, as $\{a\} = \{x \in M : \mathcal{M} \vDash x = a\}$. So $\mathcal{M}$ has uncountably many definable subsets which are not parameter-free definable.

Finally, it is worth mentioning a particular aspect of definability in second-order logic. Consider the natural numbers $\mathcal{N}$ in the full semantics, and consider the set $\{(n, A) \in N \times \wp(N) : \mathcal{N} \vDash A(n)\}$ consisting of all pairs of numbers and sets of numbers such that the number is in the set. It obviously makes good sense to say that this set is definable, even though it is not a subset of $N \times N$ but rather of $N \times \wp(N)$. So, in the case of second-order logic, we expand the notion of definability to include both subsets of products of the *second-order* domain, and subsets of products of the first-order domain and the second-order domain. This point holds for both the Henkin and the full semantics.

1.A First- and second-order arithmetic

We have laid down the syntax and semantics for the logics which occupy us throughout this book. However, we will frequently discuss certain specific mathematical theories. So, for ease of reference, in this appendix we lay down the usual first- and second-order axioms of arithmetic. We cover set theory in the next appendix, and reserve all philosophical commentary for later chapters.

Definition 1.9: *The theory of Robinson Arithmetic, Q, is given by the universal closures of the following eight axioms:*

(Q1) $S(x) \neq 0$
(Q2) $S(x) = S(y) \rightarrow x = y$
(Q3) $x \neq 0 \rightarrow \exists y\, x = S(y)$
(Q4) $x + 0 = x$
(Q5) $x + S(y) = S(x + y)$
(Q6) $x \times 0 = 0$
(Q7) $x \times S(y) = (x \times y) + x$
(Q8) $x \leq y \leftrightarrow \exists z\, x + z = y$

The theory of Peano Arithmetic, PA, *is given by adding to Robinson Arithmetic the following Induction Schema:*

$$[\varphi(0) \wedge \forall y\, (\varphi(y) \rightarrow \varphi(S(y)))] \rightarrow \forall y \varphi(y)$$

While PA obviously formalises an important part of number-theoretic practice, it was axiomatised only in 1934.[32] We now turn to second-order arithmetic:

Definition 1.10: *The theory of* second-order Peano arithmetic, PA_2, *is given by axioms (Q1)–(Q3) of Definition 1.9, the Comprehension Schema of §1.11, and the following mathematical Induction Axiom:*

$$\forall X([X(0) \wedge \forall y(X(y) \rightarrow X(S(y)))] \rightarrow \forall y X(y))$$

With the exception of the Comprehension Schema, the axioms of PA_2 were first explicitly written down by Dedekind.[33] The Choice Schema is typically not built into axiomatisations of PA_2, although it is valid on the standard semantics.[34]

Note that the signature of PA_2 is just $\{0, S\}$, whereas the signature of the first-order theory PA is $\{0, S, <, +, \times\}$. However, in the setting of PA_2, order, addition and multiplication are explicitly definable in the sense of Definition 1.8. For instance, the graph of the addition function is the unique three-place relation which is the union of all three-place relations satisfying the following condition, which intuitively describes an initial segment of the graph of addition:

$$\Phi(B) := \forall x B(x, 0, x) \wedge \forall x \forall y \forall w [B(x, S(y), w) \rightarrow \exists z\, (w = S(z) \wedge B(x, y, z))]$$

By Comprehension, there is a three-place relation A satisfying $A(a, b, c)$ iff $\exists B(\Phi(B) \wedge B(a, b, c))$. If we then define $a + b = c$ by $A(a, b, c)$ we can easily show by induction that this satisfies axioms (Q4)–(Q5) of Definition 1.9. An analogous definition can be presented in second-order logic for a formula which satisfies axioms (Q6)–(Q7). Finally, obviously (Q8) allows $\leq$ to be explicitly defined in terms of addition and first-order logic.

[32] Hilbert and Bernays (1934). For contemporary references on PA and its subsystems, see e.g. Kaye (1991) and Hájek and Pudlák (1998).

[33] Dedekind (1888).

[34] A contemporary reference on PA_2 and its subsystems is Simpson (2009).

1.B First- and second-order set theory

We now turn to set theory. The signature of set theory consists just of the binary relation $\in$, where we read $x \in y$ as 'x is a member of y'. We start with the following axioms, which we state slightly informally, leaving the reader to transcribe them into sentences of first-order logic if she wishes. Here and throughout, $(\forall y \in x)\varphi$ abbreviates $\forall y(y \in x \rightarrow \varphi)$ and $(\exists y \in x)\varphi$ abbreviates $\exists y(y \in x \wedge \varphi)$.

Extensionality. *For all x and y, we have: $x = y$ iff $\forall z\,(z \in x \leftrightarrow z \in y)$*
Pairing. *For all x and y, there is a unique set, $\{x,y\}$, such that for all z: $z \in \{x,y\}$ iff either $z = x$ or $z = y$*
Union. *For all x, there is a unique set, $\bigcup x$, , such that for all z: $z \in \bigcup x$ iff $(\exists y \in x)z \in y$*
Power Set. *For all x, there is a unique set, $\wp(x)$, such that for all z: $z \in \wp(x)$ iff $z \subseteq x$*
Separation Schema. *For all x and $\bar{v}$ there is a unique set, $\{y \in x : \varphi(y,\bar{v})\}$, such that for all z: $z \in \{y \in x : \varphi(y,\bar{v})\}$ iff both $z \in x$ and $\varphi(z,\bar{v})$*

In the Separation Schema, there is one axiom for each formula $\varphi(y,\bar{v})$ in the signature. It is worth noting that the uniqueness claims in Pairing, Union, Power Set, and the Separation Schema are redundant, given Extensionality,[35] and that the left-to-right directions of the biconditionals in Pairing, Union, and Power Set are redundant, given the Separation Schema. For instance, suppose that for all x and y there is some v such that if $z = x$ or $z = y$ then $z \in v$. Then $\{z \in v : z = x \vee z = y\}$ exists by Separation and is obviously equal to $\{x,y\}$.

Using these axioms, we define $\varnothing$ as the unique set with no members; the empty set. Whilst there are *philosophical* discussions to have about $\varnothing$'s existence,[36] there are no *technical* discussions to be had. The usual background axioms for first-order logic assert that there exists at least one object x, and applying Separation to the formula $z \neq z$ we obtain a set $\varnothing$ such that, $\forall z\,(z \in \varnothing \leftrightarrow (z \in x \wedge z \neq z))$, from which it follows by elementary logic that $\forall z\, z \notin \varnothing$. The uniqueness of the empty set then follows from Extensionality.

The intersection of x, written $\bigcap x$, is the set whose members elements are exactly those which are members of every element of x. This exists whenever x is non-empty, since $\bigcap x = \{y \in \bigcup x : (\forall z \in x)y \in z\}$, which exists by Union and Separation. The usual binary operations of union $x \cup y$ and intersection $x \cap y$ can then be defined via $x \cup y = \bigcup\{x,y\}$ and $x \cap y = \bigcap\{x,y\}$. Finally, the singleton $\{x\}$ is defined to be $\{x,x\}$ and is the set whose unique member is x.

We define the successor $s(x)$ of x to be the set $x \cup \{x\}$, so that $z \in s(x)$ iff either $z = x$ or $z \in x$. This notation allows us to state another axiom:

Infinity. *There is a set w such that $\varnothing \in w$ and for all x, if $x \in w$ then $s(x) \in w$*

[35] For philosophical commentary on uniqueness, see Potter (2004: 258–9).
[36] See e.g. Oliver and Smiley (2006: 126–32).

The empty set $\varnothing$ plays a role in set theory similar to the role zero plays in arithmetic, and the successor function s in set theory is similar to the successor function S from the axioms of Definition 1.9. In these terms, the Infinity Axiom says that there is a set which contains the ersatz of zero and is closed under the ersatz of successor.

Using the intersection operation, defined above, we can also state another axiom, whose role is to rule out infinite descending membership chains:

Foundation. *For every non-empty set x there is some $z \in x$ such that $z \cap x = \varnothing$*

After all, if an infinite chain $\ldots \in x_n \in \ldots \in x_2 \in x_1 \in x_0$ existed, then the non-empty set $x = \{x_0, x_1, x_2, \ldots, x_n, \ldots\}$ would violate Foundation.

Introducing the usual notation $\exists! x\varphi$ to abbreviate $\exists x \forall v(\varphi \leftrightarrow x = v)$, for any variable v not occurring in φ, we lay down an axiom schema which, intuitively, states that the image of any set under a function is a set:

Replacement Schema. *For all w and all $\bar{v}$: if $(\forall x \in w)\exists! y\varphi(x, y, \bar{v})$, then $\exists z(\forall x \in w)(\exists y \in z)\varphi(x, y, \bar{v})$*

Finally, we lay down an axiom stating that any set can be equipped with a binary relation that satisfies the axioms of a well-order:

Choice. *Any set can be well-ordered*

A well-order is a linear order such that any non-empty set of ordered elements has a least element. (The axioms of a linear order were given in §1.12.) Note that Choice, here, is a single axiom, expressed in first-order logic with an additional primitive, $\in$. This single Axiom should *not* be confused with the Choice Schema for second-order logic, as laid down in §1.11, which yields infinitely many second-order sentences. That said, there is evidently a connection between the Axiom and the Schema: the Axiom of Choice (in our model theory) entails that the full semantics for second-order logic always satisfies the Choice Schema, since one can use a well-order of the underlying domain of the model (or one of its finite products) to obtain the relevant witnesses for the Choice Schema.

Having discussed the axioms, we can finally define some theories:[37]

Definition 1.11: *The axioms of* first-order Zermelo–Fraenkel set theory, ZF, *are Extensionality, Pairing, Union, Power Set, Infinity, Foundation, the Separation Schema, and the Replacement Schema. The theory* ZFC *adds Choice to* ZF.

We can form second-order versions of these theories by replacing the first-order schemas with appropriate second-order sentences. In particular, we replace the

[37] A contemporary reference for ZFC is e.g. the monograph Kunen (1980).

Separation and Replacement *Schemas* with simple *Axioms*, i.e. individual sentences of second-order logic with an additional primitive, $\in$:

Separation. $\forall F \forall x \exists y \forall w\, [w \in y \leftrightarrow (w \in x \wedge F(w))]$
Replacement. $\forall G \forall w\, [(\forall x \in w)\exists! y G(x,y) \rightarrow \exists z (\forall x \in w)(\exists y \in z) G(x,y)]$

We then define:

Definition 1.12: *The theory of* second-order Zermelo–Fraenkel set theory with Choice, ZFC_2, *is formed by taking the axioms of first-order* ZFC, *and replacing the Separation Schema with the Separation Axiom, and the Replacement Schema with the Replacement Axiom, and by adding on the Comprehension Schema.*

As with second-order arithmetic, the Choice Schema is not built into these theories, and should not be confused with the (set-theoretic) Axiom of Choice. The theory ZFC_2 is sometimes also called *Kelly–Morse set theory*.[38] While second-order set theory is less widely used than first-order set theory, it plays an important role in the foundations and philosophy of set theory. We discuss this in Chapters 8 and 11.

Occasionally, but especially from Chapter 7 onwards, we invoke elementary considerations about ordinals and cardinals. As is usual, we reserve $\alpha, \beta, \gamma, \delta$ for ordinals. An *ordinal* is defined to be a transitive set which is well-ordered by membership, where x is transitive iff every member of x is a subset of x. The membership relation on ordinals is usually just written with $<$, and it is provable in very weak fragments of ZFC that $<$ well-orders the ordinals. The successor operation $s(\alpha) = \alpha \cup \{\alpha\} = \alpha + 1$ on ordinals is such that $\alpha < s(\alpha)$ and there is no ordinal β with $\alpha < \beta < s(\alpha)$. We define $0 = \varnothing$, then $1 = s(0), 2 = s(1), 3 = s(2), \dots$, and $\omega = \{0, 1, 2, 3, \dots\}$. A limit ordinal is an ordinal β such that $\beta \neq 0$ and $\beta \neq s(\gamma)$ for any ordinal γ; and ω is the least limit ordinal.

A *cardinal* is an ordinal which is not bijective with any smaller ordinal. The finite ordinals $0, 1, 2, \dots$ and ω are all cardinals. The aleph sequence provides the standard enumeration of infinite cardinals: $\aleph_0 = \omega$; $\aleph_{\alpha+1}$ is the least cardinal greater $\aleph_\alpha$; and when α is a limit ordinal, the cardinal $\aleph_\alpha$ is the least upper bound of $\{\aleph_\beta : \beta < \alpha\}$. Hence $\aleph_\omega$ is the least ordinal which is greater than $\aleph_0, \aleph_1, \aleph_2, \dots$ and it too is a cardinal. We reserve κ, λ for cardinals, and we use $|X|$ for the *cardinality* of the set X, that is $|X| = \kappa$ iff X is bijective with κ but with no smaller ordinal. We frequently invoke the facts that $|X \times Y| = \max\{|X|, |Y|\}$ when one of $|X|, |Y|$ is infinite, and that the union of $\leq \kappa$-many sets of cardinality $\leq \kappa$ itself has cardinality $\leq \kappa$ when κ is infinite.[39]

---

[38] See Monk (1969) for an axiomatic development of set theory in this framework.

[39] These elementary facts about cardinality can be found in any set-theory textbook, such as Hrbáček and Jech (1999) or the beginning chapters of Kunen (1980) or Jech (2003).

1.C Deductive systems

In several places in this book, we will need to refer to a deductive system for first-order and second-order logics. Many different but provably equivalent deductive systems are possible, and we could compare and contrast their relative technical and philosophical merits. However, deduction is not the focus of this book, so we will simply set down a system of natural deduction without much comment.[40]

To be clear: we do not expect anyone to be able to learn how to use or manipulate natural deductions just by reading this appendix. Equally, we did not expect that anyone could learn how to do arithmetic or set theory just by reading the previous two appendices. The aim is just to lay down a particular system, so that we can refer back to it later in this book.

First, we lay down rules for the sentential connectives. In the rules ¬E, ∨E, and →I, an assumption is *discharged* at the point when the rule is applied. We mark this using square brackets, and a cross-referencing index, *n*:

$$\frac{\bot}{\varphi}\ \text{Ex} \qquad \frac{\varphi \quad \neg\varphi}{\bot}\ \text{Raa}$$

$$\frac{\begin{matrix}[\varphi]^n \\ \vdots \\ \bot\end{matrix}}{\neg\varphi}\ \neg\text{I}, n \qquad \frac{\begin{matrix}[\neg\varphi]^n \\ \vdots \\ \bot\end{matrix}}{\varphi}\ \neg\text{E}, n$$

$$\frac{\varphi \quad \psi}{(\varphi \wedge \psi)}\ \wedge\text{I} \qquad \frac{(\varphi \wedge \psi)}{\varphi}\ \wedge\text{E} \qquad \frac{(\varphi \wedge \psi)}{\psi}\ \wedge\text{E}$$

$$\frac{\varphi}{(\varphi \vee \psi)}\ \vee\text{I} \qquad \frac{\psi}{(\varphi \vee \psi)}\ \vee\text{I} \qquad \frac{(\varphi \vee \psi) \quad \begin{matrix}[\varphi]^n \\ \vdots \\ \chi\end{matrix} \quad \begin{matrix}[\psi]^n \\ \vdots \\ \chi\end{matrix}}{\chi}\ \vee\text{E}, n$$

$$\frac{\begin{matrix}[\varphi]^n \\ \vdots \\ \psi\end{matrix}}{(\varphi \rightarrow \psi)}\ \rightarrow\text{I}, n \qquad \frac{\varphi \quad (\varphi \rightarrow \psi)}{\psi}\ \rightarrow\text{E}$$

We now consider the rules for first-order quantifiers. These rules are subject to the following restrictions: t can be any term; in ∀I, c must not occur in any undischarged assumption on which $\varphi(c)$ depends; in ∃I one can replace any/all occurrences of t with x, but in ∀I one must replace *all* occurrences of c with x, and in both of these rules x should not already occur in $\varphi(c)$; finally, in implementing ∃E, c must not occur in $\exists x\varphi(x)$, in ψ, or in any undischarged assumption on which ψ depends, except for $\varphi(c)$.

[40] It is essentially based on Prawitz (1965).

$$\frac{\varphi(c)}{\forall x \varphi(x)}\ \forall\text{I} \qquad \frac{\forall x \varphi(x)}{\varphi(t)}\ \forall\text{E}$$

$$\frac{\varphi(t)}{\exists x \varphi(x)}\ \exists\text{I} \qquad \frac{\exists x \varphi(x) \quad \begin{matrix}[\varphi(c)]^n \\ \vdots \\ \psi\end{matrix}}{\psi}\ \exists\text{E}, n$$

To complete the rules for first-order logic, we have the rules for identity. Note that adopting the rule =I is equivalent to treating every instance of $t = t$ as an *axiom*, since it is licensed on any (including no) assumptions:

$$\frac{}{t = t}\ {=}\text{I} \qquad \frac{t_1 = t_2 \quad \varphi(t_1)}{\varphi(t_2)}\ {=}\text{E} \qquad \frac{t_2 = t_1 \quad \varphi(t_1)}{\varphi(t_2)}\ {=}\text{E}$$

To move to a deduction system for second-order logic, we simply add rules for the quantifiers, exactly analogous to the first-order case. So, for relation-variables we have (with similar restrictions as before):

$$\frac{\varphi(R^m)}{\forall X^m \varphi(X^m)}\ \forall_2\text{I} \qquad \frac{\forall X^m \varphi(X^m)}{\varphi(R^m)}\ \forall_2\text{E}$$

$$\frac{\varphi(R^m)}{\exists X^m \varphi(X^m)}\ \exists_2\text{I} \qquad \frac{\exists X^m \varphi(X^m) \quad \begin{matrix}[\varphi(R^m)]^n \\ \vdots \\ \psi\end{matrix}}{\psi}\ \exists_2\text{E}, n$$

The case of function symbols is exactly similar. Finally, to ensure that our deduction system aligns with *faithful* Henkin models, we also allow as axioms any instance of the Comprehension or Choice schemas, i.e. we add these rules:

$$\frac{}{\exists X^n \forall \bar{v}(\varphi(\bar{v}) \leftrightarrow X^n(\bar{v}))}\ \text{Comp}$$

$$\frac{}{\forall X^{n+1}\left(\forall \bar{v} \exists y\, X^{n+1}(\bar{v}, y) \rightarrow \exists p^n \forall \bar{v}\, X^{n+1}(\bar{v}, p^n(\bar{v}))\right)}\ \text{Choice}$$

These are all the rules for our deduction systems for sentential, first-order and second-order logic. When we have a deduction whose only undischarged assumptions are members of T and which ends with the line φ, we write $T \vdash \varphi$.

2

Permutations and referential indeterminacy

In Chapter 1, we introduced some of the most basic ideas in model theory: structures, signatures, and satisfaction. In this chapter, we introduce the fundamental notion of an *isomorphism*. This provides the technical basis for several philosophical issues made famous by Benacerraf and Putnam, concerning both the 'intuitive' idea of a mathematical structure and referential indeterminacy.

2.1 Isomorphism and the Push-Through Construction

One of the most fundamental ideas in model theory is *isomorphism*. We come to the idea of isomorphism via the notion of a map which 'preserves structure'.

Definition 2.1: *Let $\mathcal{M}$ and $\mathcal{N}$ be $\mathscr{L}$-structures. A bijection $h : M \longrightarrow N$ is an* isomorphism *from $\mathcal{M}$ to $\mathcal{N}$ iff: for any $\mathscr{L}$-constant symbol c, any n-place $\mathscr{L}$-relation symbol R, any n-place $\mathscr{L}$-function symbol f, and all $a_1, \ldots, a_n$ from M:*

$$h(c^{\mathcal{M}}) = c^{\mathcal{N}}$$
$$(a_1, \ldots, a_n) \in R^{\mathcal{M}} \text{ iff } (h(a_1), \ldots, h(a_n)) \in R^{\mathcal{N}}$$
$$h(f^{\mathcal{M}}(a_1, \ldots, a_n)) = f^{\mathcal{N}}(h(a_1), \ldots, h(a_n))$$

When there is an isomorphism from $\mathcal{M}$ to $\mathcal{N}$, we say that $\mathcal{M}$ and $\mathcal{N}$ are isomorphic, *and write $\mathcal{M} \cong \mathcal{N}$.*

We continue to use overlining to discuss tuples, as introduced §1.2. So $\bar{a}$ will be some sequence of elements $(a_1, \ldots, a_n)$. We also introduce some notation to allow tuples to interact easily with functions. The idea is to 'push h through' sets of elements of M and N, through sets of sets of elements of M and N, and so on:

Definition 2.2: *Let $h : M \longrightarrow N$ be any function and let $\bar{a}$ be from M. We define $\hbar(\bar{a}) = (h(a_1), \ldots, h(a_n))$. For each $X \subseteq M^n$, we define $\hbar(X) = \{\hbar(\bar{a}) : \bar{a} \in X\}$. Likewise, for each $Y \subseteq \wp(M^n)$ we define $\hbar(Y) = \{\hbar(X) : X \in Y\}$.*

In these terms, we can rewrite the last part of the definition of an isomorphism as $h(f^{\mathcal{M}}(\bar{a})) = f^{\mathcal{N}}(\hbar(\bar{a}))$. Where no ambiguity can arise, we sometimes simply write $h(\bar{a})$ rather than $\hbar(\bar{a})$. Now, we have explicitly written out the definition of $\hbar$ on the first couple of levels of the set-theoretic hierarchy above M, but the definition generalises naturally to higher levels. In each case, we simply define the action of $\hbar$ on a higher-level object X as the set which collects together the action of $\hbar$ on all of X's members.

There are many equivalent ways to define the notion of an isomorphism:

Theorem 2.3: *For any $\mathscr{L}$-structures $\mathcal{M}$ and $\mathcal{N}$ and any bijection $h : M \longrightarrow N$, the following are equivalent:*

(1) *h is an isomorphism from $\mathcal{M}$ to $\mathcal{N}$*
(2) *$\mathcal{M} \vDash \varphi(\bar{a})$ iff $\mathcal{N} \vDash \varphi(\hbar(\bar{a}))$, for all $\bar{a}$ from M^n and all atomic $\mathscr{L}$-formulas $\varphi(\bar{v})$ with free variables displayed*
(3) *$\mathcal{M} \vDash \varphi(\bar{a})$ iff $\mathcal{N} \vDash \varphi(\hbar(\bar{a}))$, for all $\bar{a}$ from M^n and all first-order $\mathscr{L}$-formulas $\varphi(\bar{v})$ with free variables displayed*
(4) *$\mathcal{M} \vDash \varphi(\bar{a})$ iff $\mathcal{N} \vDash \varphi(\hbar(\bar{a}))$, for all $\bar{a}$ from M^n and all second-order $\mathscr{L}$-formulas $\varphi(\bar{v})$ with free variables displayed, with consequence read either via the full or the Henkin semantics for second-order logic (see §§1.10–1.11)*

The proof of this result involves a lengthy induction on complexity of formulas, which we relegate to §2.B. But Theorem 2.3 has an immediate, important corollary. Since sentences are just formulas with no free variables, the entailment (1) $\Rightarrow$ (3) shows that isomorphic structures make exactly the same first-order sentences true. This idea is significant enough to merit some new terminology.

Definition 2.4: *Let $\mathcal{M}$ and $\mathcal{N}$ be $\mathscr{L}$-structures. We say that $\mathcal{M}$ and $\mathcal{N}$ are* elementarily equivalent, *written $\mathcal{M} \equiv \mathcal{N}$, iff they satisfy exactly the same $\mathscr{L}$-sentences, i.e. $\mathcal{M} \vDash \varphi$ iff $\mathcal{N} \vDash \varphi$, for all $\mathscr{L}$-sentences φ.*

With this notation, the corollary of Theorem 2.3 which we just observed becomes:

Corollary 2.5: *If $\mathcal{M} \cong \mathcal{N}$, then $\mathcal{M} \equiv \mathcal{N}$.*

The *converse* to Corollary 2.5 is false. However, showing this requires a slightly different set of tools, and so we defer discussion of this point until Chapter 4.

Isomorphic—and so elementarily equivalent—structures are very easy to construct. Indeed, given any structure and any bijection whose domain is the structure's underlying domain, we can treat that bijection as an isomorphism. This is, in fact, one of the most basic constructions in model theory.

The Push-Through Construction. *Let $\mathcal{L}$ be any signature, let $\mathcal{M}$ be any $\mathcal{L}$-structure with domain M, and let $h : M \longrightarrow N$ be any bijection. We use h to define an $\mathcal{L}$-structure, $\mathcal{N}$, with domain N, by defining $s^{\mathcal{N}} = \hbar(s^{\mathcal{M}})$ for each $\mathcal{L}$-symbol s.*[1] *So, for any $\mathcal{L}$-constant symbol c, any n-place $\mathcal{L}$-relation symbol R, any n-place $\mathcal{L}$-function symbol f, and all $\bar{a}$ from M^n:*

$$c^{\mathcal{N}} = h(c^{\mathcal{M}})$$
$$R^{\mathcal{N}} = \hbar(R^{\mathcal{M}}) = \{\hbar(\bar{a}) : \bar{a} \in R^{\mathcal{M}}\}$$
$$f^{\mathcal{N}} = h \circ f^{\mathcal{M}} \circ \hbar^{-1}, \text{ so that } f^{\mathcal{N}}(\hbar(\bar{a})) = h(f^{\mathcal{M}}(\bar{a}))$$

In this, $\hbar^{-1}(\bar{b}) = \bar{a}$ iff $\hbar(\bar{a}) = \bar{b}$. We may write $\hbar : \mathcal{M} \longrightarrow \mathcal{N}$ to indicate that we are considering the function built from h which induces $\mathcal{L}$-structure.

As mentioned, this Construction is an extremely simple way to generate new structures from old ones. But it has a surprising number of rich philosophical consequences, which we explore in this chapter and the next.

2.2 Benacerraf's use of Push-Through

Since the Push-Through Construction makes isomorphic copies of structures *so* easy to come by, it is no surprise that, for many mathematical purposes, it seems not to matter which of two isomorphic models one works with.

The most famous philosophical statement of this point is due to Benacerraf. He focussed specifically on the fact that, when doing arithmetic, it makes no difference whether we think of the natural numbers as Zermelo's finite ordinals:

$$\varnothing, \{\varnothing\}, \{\{\varnothing\}\}, \{\{\{\varnothing\}\}\}, \ldots$$

or as von Neumann's finite ordinals:[2]

$$\varnothing, \{\varnothing\}, \{\varnothing, \{\varnothing\}\}, \{\varnothing, \{\varnothing\}, \{\varnothing, \{\varnothing\}\}\}, \ldots$$

Benacerraf maintained:

For arithmetical purposes, the properties of numbers which do not stem from the relations they bear to one another in virtue of being arranged in a progression are of no consequence whatsoever.[3]

[1] We can also define the construction when $\mathcal{M}$ is a Henkin $\mathcal{L}$-structure, as in Definition 1.6. The idea is to keep pushing h through the range of $\mathcal{N}$'s second-order variables. So we set $N_n^{\text{rel}} = \hbar(M_n^{\text{rel}})$ and $N_n^{\text{fun}} = \hbar(M_n^{\text{rel}})$. Note that there is no guarantee that $R^{\mathcal{N}} = \hbar(R^{\mathcal{M}})$ should be a member of M_n^{rel}, even when $M = N$; the existence of $R^{\mathcal{N}}$ is guaranteed by the *ambient* set-theoretic framework within which we are working. Similarly, we should not expect that $N_n^{\text{rel}} = M_n^{\text{rel}}$ or that $N_n^{\text{fun}} = M_n^{\text{fun}}$.

[2] By our definition of $s(x)$ in §1.B, the finite von Neumann ordinals are the sets $s^n(\varnothing)$ for each n.

[3] Benacerraf (1965: 69-70).

Otherwise put: for arithmetical purposes, all we require is that the progression is both long enough and also appropriately structured. Since we can always use the Push-Through Construction to induce the appropriate structure, this reduces to the requirement that we have *enough* things.

Such observations suggest that mathematicians—and so, perhaps, philosophers—can typically focus only on 'mathematical structure', informally construed. Moreover, this attitude is reflected in some parts of mathematical practice. Mathematical discourse is rich with apparent definite descriptions of mathematical structures, like '*the* natural number structure' or '*the* Klein four-group'; but, given the Push-Through Construction and related issues, it is equally rich with the idea that we only care about the identity of such entities 'up to isomorphism'.

This motivates a compelling idea: *mathematical structures, as discussed informally by mathematicians, are best explicated by isomorphism types.* An *isomorphism type* is just a class of isomorphic models.[4]

We discuss this idea in much more detail in Part B, under the name *modelism.* For now, though, we should perhaps stress that this explication need not be presented as an *ontological* hypothesis. The point of treating informal-structures as isomorphism types is that it will allow us to deploy model theory when discussing philosophical issues that arise within mathematics. The idea, then, is to think of model theory as a branch of *applied mathematics,* whose target area of application is mathematical discourse and mathematical practice (cf. §1.5).

In particular, if a mathematician claims that the theory of arithmetic picks out '*the* natural number structure', the idea will be that the theory of arithmetic picks out a particular equivalence class of isomorphic models. This is a very natural claim to make, in the case of arithmetic. But we should emphasise that not every mathematical theory is like the arithmetic in this respect. Group theory does not aim to pick out any particular mathematical structure (informally construed): the whole *point* of group theory is that it can be applied to many different mathematical areas, such as modular arithmetic, certain classes of permutations, and so forth. Similar points hold for theories governing rings, fields, topologies, and so on. Following Shapiro, we call these kinds of theories *algebraic.*[5] These are to be contrasted with *univocal* theories, like arithmetic, which aim to 'describ[e] a certain definite mathematical domain'[6] or to 'specify *one particular interpretation*',[7] loosely speaking.

This kind of view can, however, easily lead to the view that mathematical language displays a high degree of referential indeterminacy, even for univocal theo-

[4] Unfortunately, there is an infelicity here. A model theorist's structures are always relative to a specific *signature*. Consequently, the natural numbers in the signature $\{0, S\}$ induce a *different* equivalence class of isomorphic structures than the natural numbers in the signature $\{S\}$. We revisit this in Chapter 5.

[5] Shapiro (1997: 40–1).

[6] Grzegorczyk (1962: 39).

[7] Kline (1980: 273).

ries. Suppose that we do, indeed, hold that when we talk about 'the natural numbers', we are really just discussing a particular isomorphism type. Importantly, no single object in that isomorphism type can be thought of as *the* natural number 27, for example. Rather, every structure in that isomorphism type has its own 27th element. Moreover, by considering the Push-Through Construction, every object is the 27th element of *some* model of arithmetic. So, when we consider the arithmetical term '27', there is no *single* object in the isomorphism type for that term to pick out.[8] Indeed, on this view, if '27' refers at all, then it surely refers to all of 'the 27s' of all of the isomorphic models equally, i.e. it refers to *every* object equally. And this is just to say that the our arithmetical vocabulary is radically referentially indeterminate.

2.3 Putnam's use of Push-Through

We just used the Push-Through Construction to motivate a focus on 'mathematical structure' (informally construed). In turn, this led us to consider referential indeterminacy within mathematics. But an equally celebrated use of the Push-Through Construction, made famous by Putnam, leads to referential indeterminacy more directly.[9]

The permutation argument

Philosophers sometimes use model theory to explicate the intuitive notions of reference and truth. To take a toy example, consider a simple theory, stated in natural English, consisting of three sentences: 'Ajax is a cat', 'Betty is a cat', and 'Chad is not a cat'.[10] We might formalise this using a one-place predicate, C, and three constant symbols, c_1, c_2, c_3, as the following theory: $\{C(c_1), C(c_2), \neg C(c_3)\}$. We can then provide a model of this theory, $\mathcal{M}$, as follows:

$$M = \{\text{Ajax, Betty, Chad}\} \qquad c_1^{\mathcal{M}} = \text{Ajax}$$
$$C^{\mathcal{M}} = \{\text{Ajax, Betty}\} \qquad c_2^{\mathcal{M}} = \text{Betty}$$
$$c_3^{\mathcal{M}} = \text{Chad}$$

Now we think of $\mathcal{M}$ as explicating the reference relation. Its formal signature tracks our natural language vocabulary, with, for example, 'c_1' acting as a surrogate for the English word 'Ajax', and 'C' acting as a surrogate for our predicate '... is a cat'. And,

[8] Few formal languages have '27' as a primitive term, but we can just use the term $S^{27}(0)$, as defined in (*numerals*) of §1.13.

[9] The idea has a long history; for details, see Button (2013: 14fn.1–2, 18fn.7, 27fn.2).

[10] For the toy, see Putnam (1977: 484) and Button (2013: 14–15).

viewed thus, $\mathcal{M}$ explicates reference as follows: the name 'Ajax' refers to Ajax, the name 'Betty' refers to Betty, the name 'Chad' refers to Chad, and the predicate '... is a cat' picks out the only (relevant) cats, namely Ajax and Betty. Similarly, the fact that $\mathcal{M} \vDash C(c_1)$ is taken to explicate why 'Ajax is a cat' is *true*.

With this explication as a backdrop, Putnam used model theory to raise philosophical questions about reference. Suppose we start with a model which makes true everything which, intuitively, *should* be true. (The theory this model satisfies will, of course, be much more complicated than our toy theory, but the idea is the same.) Putnam now uses a Push-Through Construction to generate a distinct but isomorphic model. Since the two models are isomorphic, they make true exactly the same sentences; but since they are distinct, they differ on the (explicated) reference of some symbol. And this raises a philosophical question: What makes one model, rather than the other, a better explication of the reference relation? More briefly: *What, if anything, fixes reference?*

Putnam focussed specifically on the case where we generate a new model by permuting the underlying domain of the given model, rather than by substituting in new objects; hence his argument is called the *permutation argument*. We can illustrate this with our toy example. Let h be a permutation on our earlier domain, M, such that $h(\text{Ajax}) = \text{Betty}$ and $h(\text{Betty}) = \text{Chad}$ and $h(\text{Chad}) = \text{Ajax}$. Then the model we generate by the Push-Through Construction, $\mathcal{N}$, has the same domain as our original model, but differs on the interpretation of every symbol:

$$N = \hbar(M) = M \qquad c_1^{\mathcal{N}} = h(c_1^{\mathcal{M}}) = \text{Betty}$$

$$C^{\mathcal{N}} = \hbar(C^{\mathcal{M}}) = \{\text{Betty}, \text{Chad}\} \qquad c_2^{\mathcal{N}} = h(c_2^{\mathcal{M}}) = \text{Chad}$$

$$c_3^{\mathcal{N}} = h(c_3^{\mathcal{M}}) = \text{Ajax}$$

So, comparing the rival explications of reference provided by $\mathcal{M}$ and $\mathcal{N}$, we must now ask why our predicate '... is a cat' picks out Ajax and Betty (as $\mathcal{M}$ suggests) rather than Betty and Chad (as $\mathcal{N}$ suggests).

The issue generalises rapidly. By Pushing-Through, we can see that any name could be taken to refer to *anything*, that any one-place predicate could be taken to pick out *any* collection of things (provided only that there are enough of them), and similarly for all the other expressions of our language. We will stare into the abyss of *radical referential indeterminacy*, where every word refers equally to everything, which is just to say that nothing refers at all.[11]

Preferable models

A natural reaction to Putnam's permutation argument, and to this threat of radical referential indeterminacy, is to claim that some models are simply more *prefer-*

[11] Concerning why this is an abyss, see Button (2013: 59–60).

able than others as candidates for explicating reference. Howsoever 'preferability' is spelled out, the crucial thought is that isomorphic models can differ as regards their preferability.[12]

In response, Putnam insisted that anyone advancing this line must outline the required notion of preferability and explain why preferable models (so construed) are better at explicating reference.[13] In doing so, Putnam maintained, his opponent would have to provide a *theory* of preferability. But the permutation argument establishes that *any* theory has multiple models, *including* this theory of preferability (taken together, if we like, with anything else we want to say). So Putnam maintained that the theory of preferability is *just more theory*—more grist to be Pushed-Through the permutation-mill—and so cannot help to pin down reference. For obvious reasons, Putnam's general argumentative strategy here is called the *just-more-theory manoeuvre.*

This manoeuvre has been widely criticised.[14] It is not hard to see why. To continue with our toy example: $\mathcal{M}$ suggested that the predicate '...is a cat' picks out the cats, Ajax and Betty; whereas our permuted model, $\mathcal{N}$, suggested that the same predicate picks out a mixture of cats and non-cats, namely Betty and Chad. This permuted model therefore seems to ignore the *causal* relationships that link our use of the word 'cat' to the cats. This leads to the thought that a preferable model must (among other things) respect certain causal constraints on language–object relations. In this context, the just-more-theory manoeuvre amounts to noting that we can reinterpret the word 'causation'. But if causation *does* fix reference, then this *re*interpretation is just a *mis*interpretation, with no philosophical significance.

We just considered glossing preferability in terms of causation. We will discuss an alternative account of preferability, pioneered by Lewis, in §2.A. But the above is not intended as a final word *at all.* When it comes to Putnam's permutation argument, as it applies generally, it is only the *beginning* of the story.[15]

Referential indeterminacy for moderate objects-platonism

In this book, however, we tend to be less concerned with physical objects, like cats and cherries, and more concerned with *mathematical* objects. So we will now focus on the mathematical case.

It is easy to see that Putnam's permutation argument and his just-more-theory manoeuvre are more immediately threatening in mathematical contexts. In brief:

---

[12] See Merrill (1980: 80) and Lewis (1984: 227–8).

[13] Putnam (1977: 486–7, 1980: 477, 1981: 45–8, 1983: ix). For partial anticipations of this response, see Winnie (1967: 228–9), Field (1975: 383–4), and Wallace (1979: 309–11).

[14] See e.g. Lewis (1984: 225) and Bays (2001: 342–8, 2008: 197–207); for more references, see Button (2013: 29fn.8).

[15] For a much fuller story, see Button (2013: 3, 27–73). We should emphasise that Putnam never *advocated* radical referential indeterminacy, but used it as a kind of reductio against his opponents.

certain rather natural positions in the philosophy of mathematics are bound, by their own lights, to treat the notion of a preferable reference candidate as *just more theory*. To explain this point, we will introduce a particular position within the philosophy of mathematics, which we call *moderate objects-platonism*, which must accept that mathematical language is radically referentially indeterminate. (But, as we will explain in §2.5, we do not think that this is necessarily a *problem*.)

Moderate objects-platonism has essentially two components: moderation and objects-platonism. We begin with the latter. As we use the phrase, *objects-platonists* believe that mathematical entities are genuine objects. When they hear someone say, for example, 'there are infinitely many prime numbers', they take that as a face-value existential claim. *There really are prime numbers*, they claim, *and indeed infinitely many of them*. They do not want to paraphrase away this claim, to remove the apparently existential commitment (i.e. they are not nominalists). They do not think that this is a claim which is only to be made in the context of some 'pretence' that mathematicians like to indulge in (i.e. they are not fictionalists). They take the claim it as it stands. Furthermore, objects-platonists have a minimal thesis concerning the *nature* of these mathematical objects, namely: they are abstract entities, neither spatial nor temporal, and so in particular they are not our *creations*. (This does not mean that they refuse to talk of 'constructing' mathematical entities; but they always regard such talk as a helpful heuristic, rather than a metaphysical claim.) Our moderate objects-platonist is an *objects-platonist* in exactly this sense.

We now consider *moderation*. The use of this expression in this context is due to Putnam, and it is very much a term of art, indicating a particular kind of *naturalism*. Now, the word 'naturalism' has come to mean many things in philosophy,[16] but the moderate objects-platonist buys into a very *specific* brand of naturalism. At a very rough first pass, the moderate has an image of human beings as closer to apes than angels. Consequently, when she speaks of human *faculties*, she wants those faculties to be the kinds of things which could plausibly both have evolved within a species, and also could have developed within an individual creature as it grew from a foetus into an adult. Broadly speaking, then, she is closer aligned to empiricism than to rationalism. And, in particular, the moderate rejects all appeals to faculties of *mathematical intuition*, or anything similar, within her philosophy of mathematics. Our moderate objects-platonist is a *moderate*, in exactly this sense.

Admittedly, this is a somewhat rough sketch of moderate objects-platonism. But we doubt that a more precise statement is either possible or desirable. For, on the one hand, moderate objects-platonism is not so much a declared and precise position, as a zeitgeisty *default* for contemporary would-be platonists. And, on the other hand, we do not need to offer anything more precise in order to show that moderate objects-platonists face certain issues concerning the determinacy of reference.

---

[16] See, in particular, Maddy (2005).

Here is how those issues arise. Our moderate objects-platonist regards mathematical objects as abstract entities which we do not create. Still, she thinks that we talk about them. But, since they are abstract, she accepts that we cannot fix reference to mathematical entities by *seeing* them, *pointing* to them, or *interacting* with them in any way; her moderation rules that out. Indeed, she will follow Putnam in rhetorically asking:

> What neural process, after all, could be described as the perception of a mathematical object? Why of *one* mathematical object rather than another?[17]

A better thought is that we come to refer to mathematical entities after some process of mathematical education. But, for a moderate, there is only so much that education can achieve. As Dummett quipped in this context, 'No amount of training will teach a chimpanzee to talk.' Since the moderate sees humans as closer to apes than angels, she cannot believe that the process of education 'awakens' something 'latent in the child's mind'.[18] If learning some mathematical theory is what allows us to refer to specific mathematical entities, then the theory cannot merely gesture in roughly the right direction, with the slack picked up by some innate faculties (as it were). Rather, the theories *themselves* must precisely pin down the mathematical entities. In a brief slogan: for the moderate objects-platonist, there can be no 'reference by acquaintance' to mathematical objects; 'reference by description' is her only hope.

Putnam's permutation argument kicks in at exactly this point. If a theory has any models, then it has many distinct isomorphic models, just by the Push-Through Construction, and every different model will differ on the explicated reference of some expression. So the moderate objects-platonist *cannot* hope that laying down a formal theory will enable her to refer determinately to mathematical objects. Her only hope has been dashed.

The moderate objects-platonist might, perhaps, try to respond by insisting that some structures are *preferable* over others as reference candidates. But her moderation renders this move entirely ineffective. Given her moderation, she cannot think that the preferability of one model over another consists in allowing our 'mathematical gaze' to fall upon certain objects rather than others, or in singling out some specific entity in mathematical intuition. But it then becomes extremely unclear what preferability *could* consist in. It might be reasonable to say, for example, that a preferable model of PA should have a countable domain. (We discuss this further in Chapter 7; recall that PA is the usual axiomatisation of first-order arithmetic, outlined in §1.A.) But this will not address the permutation argument at all: by the Push-Through Construction, if there are any countable models of PA, then there are (absolutely infinitely) many of them. No; if preferability is to *fix* reference, it

---

[17] Putnam (1980: 471); Putnam's emphasis.

[18] Dummett (1963: 189–90).

must be that some particular entity is the *preferred* referent of '1', that some other particular entity is the *preferred* referent of '2', and so forth. But this just seems incompatible with the moderate objects-platonist's own naturalistic constraints on what preferability could amount to.

Simply put, there is a dilemma: any notion of preferability which could fix mathematical reference is incompatible with moderation.[19] As such, *moderate* objects-platonists must embrace radical referential indeterminacy. (But an *im*moderate objects-platonist might yet manage to insist on determinacy of reference.)

2.4 Attempts to secure reference in mathematics

We have argued that objects-platonists must chose between moderation and determinacy of reference. In this section, we illustrate the point by considering three different positions in the philosophy of mathematics which have attempted to *avoid* referential indeterminacy.

Shapiro's ante rem structuralism

In §2.2, we discussed a mathematical-cum-philosophical focus on 'mathematical structure'. Shapiro has embraced this, whilst attempting to secure the *determinacy* of reference for mathematical terms, by advancing a position which he calls *ante rem structuralism*.

At its most basic, this position holds that mathematics is about 'structures'. However, the kinds of thing which Shapiro calls 'structures' are, emphatically, *not* structures in the sense of Definition 1.2. So, to avoid ambiguity, we will call the entities which Shapiro postulates *ante-structures*. According to Shapiro, ante-structures are abstract entities; they consist of *places*, with certain intra-structural relations holding between them.[20] These abstract entities should be thought of along the lines of universals, or platonic forms, which are abstract but can be multiply realised. Shapiro calls such realisations *systems*, and a *place-holder* is an object which, on that realisation, instantiates a particular place in the realised ante-structure.[21] Shapiro's *ante rem* structuralism is then the claim that ante-structures exist independently from the systems realising them.[22]

In some senses, Shapiro's ante-structures are rather like the isomorphism types which, in §2.2, we considered using to explicate the informal notion of mathemati-

[19] Cf. Hodes (1984: 127, 133–5) and McGee (1997: 35–8). Button (2013: chs.3–7) argues that the same dilemma applies in a rather more general context, against any *external realist*.

[20] Shapiro's ante-structures can be compared to Resnik's (1981, 1997) *patterns*, and Shapiro's places can be compared to Resnik's *positions*.

[21] Shapiro (1997: 73–4).

[22] Shapiro (1997: 9, 84–5, 109).

cal 'structure'. But there is a crucial difference. As mentioned at the end of §2.2, no single object in the isomorphism type of the natural numbers can be thought of as *the* natural number 27; rather, every structure in that isomorphism type has its own 27th element. By contrast, according to Shapiro, the ante-structure for arithmetic—*the* natural number ante-structure—contains *places*. So some particular entity is *the* 27-place in this ante-structure. And Shapiro maintains that the expression '27' determinately refers to *this* position.[23] This is how Shapiro aims to secure the determinacy of reference.

However, since places in ante-structures are abstract objects, we must ask Shapiro *why* the expression '27' refers to the 27-position in the natural number structure, rather than to any *other* abstract entity. As others before us have noted, we can make the question sharp by noting that the Push-Through Construction allows us to create a model of arithmetic according to which '27' refers to *any* entity we like, including any other place in the natural number ante-structure.[24] Indeed, as it stands, the problem is a simple one. The Push-Through Construction generates *too many* referential candidates, and postulating new types of abstract objects only *adds* to the embarrassment of potential referential riches.

In response, Shapiro might emphasise that the Push-Through Construction only generates a structure in the sense of Definition 1.2, and does not generate one of his *ante*-structures. He may then go on to argue that ante-structures are *preferable* reference candidates for our mathematical language. However, this would simply return us to the dialectic of §2.3: given that all the entities involved are abstract, it is entirely unclear *why* ante-structures, rather than $\mathscr{L}$-structures, should count as *preferable*. As in §2.3, this problem looks simply unanswerable if Shapiro's ante rem structuralism is supposed to be a version of *moderate* objects-platonism. Conversely, if Shapiro's ante rem structuralism is *immoderate*, it is unclear why Shapiro's ante rem structuralism—rather than his lack of moderation—is the 'active ingredient' in pinning down reference.

Putnam's internal realism (constructivist reading)

Putnam's own response to his permutation arguments was not structuralist, nor did it involve embracing referential indeterminacy. Instead, he abandoned moderate objects-platonism in favour of a position he called 'internal realism'. Famously, he claimed that 'Models are not lost noumenal waifs looking for someone to name them; they are constructions within our theory itself, and they have names from birth.'[25] And he later supplemented this pregnant idea with the claim:

---

[23] Shapiro (1997: 14, 55fn.15, 141fn.8).

[24] See Balaguer (1998b: 80–4), Hellman (2001: 193–6, 2005: 546), and McGee (2005: 151).

[25] Putnam (1980: 482).

If, as I maintain, 'objects' themselves are as much made as discovered, as much products of our conceptual invention as of the 'objective' factor in experience [...], then of course objects intrinsically belong under certain labels; because those labels are the tools we used to construct a version of the world with such objects in the first place. But *this* kind of 'Self-Identifying Object' is not mind-independent.[26]

In this last quote, Putnam speaks of 'objects' in general, not just mathematical objects. However, the basic thought at this time seems to be as follows: we name the objects at the same time as we construct them, and this is what makes some models preferable as explications of reference. Evidently, though, this involved rejecting moderate objects-platonism, for our objects-platonist denies that we *construct* mathematical entities (in any serious sense).

Putnam did not, however, persist with this constructivist imagery for very long. His 'internal realism' moved on in rather different directions,[27] and he made very few further remarks about the specific topic of reference to mathematical entities. In Chapters 9–12, we will extract some ideas from Putnam's 'internal realist' writings which do *not* require any constructivist imagery (but which, equally, do not secure determinacy of mathematical reference). Here, we simply want to note the availability of a constructivist position, which might secure referential determinacy.

Syntactic Priority

Finally, we consider a version of objects-platonism which embraces a principle known as the *Syntactic Priority Thesis*. This Thesis has come to be associated with Hale and Wright's neo-logicism,[28] but its simplest formulation is due to Dummett:

If a word functions as a proper name, then it *is* a proper name [...] If its syntactical function is that of a proper name, then we have fixed the sense, and with it the reference, of a proper name.[29]

Moreover, some incarnations of the Syntactic Priority Thesis were supposed to *block* the indeterminacy of reference. For example, Hale claims that 'when the truth values of all sentences containing a singular term have been fixed, there can be no *further* question about which object it stands for.'[30]

[26] Putnam (1981: 54).

[27] For details, see Button (2013: chs.9–11, 18–19).

[28] Wright (1983: 14, 24, 51–3, 171fn.5).

[29] Dummett (1956: 494). Duke (2012: chs.3–4) emphasises the importance of this early article in Dummett's later thought e.g. Dummett (1981: 497). Dummett developed this from Frege's (1980: x) 'context principle', and Wright (1983: 14, 171fn.5) is clear that he regards the Syntactic Priority Thesis as deriving explicitly from Dummett.

[30] Hale (1987: 229–30); and on the previous page he insists that anyone who runs Putnam's permutation argument must deny the context principle.

At first glance, Hale's claim might seem bizarre. After all, by Corollary 2.5, all models generated by the permutation argument are elementarily equivalent, so that the 'truth values of all sentences' are undisrupted by the reinterpretation, even though they differ on the explicated reference of any terms you like. What this shows, though, is just that those who advance the Syntactic Priority Thesis as a cure for referential indeterminacy must have a somewhat different conception of *reference* in mind.[31]

Whatever exactly that different conception amounts to, though, it seems implausible that any purely *syntactic* test could settle whether or not a term refers determinately. This point has been well made by Rosen.[32] Syntax does not seem to tell me that 'bald' is a vague predicate; but, if we think of predicates as referring at all, we might well say that 'bald' refers somewhat indeterminately. We may happily speak of '*the* third sock' that was mistakenly included in the pair of socks we bought, even though all three socks are qualitatively alike, so that '*the* third sock' does not pick out any particular sock.[33] You will never meet *the* man on the Clapham omnibus, Joe Sixpack, or Otto Normalverbraucher, even though all of these characters were introduced using grammatically singular expressions. Perhaps most pertinently: a specification of a legal procedure may discuss 'the plaintiff', in the singular, but the case at hand will determine who the plaintiff is. So it may be with numbers: number theory discusses 'the number 27', in the singular, but context may determine what the number 27 is. We revisit some of these examples in §9.4 and §15.2. For now, our point is just that syntax *alone* provides little clue as to whether reference is determinate.

2.5 Supervaluationism and indeterminacy

The considerations of §§2.2–2.4 have led many philosophers simply to *embrace* the idea that mathematical terms refer indeterminately.[34] To close the chapter, though, we will explain why this seems to be entirely *compatible* with moderate objects-platonism. In particular, we will sketch a *supervaluational* semantics which mod-

[31] See Wright (1983: 82–3), in particular the discussion of the point that: 'After all, what we state metalinguistically by "a has reference" is just the object-language "$(\exists x)x = a$"'. Cf. also Dummett (1981: 508).

[32] Rosen (2003: 231ff). Hale and Wright (2003: 253) respond to Rosen by toning down what the Syntactic Priority Thesis could hope to show. For further criticism of what the neo-Fregean can hope to learn from studying syntax, see Trueman (2014).

[33] Cf. Shapiro (2012: 393, 396).

[34] For example: McGee (1997: 36) regards the question of 'how mathematical terms come to have determinate referents' as 'insoluble'. Balaguer (1998b: 84, 1998a: 73) says that platonists 'have to claim that while [theories like PA] truly describe collections of abstract mathematical objects, they do not pick out *unique* collections of such objects'. MacBride (2005: 581) similarly writes that 'indeterminacy appears to be an ineliminable aspect of reference to mathematical objects'. Indeed, the prospect of referential indeterminacy is also sometimes used as evidence for the conclusion that mathematical language does not refer at all (cf. Hodes 1984: 139).

erate objects-platonists can adopt, but which allows for referential indeterminacy.

The core idea behind supervaluationism is to define truth (simpliciter) in terms of what happens in all appropriate models.[35] More specifically, we schematically stipulate that, for each sentence φ in the relevant language:

(a) φ is *true* iff every appropriate model satisfies φ
(b) φ is *false* iff every appropriate model satisfies $\neg\varphi$
(c) φ is *indeterminate* otherwise

Now, this stipulation is schematic, since exactly which models count as 'appropriate' will depend upon the purpose at hand. But the crucial point is that the appropriate models can be $\mathscr{L}$-structures, exactly as defined in Definition 1.2, even whilst the resulting semantics allows for *indeterminacy* of truth value. Let us see this in action.

For everything we have said in this chapter, the moderate objects-platonist may be able to insist that the *appropriate models of arithmetic* are all isomorphic. We will extensively criticise this suggestion in Chapter 7; but let us grant it for now. In that case, it is easy to show that the supervaluational semantics will entail both that:

(i) no arithmetical sentence is *indeterminate*; and
(ii) no arithmetical expression refers determinately

Concerning (i): let φ be any arithmetical sentence, and let $\mathcal{M}$ be any appropriate model of arithmetic. Either $\mathcal{M} \vDash \varphi$ or $\mathcal{M} \vDash \neg\varphi$. Now suppose also that $\mathcal{N}$ is an appropriate model. Since all appropriate models are (by hypothesis) isomorphic, $\mathcal{M} \cong \mathcal{N}$. So, by Corollary 2.5, $\mathcal{M} \equiv \mathcal{N}$. So if $\mathcal{M} \vDash \varphi$ then $\mathcal{N} \vDash \varphi$; and if $\mathcal{M} \vDash \neg\varphi$ then $\mathcal{N} \vDash \neg\varphi$. Generalising: either every appropriate model satisfies φ, so that φ is *true* by (a); or every appropriate model satisfies $\neg\varphi$, so that φ is *false* by (b).

Concerning (ii): let a and b be two different abstract entities. Then, by a Push-Through Construction, there is an appropriate model where '0' picks out a, and a different appropriate model where '0' picks out b. So, if we expand the notion of an *appropriate model* to accommodate semantic vocabulary,[36] then by (c) it will be indeterminate what '0' refers to. Generalising the point, *no* arithmetical word will refer determinately to any particular thing.

It is worth noting, though, that the use of supervaluational semantics does not stop at the level of arithmetical vocabulary, but percolates upwards through the model theory itself. To illustrate the point, let us continue with our example of two abstract objects, a and b. Because the moderate objects-platonist cannot refer determinately to *any* abstract entity, she cannot refer determinately to a rather than

[35] While natural in any discussion of indeterminacy (such as in vagueness), it is rarer to see this kind of supervaluational approach explicitly developed in the philosophy of mathematics. Some exceptions are McGee (2005: 151), Hodes (1990: 365, 1991: 158), and Walsh and Ebels-Duggan (2015: §8).

[36] There are many ways to do this. One is to expand the notion of an appropriate model, so that it includes semantic as well as arithmetical vocabulary. Another is to expand the supervaluational semantics to open-formulas, offering e.g.: $\varphi(v)$ is *true of* a iff every appropriate model has $\varphi(a)$. Since we have appropriate models $\mathcal{M}$ and $\mathcal{N}$ with $0^{\mathcal{M}} = a$ and $0^{\mathcal{N}} = b$, we can gloss this in terms of referential indeterminacy.

b. So she cannot tell us *which* structure labels which entity with '0'. In short, she cannot refer determinately to either of the two structures, any more than she can refer determinately to either of the original abstract entities.

We see no intrinsic difficulty with this idea. However, it is worth observing that the supervaluational framework yields an interesting mismatch between our natural vernacular and our formal semantics. In the example just given: within each structure, '0' refers determinately to *some* particular abstract entity (albeit a different entity in each case). However, the moderate objects-platonist must accept that *our* word '0' just *could not* pick out one abstract entity rather than any other. So: neither of the two structures, by itself, provides a faithful model of how our natural language works. The 'faithful model', if there is one, comes from considering *all* structures simultaneously.

Moreover, as Lavine notes, all of these the issues concerning names carry over to quantifiers.[37] This is most easily seen in the case of the Robinsonian and Hybrid approaches to semantics, where we handled quantifiers by 'adding new names' to the structure (see §1.5 and §1.7). More precisely: we added new constant symbols which, as usual, must each pick out exactly *one* element of the domain. But, if the initial structure's domain consisted entirely of abstract objects, then the moderate objects-platonist must accept that these new constant symbols cannot be thought of as names that any human could wield, for, according to her, no human could refer determinately to any abstract objects. As such, the notion of 'adding new names' must be meant in a slightly more technical sense than it may have first appeared. Still, to provide a formal semantics, we require only *that these augmented structures exist*; we do not require that we can determinately refer, in natural language, to the entities in their domain, or to the augmented structures themselves.

In sum: moderate objects-platonists must embrace the radical referential indeterminacy of mathematical language. They can make explicate this using a supervaluational semantics. And the indeterminacy of reference percolates through every level of that formal semantics.

2.6 Conclusion

The philosophical content of this chapter has been generated just by repeated applications of the Push-Through Construction. The Construction showed that isomorphism is extremely easy to come by. And this led to two main thoughts.

First: the informal notion of 'mathematical structure' can, perhaps, be fruitfully explicated via isomorphism types, i.e. classes of isomorphic structures.

[37] Lavine (2000: 20–6). Lavine also notes that the Tarskian approach of §1.3 faces the same issue, since moderate objects-platonists must accept that they cannot pick out any particular variable-assignment over abstract entities.

Second: moderate objects-platonists cannot plausibly maintain that mathematical terms refer determinately. Crudely: moderation rules out any plausible mechanism of reference to abstract entities that were not our own creations.

The two uses of the Push-Through Construction which brought us these morals are clearly linked both technically and philosophically. They are also, however, linked *historically*. Benacerraf's famous 1965-paper grew out of his 1960-dissertation,[38] where he writes:

> One day in conversation, Putnam made very suggestive remarks in the course of a discussion of the question 'Can, or should, the numbers be identified with sets of sets?' His point was to reject the question, arguing that it arises from a distinction between number words and numbers parallel to that between, say, furniture words and furniture, and that in the former case the distinction was unwarranted.[39]

Moreover, some of Putnam and Benacerraf's considerations were motivated by a reading of chapter 2 of Cassirer's 1910-book, in which Cassirer defends Dedekind against Frege on the foundations of arithmetic. Cassirer writes:

> It is a fundamental characteristic of the ordinal theory [of natural number] that in it the individual number never means anything by itself alone, that a fixed value is only ascribed to it by its position in the total system.[40]

To our ears, this both harkens back to Dedekind,[41] and heralds modern structuralists like Shapiro and Parsons.

2.A Eligibility, definitions, and Completeness

In §2.3, we discussed responses to Putnam's permutation argument which invoke *preferability*. In this appendix, we consider a recent discussion of referential indeterminacy which links Lewis's views on preferability to a result which we will encounter in Chapter 4. Consequently, some readers may want to read ahead before returning to this appendix. And this appendix can be safely omitted, at least on a first reading, since nothing said here affects the conclusions of this chapter.

Lewis famously held that basic physical properties are maximally *eligible* as the referents of our predicates.[42] Subsequent authors have suggested, more broadly, that certain fundamental properties (which need not be physical) are maximally eligible. In this appendix, we use the word 'eligible' in this sense, without wishing

---

[38] The dissertation Benacerraf (1960) is listed as item B1 in Benacerraf's bibliography in Morton and Stich (1996: 263) and its relation to his papers is discussed in Benacerraf (1996: 24).

[39] Benacerraf (1960: 162).

[40] Cassirer (1910: 47–8).

[41] Dedekind (1888: ¶¶73, 134).

[42] Lewis (1984: 227–8).

ourselves to endorse either the notion of fundamentality at work, or the thought that eligibility (so construed) has much to do with reference.

As outlined, eligibility is first and foremost a feature of the possible *referents* of expressions, such as the set of electrons. But if eligibility is to block Putnam's permutation argument, then we need a way to measure the eligibility of entire *structures*. The simplest way to do this would be if the eligibility of a model were a function of the eligibility of its referents. Lewis made some suggestions in this direction himself, and these have recently been developed by Williams.[43] His idea is as follows. We first assume that our model-theoretic *metalanguage* has primitive predicates for each of the maximally eligible referents.[44] To compute the eligibility of a structure, we then simply consider the length of the metalinguistic definition of the structure. Or rather, because a structure has multiple definitions if it has any, we say: *The eligibility of $\mathcal{M}$ is the reciprocal of the length of the shortest possible definition of $\mathcal{M}$ in the metalanguage.* So, the shorter the definition, the more eligible the structure.

To see how this is supposed to deal with Putnam's permutation argument, let $\mathcal{M}$ be any structure, let h be any bijection with domain M, and use the Push-Through Construction to generate a structure $\mathcal{N}$ isomorphic to $\mathcal{M}$. To determine whether $\mathcal{M}$ or $\mathcal{N}$ is more eligible, we must compare the lengths of their shortest possible definitions. Let D be a shortest definition of $\mathcal{M}$. Crucially, there is no guarantee that any of $\mathcal{N}$'s definitions is as short as D. Indeed, the only definition of $\mathcal{N}$ which is guaranteed to exist, in general, involves inserting the symbol 'h' into D several times, and this definition is of course strictly longer than D itself. In short, and as desired: isomorphic models need not be equally eligible.

This blunts some of the force of Putnam's permutation argument. However, having proposed the idea, Williams goes on to argue that it is ultimately *ineffective* in dealing with the permutation argument, claiming that Gödel's Completeness Theorem can be used to rekindle the threat of referential indeterminacy.[45] (We state and prove the Completeness Theorem in §4.A.) His reasoning is as follows. Suppose we start with a theory of the physical world, T, in a finite signature, which is consistent with (a formalisation of) the claim 'there are exactly n things'. By the Completeness Theorem 4.24, there is a model $\mathcal{N}$ of T whose domain has exactly n elements. Following Williams, we can take these elements to be the natural numbers 1 through n. Since $\mathcal{N}$ is finite, 'brute force' guarantees that there will be a finite description of $\mathcal{M}$: just explicitly list each and every aspect of the interpretation. But now, according to Williams, there is a genuine threat: if the natural numbers are fundamental entities, then this 'brute force' description might be very short, so that $\mathcal{N}$ is extremely eligible; but $\mathcal{N}$ is clearly 'unintended', for its domain consists of

[43] Williams (2007); see also also Hale (1987: 222) and Hale and Wright (1997: 438). Many thanks to Williams for discussion on the material in this section.

[44] Metaphysicians may think of this as Ontologese, if they like; see e.g. Sider (2011: 171–3).

[45] Williams (2007: 381–93).

abstract numbers and not physical things. In summary: in this case, an 'eligibility benchmark' can be established by brute force, and we have no guarantee that the genuinely 'intended' model meets the benchmark.[46]

In response, we note that Williams's observation here has very little to do with the Completeness Theorem. The *Push-Through Construction* is what allows us to take $\mathcal{N}$'s domain to be the natural numbers 1 through n.[47] The Completeness Theorem first gives us a model, $\mathcal{M}$, with n elements; since it has n elements, there is a bijection $h : M \longrightarrow \{1, ..., n\}$; and now we can generate $\mathcal{N}$ using the Push-Through Construction. So it is just the Push-Through Construction which enables us to say that brute force will establish an 'eligibility benchmark' for any theory in a finite signature with a finite model. Williams's appeal to the Completeness Theorem adds only this: *if* it is consistent to add to a theory the claim 'there are exactly n objects', *then* that theory has a model with exactly n elements. *Pace* Williams, we cannot see why this increases the threat of referential indeterminacy.

Moreover, there is a serious risk of getting tangled in the weeds here. The entire discussion trades on measuring a structure's eligibility by the length of its metalinguistic definition. This is a poor measure for several reasons, but a simple one is that it only yields results for structures which are explicitly definable in the metalanguage using finitely many sentences. This is, of course, why we could only establish an 'eligibility benchmark' by brute force for theories in finite signatures with finite models.[48] But if we want to consider theories with infinite models—and we often will—then all bets are off again.

Indeed, all we are really entitled to infer from the preceding discussion is this: if the eligibility of a model is a function of the eligibility of its referents, then *maybe* the 'intended' model will fail to be maximally eligible. And maybe not. Not enough has been said about *eligibility* to have any real clue.

2.B Isomorphism and satisfaction

This appendix presents the inductive proof of Theorem 2.3. We leave the case of (4) as an exercise for the reader. During the proof, we work with multiple n_i-tuples, and we subscript these; so for example $\bar{a}_1 = (a_{1,1}, ..., a_{1,n_1})$ and $\bar{a}_2 = (a_{2,1}, ..., a_{2,n_2})$.

Theorem (2.3): *For any $\mathscr{L}$-structures $\mathcal{M}$ and $\mathcal{N}$ and any bijection $h : M \longrightarrow N$, the following are equivalent:*

---

[46] Williams (2007: 388).

[47] So when Williams (2007: 381) says that nothing stops us from 'setting up the model ... with any domain we choose, so long as the size is appropriate', he implicitly invokes the Push-Through Construction.

[48] This may be partially mitigated by using more sophisticated measures of eligibility than the mere length of a structure's definition, e.g. perhaps Kolmogorov complexity (cf. Williams 2007: 377fn.30). But there are hard limits to how far we can hope that the eligibility of structures can be a function of definitions.

(1) *h is an isomorphism from $\mathcal{M}$ to $\mathcal{N}$*
(2) *$\mathcal{M} \vDash \varphi(\bar{a})$ iff $\mathcal{N} \vDash \varphi(\hbar(\bar{a}))$, for all $\bar{a}$ from M^n and all* atomic *$\mathscr{L}$-formulas $\varphi(\bar{v})$ with free variables displayed*
(3) *$\mathcal{M} \vDash \varphi(\bar{a})$ iff $\mathcal{N} \vDash \varphi(\hbar(\bar{a}))$, for all $\bar{a}$ from M^n and all first-order $\mathscr{L}$-formulas $\varphi(\bar{v})$ with free variables displayed*
(4) *$\mathcal{M} \vDash \varphi(\bar{a})$ iff $\mathcal{N} \vDash \varphi(\hbar(\bar{a}))$, for all $\bar{a}$ from M^n and all second-order $\mathscr{L}$-formulas $\varphi(\bar{v})$ with free variables displayed, with consequence read either via the full or the Henkin semantics for second-order logic (see §§1.10–1.11)*

Proof. *(1) ⇒ (2).* Suppose (1). We first prove that, where $t(\bar{v})$ is a term with free variables displayed, we have $h(t^{\mathcal{M}}(\bar{a})) = t^{\mathcal{N}}(\hbar(\bar{a}))$. This is by induction on complexity of terms. The only interesting case is when $t = f(t_1, \ldots, t_k)$ for some $\mathscr{L}$-function symbol f and $\mathscr{L}$-terms $t_1, \ldots, t_k$, in which case, by (1):

$$\begin{aligned} h(f^{\mathcal{M}}(t_1^{\mathcal{M}}(\bar{a}_1), \ldots, t_k^{\mathcal{M}}(\bar{a}_k))) &= f^{\mathcal{N}}(h(t_1^{\mathcal{M}}(\bar{a}_1)), \ldots, h(t_k^{\mathcal{M}}(\bar{a}_k))) \\ &= f^{\mathcal{N}}(t_1^{\mathcal{N}}(\hbar(\bar{a}_1)), \ldots, t_k^{\mathcal{N}}(\hbar(\bar{a}_k))) \end{aligned}$$

We now turn our attention to atomic $\mathscr{L}$-sentences. Where R is an $\mathscr{L}$-relation symbol, by (1) and the above, we have:

$$\begin{aligned} \mathcal{M} \vDash R(t_1(\bar{a}_1), \ldots, t_n(\bar{a}_n)) &\text{ iff } (t_1^{\mathcal{M}}(\bar{a}_1), \ldots, t_n^{\mathcal{M}}(\bar{a}_n)) \in R^{\mathcal{M}} \\ &\text{ iff } (h(t_1^{\mathcal{M}}(\bar{a}_1)), \ldots, h(t_n^{\mathcal{M}}(\bar{a}_n))) \in R^{\mathcal{N}} \\ &\text{ iff } (t_1^{\mathcal{N}}(\hbar(\bar{a}_1)), \ldots, t_n^{\mathcal{N}}(\hbar(\bar{a}_n))) \in R^{\mathcal{N}} \\ &\text{ iff } \mathcal{N} \vDash R(t_1(\hbar(\bar{a}_1)), \ldots, t_n(\hbar(\bar{a}_n))) \end{aligned}$$

The case of identity is similar, here invoking the fact that h is a bijection; we leave it to the reader.

(2) ⇒ (3). We prove (3) by induction on complexity of formulas. The base case holds by assuming (2). The remaining cases are given by the recursion clauses for satisfaction, together with the induction hypothesis (for the middle biconditional each time). In the case of negation:

$$\mathcal{M} \vDash \neg\varphi(\bar{a}) \text{ iff } \mathcal{M} \nvDash \varphi(\bar{a}) \text{ iff } \mathcal{N} \nvDash \varphi(\hbar(\bar{a})) \text{ iff } \mathcal{N} \vDash \neg\varphi(\hbar(\bar{a}))$$

The case of conjunction is similar, and we leave it to the reader. In the case of universal quantification:

$$\begin{aligned} \mathcal{M} \vDash \forall x \varphi(\bar{a}, x) &\text{ iff } \mathcal{M} \vDash \varphi(\bar{a}, b) \text{ for all } b \in M \\ &\text{ iff } \mathcal{N} \vDash \varphi(\hbar(\bar{a}), d) \text{ for all } d \in N \\ &\text{ iff } \mathcal{N} \vDash \forall x \varphi(\hbar(\bar{a}), x) \end{aligned}$$

The middle step in this last case invokes the fact that h is a bijection.

(3) ⇒ (1). Clearly (3) ⇒ (2). So, assuming (2), we first show that relation symbols are preserved. With $a_1, \ldots, a_n \in M$, we have:

$$\begin{aligned}(a_1, \ldots, a_n) \in R^{\mathcal{M}} &\text{ iff } \mathcal{M} \vDash R(a_1, \ldots, a_n)\\ &\text{ iff } \mathcal{N} \vDash R(h(a_1), \ldots, h(a_n))\\ &\text{ iff } (h(a_1), \ldots, h(a_n)) \in R^{\mathcal{N}}\end{aligned}$$

We next show that function symbols are preserved. Let $\bar{a}$ be from M and suppose $b = f^{\mathcal{M}}(\bar{a})$. Since $\mathcal{M} \vDash f(\bar{a}) = b$, we have $\mathcal{N} \vDash f(\hbar(\bar{a})) = h(b)$ by (2). Hence $f^{\mathcal{N}}(\hbar(\bar{a})) = h(b) = h(f^{\mathcal{M}}(\bar{a}))$. The case of constant symbols is similar. □

3

Ramsey sentences and Newman's objection

Whilst the main focus of this book is the philosophy of mathematics, Chapter 2 puts us in an excellent position to consider a particular topic in the philosophy of science. In particular, in this chapter, we will consider some recent discussions of Ramsey sentences and Newman's objection.

In 1928, Newman presented a now famous objection to Russell's (1927) theory of causation. The simplest form of his objection is just an application of the Push-Through Construction of §2.1. In this simple form, Newman's objection has a rather specific target. However, as subsequent commentators have noted, Newman's objection can be directed at a wider range of positions. The aim of this chapter is to develop several versions of the Newman objection, and to relate them to different notions of *conservation*. Our main observation is that the best version of Newman's objection is a slightly spruced-up application of the Push-Through Construction, and that the dialectic surrounding this objection should precisely parallel the dialectic surrounding Putnam's permutation argument.

3.1 The o/t dichotomy

The focus of this chapter is a certain kind of *structuralism* within the philosophy of science. So we must start by motivating that version of structuralism. It begins with the idea is that our scientific theorising is split into two parts. The *okay* part is in good standing, and poses no particular philosophical puzzles. Unfortunately, there is also a *troublesome* part.

Such a dichotomy might arise by treating the *observational* as okay but the *theoretical* as troubling. Consider, in particular, the following view. Some statements are observation statements. Their role is simply to make claims about observations, and nothing beyond that. There is no mystery (on this view) concerning how observation statements work: to determine whether they are true, one simply makes a relevant observation, and thereby checks whether the world is as the observation statement claims. An example of an observation statement might be 'a cat is on the mat', and to check whether this is true, you simply *look and see*. But, continuing with this view, some statements are theoretical statements. An example might be 'carbon has fifteen isotopes, only two of which are stable'. And there may be some initial mystery concerning how we could so much as *check* whether the world is as the theoretical statement claims.

For those who want to effect such a dichotomy between the observational and the theoretical, the dichotomy shines through in the *vocabulary*. The predicate '... is a cat' is an observation-predicate, but the predicate '... is a carbon isotope' is not. To keep track of this, in what follows we let $\mathscr{V}$ be the signature which we use to frame our scientific theories. (Recall from Definition 1.1 that a signature is a formal, regimented *vocabulary*.) Then $\mathscr{V}_o$ is a privileged sub-signature of $\mathscr{V}$, which is treated as $\mathscr{V}$'s *observational* vocabulary, and $\mathscr{V}_t$, i.e. $\mathscr{V} \setminus \mathscr{V}_o$, is its *theoretical* vocabulary.

Now, we have motivated an okay/troublesome dichotomy—more briefly, an o/t dichotomy—by considering a dichotomy between the observational and theoretical. We do not want to suggest that this is the only way to generate an o/t dichotomy.[1] Equally, we do not want to suggest that there *is* an important o/t dichotomy (we revisit this in §3.4). We just want to use model theory to explore what happens *if* one effects some o/t dichotomy.

3.2 Ramsey sentences

Suppose, then, that we effect an o/t dichotomy. Our next question is how we should handle this within our philosophising. This is the context in which Ramsey sentences present themselves.

Ramsey sentences are sentences of second-order logic. We outlined second-order logic in §§1.9–1.11; here are the main points. Syntactically, second-order logic allows us to quantify into 'predicate position'. Semantically, matters are more subtle; but we can afford to relegate the subtleties to footnotes, since all of the results of this chapter go through with either of the two most common semantics for second-order logic, namely, *full* semantics and *faithful Henkin* semantics. (For more, see §§1.9–1.11, and footnotes 17, 23, 24, and 30 of this chapter.)

Ramsey sentences are defined *from* theories. So we start by making some assumptions about the theory, T, whose Ramsey sentence we want to define. (These assumptions hold good throughout this chapter, but no further.)

First, we assume that our theory, T, is given in the signature $\mathscr{V}$, which has an appropriate o/t dichotomy. So, $\mathscr{V}_o$ and $\mathscr{V}_t$ are disjoint, but together exhaust $\mathscr{V}$.

Second, we assume that $\mathscr{V}$ is relational, i.e. that it contains only predicates. This is no loss, as we can simulate constants and functions using relations and identity.

Third, we assume that T is a *finite* set of sentences. This is a bigger assumption.

[1] Lewis (1970), for example, considered the *old/theoretical* distinction. And in the philosophy of mathematics, there is Hilbert's (1925: 179) analogous real/ideal distinction. In §4.7, we consider the idea that *infinitesimals* are ideal.

However, once we have made it, we can harmlessly treat T as a *single* sentence, by conjoining all of its finitely many members. Moreover, since T is finite, only finitely many $\mathscr{V}_o$-symbols occur in T. And this allows us to construct T's Ramsey sentence. Intuitively, we do this by treating all of T's theoretical predicates as variables—i.e. we treat the $\mathscr{V}_t$ relation symbols as relation-variables—and then bind them with existential quantifiers:

Definition 3.1: *Let T be a finite $\mathscr{V}$-theory, and let $R_1, \ldots, R_n$ be the only $\mathscr{V}_t$-predicates appearing in T. Where $\hat{T}$ is the conjunction of the sentences in T, and where $\hat{T}[\overline{X}/\overline{R}]$ is the result of replacing each instance of the predicate R_i in $\hat{T}$ with the variable X_i, we say that T's* Ramsey sentence *is*

$$\text{Ramsey}(T) := \exists X_1 \ldots \exists X_n \hat{T}[\overline{X}/\overline{R}]$$

Consequently, Ramsey(T) is a second-order $\mathscr{V}_o$-sentence; that is, the only symbols occuring in Ramsey(T) are logical expressions and $\mathscr{V}_o$-symbols. (And note that T's Ramsey sentence is only *definable* when T is finite.)

3.3 The promise of Ramsey sentences

Ramsey sentences were first formulated in Ramsey's posthumously published 'Theories'.[2] Carnap brought them to prominence in the late 1950s and 1960s, relating them to his long-running concerns about the meaning of theoretical terms.[3] In this section, we will outline a contemporary motivation for such concerns.

As noted in §2.3, it is plausible that our (causal) acquaintance with cats helps to explain why our word 'cat' refers to cats rather than to cherries. In this case, however, we are likely to treat '... is a cat' as an observational predicate. It is much less clear what could pin down the reference of *theoretical* predicates. Crudely put, we cannot point at electrons in the same way as we point at cats. And so it may seem that our knowledge of theoretical predicates will have to come via description—in terms of their impact on our more immediate observations—rather than by acquaintance.[4]

If we buy this line of thought, then Ramsey(T) may start to recommend itself to us. After all, Ramsey(T) is a $\mathscr{V}_o$-sentence, and so it contains no theoretical vocabulary. As such, it is not obviously vulnerable to the concerns just mentioned.[5] Moreover, and crucially, not much is lost by considering Ramsey(T) rather than T

---

[2] Ramsey (1931).

[3] Carnap (1958: 245, 1959: 160–5, 1966: 248, 252, 265ff). For an account of the history, see Psillos (1999: 46–9).

[4] See Maxwell (1971), Zahar and Worrall (2001: 239, 243), and Cruse (2005: 562–3), and the discussion of moderate objects-platonism in Chapter 2.

[5] See Carnap (1958: 242, 245, 1966: 251, 269).

itself, for it is easy to show that the two are *observationally equivalent* in the following sense (for the proof, see Proposition 3.5, below):[6]

Proposition 3.2: *Let T be a finite $\mathscr{V}$-theory. Then $T \vDash \varphi$ iff* Ramsey(T) $\vDash \varphi$, *for all $\mathscr{V}_o$-sentences φ.*

This result led Carnap to claim that Ramsey(T) expresses T's 'factual content'.[7]

Suppose we follow Carnap's initial motivations: there are difficulties with the $\mathscr{V}_o$-vocabulary; but Ramsey(T) does not share those difficulties; and indeed Ramsey(T) has the same observational content as T, in the sense of Proposition 3.2. But suppose we *also* maintain that Ramsey(T) expresses some additional content, beyond its observational consequences. That additional content is existential: there *are* some theoretical objects and relations, whose behaviour is specified by Ramsey(T). It is worth noting that, for those who accept an o/t dichotomy, Ramsey(T) will seem to provide the fullest thesis that one could plausibly hope to maintain concerning a *troublesome* realm within which determinate reference is bound to be problematic.[8] And this line of thought culminates in what we call *ramsified realism*.[9] This holds that, for a physical theory T,

(a) Ramsey(T) expresses T's genuine content, but
(b) the content expressed by Ramsey(T) goes significantly beyond T's mere observational consequences, since Ramsey(T) makes substantial—and hopefully true—claims about the theoretical.

3.4 A caveat on the o/t dichotomy

In what follows, we will use model-theoretic results to raise problems for ramsified realism. First, though, we should emphasise that there are plenty of problems with ramsified realism which have nothing to do with model theory.

---

[6] Carnap (1958: 245, 1959: 162–4, 1963: 965, 1966: 252), Psillos (1999: 292 n.6), Worrall (242–3 2007: 150), and Zahar and Worrall (2001).

[7] Carnap (1959: 164–5). Carnap's own interest in Ramsey sentences was sharpened by a further observation: Ramsey(T) $\wedge$ (Ramsey(T) $\rightarrow \hat{T}$) $\vDash \hat{T}$ and $T \vDash$ Ramsey(T) $\wedge$ (Ramsey(T) $\rightarrow \hat{T}$). Consequently, Carnap maintained that physical theories can be exclusively and exhaustively factored into a purely *factual* part, Ramsey(T), and a purely *analytic* part, the conditional (Ramsey(T) $\rightarrow \hat{T}$), which records the theory's 'meaning postulates' (1958: 246, 1959: 163–4, 1963: 965, 1966: 270–2). So Carnap's own interest in Ramsey sentences is tied to his long-standing interest in analyticity. Admittedly, most contemporary philosophers are less interested in analyticity than Carnap was; so the conditional (Ramsey(T) $\rightarrow \hat{T}$), which is now called T's *Carnap sentence*, will not detain us.

[8] See Worrall (2007: 148) and Demopoulos (2011: 200).

[9] Psillos (1999: ch.3) seems to suggest that Carnap himself was a ramsified realist. For a contrasting view, see Demopoulos (2003: 384–90, 2011: 195–7). We cannot pursue this in detail, but we note that Carnap (esp. 1966: 256) may well have simply *embraced* the conclusions of the various Newman-style objections against ramsified realism.

Crucially, ramsified realism only makes sense given that one has accepted some o/t dichotomy. And it is worth emphasising that this really is a *dichotomy*. The o/t distinction is *exhaustive*, in that $\mathcal{V}_o \cup \mathcal{V}_t = \mathcal{V}$. It is *exclusive*, in that $\mathcal{V}_o \cap \mathcal{V}_t = \varnothing$. Then all and only the troublesome vocabulary is 'ramsified away' (i.e. replaced with an existentially bound relation-variable).

It is far from clear that natural language—whether mundane or scientific—really displays such a dichotomy. To take an example from Cruse, consider the predicate 'x is bigger than y'.[10] Some instances of this are *observational*: Big Ben is clearly bigger than the tourist next to it. However, some instances seem straightforwardly *theoretical*: protons are bigger than electrons. But some instances seem to be *mixed*: Big Ben is bigger than a proton. So the predicate 'x is bigger than y' resists categorisation as (dichotomously) either observational or theoretical.

If we insist on an o/t dichotomy, then the only way to handle this natural-language predicate will be to split it in half. We will use a $\mathcal{V}_o$-predicate, B_1, to formalise the observational instances, and a $\mathcal{V}_t$-predicate, B_2, to formalise the theoretical instances. When it comes to the mixed instances, we must make a choice as to whether to formalise them with B_1 or B_2. However, it is likely that we will use B_2, since we cannot directly observe the size of the proton, so that these judgments of comparative size must be (somewhat) troublesome. In this manner, we will retain the o/t dichotomy and 'ramsify away' B_2 but not B_1. How (un)satisfying this is will, of course, depend upon whether we think that there really are two dichotomously different relations here.

This simple example illustrates a fundamental point. It is one thing to believe that there is a *distinction* between the observational and the theoretical; it is another to believe that there is a *dichotomy*, in the sense endorsed by ramsified realism. For our part, we believe that any distinction we should draw between between the observational and the theoretical is likely to be, not a once-and-for-all dichotomy, but rough and ready, porous, and context-sensitive.[11] So, *we* are not ramsified realists.

However, to pursue this any further would take us deep into issues at the intersection of philosophy of science and philosophy of perception. We raise the point simply to highlight that the very idea of an o/t dichotomy is quite contentious, and for reasons which have nothing to do with model theory.

3.5 Newman's criticism of Russell

With this caveat out of the way, we will consider various model-theoretic results which raise problems for ramsified realists (who *do* embrace an o/t dichotomy).

---

[10] Cruse (2005: 561).

[11] This is the line pushed in Button (2013: 50–1), and it draws in various ways on Maxwell (1962: 7–8, 14–15), Putnam (1987: 1, 20–1, 26–40, 1994: 502ff), and Okasha (2002: 316–9).

These problems came to contemporary prominence when Demopoulos and Friedman related Ramsey sentences to Newman's criticism of Russell.[12]

As Newman understood Russell's position in *The Analysis of Matter*, Russell was committed to the doctrine that we could only know the *structure* of the external world. Newman's objection to this view was straightforward. For any structure $\mathcal{W}$,

> Any collection of things can be organised so as to have the structure $\mathcal{W}$, provided there are the right number of them. Hence the doctrine that *only* structure is known involves the doctrine that *nothing* can be known that is not logically deducible from the mere fact of existence, except ('theoretically') the number of constituting objects.[13]

Note that Newman's initial claim is *just* a restatement of the Push-Through Construction from §2.1. Indeed Newman gestures at the Construction himself.[14]

Newman's objection raises serious problems for anyone who thinks that we can know only the structure of the external world. But this seems to be an extremely restricted target.[15] For example, our ramsified realist accepts an o/t dichotomy, and glosses it in terms of the observational and the theoretical. Given the troublesome nature of the theoretical, our ramsified realist might concede that our knowledge of the external *theoretical* realm is limited to its structure. But she will surely claim that we can know more about the external *observational* realm than its mere structure. So the obvious question to ask is whether Newman's objection against Russell can be turned into an objection against ramsified realism.

Over the next few sections, we will try to formulate the strongest possible Newman-style objection against ramsified realism. It arises by combining two simple thoughts: one relating to the Push-Through Construction; the other relating to various model-theoretic notions of *conservation*.

3.6 The Newman-conservation-objection

We start by developing the notion of *conservation*, and proving that *Ramsey sentences are object-language statements of conservation.* This is essentially our Proposition 3.5, below. We then use this to present an interesting, but ultimately ineffectual, Newman-style objection against ramsified realism.

Suppose that S is a theory which has been formulated in a purely observational language, and that T is a wider theory of physics. Suppose, too, that we have a guarantee that no more observational consequences follow from T than follow from S alone. Obviously, the notions of 'guarantee' and 'following from' need to be made

---

[12] See Demopoulos and M. Friedman (1985), Newman (1928), and Russell (1927).

[13] Newman (1928: 144), with a slight change to typography.

[14] Newman (1928: 145–6).

[15] See Zahar and Worrall (2001: 238–9, 244–5), Worrall (2007: 150), and Ainsworth (2009: 143–4).

more precise. But if they are suitably rich, this guarantee will allow us to become instrumentalists (or fictionalists) about everything in T which goes beyond S. That is: we could in good conscience employ the full power of T, without having to insist that its distinctively theoretical claims are *true*; they would simply be an expedient way to allow us to reason our way around observables. (Note, in passing, that this line of thought seems to pull us away from ramsified realism, since ramsified realists wanted to incur some *substantive* theoretical commitments; this was the point of clause (b) of §3.3.)

To tidy up the line of thought that points us towards instrumentalism, we can employ the following definition:

Definition 3.3: *Let T be an $\mathscr{L}^+$-theory and S be an $\mathscr{L}$-theory, with $\mathscr{L}^+ \supseteq \mathscr{L}$. T is* consequence-conservative *over S iff: if $T \vDash \varphi$ then $S \vDash \varphi$, for all $\mathscr{L}$-sentences φ.*

The relation $T \vDash \varphi$ in this definition is the model-theoretic consequence relation from §1.12, defined by: if $\mathcal{M} \vDash T$ then $\mathcal{M} \vDash \varphi$, for all structures $\mathcal{M}$. The immediate interest in consequence-conservation is as follows. Suppose that S is the set of all *true* observation sentences, expressed in some suitably rich vocabulary. To say that T is consequence-conservative over S is, then, to say that T is perfectly 'observationally reliable'. After all, any observation sentence entailed by T is entailed by S, and hence true by assumption.

Here, though, is a second way to think about 'observational reliability'. Imagine that we can turn any model of S into a model of T, just by interpreting a few more symbols in T's vocabulary. Then any configuration of observational matters that makes S true is *compatible* with the truth of T. Intuitively, then, T cannot make any false judgements about observational matters; so, again, T is 'observationally reliable'.

Formalising these intuitive ideas yields a second notion of conservation. We already formalised the idea of interpreting new symbols, when introducing Robinsonian semantics (see Definition 1.4). So we can offer the following:

Definition 3.4: *Let T be an $\mathscr{L}^+$-theory and S be an $\mathscr{L}$-theory, with $\mathscr{L}^+ \supseteq \mathscr{L}$. T is* expansion-conservative *over S iff: for any $\mathscr{L}$-structure $\mathcal{M}$ such that $\mathcal{M} \vDash S$, there is an $\mathscr{L}^+$-structure $\mathcal{N}$ which is a signature expansion of $\mathcal{M}$ and such that $\mathcal{N} \vDash T$.*

It is natural to ask how these two notions of conservation relate to each other. In fact, the Ramsey sentence allows us to demonstrate that they align perfectly:[16]

[16] We have not found Proposition 3.5 stated in the literature on Ramsey sentences, though we note three near misses. First: as just noted, Carnap's Proposition 3.2 is an immediate corollary of Proposition 3.5. Second: Worrall (2007: 151–2) suggests that Demopoulos and M. Friedman (1985) might want to invoke something like Proposition 3.5. Third: Ketland (2009: 42) obtains a restricted corollary of Proposition 3.5.

Proposition 3.5: *Let T be any finite $\mathscr{V}$-theory, and let S be any $\mathscr{V}_o$-theory. The following are equivalent:*

(1) *T is consequence-conservative over S*
(2) *T is expansion-conservative over S*
(3) $S \vDash \text{Ramsey}(T)$

Proof. (3) ⇒ (2). Suppose $S \vDash \text{Ramsey}(T)$. Let $\mathcal{M} \vDash S$. Since $\mathcal{M} \vDash \text{Ramsey}(T)$, we can choose suitable witnesses for Ramsey(T)'s initial existential quantifiers, and then take these witnesses as the interpretations of the $\mathscr{V}_t$-predicates, giving us a $\mathscr{V}$-structure $\mathcal{N} \vDash T$ which is a signature expansion of $\mathcal{M}$.

(2) ⇒ (1). Let φ be any $\mathscr{V}_o$-sentence such that $T \vDash \varphi$. Suppose $\mathcal{M} \vDash S$. As T is expansion-conservative over S, there is a $\mathscr{V}$-structure $\mathcal{N} \vDash T$ which is a signature expansion of $\mathcal{M}$. Since $\mathcal{N} \vDash T$, we have $\mathcal{N} \vDash \varphi$, and hence $\mathcal{M} \vDash \varphi$ since φ is a $\mathscr{V}_o$-sentence and $\mathcal{N}$ and $\mathcal{M}$ agree on the interpretation of the $\mathscr{V}_o$-vocabulary. Since $\mathcal{M}$ was arbitrary, $S \vDash \varphi$.

(1) ⇒ (3). $T \vDash \text{Ramsey}(T)$,[17] and Ramsey($T$) is a $\mathscr{V}_o$-sentence. So if T is consequence-conservative over S, then $S \vDash \text{Ramsey}(T)$. □

Informally glossed, this Proposition states that Ramsey sentences are object-language statements of conservation (in both senses). Additionally, Proposition 3.2—which told us that T and Ramsey(T) are observationally equivalent—is an easy corollary of Proposition 3.5.[18] However, it should be emphasised that this Proposition crucially relies upon the use of *second-order* logic: not only does condition (3) involve Ramsey sentences, which are by definition second-order, but the equivalence between (1) and (2) *fails* for first-order theories (see §3.B).

This Proposition now allows us to formulate a neat (if ultimately ineffectual) Newman-style objection against ramsified realism:

Newman-conservation-objection. *Instrumentalists and realists can agree that their favourite theories should be $\mathscr{V}_o$-sound, i.e. that all of their $\mathscr{V}_o$-consequences should be true.*[19] *Where S is the set of all true $\mathscr{V}_o$-sentences, to say that T is $\mathscr{V}_o$-sound is to say that T is consequence-conservative over S. By Proposition 3.5, this is equivalent to the claim that $S \vDash \text{Ramsey}(T)$. So* Ramsey($T$) *expresses no truth-evaluable content beyond T's observational consequences. So clause (b) of §3.3—which states precisely that the content expressed by* Ramsey(T) *goes* beyond *T's mere observational consequences—is false, and ramsified realism fails.*

[17] Assuming only that our structures satisfy a few instances of the Comprehension Schema (see §1.11).

[18] For the left-to-right direction of Proposition 3.2, just take $S = \text{Ramsey}(T)$, so that condition (3) of Proposition 3.5 is trivially satisfied, and then apply the equivalence between condition (3) and condition (1) guaranteed by Proposition 3.5. The right-to-left direction of Proposition 3.2 follows since $T \vDash \text{Ramsey}(T)$, as we had occasion to note in the proof of Proposition 3.5.

[19] Ketland (2009: Definition D) calls $\mathscr{V}_o$-soundness 'weak O-adequacy'.

3.7 Observation vocabulary versus observable objects

The ramsified realist has only one possible response to the Newman-conservation-objection: she must deny that instrumentalists are committed to the $\mathcal{V}_0$-soundness of physical theories.

As Worrall explains, this is much less strange than it initially sounds.[20] Suppose T entails the existence of an object which falls under *no* $\mathcal{V}_0$-predicate. If we believe that our $\mathcal{V}_0$-vocabulary is adequate to deal with any observable object we encounter, then we can characterise this by saying that T entails a $\mathcal{V}_0$-sentence which says, in effect, 'there is an unobservable object'. So, in order for T to be $\mathcal{V}_0$-sound, there must be unobservable objects. But instrumentalists need not be committed to the existence of any unobservables. So instrumentalists should not be regarded as being committed to the $\mathcal{V}_0$-soundness of physical theories after all.

To deal with this point, in what follows we will assume that our observational vocabulary $\mathcal{V}_0$ includes a primitive one-place predicate, O, which is intuitively to be read as '... is an observable object'.[21] In this new setting, we should not count just *any* $\mathcal{V}_0$-sentence as an *observation* sentence. Rather, the observation sentences will be just those sentences which, intuitively, are restricted to telling us *about* the observable objects. Specifically, the observation sentences are the $\mathcal{V}_0$-sentences whose quantifiers are restricted to O. Call these the $\mathcal{V}_0^O$-sentences. We define these precisely as follows.

Definition 3.6: *Where φ is any $\mathcal{V}_0$-formula, we recursively define:*

$$\begin{aligned}
\varphi^O &:= \varphi, \textit{ if } \varphi \textit{ is atomic} \\
(\varphi \wedge \psi)^O &:= (\varphi^O \wedge \psi^O) \\
(\neg\varphi)^O &:= \neg(\varphi^O) \\
(\exists x\varphi)^O &:= \exists x(O(x) \wedge \varphi^O) \\
(\exists X^n\varphi)^O &:= \exists X^n \left(\forall \bar{v}\left[X^n(\bar{v}) \rightarrow (O(v_1) \wedge \ldots \wedge O(v_n))\right] \wedge \varphi^O\right)
\end{aligned}$$

The set of $\mathcal{V}_0^O$-formulas is then $\{\varphi^O : \varphi$ is a $\mathcal{V}_0$-formula$\}$.

We can now easily accommodate Worrall's point. The sentence 'there are unobservable objects', i.e. $\exists x \neg O(x)$, uses an unrestricted existential quantifier, and so is a $\mathcal{V}_0$-sentence but not a $\mathcal{V}_0^O$-sentence. Moreover, in these terms, instrumentalists and realists will agree only on the fact that T should entail true $\mathcal{V}_0^O$-sentences, i.e. that theories should be $\mathcal{V}_0^O$*-sound*.[22] But it is clear that $\mathcal{V}_0^O$-soundness is strictly

---

[20] Worrall (2007: 152).

[21] This predicate allows us to simulate Ketland's (2004: 289ff, 2009: 38ff) use of a two-sorted language.

[22] This is essentially Ketland's (2004: Definition D) notion of 'weak empirical adequacy'.

weaker than $\mathscr{V}_o$-soundness, and it is only obvious that *realists* have any motivation for insisting on the latter. So the Newman-conservation-objection fails.

3.8 The Newman-cardinality-objection

Our aim, now, is to present a version of Newman's objection which can handle the fact that the quantifiers range over observables and unobservables alike. The essential idea behind the objection is quite simple. In concessive spirit, we will grant to the ramsified realist that there are no problems whatsoever with regard to the $\mathscr{V}_o^O$-claims. In particular, they have an *intended model.* But we will then ask the ramsified-realist what more it takes for her 'troublesome' claims to be true. Model theory threatens that she must answer, only, that there be *enough* unobservables.

To implement this strategy, let C be the *correct* model of the $\mathscr{V}_o^O$-sentences. Now, in allowing that there *is* such a model, we are simply waiving any Putnam-esque concerns about the reference of $\mathscr{V}_o$-expressions, such as those from §2.3. To do this, though, is just to allow the ramsified realist (for the sake of argument) that this is all *okay*.

Since C is the intended model of the $\mathscr{V}_o^O$*-sentences,* it is a $\mathscr{V}_o$-structure whose domain $C = O^C$. Since C is *correct,* C comprises all and only the actual observable entities in the physical universe, and any given entities stand in a certain observation relation to each other iff C represents them as so doing.

We now turn to the question of what it takes for T, or Ramsey(T), to be *true.* Since we have granted the ramsified realist the existence of an *intended model* of the $\mathscr{V}_o^O$-sentences, at the very least T or Ramsey(T) must be *compatible* with C. Somewhat more precisely, there must be a model of T whose 'observable part' is just C itself. But, to make this fully precise, we must explain the idea of a 'part' of a structure. We do this by introducing the notion of a substructure:

Definition 3.7: *Let $\mathcal{M}$ and $\mathcal{N}$ be $\mathscr{L}$-structures, with $M \subseteq N$. Then we say that $\mathcal{M}$ is a* substructure *of $\mathcal{N}$ iff: for any $\mathscr{L}$-constant symbol c, any n-place $\mathscr{L}$-relation symbol R, and any n-place $\mathscr{L}$-function symbol f:*[23]

$$c^{\mathcal{M}} = c^{\mathcal{N}}$$
$$R^{\mathcal{M}} = R^{\mathcal{N}} \cap M^n$$
$$f^{\mathcal{M}} = f^{\mathcal{N}}|_{M^n}$$

[23] The notation $g|_X$ indicates g's restriction to X, implemented set-theoretically as $g|_X := \{(\bar{x}, g(\bar{x})) \in g : \bar{x} \in X\}$. When we consider Henkin $\mathscr{L}$-structures, as in §1.6, the definition of *substructure* requires two further clauses (see Shapiro 1991: 92) for each $n < \omega$: (i) if $Q \subset M_n^{\text{rel}}$ then there is some $R \in N_n^{\text{rel}}$ such that $Q = R \cap M^n$; and (ii) if $g \in M_n^{\text{fun}}$ then there is some $h \in N_n^{\text{fun}}$ such that $g = h|_{M^n}$.

Now, where $\mathcal{M}$ is a $\mathscr{V}_o$-structure, we say that *$\mathcal{M}$'s observable part*, $\text{Ob}(\mathcal{M})$, is the $\mathscr{V}_o$-reduct of the substructure of $\mathcal{M}$ whose domain is the observables in $\mathcal{M}$, i.e. $O^{\mathcal{M}}$.[24] Since we have assumed that $\mathscr{V}$ only contains relation symbols, this amounts to the following condition: $\text{Ob}(\mathcal{M})$'s domain is $O^{\mathcal{M}}$, and $R^{\text{Ob}(\mathcal{M})} = R^{\mathcal{M}} \cap (O^{\mathcal{M}})^n$ for each n-place $\mathscr{V}_o$-predicate, R.[25]

We now say that T is *C-alright* iff there is a $\mathscr{V}$-structure $\mathcal{M} \models T$ such that $\text{Ob}(\mathcal{M}) = C$.[26] So, intuitively, if T or $\text{Ramsey}(T)$ is true then they must (at least) be C-alright.

To run a Newman-style argument in this setting, we simply need to combine two ideas that we have encountered already. The first idea comes from the Push-Through Construction of §2.1, which was also the basis for Newman's original objection to Russell in §3.5. In the present setting, the important point is that *any* object will do as an 'unobservable'; so we can push-through the 'unobservables' as much as we like, so long as we leave the 'observables' undisturbed.[27]

The second idea relates to a trivial corollary of Proposition 3.5, that T is expansion-conservative over $\text{Ramsey}(T)$. This lets us move freely between models of T and models of $\text{Ramsey}(T)$, by signature-expansion and reduction.

Combining these two ideas, we obtain just the result we need:[28]

Proposition 3.8: *Let T be any $\mathscr{V}$-theory, with $O \in \mathscr{V}_o$. Let $\mathcal{A}$ be any $\mathscr{V}$-structure satisfying $\forall x O(x)$. Then the following are equivalent:*

(1) *There is a $\mathscr{V}$-structure $\mathcal{M}$ such that $\mathcal{M} \models T$ and $\text{Ob}(\mathcal{M}) = \mathcal{A}$*
(2) *There is a cardinal κ with the following property: if U is a set of cardinality κ with $A \cap U = \varnothing$, then there is a $\mathscr{V}_o$-structure $\mathcal{P}$ with domain $A \cup U$, with $\text{Ob}(\mathcal{P}) = \mathcal{A}$ and with $\mathcal{P} \models \text{Ramsey}(T)$*

---

[24] When we consider Henkin $\mathscr{L}$-structures, we also need to stipulate that $\text{Ob}(\mathcal{M})^{\text{rel}}_n := \{X \cap (O^{\mathcal{M}})^n : X \in M^{\text{rel}}_n\}$ and $\text{Ob}(\mathcal{M})^{\text{fun}}_n := \{g|_{(O^{\mathcal{M}})^n} : g \in M^{\text{fun}}_n \text{ and range}(g|_{(O^{\mathcal{M}})^n}) \subseteq O^{\mathcal{M}}\}$.

[25] For similar notions, see Przełęckie (1973: 287), Hodges (1993: 202), and Lutz (2014: §4.3).

[26] There is a delicacy here. So far, we have made no particular assumptions about how the observable *entities* interact with the observation *relations*. Two plausible assumptions are: (a) observable objects never fall under $\mathscr{V}_t$-predicates, or (b) unobservable objects never fall under $\mathscr{V}_o$-predicates. The legitimacy and consequences of such assumptions are touched upon in the literature, in particular in connection with predicates like 'x is bigger than y' which, as discussed in §3.4, resist categorisation as (dichotomously) either observational or theoretical. (See Cruse 2005: 561ff; Ainsworth 2009: 145–6, 155–60; Ketland 2004: 292, 2009: 38–9.) However, *provided* that some o/t dichotomy has been embraced, we think that we can sidestep this debate. A philosopher who rejects (b) might insist on expanding C, so that it includes all unobservable entities which fall under some observation relation. If we concede this point, then we will want to expand the definition of $\text{Ob}(\mathcal{M})$ accordingly. (It is harder to see how rejecting (a) would force us to change anything in our setup.) But making these changes this will not affect the fundamental idea behind Proposition 3.8: we can always arbitrarily permute away the entities falling outside $\text{Ob}(\mathcal{M})$'s domain.

[27] This idea is in Winnie (1967: 226–7). Putnam (1981: 218, Second Comment) mentions it in his fullest presentation of his permutation argument. See also Button (2013: 41–2).

[28] Again, as is standard, we use κ to denote a cardinal; see the end of §1.B for a brief review of cardinals.

Proof. (1) ⇒ (2). Let $\mathcal{M}$ be as in (1), so $O^{\mathcal{M}} = O^{\mathcal{A}} = A$. Let κ be the cardinality of the unobservables in $\mathcal{M}$, i.e. $\kappa = |M \setminus A|$. Let U be any set of cardinality κ with $A \cap U = \emptyset$; so there is a bijection $g : (M \setminus A) \longrightarrow U$. Define a bijection $h : M \longrightarrow (A \cup U)$ by setting $h(x) = x$ if $x \in A$, and $h(x) = g(x)$ otherwise. Now use a Push-Through Construction, applying h to $\mathcal{M}$ to define a $\mathscr{V}$-structure, $\mathcal{N}$, and simply let $\mathcal{P}$ be $\mathcal{N}$'s $\mathscr{V}_0$-reduct.

(2) ⇒ (1). Let $\mathcal{P}$ be as in (2). As T is expansion-conservative over Ramsey(T), there is a $\mathscr{V}$-structure $\mathcal{M}$ which is a signature expansion of $\mathcal{P}$ and $\mathcal{M} \vDash T$. □

The significance of this result emerges, when we substitute the intended model C of the $\mathscr{V}_0^O$-sentences for $\mathcal{A}$, to obtain a new objection to ramsified realism:

Newman-cardinality-objection. *Instrumentalists and realists can agree that, at a minimum, T must be C-alright. But Proposition 3.8 entails that T is C-alright* iff *there is some cardinal κ such that we can obtain a model of* Ramsey(T) *just by adding κ 'unobservables' to C. Otherwise put:* Ramsey(T) *is* true *provided both that T is C-alright and there are 'sufficiently many' unobservables. And so, we are back at Newman's criticism: 'the doctrine that* only *structure is known', when it comes to the theoretical, 'involves the doctrine that* nothing *can be known' about the theoretical, except 'the number of' theoretical objects.*[29]

This is the most powerful version of the Newman objection, as generalised against ramsified realism.[30] Maybe some ramsified realists can learn to embrace the idea that the purpose of physics is to get everything right at the level of observables and additionally tell us the mere *cardinality* of the theoretical realm; but we doubt that this view will find many takers.

We can summarise the problem as follows. Proposition 3.8 uses model theory to generate a 'trivialising' structure, $\mathcal{P}$, from a given structure, $\mathcal{M}$. The existence of $\mathcal{P}$ is guaranteed by the standard set-theoretic axioms which we assume whenever we do model theory. So the ramsified-realist needs to explain why $\mathcal{P}$ is somehow *unintended* (cf. §2.3). To close the chapter, we consider two attempts to do this.[31]

3.9 Mixed-predicates again: the case of causation

When we discussed the permutation argument in §2.3, we noted that someone might complain that models obtained by the Push-Through Construction can fail

---

[29] Newman (1928: 144).

[30] Our *Newman-cardinality-objection* is close to Ketland's preferred version of the Newman objection (See also Ainsworth 2009: 144–7). Ketland's objection falls out of his Theorem 6 (2004: 298–9), which is close to our Proposition 3.8. However, Ketland invokes *full* second-order logic, whereas our Proposition 3.8 holds with both full and faithful Henkin semantics (see §3.2, and also footnote 1 from Chapter 2).

[31] For an excellent survey of responses, see Ainsworth (2009: 148ff).

to respect causal connections between words and the world. The ramsified realist might offer a similar response, concerning our trivialising model, $\mathcal{P}$.

By construction, $\mathcal{P}$ will respect all the causally-constrained reference relations between *okay* vocabulary and *okay* entities. The only possible concern, then, must be about the relationship between *troublesome* vocabulary and *troublesome* entities. We want to argue for a conditional claim: *if* our realist took the o/t dichotomy seriously in the first place, then she cannot make any hay here.

To see why, we must revisit §3.4. There, we considered the predicate 'x is bigger than y'. and noted that, prima facie, there are *observable* instances of this relation, *theoretical* instances of this relation, and *mixed* instances of this relation. The same point applies to the predicate 'x causes y'. Waiving Humean concerns, there are observational instances of causation: the striking of the match causes the fire. There are also theoretical instances of causation: consider the microscopic events involved in a chemical reaction. And then there are *mixed* instances: the decay of a radium atom causes the Geiger counter to click.

As in §3.4, such considerations might lead us to doubt the very idea that there is a sharp o/t *dichotomy*. And if we deny that there is any such dichotomy, then nothing will prevent us from going on to insist that the observable instances (in some rough and ready sense) of causation give us sufficient handle on the general notion of causation for us to see why $\mathcal{P}$ may be unintended.[32] However, to abandon the o/t dichotomy is *precisely* to abandon the position against which the Newman-cardinality-objection was targeted.

If, instead, we continue to insist on an o/t dichotomy, then we have no option but to 'split' the 'mixed' claims about causation into the okay ones and the troublesome ones. (Compare the fate of 'x is bigger than y' in §3.4.) But once we have done *that*—and taken seriously that this is a dichotomy and not a continuous transition from more to less observable—then it is profoundly unclear what would allow us to maintain that we have sufficient handle on the 'troublesome side' of causation, even to *articulate* the idea that $\mathcal{P}$ has failed to respect the (troublesome) causal relationship between troublesome vocabulary and troublesome entities. For the 'troublesome side' of causation is *just more* $\mathscr{V}_t$*-vocabulary*, to be ramsified away.[33]

3.10 Natural properties and just more theory

An alternative response is for the ramsified-realist to insist that some properties are more *natural* than others.

The existence of the trivialising structure, $\mathcal{P}$, is guaranteed by the standard set-theoretic axioms which we assume whenever we do model theory. The ramsified

[32] See the references in footnote 11.

[33] This is essentially the point of Button (2013: 50–1).

realist may, then, maintain that certain properties and relations are *natural*, whilst others are *mere artefacts* of a model theory that might be suitable for pure mathematics, but which is unsuitable for the philosophy of science. If she can uphold this point, then she can reject the significance of Proposition 3.8 and so answer the Newman-cardinality-objection.[34]

The obvious question is whether ramsified realism is consistent with a belief in such natural kinds. To believe that some properties or relations are natural is to postulate a higher-order property of properties, namely, *naturalness*. Syntactically, this will be introduced via a second-order predicate, N, such that we can say of a first-order predicate, R, such as '... is an electron', something of the form $\mathrm{N}(R)$, i.e. roughly '*electronhood* is a natural property'. Since, though, our ramsified realist believes in an o/t dichotomy, we must ask her whether N falls on the o-side or the t-side.[35] Certainly N cannot belong to the *observational* vocabulary—even if *electronhood* is natural, one cannot simply *observe* that it is—so N must belong with the *theoretical* vocabulary, i.e. $\mathrm{N} \in \mathscr{V}_{\mathrm{t}}$. As such, the ramsified realist is duty bound to ramsify away the higher-order property of *naturalness*. That is, she must replace N with an existentially bound (third-order) variable Y, so that her new Ramsey sentence for T is:

$$\exists Y \exists X_1 \dots \exists X_n \hat{T}[Y/\mathrm{N}, \overline{X}/\overline{R}]$$

But now, given even remotely permissive comprehension principles for third-order logic, the invocation of a *ramsified* property of *naturalness* will impose no constraint whatsoever upon our structures.[36]

The shape of this problem should look familiar: in the last few paragraphs, we have essentially recapitulated the dialectic surrounding Putnam's permutation argument from §2.3. Our ramsified realist wanted to maintain that a Ramsey sentence is made true (if at all) only by the existence of a structure whose second-order entities are *natural*. This is just a version of the idea that certain referential candidates are *preferable*, as considered in §2.3 in response to Putnam's permutation argument.[37] And, as in §2.3, any defender of this notion of *preferability* must face up to Putnam's just-more-theory manoeuvre. In the present context, the allegation will be that to invoke 'naturalness' is just more *theory*.

---

[34] See Psillos (1999: 64–5), Ketland (2009: 44), and Ainsworth (2009: 167–9).

[35] Brian King suggested to us that the ramsified realist might decline to answer the question, on the grounds that N is not itself an expression from the sciences, but an expression from some metasemantic theory. Whilst the reply is possible in principle, it is hard to see how it could be developed. A ramsified realist who offers this reply must explain why her initial motivations for embracing the o/t dichotomy stop short of rendering *naturalness* troublesome. This is particularly difficult if those initial motivations were essentially epistemological, for our epistemological situation with regard to *naturalness* is worse than our epistemological situation with regard to *electronhood* (for example).

[36] For a similar argument, see Ainsworth (2009: 160–2, 169); also Demopoulos and M. Friedman (1985: 629), Psillos (1999: 64–6), and Ketland (2009: 44).

[37] See Ainsworth (2009: 162–3).

Now, in §2.3, we noted that there are positions according to which the just-more-theory manoeuvre simply looks hopelessly question-begging. (If causation fixes reference, then permuting the reference of 'causation' is besides the point.) However, we also noted that the just-more-theory manoeuvre poses real difficulties for certain other positions, such as moderate objects-platonism. (To the moderate objects-platonist, anything compatible with moderation looks like more mere theory.) The same point arises here. Belief in intrinsically natural kinds may be a defensible philosophical thesis. However, the present allegation is that the ramsified realist is bound, *by her own lights*, to treat 'naturalness' as just more theory, in the precise sense that it belongs to theoretical-vocabulary rather than the observational-vocabulary, and hence *must* be ramsified away.

At this point, though, we see that there are no *shortcuts* to the philosophical significance of the Newman-cardinality-objection against ramsified realism. Its significance can only be settled by delving deeply into the very idea of an o/t dichotomy. And *that* would take us far away from model theory, and so far beyond the scope of this book.

So, to close the chapter, we simply note the deep connections between the permutation argument and the Newman objection. First: ramsified realism itself can be motivated by permutation-style worries about the reference of theoretical vocabulary (see §3.2). Second: the Push-Through Construction is used in Putnam's permutation argument, in Newman's original argument, and in the Newman-cardinality-objection.[38] Third: similar responses are available to both arguments, and such responses must confront the just-more-theory manoeuvre head-on.[39]

3.A Newman and elementary extensions

We have presented and discussed our favourite version of the Newman objection. However, in this appendix, we consider a Newman-esque objection in a first-order context. Before starting, we should issue a caution. The appendix draws on technical results which we will first encounter in Chapter 4. As such, some readers may want to read ahead before returning to this appendix. And the material in this appendix can, indeed, be safely omitted, since the *best* version of the Newman objection is the Newman-cardinality-objection of §3.8.

In the next chapter, we prove the following:[40]

[38] See Demopoulos and M. Friedman (1985: 629–30), Lewis (1984: 224fn.9), Ketland (2004: 294–5, 298–9), and Hodesdon (forthcoming: §5).

[39] See Hodesdon (forthcoming: §6).

[40] Actually, in Chapter 4 we state the result in terms of deduction-conservativeness (see Definition 4.17) rather that consequence-conservativeness. However, for first-order theories, these notions are identical, by the Soundness and Completeness of first-order logic.

Proposition (4.18): *Let T be a first-order $\mathscr{L}^+$-theory and S be a first-order $\mathscr{L}$-theory, with $\mathscr{L}^+ \supseteq \mathscr{L}$. The following are equivalent:*

(1) *T is consequence-conservative over S.*
(2) *For any $\mathscr{L}$-structure $\mathcal{M} \vDash S$, there is an $\mathscr{L}^+$-structure $\mathcal{N}$ which satisfies T and whose $\mathscr{L}$-reduct is an elementary extension of $\mathcal{M}$.*

This result connects a notion of conservation (see Definition 3.3) with the existence of an elementary extension (see Definition 4.3), i.e. typically to a structure with an enlarged *ontology*. It is worth explicitly contrasting this first-order result with our second-order Proposition 3.5, which connected a notion of conservation with a structure with no new ontology, but merely a richer *ideology*.

Despite the differences, Demopoulos has recently used Proposition 4.18 to present a Newman-esque objection.[41] His objection does not consider Ramsey sentences, or target ramsified realism, *per se*. Rather, Demopoulos's target is a realist who believes that physical theories have a kind of content which goes beyond the observational claims, but which is in some sense 'merely structural'. Demopoulos criticises this realist via something very much like the Newman-conservation-objection of §3.6, but employing the first-order Proposition 4.18 rather than the second-order Proposition 3.5. As we understand it, Demopoulos' objection runs as follows:

Newman-extension-objection. *As in the Newman-conservation-objection of §3.6, instrumentalists and realists of all stripes can surely agree that their theory, T, should be consequence-conservative over the set of all true $\mathscr{V}_0$-sentences, S. Additionally, instrumentalists and realists of all stripes can surely agree that there is a model, say C, of S. So, by (1) $\Rightarrow$ (2) of Proposition 4.18, there is a model of T (indeed, an elementary extension of C). So the realist does not, in fact, manage to incur any substantial commitments by affirming T.*

Now, like the Newman-conservation-objection of §3.6, this Objection invokes the idea that the instrumentalist must be committed to $\mathscr{V}_0$-soundness. However, as we pointed out in §3.7, that is a mistake. If there is a one-place $\mathscr{V}_0$-predicate, O, to be read as '... is an observable object', then $\exists x \neg O(x)$ is a $\mathscr{V}_0$-sentence which is probably entailed by T, but which the instrumentalist need not accept. Consequently, instrumentalists need not be committed to the $\mathscr{V}_0$-soundness of physical theories after all. Exactly the same point applies to the Newman-extension-objection.

However, it is worth pointing out a *further* defect with the Newman-extension-objection. The Objection involves the claim that there is an intended model, C, of the true $\mathscr{V}_0$-sentences. When we apply Proposition 4.18 to C, we obtain a structure,

[41] Demopoulos (2011: 186–90).

$\mathcal{N}$, whose $\mathscr{V}_0$-reduct is an elementary extension of $\mathcal{C}$. Crucially, this elementary extension will (probably) *enlarge* the ontology; that is, $N \setminus C$ will be non-empty. And if the objection is to hit its target, then we should think of C as all that is *observable*, so that the entities in $N \setminus C$ are all deemed *theoretical*.[42] But Proposition 4.18 provides no guarantee that we might not find some new element $a \in N \setminus C$ such that $\mathcal{N} \vDash O(a)$. On the one hand, such an element would be *observable*, at least according to $\mathcal{N}$; on the other hand, it should be(merely) *theoretical*, since it is in $N \setminus C$. It is unclear why such a confused structure should command our attention.

We can illustrate this concern via a toy example, which employs a lovely result from the model theory of PA. The result is that PA cannot define the notion of a 'standard number' (the notation $S^n(0)$ was defined in (*numerals*) of §1.13):

Lemma 3.9 (Overspill): *Let $\varphi(v)$ be any one-place formula in the signature of* PA. *Let $\mathcal{M}$ be a non-standard model of* PA. *If $\mathcal{M} \vDash \varphi(S^n(0))$ for all $n < \omega$, then there is some non-standard element b such that $\mathcal{M} \vDash \varphi(b)$.*

Proof. Suppose $\mathcal{M} \vDash \varphi(S^n(0))$ for all n in the metatheory. Let a be a non-standard element of $\mathcal{M}$. If $\mathcal{M} \vDash \varphi(a)$, then we are done. Otherwise, $\mathcal{M} \vDash \neg\varphi(a)$. Since $\mathcal{M} \vDash$ PA, it satisfies the least-number-principle,[43] so let a_0 be the least element of $\mathcal{M}$ satisfying $\neg\varphi(x)$. Since a_0 is non-standard, $a_0 \neq 0$, so there is some b such that $\mathcal{M} \vDash S(b) = a_0$. Then b is also non-standard and since $\mathcal{M} \vDash b < a_0$, we have that $\mathcal{M} \vDash \varphi(a)$. □

To see the significance of this result, suppose that the observable objects happen to 'line up' nicely, so that they can be enumerated. So we can treat the (standard) natural numbers as *proxies* for the genuinely observable objects, and regard the (standard) natural number structure as a proxy for our intended model of the observables, $\mathcal{C}$. Then when $\mathcal{N}$ is a *proper* elementary extension of $\mathcal{C}$, it must add some new elements to the domain. In our toy, this will amount to adding some *nonstandard* numbers. Suppose, furthermore, that we have defined a one-place predicate 'O' in the language of PA, which we can usefully treat as a proxy for the predicate '... is an observable' in $\mathcal{C}$. Then, by the Overspill Lemma 3.9, some of these new nonstandard elements in $N \setminus C$ must satisfy the predicate 'O' in $\mathcal{N}$.

In sum, the Newman-extension-objection mishandles observable entities in at least two ways. This is why we favour the Newman-cardinality-objection.

---

[42] Hence Demopoulos writes: 'Let $\mathcal{C}$ be the $\mathscr{V}_0$-structure whose domain is the domain of observable events.... We must show that the domain C of $\mathcal{C}$ has an extension $\mathcal{N}$ which is the domain of a model of T, where N is the set of observable and theoretical events' (Demopoulos 2011: 186–7, notation changed to match main text).

[43] Roughly: if some elements are φs, then there is a $<$-least φ. See e.g. Kaye (1991: 44–5).

3.B Conservation in first-order theories

Theorem 3.5 shows that consequence-conservation and expansion-conservation align for second-order theories. In this appendix, we prove that this alignment *fails* for first-order theories. In particular, while expansion-conservation still implies consequence-conservation for first-order theories, the converse fails. This is precisely the content of Proposition 3.11, below.

As with §3.A: the topic of this appendix means that it belongs in this chapter, even though the result requires some ideas from later chapters. In particular, our proof of Proposition 3.11 uses both non-standard models of PA (see Chapter 4) and the Löwenheim–Skolem Theorem 7.2 (see Chapter 7). Still, our proof requires only basic model-theoretic tools.[44]

We say that a theory T is *complete* iff either $T \vDash \varphi$ or $T \vDash \neg\varphi$ for every sentence φ in T's signature. Then the theory of arithmetic in the signature $\{0, S\}$ is complete. More precisely, let SA, for *successor arithmetic*, be the theory whose axioms are just (Q1) and (Q2) from Definition 1.9, i.e., $\forall x\, S(x) \neq 0$ and $\forall x \forall y(S(x) = S(y) \rightarrow x = y)$. Now:

Proposition 3.10: SA *is complete.*

Proof. Where $\mathcal{M}$ is a model of SA, say that a *chain* in $\mathcal{M}$ is a subset $Z \subseteq M$ such that for every $a, b \in Z$ there is some natural number n in the metatheory such that either $(S^n(a))^{\mathcal{M}} = b$ or $(S^n(b))^{\mathcal{M}} = a$, using the (*numerals*) notation from §1.13.

Let $\mathcal{M}_0$ and $\mathcal{M}_1$ be two models of SA of a fixed uncountable cardinality, κ. Since they are of the same cardinality κ, their domains can be written as the disjoint unions $M_i = N_i \cup \bigcup_{\alpha<\kappa} Z_{i,\alpha}$ where the N_i component is isomorphic to the standard natural numbers in the signature $\{0, S\}$, and each $Z_{i,\alpha}$ component is isomorphic to the integers in the signature $\{S\}$. By mapping N_0 to N_1 and $Z_{0,\alpha}$ to $Z_{1,\alpha}$, we have that $\mathcal{M}_0 \cong \mathcal{M}_1$. Since any incompleteness in SA would have to be registered on models of size κ by the Löwenheim–Skolem Theorem 7.2, SA is complete. □

We can now use this result to offer an example of consequence-conservation without expansion-conservation.

Proposition 3.11: PA *is consequence-conservative over* SA *but not expansion-conservative over* SA.

Proof. First observe that that PA's signature strictly expands SA's.

[44] As such, our proof is much simpler than the one usually offered in the literature, which requires familiarity with formal theories of truth. See Halbach (2011: Theorem 8.12 p.80, Theorem 8.31 p.98).

Consequence-conservation. Let φ be a first-order sentence in the signature $\{0, S\}$ such that PA $\vDash \varphi$. If SA $\vDash \neg\varphi$, then also PA $\vDash \neg\varphi$, since PA has all of SA's axioms; but then, absurdly, PA would be inconsistent; so SA $\nvDash \neg\varphi$. Since SA is complete, SA $\vDash \varphi$, as required.

Failure of expansion-conservation. Let $\mathcal{M}$ be a model of SA given by a copy of the natural numbers followed by a single disjoint copy of the integers, with zero and successor interpreted in the ordinary way. So we may write $\mathcal{M}$'s domain as $M = N \cup Z$, where the N component is isomorphic to the standard natural numbers in the signature $\{0, S\}$, and the Z component is a chain.

For reductio, suppose there is a model $\mathcal{M}^*$ of PA whose $\mathscr{L}$-reduct is just $\mathcal{M}$. Let a be an element of Z. Now, PA proves that every number is even or odd. Begin by supposing that $\mathcal{M}^* \vDash b + b = a$ for some b in M. Then b too is in Z rather than N. Since a and b are part of the same chain, $\mathcal{M} \vDash S^k(b) = a$ for some $k \in \omega$. So $\mathcal{M}^* \vDash S^k(0) + b = S^k(b) = a = b + b$. Subtracting b from both sides, $\mathcal{M} \vDash S^k(0) = b$, contradicting the fact that b is in Z rather than N. And a similar contradiction arises from supposing $\mathcal{M}^* \vDash b + b + 1 = a$ for some b in M, completing the reductio. □

4
Compactness, infinitesimals, and the reals

In this chapter, we introduce the Compactness Theorem. This provides us with the tools to investigate non-standard models. These non-standard models will occupy us in Part B of this book. There, they will be treated as some kind of *foe*. But, in this chapter, non-standard models come as *friends*.

Arguably the most famous application of the Compactness Theorem lies in Robinson's development of non-standard analysis. Robinson's aim was to resuscitate *infinitely small quantities*, as used in the historical calculus of the seventeenth and early eighteenth century. In this chapter, we suggest that Robinson's attempt to 'fully vindicate' the historical calculus can be extended beyond even the point that Robinson himself realised.[1] However, this is not merely of historic interest: it provides a stunning example of how model-theoretic methods can help to articulate and defend varieties of instrumentalism in the philosophy of mathematics.

The ideas of this chapter focus on a single mathematical structure, namely the real numbers and some of its model-theoretic extensions. While most of our efforts in this chapter concern non-standard analysis, we close the chapter with a brief discussion of other model-theoretic perspectives on the real numbers.

4.1 The Compactness Theorem

Here is one of the most fundamental results in model theory:

Theorem 4.1 (Compactness Theorem): *Let T be a first-order theory. Then T has a model iff every finite subtheory of T has a model.*

Proof. Left-to-right is trivial, since any model of T is a model of each of T's finite subtheories. *Right-to-left* holds by assuming that every finite subtheory of T has a model. Then T is consistent, by the soundness of the deductive system and Proposition 4.20. So T has a model by Gödel's Completeness Theorem 4.24. □

This quick proof of Compactness uses Gödel's Completeness Theorem. No doubt

[1] The phrase 'fully vindicate' is in A. Robinson (1966: 2). Elsewhere, Robinson was more circumspect about the aims of the project: he says that his theory 'provides a satisfactory framework for the development' of Leibniz's notion of infinitesimals (1968: 70), and that it is 'a natural approach to [...] a calculus involving infinitesimals' (1961: 433).

most readers will have encountered this result before; but, for the sake of completeness, we prove it in §4.A. However, given its centrality, we also offer two more proofs of the Compactness Theorem in this book: a more model-theoretic proof in §4.B, and a set-theoretic-cum-algebraic proof in §13.C.

To demonstrate the power of the Compactness Theorem, we will use it to prove the existence of a non-standard model of arithmetic. Let T be the theory of 'true arithmetic', i.e. the theory containing all the sentences in the signature of PA which are true on the natural numbers (PA is set out in Definition 1.9 of §1.A). Let c be a new constant symbol, and consider the following expanded theory where the term $S^n(0)$ is defined as in (*numerals*) of §1.13:

$$T^* = T \cup \{S^n(0) < c : n < \omega\} \qquad \textit{(non-standard)}$$

Let T_0^* be any finite subtheory of T^*. Since T_0^* is finite, it contains at most finitely many sentences of the form $S^n(0) < c$; so let m be the largest number n such that $S^n(0) < c$ appears in T_0^* (or 0 if there are no instances). Then consider the model $\mathcal{N}^+$ which is a signature expansion of $\mathcal{N}$ in which c is interpreted to name $m + 1$. Clearly $\mathcal{N}^+$ is a model of T_0^*. And since T_0^* was an arbitrary finite subtheory of T^*, the Compactness Theorem 4.1 entails that T^* itself has a model, $\mathcal{N}^*$.

In $\mathcal{N}^*$, the interpretation of c must differ from the interpretation of each $S^n(0)$, as n ranges over natural numbers in the meta-theory. So this element must be *infinitary*, for according to $\mathcal{N}^*$ it is larger than any number picked out by a numeral. It will be useful to have a term for such models and elements:

Definition 4.2: *An element b of a model $\mathcal{M}$ of* PA *is* non-standard *iff $\mathcal{M} \models b \neq S^n(0)$ for each $n < \omega$, where n ranges over natural numbers in the meta-theory. A* non-standard model *of* PA *is a model which contains a non-standard element.*

In these terms, $c^{\mathcal{N}^*}$ is a non-standard element in the non-standard model $\mathcal{N}^*$.

Our model $\mathcal{N}^*$ also illustrates an important model-theoretic notion. We first introduce the notion, and then show how it applies to $\mathcal{N}^*$.[2]

Definition 4.3: *For any $\mathcal{L}$-structures $\mathcal{A}$ and $\mathcal{B}$ such that $A \subseteq B$, we say that $\mathcal{B}$ is an* elementary extension *of $\mathcal{A}$ iff for any elements $\bar{a}$ from A and any $\mathcal{L}$-formula $\varphi(\bar{x})$ we have $\mathcal{A} \models \varphi(\bar{a})$ iff $\mathcal{B} \models \varphi(\bar{a})$. Equivalently, we say that $\mathcal{A}$ is an* elementary substructure *of $\mathcal{B}$, or simply write $\mathcal{A} \preceq \mathcal{B}$.*

A closely related notion drops the requirement that $A \subseteq B$:

---

[2] See Hodges (1993: 54) and Marker (2002: 44).

Definition 4.4: *For any $\mathcal{L}$-structures $\mathcal{A}$ and $\mathcal{B}$, we say that a map $h : \mathcal{A} \longrightarrow \mathcal{B}$ is* an elementary embedding *iff for any elements $\bar{a}$ from A and any $\mathcal{L}$-formula $\varphi(\bar{x})$ we have $\mathcal{A} \vDash \varphi(\bar{a})$ iff $\mathcal{B} \vDash \varphi(\bar{h}(\bar{a}))$.*

As in Definition 2.2, if $\bar{a} = (a_1, \ldots, a_n)$ then $\bar{h}(\bar{a}) = (h(a_1), \ldots, h(a_n))$. The name 'embedding' comes from fact that if $\mathcal{B} \vDash h(a) = h(b)$ then $\mathcal{A} \vDash a = b$, so that elementary embeddings are always injective. Finally, if $h : \mathcal{A} \longrightarrow \mathcal{B}$ is an elementary embedding, then $\mathcal{B}$ is isomorphic to an elementary extension of $\mathcal{A}$, by a simple Push-Through Construction. Up to isomorphism, then, elementary extensions and elementary embeddings are two viewpoints on the same thing.

We can now apply these notions to our model $\mathcal{N}^*$ from before. Where $\mathcal{N}$ is again the natural numbers, the map $h : \mathcal{N} \longrightarrow \mathcal{N}^*$ given by $h(n) = (S^n(0))^{\mathcal{N}^*}$ is an elementary embedding. For, given any one-place formula $\varphi(x)$, we have:

$$\mathcal{N} \vDash \varphi(n) \text{ iff } \mathcal{N} \vDash \varphi(S^n(0)) \text{ iff } \mathcal{N}^* \vDash \varphi(S^n(0)) \text{ iff } \mathcal{N}^* \vDash \varphi(h(n)) \qquad (elem)$$

The first biconditional holds because the term $S^n(0)$ picks out n on $\mathcal{N}$, the second biconditional holds because $\mathcal{N}$ and $\mathcal{N}^*$ both satisfy the same complete theory T in their common signature and $\varphi(S^n(0))$ is a sentence in this signature, and the final biconditional holds thanks to the definition of h. And obviously the same argument as given for (*elem*) works for formulas in any number of free variables. So h is an elementary embedding and our model $\mathcal{N}^*$ is isomorphic to an elementary extension of the natural numbers. But it is not isomorphic to the natural numbers *themselves*. (This refutes the converse of Corollary 2.5.)

4.2 Infinitesimals

Armed with the general idea of a non-standard model, we now turn to the development of Robinson's non-standard analysis. The aim, recall, is to resuscitate infinitesimals, so we must start by saying what they are:

Definition 4.5: *An* infinitesimal *is a quantity η such that $0 < |\eta| < \frac{1}{n}$ for every natural number $n > 0$.*

Here, and throughout the chapter, we use vertical bars to indicate *absolute value.* (Note that many other texts treat zero itself as an infinitesimal.[3] However, by adopting the convention that zero is not infinitesimal, we can state our results and discuss certain philosophical questions more concisely.)

Now, let $\mathcal{R}$ be any expansion of the real numbers as a linear order with addition and multiplication and constants for zero and one. Let T be the complete theory

[3] See Hurd and Loeb (1985: 25) and Goldblatt (1998: 50).

of $\mathcal{R}$ in this signature. By using the Compactness Theorem 4.1 exactly as in §4.1, we can create a model of T with infinitesimals. In particular, if η is the new constant symbol, we let

$$T^* = T \cup \{(0 < |\eta| < \tfrac{1}{n}) : n < \omega\}$$

Let T_0^* be any finite subtheory of T; let m be the greatest n with $(0 < |\eta| < \frac{1}{n})$ in T_0^*; then the expansion of $\mathcal{R}$ in which η is interpreted as $\frac{1}{m+1}$ is a model of T_0^*. Since T_0^* was an arbitrary finite subtheory of T^*, by the Compactness Theorem 4.1 the theory T^* has a model.

An argument similar to that of §4.1 also shows that T^* has a model $\mathcal{R}^*$ which is an elementary extension of $\mathcal{R}$. For, in the previous paragraph, there were no restrictions on the type of signature which we could interpret on the real numbers, other than that it contained $\{0, 1, <, +, \times\}$. So, we can assume that each real number r is denoted by a constant c_r in the structure $\mathcal{R}$.[4] As above, there is a model $\mathcal{R}^*$ of T^* in this expanded signature. There is therefore a natural map $h : \mathcal{R} \longrightarrow \mathcal{R}^*$ given by sending each real number r to the interpretation of c_r in the structure $\mathcal{R}^*$. So for each n-place formula $\varphi(\bar{x})$ in the signature, we have the following (compare this with (*elem*) of §4.1):

$$\mathcal{R} \vDash \varphi(\bar{r}) \text{ iff } \mathcal{R} \vDash \varphi(\overline{c_r}) \text{ iff } \mathcal{R}^* \vDash \varphi(\overline{c_r}) \text{ iff } \mathcal{R}^* \vDash \varphi(\hbar(\bar{r}))$$

Accordingly, h is an elementary embedding, so that $\mathcal{R}^*$ is isomorphic to an elementary extension of $\mathcal{R}$ by a Push-Through Construction. Hence, $\mathcal{R}$ has an elementary extension containing infinitesimals.

In §4.1, we used Compactness and sentences of the form $n < c$ to obtain infinitary elements. Our method for obtaining infinitesimals involved considering sentences of the form $(0 < |\eta| < \frac{1}{n})$. This suggests, correctly, that infinitary elements are the reciprocals of infinitesimals, and vice versa. Indeed, we define:

Definition 4.6: *A quantity b is* infinite *iff $n < |b|$ for all natural numbers n. A quantity b is* finite *iff there is a natural number n such that $|b| \leq n$.*

It is now easy to see b is infinitesimal iff $\frac{1}{b}$ is infinite. It follows that any elementary extension of $\mathcal{R}$ containing infinitesimals will also contain infinite elements, for division is defined everywhere in $\mathcal{R}$ (except on 0). It is also easy to see that any quantity is either: (i) infinite, (ii) finite and not infinitesimal, or (iii) infinitesimal.

[4] Using the notation of §1.5, we could do this by moving from $\mathcal{R}$ to $\mathcal{R}^\circ$; but this would introduce more superscripts than necessary.

4.3 Notational conventions

Over the next few sections, we use infinitesimals to develop a non-standard approach to real analysis. Readers with no prior familiarity with real analysis will struggle with this material; they may wish to take it on trust that we *can* develop an entirely rigorous theory of infinitesimals, and skip directly to §4.7, where we discuss the largely philosophical issue of instrumentalism about infinitesimals. But readers who are sticking with us should be forearmed with some nomenclature.

In what follows, we work with elementary extensions $\mathcal{R}^*$ of $\mathcal{R}$. The signature of $\mathcal{R}$ will vary from application to application. Since the first-order part of $\mathcal{R}$ is simply the real numbers, we simply call elements of $\mathcal{R}$ *reals*. For lack of any better term, we continue to call the elements of $\mathcal{R}^*$ *quantities*. So, on this definition, all reals are quantities.

We now define an important equivalence relation, of 'almost-equality':

$$a \approx b \text{ iff } a - b \text{ is infinitesimal or zero}$$

The notation $a \approx b$ also allows us to define the standard part of a finite quantity. In particular, if a is finite, then we define its *standard part* to be the unique real number r such that $r \approx a$.[5] For instance, if η is infinitesimal, then the standard part of η is zero, while the standard part of $5 + \eta$ is just 5.

Our final piece of notation relates to definability (as laid down in §1.13). If $f : \mathcal{R} \longrightarrow \mathcal{R}$ is any function which is $\mathcal{R}$-definable by a formula $\varphi(x, y)$, then by the elementary equivalence of $\mathcal{R}$ and $\mathcal{R}^*$, the formula $\varphi(x, y)$ as interpreted in $\mathcal{R}^*$ defines a function from $\mathcal{R}^*$ to $\mathcal{R}^*$. In what follows, we denote this function by $f^* : \mathcal{R}^* \longrightarrow \mathcal{R}^*$. We proceed similarly with n-place functions and n-place relations. However, for certain very frequently used functions, like absolute value, we sometimes abuse notation and write $|x|$ instead of the more cumbersome $|x|^*$.

4.4 Differentials, derivatives, and the use of infinitesimals

We will start by showing how to use infinitesimals to define both the derivative and its historical predecessor, the differential. The very first textbook on the calculus, l'Hôpital's *Analysis of the Infinitely Small*, contains this definition at its outset: 'The infinitely small portion by which a variable quantity continually increases or decreases is called the *Differential*.'[6] In this vein, we offer the following:

Definition 4.7: *Let η be an infinitesimal quantity, let f be a function on quantities, and let t be any quantity. The* differential *is then* $(d_\eta f)(t) = f(t + \eta) - f(t)$

[5] The standard part was first defined in A. Robinson (1966: 57). It takes some effort to verify its existence and uniqueness; see e.g. Hurd and Loeb (1985: 26) and Goldblatt (1998: 53).

[6] l'Hôpital (2015: Definition II, p.2).

That is, the differential $(d_\eta f)(t)$ tells us how much a 'variable quantity' changes when its argument is infinitesimally increased from t to $t + \eta$.

The differential was the primary concept in the historical calculus. But we can define the now more familiar notion of the derivative in terms of the differential.

Definition 4.8: *Let η, f and t be as in Definition 4.7 and let $L \in \mathcal{R}$. Then f has* derivative *L at t iff $\frac{(d_\eta f)(t)}{\eta} \approx L$ for all infinitesimals η.*

For example, the derivative of $f(t) = t^2$ is $2t$:

$$\frac{(d_\eta f)(t)}{\eta} = \frac{(t+\eta)^2 - t^2}{\eta} = \frac{t^2 + 2t\eta + \eta^2 - t^2}{\eta} = \frac{2t\eta + \eta^2}{\eta} = 2t + \eta \approx 2t$$

When t is real and η is infinitesimal, both the numerator $2t\eta + \eta^2$ and the denominator η are infinitesimal; but $2t$ is not. As Euler put it, in the differential calculus 'the work involved is not finding the differentials themselves, which are both equal to zero, but rather in finding their geometric ratio.'[7]

The computation of $f(t) = t^2$, however, led to many criticisms of the use of infinitesimals. To some early readers of the calculus, it seemed that the last step was justified by supposing that the infinitesimal η was *equal* to zero, i.e. that $2t + \eta = 2t$. (Look again at Euler's claim, that the differentials 'are both equal to zero'.) This, of course, contradicts the assumption that η was *non-zero*, which we needed in order to divide by η at the outset of the calculation. Berkeley put the point as follows:

I admit that signs may be made to denote either any thing or nothing: and consequently that in the original notation $t + \eta$, η might have signified either an increment or nothing. But then which of these soever you make it signify, you must argue consistently with such its signification, and not proceed upon a double meaning: Which to do were a manifest sophism.[8]

That is, Berkeley saw the original versions of the calculus as haphazardly assuming that a quantity η was non-zero at the outset of a calculation, and then *later in that same calculation* explicitly violating this assumption.

Fortunately, Robinson's distinction between $x = y$ and $x \approx y$ dispels this worry. We do not say that $2t + \eta = 2t$, only that $2t + \eta \approx 2t$. As Robinson said:

[...] instead of claiming that two quantities which differ only by an infinitesimal amount, e.g. x and $x + \eta$, are actually equal, we find only that they are equivalent in a well-defined sense, $x + \eta \approx x$ and thus can be substituted for one another in some relations but not in others.[9]

---

[7] Euler (1755: 66).

[8] Ewald (1996: v.1 p.69), variables changed to match preceding text.

[9] A. Robinson (1967: 34), variables and notation changed to match main text. See also (1966: 266).

In short, the distinction between = and ≈ gives firm-footing to the notion of a derivative, as presented in terms of infinitesimals via Definition 4.8.

However, it is still not clear how to relate *that* notion of a derivative, with the notion of a derivative *as we currently understand it*, namely as:

Definition 4.9: *For any function* $f : \mathbb{R} \longrightarrow \mathbb{R}$ *and any* $t \in \mathbb{R}$, *we define* $f'(t) = \lim_{x \to 0} \frac{f(t+x)-f(t)}{x}$, *when this limit exists.*

This is where our model-theoretic machinery starts to pay its way. As Robinson noted, these two notions provably align:[10]

Proposition 4.10: *If* $f : \mathcal{R} \longrightarrow \mathcal{R}$ *is* $\mathcal{R}$*-definable and* $t, L \in \mathcal{R}$, *then the following are equivalent:*

(1) $f'(t) = L$ *in the sense of Definition 4.9*
(2) f^* *has derivative L at t in the sense of Definition 4.8*

Proof. (1) ⇒ (2). Suppose that $f'(t) = L$. Let η be infinitesimal and let $\varepsilon > 0$ be a positive real number. We must show that $\left|\frac{(d_\eta f^*)(t)}{\eta} - L\right| < \varepsilon$. Applying Definition 4.9 to $\varepsilon > 0$, we obtain a real $\delta > 0$ such that for all $x \in \mathcal{R}$ if $0 < |x| < \delta$ then $\left|\frac{f(t+x)-f(t)}{x} - L\right| < \varepsilon$. By elementarity,[11] this holds in $\mathcal{R}^*$. Since η is infinitesimal, of course $0 < |\eta| < \delta$ and so $\left|\frac{(d_\eta f^*)(t)}{\eta} - L\right| < \varepsilon$, as required.

(2) ⇒ (1). Suppose that $\frac{(d_\eta f^*)(t)}{\eta} \approx L$ for all infinitesimal η. Let $\varepsilon > 0$ be a real. Suppose that $\delta > 0$ is infinitesimal with $0 < |\eta| < \delta$, so that η is also infinitesimal. Then by hypothesis, $\left|\frac{(d_\eta f^*)(t)}{\eta} - L\right| < \varepsilon$. Hence, in $\mathcal{R}^*$ there is $\delta > 0$ such that if $0 < |\eta| < \delta$ then $\left|\frac{(d_\eta f)(t)}{\eta} - L\right| < \varepsilon$. By elementarity, this also holds in $\mathcal{R}$. □

4.5 The orders of infinite smallness

We have just glimpsed the power of Robinson's non-standard analysis. Moreover, there is clearly *some* resemblance between Robinson's notion of an infinitesimal and its historical antecedents. But Robinson claimed at times to have '*fully* vindicated' Leibniz's use of infinitesimals (our emphasis). And this claim is open to criticism.

In particular, Bos has argued that there are aspects of the historical usage of infinitesimals which have no counterpart in Robinson's non-standard analysis. Consequently, Bos writes that the early calculus and non-standard analysis simply

---

[10] A. Robinson (1961: 436, 1966: 68, 1967: 31).

[11] I.e. appealing to the following: since $\mathcal{A} \preceq \mathcal{B}$, when $\bar{a}$ are from A we have $\mathcal{A} \vDash \varphi(\bar{a})$ iff $\mathcal{B} \vDash \varphi(\bar{a})$.

present us with two different conceptions of 'the structure of the set of infinitesimals.'[12] In this section, we explain Bos's reasons for making this claim, and we will respond to Bos on Robinson's behalf in §4.6. However, we want to emphasise that Bos's objection poses no threat to non-standard analysis as a part of mathematics or logic (and nor was it meant to). Instead, Bos's criticism challenges Robinson's claims about the broader intellectual significance of his non-standard analysis.

As a way into Bos's objection, we consider the proof of the product rule for derivatives. Let f and g be two functions, let t be a point of evaluation, and let η be an infinitesimal. For readability, we will (when convenient) write the differential of h as dh rather than $(d_\eta h)(t)$, and the value $h(t)$ as h. Then the traditional proof of the product rule began thus:

$$\begin{aligned} d(f \cdot g) &= (f \cdot g)(t+\eta) - (f \cdot g)(t) \\ &= f(t+\eta) \cdot g(t+\eta) - f(t) \cdot g(t) \\ &= (df + f(t)) \cdot (dg + g(t)) - f(t) \cdot g(t) \\ &= df \cdot dg + df \cdot g(t) + f(t) \cdot dg + f(t) \cdot g(t) - f(t) \cdot g(t) \\ &= f \cdot dg + g \cdot df + df \cdot dg \qquad \textit{(prod-comp)} \end{aligned}$$

Robinsonian analysis provides us with robust tools to continue this proof. We first divide throughout by η, obtaining $\frac{d(f \cdot g)}{\eta} = f \cdot \frac{dg}{\eta} + g \cdot \frac{df}{\eta} + df \cdot \frac{dg}{\eta}$. Now, assuming that f is differentiable at t,[13] we can prove that df is infinitesimal:

Proposition 4.11: *Suppose that $f : \mathcal{R} \longrightarrow \mathcal{R}$ is $\mathcal{R}$-definable and differentiable at $t \in \mathcal{R}$, and that η in $\mathcal{R}^*$ is infinitesimal. Then $(d_\eta f)(t) = df$ is infinitesimal.*

Proof. By Proposition 4.10, for some $L \in \mathcal{R}$ we have $\frac{f(t+\eta)-f(t)}{\eta} \approx L$. So $\left|\frac{f(t+\eta)-f(t)}{\eta} - L\right| < \frac{1}{n}$ for all $n > 0$. Since $\eta \neq 0$, we may multiply through by $|\eta|$ to obtain $|(f(t+\eta) - f(t)) - L \cdot \eta| < \frac{|\eta|}{n} < \frac{1}{n}$. Since $n > 0$ was arbitrary, $f(t+\eta) - f(t) \approx L \cdot \eta \approx 0$, where the last equivalence holds because L is in $\mathcal{R}$ and η is infinitesimal. □

If g is differentiable at t, then $\frac{dg}{\eta}$ is finite by Proposition 4.10, so that $df \cdot \frac{dg}{\eta}$ is infinitesimal, so that now $\frac{d(f \cdot g)}{\eta} \approx f \cdot \frac{dg}{\eta} + g \cdot \frac{df}{\eta}$. By Proposition 4.10 again, both $\frac{dg}{\eta} \approx g'$ and $\frac{df}{\eta} \approx f'$, so that $\frac{d(f \cdot g)}{\eta} \approx f \cdot g' + g \cdot f'$. Invoking Proposition 4.10 one last time, we obtain the now-familiar product rule:

$$(f \cdot g)' = f \cdot g' + g \cdot f' \qquad \textit{(prod-rule)}$$

[12] Bos (1974: 84).

[13] The assumption is necessary. Let $f(x) = x^{-1}$ if $x \neq 0$, and let $f(x) = 0$ otherwise. Then $d_\eta f(0) = f(\eta) - f(0) = \eta^{-1}$, which is infinite.

But that is how things continue in the *Robinsonian* setting, and it involves noting the difference between $\approx$ and $=$. Historical practitioners of the calculus, however, obtained the product rule just by *discarding* the term $df \cdot dg$ at the end of the calculation (*prod-comp*).

In an attempt to justify discarding this term, l'Hôpital wrote that 'this is because the quantity $df \cdot dg$ is infinitely small with respect to the other terms $g \cdot df$ and $f \cdot dg$.'[14] But this explanation is dangerously quick. The product of an infinitesimal with a real number is itself an infinitesimal, and hence the two terms $f \cdot dg$ and $g \cdot df$ are *both* infinitesimals since dg and df are infinitesimals (as we just saw). So l'Hôpital's point cannot have been that the term $df \cdot dg$ is an infinitesimal while the other two terms were not. Equally, l'Hôpital's point cannot have been that we can *always* discard infinitesimals, for then we could discard all three terms. Finally, while the absolute value of $df \cdot dg$ is strictly less than that of the other two terms, this does not itself explain why we should be allowed to discard $df \cdot dg$, any more than we could hope to obtain a true identity by discarding the term 3 in the equation $15 = 7+5+3$. Rather, l'Hôpital's point must be something more like the following: $df \cdot dg$ is somehow of a *different order of smallness* than the other two terms. But Robinsonian analysis has not (yet) given us any way to make sense of this notion.

The concern just voiced about the product rule is not anachronistic; Berkeley made essentially the same complaint. Working geometrically, Berkeley conceived of f and g as the lengths of two lines A and B, and of df and dg as the small lengths a and b by which they are increased. Then $(A + a)(B + b) - AB$ represents the difference in area of a rectangle when its sides are increased from A and B to $A + a$ and $B + b$. Berkeley noted that some authors had suggested that this difference was the quantity $aB + bA$. But Berkeley insisted that:

> [...] $aB+bA+ab$ will be the true increment of the rectangle, exceeding that which was obtained by the former illegitimate and indirect method by the quantity ab. And this holds universally be the quantities a and b what they will, big or little, finite or infinitesimal, increments, moments, or velocities. Nor will it avail to say that ab is a quantity exceedingly small: since we are told that 'in mathematical matters errors, however small, are not to be contemned.'[15]

Note that this a slightly different kind of objection than we discussed in §4.4. There it seemed that Berkeley was (rightly) objecting to a lack of care in distinguishing between $x = y$ and $x \approx y$. Here, Berkeley is drawing attention to the fact that the early calculus seemed simply to discard certain infinitesimals from sums of infinitesimals, apparently because some infinitesimals were sufficiently smaller than others to license this.

---

[14] l'Hôpital (2015: 4), variables changed to match preceding text.

[15] Ewald (1996: v.1 p.66).

These issues about the product rule are far from isolated, and Bos draws attention to a host of other similar cases. For instance, he cites an argument of Bernoulli that the quotient of an infinitesimal of order one and an infinitesimal of order three should be an infinitely large quantity of order two.[16] Likewise, he cites Euler's argument that the natural logarithm of an infinitesimal is of a different order of smallness than all of the nth roots of the infinitesimal.[17]

The essential problem is this. Proofs in the historical calculus appealed to a notion of *different orders of smallness*. If this notion cannot be accommodated within non-standard analysis, then there is a real concern that non-standard analysis and the historical calculus employ different notions of *infinitesimal*.

4.6 Non-standard analysis with a valuation

Fortunately, as we now show, a natural notion of different orders of smallness *is* available within Robinson's framework.

We will define a function, v, which allows us to compare different orders of smallness. So when a and b are of the *same* order of smallness, we write $v(a) = v(b)$; and when a is *much* smaller than b, and so of a greater order of smallness, we write $v(a) > v(b)$.[18] Here is the formal implementation of this intuitive idea:

Theorem 4.12: *Let the natural logarithm function,* ln, *be definable in* $\mathcal{R}$, *and let* $\mathcal{R}^* \succeq \mathcal{R}$. *Then there is an ordered abelian group* $\mathcal{G}$ *and a surjective map* $v : R^* \longrightarrow (G \cup \{\infty\})$ *satisfying the following properties:*

(1) $v(a) = \infty$ *iff* $a = 0$
(2) $v(ab) = v(a) + v(b)$
(3) $v(a + b) \geq \min(v(a), v(b))$
(4) $0 < v(a) < \infty$ *iff* a *is infinitesimal*
(5) $v(a) < 0$ *iff* a *is infinite*
(6) $v(a) = 0$ *iff* a *is finite, non-zero, and not infinitesimal*
(7) *If* $|a| \leq |b|$, *then* $v(a) \geq v(b)$

The function v in this theorem is a *valuation map*. Axioms (1)–(3) are the axioms for a *valued field*,[19] while axioms (4)–(7) indicate how v acts differently on infinitesimals, finite and infinite quantities. The notion of an 'ordered abelian group' just

[16] Bos (1974: 23).

[17] Bos (1974: 84) and Euler (1780: §§9–14).

[18] There is, we admit, an awkwardness in describing this situation by saying that 'a is of a *greater order of smallness* than b'. But we hope that this awkwardness will be quickly mitigated by the formalism of the valuation function.

[19] See Engler and Prestel (2005) for a standard treatment of valued fields, and see van den Dries (2014) for a discussion of traditional applications to model theory. See in particular Engler and Prestel (2005: 28)

encapsulates algebraic properties common to ordered structures with an addition function, such as the integers or the reals.[20]

To get a sense of how v behaves, we offer two quick observations. When a and b are infinitesimals, their product is much smaller than either infinitesimal, in that $v(ab) = v(a) + v(b) > v(a)$, by clauses (2) and (4) of the Theorem. By contrast, when a is infinitesimal but b is a finite, non-zero, non-infinitesimal, their product is exactly as small as a itself, since $v(ab) = v(a) + v(b) = v(a) + 0$, by clause (6).

We defer the proof of Theorem 4.12 to §4.C. But we also prove there that the group mentioned in Theorem 4.12 is unique up to isomorphism (see Proposition 4.26). And this shows that, up to isomorphism, our valuation function uniquely implements an extremely intuitive idea of 'orders of smallness'.

Once we are armed with our valuation map, v, we can set about recovering various historical arguments which invoked the idea of different orders of smallness. We start with the historical deployment of the product rule. To handle this, we use the following result, which shows that, when a function has non-zero derivative, its differential gets valuated the same as its infinitesimal:

Proposition 4.13: *Let $\mathcal{R}^*$ be as in Theorem 4.12. Let $f : \mathcal{R} \longrightarrow \mathcal{R}$ be $\mathcal{R}$-definable and also differentiable at t. Then $f'(t) \neq 0$ iff: $v((d_\eta f^*)(t)) = v(\eta)$ for all infinitesimals η.*

Proof. By Proposition 4.10 and Theorem 4.12, the following are equivalent:

(1) $f'(t) \neq 0$

(2) there is finite non-zero $L \in \mathcal{R}$ such that $\frac{(d_\eta f^*)(t)}{\eta} \approx L$ for all infinitesimals η

(3) $v\left(\frac{(d_\eta f^*)(t)}{\eta}\right) = 0$ for all infinitesimals η

(4) $v((d_\eta f^*)(t)) = v(\eta)$ for all infinitesimals η

The only aspect of this equivalence which merits comment is (3) $\Rightarrow$ (2). Supposing (3), by Theorem 4.12, for each infinitesimal η there is some $L_\eta \in \mathcal{R} \setminus \{0\}$ such that $\frac{(d_\eta f^*)(t)}{\eta} \approx L_\eta$. By our supposition that f is differentiable at t and Proposition 4.10, we have $L_\eta = L_{\eta'}$ for all infinitesimals η, η'. □

This yields a proposition which partly vindicates l'Hôpital:[21]

and van den Dries (2014: 77) for the definition of a valued field. We frequently use these three consequences of (1)–(2):

First, $v(1) = 0$; this holds since $v(1) = v(1 \cdot 1) = v(1) + v(1)$.

Second, $v\left(\frac{1}{a}\right) = -v(a)$; this holds since if $a \neq 0$ then $0 = v(1) = v\left(a \cdot \frac{1}{a}\right) = v(a) + v\left(\frac{1}{a}\right)$.

Third, $v\left(\frac{a}{b}\right) = v(a) - v(b)$; this holds since $v\left(\frac{a}{b}\right) = v\left(a \cdot \frac{1}{b}\right) = v(a) + v\left(\frac{1}{b}\right) = v(a) - v(b)$.

[20] The axioms of ordered abelian groups are given in all of the references from the previous footnote.

[21] In the statement and proof of this proposition, like in the initial discussion of the product rule in (*prod-comp*) of §4.5, for a function h we write (where convenient) the differential $(d_\eta h)(t)$ as dh, and the value $h(t)$ simply as h.

Proposition 4.14: *Suppose f and g are differentiable at t, and that the derivative of the product $f \cdot g$ at t is non-zero. Then $v(f \cdot dg + g \cdot df + df \cdot dg) = v(f \cdot dg + g \cdot df)$.*

Proof. Let us write $(f \cdot g)'(t) = L$ for some real $L \neq 0$. Let η be a positive infinitesimal. By our reasoning concerning (*prod-rule*) in the Robinsonian setting:

$$(f \cdot g)'(t) = f(t) \cdot g'(t) + g(t) \cdot f'(t) = L \approx \frac{f \cdot dg}{\eta} + \frac{g \cdot df}{\eta}$$

Since L is a non-zero real, $v(L) = v\left(\frac{f \cdot dg}{\eta} + \frac{g \cdot df}{\eta}\right) = 0$, so that

$$v(f \cdot dg + g \cdot df) = v\left(\eta \cdot \left(\frac{f \cdot dg}{\eta} + \frac{g \cdot df}{\eta}\right)\right) = v(\eta) + v\left(\frac{f \cdot dg}{\eta} + \frac{g \cdot df}{\eta}\right) = v(\eta)$$

But as $L \neq 0$, by (*prod-comp*) and Proposition 4.13 we have $v(f \cdot dg + g \cdot df + df \cdot dg) = v(d(f \cdot g)) = v(\eta)$. □

So, if we are only interested in the 'order of infinite smallness', we can indeed discard the term $df \cdot dg$ from $f \cdot dg + g \cdot df + df \cdot dg$. And this is just as l'Hôpital suggested, modulo the extra assumption that the derivative of the product is non-zero. However, the extra assumption is indispensable. To see this, let $f(x) = g(x) = x$ and let $t = 0$ be our evaluation-point: then $v(f \cdot dg + g \cdot df) = v(0 + 0) = \infty$, but $df = dg = \eta$, so that $v(d(f \cdot g)) = v(f \cdot dg + g \cdot df + df \cdot dg) = v(0 + 0 + \eta^2) = 2v(\eta) < \infty$.

We next turn to Bernoulli's result. To tackle this, we must define higher-order differentials,[22] by iterating Definition 4.7 in the obvious way:[23]

$$(d^1_\eta f)(t) = (d_\eta f)(t) \qquad (d^{n+1}_\eta f)(t) = (d^n_\eta f)(t + \eta) - (d^n_\eta f)(t)$$

We can now extend Proposition 4.13 to cover such n^{th}-order differentiables:

Proposition 4.15: *Let $\mathcal{R}^*$ be as in Theorem 4.12. Let $f : \mathcal{R} \longrightarrow \mathcal{R}$ be $\mathcal{R}$-definable and n^{th}-order differentiable at t. Then $f^{(n)}(t) \neq 0$ iff $v((d^n_\eta f^*)(t)) = nv(\eta)$ for all infinitesimals η.*

---

[22] A. Robinson (1966: 79–80).

[23] The notions of higher-order differentials obviously make sense also when η is not infinitesimal. If x ranges over non-zero real numbers, then $\lim_{x \to 0} \frac{(d^n_x f)(t)}{x^n}$ is sometimes today called *the nth-order unsymmetric Riemann derivative*. See e.g. Mukhopadhyay (2012: 6–8). When f is n-times differentiable, the usual n^{th} derivative is equal to the nth-order unsymmetric Riemann derivative (though the converse need not hold). One typically uses an intermediary notion of an nth Peano derivative to obtain these results. In particular, one has: (i) if f is n^{th}-order differentiable at t, then it is n^{th}-order Peano differentiable at t and the two derivatives are equal, but the converse does not hold (Mukhopadhyay 2012: Theorem 1.4.1 p.17), and (ii) if f is n^{th}-order Peano differentiable at t then it is n^{th}-order unsymmetric Riemann differentiable at t, and the two derivatives are equal (Mukhopadhyay 2012: Theorem 2.22.1 p.176). For another proof of these results, see Butzer and Berens (1967: §2.2 pp.95ff).

Proof. As in Proposition 4.10, if f is n^{th}-order differentiable at t, then $\frac{(d^n_\eta f^*)(t)}{\eta^n} \approx f^{(n)}(t)$ for all infinitesimals η.[24] The proof now runs as in Proposition 4.13, noting that $n\nu(\eta) = \nu(\eta^n)$. □

This result predicts an important feature of the early calculus, noted by Bos, namely that the higher-order differentials were conceived to be 'of successive different orders of infinity.'[25] Moreover, Proposition 4.15 allows us to vindicate Bernoulli's argument, mentioned in §4.5, that the quotient of an infinitesimal of order one and an infinitesimal of order three should be an infinitely large quantity of order two.

To show this, let η be infinitesimal with $\nu(\eta) = u > 0$, where 'u' is a mnemonic for 'unit'. Suppose that $t \in \mathcal{R}$, that f is once differentiable at t, that g is three-times differentiable at t, and that all these derivatives are non-zero. Abbreviate $d^n f = (d^n_\eta f)(t)$ and $d^n g = (d^n_\eta g)(t)$ and let a in $\mathcal{R}$ be non-zero. Then by Proposition 4.15 and Proposition 4.11 we have:

$$\nu\left(\frac{a \cdot df}{d^3 g}\right) = \nu(a \cdot df) - \nu(d^3 g) = \nu(a) + \nu(df) - \nu(d^3 g) = 0 + u - 3u = -2u$$

So, dividing an infinitesimal quantity $a \cdot df$ of value u by an infinitesimal quantity $d^3 g$ of value $3u$ results in a quantity $\frac{a \cdot df}{d^3 g}$ of value $-2u < 0$. Further, clause (5) of Theorem 4.12 states that negative valuations are reserved for infinite quantities; so, as Bernoulli argued, this quotient is an infinitely large quantity of order two, in reference to our unit.

Last, we turn to Euler's result, mentioned in §4.5. This result is motivated by the following observation. If η is a positive infinitesimal, then so is its square root $\eta^{\frac{1}{2}}$, and hence both get valuated as positive but non-infinite, thus:

$$\nu(\eta) = \nu\left(\eta^{\frac{1}{2}} \cdot \eta^{\frac{1}{2}}\right) = \nu\left(\eta^{\frac{1}{2}}\right) + \nu\left(\eta^{\frac{1}{2}}\right) = 2\nu\left(\eta^{\frac{1}{2}}\right)$$

Similarly, η receives a valuation which is three times that of its cube root $\eta^{\frac{1}{3}}$. It follows from this that we have a descending chain:

$$\nu(\eta) > \nu\left(\eta^{\frac{1}{2}}\right) > \nu\left(\eta^{\frac{1}{3}}\right) > \nu\left(\eta^{\frac{1}{4}}\right) > \dots$$

Euler asked whether there was something beyond all of these, of a different order of infinite smallness. He answered this in the affirmative by using the natural logarithm. We will prove his result using our valuation map, ν, but using exactly the same application of l'Hôpital's rule as Euler's original proof:[26]

---

24 One can use elementarity and the fact, mentioned in footnote 23, that if f is n-times differentiable, the usual n^{th} derivative is equal to the nth-order unsymmetric Riemann derivative. Alternatively, A. Robinson (1966: 80) gives a proof under the additional assumption that the kth-order derivatives for all $k \leq n$ exist *and* are continuous on an open interval containing the point.

25 Bos (1974: 27).

26 For Euler's proof, see Bos (1974: 85), Euler (1780: §8). There it is stated in terms of the infinite element $\frac{1}{\eta}$ rather than the infinitesimal η.

Proposition 4.16: *If* $\eta > 0$ *is infinitesimal, then* $v\left(\eta^{\frac{1}{n}}\right) > v\left(\frac{1}{\ln\eta}\right) > 0$ *for all* $n \geq 1$.

Proof. Note that $\lim_{x\to 0^+} e^{x^{\frac{1}{n}}} = 1$. Hence $e^{\eta^{\frac{1}{n}}} \approx 1$, and from this it follows that $e^{\eta^{\frac{1}{n}}\cdot\ln\eta} = e^{\eta^{\frac{1}{n}}}\cdot\eta$ is infinitesimal and hence $< e^{-1}$. So we have the inequality $e^{\eta^{\frac{1}{n}}\cdot\ln\eta} < e^{-1}$. By taking natural logarithms of both sides, we obtain $\eta^{\frac{1}{n}}\cdot\ln\eta < -1$ or $\eta^{\frac{1}{n}} < -\frac{1}{\ln\eta}$. Applying v to these positive quantities, we obtain $v\left(\eta^{\frac{1}{n}}\right) \geq v\left(-\frac{1}{\ln\eta}\right) = v\left(\frac{1}{\ln\eta}\right)$.

For reductio, suppose that $v\left(\eta^{\frac{1}{n}}\right) = v\left(-\frac{1}{\ln\eta}\right) = v\left(\frac{1}{\ln\eta}\right)$. Then:

$$v\left(\eta^{\frac{1}{n}}\cdot(-\ln\eta)\right) = v\left(\eta^{\frac{1}{n}}\right) + v(-\ln\eta) = v\left(\eta^{\frac{1}{n}}\right) - v\left(\frac{1}{-\ln\eta}\right) = 0$$

So the positive quantity $\eta^{\frac{1}{n}}\cdot(-\ln\eta)$ is finite but not infinitesimal and not zero, so that there is some real $M > 0$ such that $\frac{1}{M} \leq \eta^{\frac{1}{n}}\cdot(-\ln\eta) \leq M$. By elementarity, in $\mathcal{R}$ there is a sequence $b_k > 0$ of reals with $\lim_{k\to\infty} b_k = 0$ and $\frac{1}{M} \leq b_k^{\frac{1}{n}}\cdot(-\ln b_k) \leq M$ for all k. By compactness, this sequence has a convergent subsequence whose limit lies in the closed interval $\left[\frac{1}{M}, M\right] = \left\{z \in \mathcal{R} : \frac{1}{M} \leq z \leq M\right\}$. But by l'Hôpital's rule:

$$\lim_{x\to 0^+}\left(x^{\frac{1}{n}}\cdot(-\ln x)\right) = \lim_{x\to 0^+}\frac{-\ln x}{\frac{1}{x^{\frac{1}{n}}}} = \lim_{x\to 0^+}\frac{-\frac{1}{x}}{-\frac{1}{n}\cdot\frac{1}{x^{\frac{1}{n}+1}}} = \lim_{x\to 0^+} n\cdot x^{\frac{1}{n}} = 0$$

Hence, we have a contradiction. So in fact we must have that $v\left(\eta^{\frac{1}{n}}\right) > v\left(\frac{1}{\ln\eta}\right)$.

Finally, let us note why $v\left(\frac{1}{\ln\eta}\right) > 0$. Since $\eta > 0$ is infinitesimal, $-\ln\eta$ is infinite and hence $-\frac{1}{\ln\eta} > 0$ is infinitesimal. So $0 < v\left(-\frac{1}{\ln\eta}\right) = v\left(\frac{1}{\ln\eta}\right)$. □

This section has been somewhat heavy on technicalities, so it may help to summarise what we have shown. Using a valuation function, we have introduced a natural, intuitive notion of different orders of infinite smallness which is available within non-standard analysis. And we have showed how the valuation function vindicates several historical uses of different orders of infinitesimals.

4.7 Instrumentalism and conservation

Evidently, infinitesimals can be incorporated both rigorously and beautifully into contemporary mathematics. However, this leaves open the question of whether one should *believe* in infinitesimals. In a famous letter to Varignon, Leibniz wrote:

> To speak the truth, I am not very persuaded myself that it is necessary to consider the infinite and the infinitely small other than as ideal things or as well-founded fictions.[27]

[27] Leibniz (1849–63: vol.4 p.110).

Elaborating on this, Jesseph suggests that 'a fiction is well-founded in the Leibnizian sense when it does not lead us astray, so that indulgence in the fiction is harmless.'[28] Robinson expressed a similar view, presenting a kind of fictionalism or instrumentalism concerning *all* infinitary notions in mathematics:

> My position concerning the foundations of Mathematics is based on the following two main points or principles. [¶] (i) Infinite totalities do not exist in any sense of the word (i.e. either really or ideally). More precisely, any mention, or purported mention, of infinite totalities is, literally *meaningless*. [¶] (ii) Nevertheless, we should continue the business of Mathematics 'as usual,' i.e. we should act *as if* infinite totalities really existed.[29]

The general fictionalist or instrumentalist idea is that we should not actively *believe* in infinite or infinitesimal entities, but that we can happily *employ* them in reliable reasoning (about real numbers, in this case). There are obvious echoes here of the o/t dichotomy from Chapter 3, but in this case the finite is being treated as *okay* and the infinite(simal) is *troublesome*. (And, anticipating a little, in §4.9 we will see that some contemporary presentations of non-standard analysis contain various kinds of non-standard *sets* in addition to infinitesimals. We can consider a similar attitude there: that we should not *believe* in non-standard sets, because they are troublesome, but that we can happily employ them in reliable reasoning).

In a moment, we will explain why reasoning with infinitesimals (or non-standard sets) is, indeed, reliable. But first, we should ask whether infinitesimals *should* be thought of as more troublesome than reals. There is a genuine risk here of trafficking in mere squeamishness.

Squeamishness about infinitesimals may have been reasonable, when it seemed that we had to treat them as both *distinct* from 0 and *identical* to 0 within the very same calculation. But we quashed that worry, firmly, in §4.4.

A second source of squeamishness may be that it is difficult to 'picture' a continuum which contains infinitesimals. Between 0 and any positive real, there is a smaller positive real; but all the positive infinitesimals will have to sit *after* 0 but before *any* positive real; and it is certainly hard to form an intuitive 'picture' of this ordering. That said, one might wonder whether this is any harder than the attempt to 'picture' the (now standard, in every sense) idea that the continuum is made up of *points*. After all, the pointwise conception of the continuum allows us to define several functions which are *extremely* hard to 'picture', such as Peano's space-filling curve, Bolzano–Weierstrass's continuous everywhere but differentiable nowhere function, and Conway's Base 13 function.

---

[28] Jesseph (2008: 232). Also: 'A fiction is well-founded when it reliably enables us to investigate the properties of real things, so that indulgence in the fiction cannot lead us into error' (Jesseph 1998: 35).

[29] A. Robinson (1965: 230).

As such, we are not sure whether there are principled reasons to regard real quantities as *okay* but infinitesimals as *troublesome*. The tendency to regard infinitesimals as merely fictional may just be a legacy of history.

There are, however, excellent reasons to regard infinitesimals as *reliable*. These reasons relate to *conservation*. In Definition 3.3, we defined *consequence*-conservation. We can equally define a notion of *deduction*-conservation (noting that if the deductive system is sound and complete for the semantics, then consequence-conservation and deduction-conservation align):

Definition 4.17: *Let T be an $\mathscr{L}^+$-theory and S be an $\mathscr{L}$-theory, with $\mathscr{L}^+ \supseteq \mathscr{L}$. T is* deduction-conservative *over S iff: if $T \vdash \varphi$ then $S \vdash \varphi$ for all $\mathscr{L}$-sentences φ.*

So, if our theory of infinitesimals is deduction-conservative over our 'vanilla' theory of the reals, then anything which can be proved about the reals using the theory of infinitesimals can 'already' be proved using only our 'vanilla' theory. Moreover, we can redescribe deduction-conservation in model-theoretic terms, using reducts (from Definition 1.4) and elementary extensions (from Definition 4.3).

Proposition 4.18: *Let T be a first-order $\mathscr{L}^+$-theory and S be a first-order $\mathscr{L}$-theory, with $\mathscr{L}^+ \supseteq \mathscr{L}$. The following are equivalent:*

(1) *T is deduction-conservative over S.*
(2) *For any $\mathscr{L}$-structure $\mathcal{M} \vDash S$, there is an $\mathscr{L}^+$-structure $\mathcal{N}$ which satisfies T and whose $\mathscr{L}$-reduct is an elementary extension of $\mathcal{M}$.*

Proof. (1) $\Rightarrow$ (2). Assume (1), and let $\mathcal{M} \vDash S$. Let $\mathcal{M}^\circ$ be as in Definition 1.5, so that $c_a^{\mathcal{M}^\circ} = a$ for each $a \in M$, and so that the new constants $c_a^{\mathcal{M}^\circ}$ are chosen to be distinct from any constant symbol appearing in $\mathscr{L}^+$.

Where $\text{Th}(\mathcal{M}^\circ)$ is the set of sentences true on the structure $\mathcal{M}^\circ$, we now prove that $T \cup \text{Th}(\mathcal{M}^\circ)$ is consistent. Let $\delta(\overline{c_a})$ be an arbitrary conjunction of finitely many sentences in $\text{Th}(\mathcal{M}^\circ)$, where $\overline{c_a}$ are the new constants $c_{a_1}, \ldots, c_{a_n} \notin \mathscr{L}^+$ occurring in δ. Since $\mathcal{M}^\circ \vDash \delta(\overline{c_a})$, we have $\mathcal{M}^\circ \vDash \exists \overline{v}\delta(\overline{v})$. Hence $S \nvdash \forall\overline{v}\neg\delta(\overline{v})$ and $T \nvdash \forall\overline{v}\neg\delta(\overline{v})$ by (1). So $T \cup \{\delta(\overline{c_a})\}$ is consistent. Since δ was an arbitrary finite conjunction from $\text{Th}(\mathcal{M}^\circ)$, by the Compactness Theorem 4.1, $T \cup \text{Th}(\mathcal{M}^\circ)$ has a model. Call it $\mathcal{N}^*$.

By the Push-Through Construction, we can assume that $c_a^{\mathcal{N}^*} = a$ for all $a \in M$. Letting $\mathcal{N}$ be $\mathcal{N}^*$'s $\mathscr{L}^+$-reduct, an argument similar to (*elem*) shows that $\mathcal{N} \succeq \mathcal{M}$.

(2) $\Rightarrow$ (1). Suppose (1) fails, i.e. there is some $\mathscr{L}$-sentence φ such that $T \vdash \varphi$ and $S \nvdash \varphi$. Since $S \cup \{\neg\varphi\}$ is consistent, by the Completeness Theorem there is a model $\mathcal{M} \vDash S \cup \{\neg\varphi\}$. Then for any $\mathscr{L}^+$-structure $\mathcal{N}$ whose $\mathscr{L}$-reduct is an elementary extension of $\mathcal{M}$, we have $\mathcal{N} \vDash \neg\varphi$, and hence $\mathcal{N} \nvDash T$. So (2) fails. □

This allows us to describe situations where reasoning with infinitesimals is provably 'harmless'. Let S be a theory which is true on the real numbers, and suppose that T extends S with further principles governing infinitesimals (perhaps in an expanded signature). For instance, T might simply assert that there is an infinitesimal, or more elaborately, T might assert that there is a valuation map satisfying the clauses of Theorem 4.12. If S is sufficiently rich that any model of S has an elementary extension that may be expanded to a model of T,[30] then T is deduction-conservative over S by Proposition 4.18. In that case, T says no more nor less about the reals than S does. So, one can use T to reason about the reals with a clear conscience.

Conservation results are famously associated with Hilbert's programme, which aimed to show that set-theoretic and higher-order methods were conservative over a basic kind of arithmetic.[31] So it is no accident that Robinson described his own philosophy of mathematics as 'basically, close to that of Hilbert and his school.'[32] That said, Proposition 4.18 is proved model-theoretically. As such, someone who is worried that *both* infinitesimal *and* model-theoretic methods are *troublesome* will not be convinced by this kind of argument (they will likely require a purely *proof-theoretic* demonstration that infinitesimal methods are reliable). Still, anyone who accepts model theory (and so anyone who accepts set theory) will need no further proof of the reliability of infinitesimal methods.

4.8 Historical fidelity

Model theory has paid its way several times now. Nonetheless, Bos suggests that Robinson's (essential) use of *contemporary* model-theoretic methods undermines any hope of reconstructing the *historical* calculus:

> [...] the most essential part of non-standard analysis, namely the proof of the existence of the entities it deals with [i.e. infinitesimals inside elementary extensions], was entirely absent in the Leibnizian infinitesimal calculus, and this constitutes, in my view, so fundamental a difference between the theories that the Leibnizian analysis cannot be called an early form, or precursor, of non-standard analysis.[33]

[30] If T merely asserts the existence of infinitesimals, then S only needs the meagre resources to talk about e.g. zero, one, addition, multiplication and a linear order, and to show $\frac{1}{n+1} < \frac{1}{n}$ for each $n > 0$. If we want to invoke the valuation map from Theorem 4.12, then we must ensure that S includes enough principles governing the natural logarithm function to ensure that the proof of that theorem works for any model of S.

[31] For references on Hilbert's views on conservation, see Detlefsen (1996: 79, 1986: 30–1). For discussion of recent conservation results in the setting of arithmetic and their bearing on the reevaluation of Hilbert's programme, see Simpson (1988: 353ff).

[32] A. Robinson (1965: 229). Elsewhere Robinson writes that 'Leibniz's approach is akin to Hilbert's original formalism, for Leibniz, like Hilbert, regarded infinitary entities as ideal, or fictitious, additions to concrete Mathematics' (1967: 39–40).

[33] Bos (1974: 83).

But surely neither Robinson, nor anyone else, was suggesting that any kind of model theory was implicit in the early calculus. The claim is only that model theory establishes that one can *reason reliably* with infinitesimals. We have seen how this takes place. And once that *has* taken place, there is no further need to conceive of infinitesimals model-theoretically, as elements of an elementary extension of the reals. If we like, we can conceive of infinitesimals deductively, as characterised by an axiom asserting the existence of entities with an absolute value smaller than any positive rational. As Robinson put it in one place, 'we may look at our theory syntactically and may consider that what we have done is to introduce *new deductive procedures* rather than new mathematical entities.'[34] In short, if we use the model-theoretic Proposition 4.18 to show that certain infinitesimal methods are deduction-conservative over real methods, we can, thereafter, appeal to an entirely proof-theoretic conception of reliability. And, while semantics and model-theory may be twentieth-century inventions, mathematics has always been in the business of proof.

We just considered Bos's complaint about Robinson's use of model theory. A similar complaint might be raised against our use of the valuation map, as follows: *the very idea of a valuation map is so alien to the mathematics of Leibniz and his contemporaries, that its invocation cannot feature in any plausible attempt to vindicate the historical calculus.*

As before, though, we are not making the implausible claim that Leibniz et al. conceived matters in terms of valuation maps. Our point is only that we can use the valuation map to show that their notion of 'different orders of smallness' was (provably) on a reliable footing.

We do not deny, then, that invoking a valuation map involves a certain amount of *reconceptualisation* of the mathematics of a previous generation. But here, we keep good company. In their famous 1882-paper, Dedekind and Weber '*define* the "points of a Riemann surface" to be *discrete valuations*', where a valuation is *discrete* if the codomain of the valuation function is the integers.[35] In doing so, they plainly added to the concept of the Riemann surface; but no one would contemn them.[36]

Moreover, it is unclear why worries about anachronism should focus solely on the use of *model theory* (whether our deployment of a valuation map, or Robinson's appeal to Compactness). The whole approach to non-standard analysis invokes, without much comment, a presentation of analysis which was only first developed in the nineteenth century, and which supplanted the eighteenth century idea of 'continuous magnitudes such as lengths and weights' and 'their "abstract"

---

[34] A. Robinson (1966: 282).

[35] This is a quotation from Dieudonné's (1985) history of algebraic geometry. The reference for the original Dedekind–Weber paper is Dedekind and Weber (1882) and Dedekind (1930–32: 1 pp.238ff).

[36] For nuanced discussions of 'concept change' in the context of the Dedekind–Weber paper, see Schappacher (2010) and Haffner (forthcoming).

counterparts'.[37] Even mentioning first-order features common to the real numbers and their elementary extensions—e.g. 'every first-order definable bounded set has a least upper bound'—involves departing non-trivially from the *historical* calculus.

In sum, our claims about our valuation map are rather modest. The apparatus of a valuation map is certainly a conceptual addition to the calculus. But it allows us to derive some of the theorems presented by its historical practitioners. And, more generally, it suggests the reliability of certain historical considerations about 'different orders of smallness'.

4.9 Axiomatising non-standard analysis

Over the past two sections, we considered a certain shift from model theory to proof theory. The focus on proof can motivate a more *axiomatic* approach to non-standard analysis. So, too, can the desire to develop a general approach to non-standard analysis that does not depend so very heavily on model theory. In this section, we outline such approaches, characterising them by their responses to three questions concerning axiomatisation.

Axiomatising elementary extensions

The most pedestrian axiomatisation of non-standard analysis would simply add the claim 'there is an infinitesimal' to some formal theory of real numbers. However, our work in the previous sections shows that non-standard analysis thrives precisely because it provides a way to go back and forth between a setting with no infinitesimals and a setting with infinitesimals. And the pedestrian approach offers us no obvious way *internal to the theory* to 'go back' to a setting with no infinitesimals.

Hence, the first question is how to axiomatise the reals *together with* an elementary extension that possesses infinitesimals. This is really just a question of how to axiomatise elementary extensions. Expressed at this level of generality, there is a natural answer. Suppose that $\mathcal{A}$ and $\mathcal{B}$ are $\mathscr{L}$-structures and that $\mathcal{A} \preceq \mathcal{B}$. Expand the signature $\mathscr{L}$ to a signature $\mathscr{L}^+$ that contains a new unary predicate symbol $St(x)$, which is a mnemonic for 'standard'. Recall from §1.9 that $(\forall x : St)\varphi$ abbreviates $\forall x(St(x) \rightarrow \varphi)$ and $(\exists x : St)\varphi$ abbreviates $\exists x(St(x) \wedge \varphi)$. Given an $\mathscr{L}$-formula φ, let φ^{St} be the result of restrcting all the quantifiers in φ to St. Then expand $\mathcal{B}$ to an $\mathscr{L}^+$-structure $\mathcal{B}^+$ where the new predicate symbol $St(x)$ is interpreted as A, the underlying domain of $\mathcal{A}$. Since $\mathcal{B} \succeq \mathcal{A}$, the structure $\mathcal{B}^+$ satisfies every sentence of the following form, where $\varphi(\bar{x})$ is an $\mathscr{L}$-formula:[38]

[37] Epple (2003: 291).

[38] The equivalent characterisation of elementary embeddings given by the Tarski–Vaught test (Proposition 7.4 of Chapter 7) yields a slightly more elegant but equivalent axiom. Compare e.g. the 'transfer

$$(\forall \bar{x} : St)(\varphi^{St}(\bar{x}) \leftrightarrow \varphi(\bar{x})) \qquad (elem{:}object)$$

This axiomatic rendition of the notion of an elementary extension is often used as a basis of axiomatic formulations of non-standard analysis. In particular, one simply takes natural axioms T for the elementary extension, and adds every sentence of the form of (*elem:object*). From this, one can deduce φ^{St} for all axioms φ of T, since this is just an object-language expression of being an elementary extension, and elementary extensions preserve first-order sentences. Hence this axiomatic rendition provides the benefits of passing back and forth between the standard reals and an elementary extension with infinitesimals.

Axiomatising the reals

The second question facing any axiomatisation of non-standard analysis is how to axiomatise the real numbers in the first place. Some choices will trivialise the project. For instance, if S is a complete theory, then any theory T in any extension of S's signature is deduction-conservative over S, so long as T is consistent with S.[39] Conversely, if we want non-trivial conservation results, then we must seek an *in*complete axiomatisation of the reals.

An obvious option is to take the usual second-order axiomatisation of the reals, whose completeness axiom states that any non-empty bounded subsets of the reals has a least upper bound.[40] This axiomatisation interprets Robinson's Q, and so it is incomplete.[41]

But there is no reason to stop at *second*-order logic. If we want to do non-standard probability theory, for example, then we will have to work with probability measures, which are even higher-order beasts. And if we admit higher-order objects of all finite orders, it becomes very natural to view the reals as simply embedded within set theory.[42] In this case, we let T be the axioms of ZFC, plus the axioms from (*elem:object*), along with the supposition that there is an infinitesimal among the non-standard reals. If we then assume that there is a model of ZFC, we can use the methods of §4.7 to show that T is deduction-conservative over ZFC.

---

principle' in Nelson (1977: 1166). This way of axiomatising elementary extensions is also sometimes used in the setting of large cardinals: see e.g. Reinhardt's Axiom S2 in Reinhardt (1974: 192).

[39] To see this, suppose that $T \vdash \varphi$, where φ is from S's signature; since S is complete, either $S \vdash \varphi$ or $S \vdash \neg\varphi$; but the latter cannot happen since a model of both T and S would satisfy both φ and $\neg\varphi$; so $S \vdash \varphi$.

[40] For formal statements, see any real analysis text. This also arises in the context of categoricity arguments (see §7), and so formal statements of these axioms can often be found in treatments of second-order logic, such as Shapiro (1991: 83–4).

[41] For details on these sorts of considerations, see §§5.A, 7.5 and 10.3. We should emphasise that the theory is here being considered *deductively*, rather than (e.g.) using the full semantics.

[42] Hrbáček motivates the move to set theory by noting that 'the work with higher-order structures involves the type-theoretic language repugnant to most mathematicians' (1978: 1).

Axiomatising non-standardness

The final design question concerns how to axiomatise the idea that the elementary extension is a *non-standard* model, while the elementary substructure is a *standard* model. Now, unlike the predicate '*St*' that we introduced a moment ago, the italicised phrases in the previous sentence are not object-language expressions, but metatheoretic. Furthermore, the adjective 'standard' as used in the metatheory does not have a sharply defined meaning, but is instead understood by reference to paradigmatic (non-)examples. For instance, the natural numbers are the standard model of PA, and any model not isomorphic to it is non-standard (in exactly the sense of Definition 4.2); likewise, the real numbers are a standard model, while any model with infinitesimals is non-standard. This usage of the word 'standard' is connected heavily with issues surrounding the categoricity of second-order theories, which will occupy us in Part B of this book. But for now, we can simply approach the standard / non-standard distinction via these paradigmatic examples.

To axiomatise *standardness*, we simply appeal to the idea that the schemata in ZFC (or whatever rich theory we are working with) should extend to any signature whatsoever, including the signature of the elementary extension.[43] In the case of ZFC, this would involve for instance the Separation schema (see §1.B), so we would insist on the existence of any subset of a standard set which can be defined by a formula involving *any* resources we like (standard sets, infinitesimals, or anything else). If the rich theory is PA_2, this would involve the Comprehension Schema (see §1.A), and so we would insist that *any* formula determines a standard second-order subset of natural numbers.

To axiomatise the *non-standardness* of elementary extensions, it seems desirable to have a single principle that handles all appeals to compactness. Nelson suggests the following axiom schema, where '*Stfin*' restricts to standard finite sets, and φ is a formula in the signature of set theory:[44]

$$(\forall z : \mathit{Stfin})\exists x(\forall y \in z)\varphi(x,y) \leftrightarrow \exists x(\forall y : \mathit{St})\varphi(x,y) \qquad (\mathit{elem{:}nelson})$$

To illustrate how this axiom is an object-language expression of compactness, observe how it implies the existence of infinitesimals. Let $\varphi(x, y)$ be the statement 'x is a positive quantity, and if y is a positive quantity then $|x| < |y|$'. Then the left-hand side of (*elem:nelson*) is satisfied, because for each finite set of positive quantities we can find a positive quantity less than all of them. But the right-hand side says that there is a positive quantity which is less than all positive standard quantities, which is just to say that there is an infinitesimal. This lets us view Nelson's (*elem:nelson*) as an object-language expression of compactness.

[43] This kind of requirement is well-known from Feferman's discussion of schemata (1991: 8). It is also related to a point about 'open-endedness' made in the discussion of categoricity theorems (see footnote 21 in §7.10 and §13.7).

[44] See the 'principle of idealization' in Nelson (1977: 1166).

However, Nelson's (*elem:nelson*) has some potentially counterintuitive consequences. For instance, take $\varphi(x,y)$ to be simply $y \in x$. Then the left-hand side of (*elem:nelson*) is trivially satisfied, because we may simply take the witness x to be z itself, and have $(\forall z : Stfin)(\forall y \in z) y \in z$. But then Nelson's (*elem:nelson*) entails that $\exists x(\forall y : St) y \in x$, that is, that there is a set which contains all the standard sets.[45] But this consequence of Nelson's object-language expression of compactness seems rather different from our earlier arguments for infinitesimals via compactness. In this earlier argument, the idea was to show that some standard set (namely, the real numbers) has more elements than we usually think (namely, infinitesimals). By contrast, Nelson's (*elem:nelson*) implies that there is a set which is not a member of any standard set. (That said, it is not obvious that this is a fatal objection to Nelson's axiom: the push-and-pull here is similar to that surrounding Fletcher's objection, with which we close this section.)

Hrbáček's alternative axiomatisation of non-standard analysis begins by correcting for this. Hrbáček defines a set to be *internal* iff it is a member of a standard set. Evidently, then, all standard sets are internal sets, since they are members of their own singletons. Letting '*Int*' abbreviate this notion of being internal, Hrbáček considers a variant of (*elem:object*) which expresses that the internal sets are an elementary extension of the standard sets:[46]

$$(\forall \bar{x} : St)(\varphi^{St}(\bar{x}) \leftrightarrow \varphi^{Int}(\bar{x}))$$

for all formulas in the signature of set theory. Hrbáček then modifies Nelson's (*elem:nelson*), by 'bounding' it with a standard set:[47]

$$(\forall a : St)\big((\forall z : Stfin)[z \subseteq a \rightarrow (\exists x : Int)(\forall y \in z)\varphi^{Int}(x,y)] \leftrightarrow \\ (\exists x : Int)(\forall y : St)[y \in a \rightarrow \varphi^{Int}(x,y)]\big)$$

This alternative principle still allows the existence of infinitesimals, since the argument we gave earlier may be bounded with the *standard* set of reals. However, it directly blocks the derivation of the existence of a set that contains all the standard sets, since no internal set could have this property.

Nevertheless, if there are non-internal sets, then we must ask how they behave. And Hrbáček has shown that, against the background of his other axioms, it is inconsistent to claim that the non-internal sets satisfy either the Power Set Axiom or the axiom that every set can be well-ordered.[48] Referring to this result, Fletcher

45 Nelson (1977: 1167) and Hrbáček (2006: 96).

46 See 'transfer' in Hrbáček (2006: 85).

47 See 'bounded idealization' in Hrbáček (2006: 86). In this schema, φ may be a formula in the signature of set theory that may contain internal sets as parameters.

48 Hrbáček (2006: 90). Of course, in ZF the axiom that every set can be well-ordered is equivalent to the Axiom of Choice, but this equivalence no longer holds when the Power Set Axiom is dropped (cf. Zarach 1982: Theorem III). For statements of these set-theoretic axioms, see §1.B.

writes that it 'undermines confidence in one's intuitive picture of the non-standard universe on the basis of which informal arguments are created and justified'.[49]

Fletcher's objection strikes us as half wrong and half right. What seems wrong is this: non-standard analysis has a formalist bent, and formalists are unlikely to be much concerned with providing an intuitive picture of the universe of non-standard sets or non-internal sets. (Of course, formalists might care about providing intuitive pictures, if those pictures help to convince us that some theory is consistent; but consistency cannot be at issue here, since both Nelson's and Hrbáček's non-standard set theories are deduction-conservative over ZFC, so that it is consistent if ZFC is.) But what seems right in Fletcher's objection is this: as formalists, practitioners of non-standard analysis should be very much concerned with isolating efficacious modes of reasoning, so it will be problematic if non-standard set theory requires large-scale changes to the usual set-theoretic axioms. After all, as Nelson once put it, the idea was supposed to be that 'what is new [in axiomatic non-standard analysis] is only an addition, not a change'.[50]

4.10 Axiomatising the reals

There is obviously much more to say about axiomatisations of non-standard analysis. However, saying any more would take us too far afield from model theory.[51] And indeed, much contemporary work in model theory on the real numbers is *not* centred around non-standard analysis, but around first-order axiomatisations of the reals. We close this chapter by briefly surveying this work.

The complete first-order theory of the reals, in the signature with zero, one, addition, and multiplication, is computable. In addition to the usual field axioms,[52] the axioms are that -1 is not a sum of squares, that either a or $-a$ is a square for every a, and that every odd degree polynomial has a root. These axioms are true of the real numbers, and it turns out that they are complete.[53]

These algebraic resources also allow us to define the ordering on reals. In particular, a real number is positive iff its square root is real, so that we define:

$$a < b \text{ iff } (\exists x \neq 0)x^2 = b - a$$

The axioms described in the previous paragraph suffice to show that all of the ordinary properties of the ordering now hold: for example, it is linear and if $0 < c$ and

[49] Fletcher (1989: 1004).

[50] Nelson (1977: 1165).

[51] For those who want to read more on this topic, we highly recommend Hrbáček (2006).

[52] The field axioms are those axioms common to the rationals, the reals, and the complexes that indicate how zero and one interact with addition and multiplication. See any algebra textbook for the formal definition of the field axioms, e.g. Hungerford (1980: 116).

[53] See Marker (2002: §3.3, 2006: §2).

$a < b$ then $a \times c < b \times c$.[54]

An explicit description of this theory was found independently by Tarski and by Artin and Schreier. Tarski had two primary reasons for interest in this theory. First, he was interested in the fact that this axiomatisation provides a decision procedure for whether a first-order statement is true on the real numbers.[55] Second, in 1931 Tarski started studying the model theory of the real numbers as a way to convince other mathematicians that the notions of satisfaction and definability were not paradoxical.[56] Thus his work on the model theory of the real numbers serves as a mathematical supplement to his famous definition of truth.

Artin and Schreier's work is located not in model theory *per se*, but in a distinct tradition of real algebra, whose history has been exhaustively studied by Sinaceur.[57] According to Sinaceur, Hilbert's work on the foundations of geometry provides the mathematical antecedent of Artin and Schreier's work. Hilbert had attempted to ascertain which parts of geometry depend essentially upon 'higher-order' postulates, like the completeness of the real field or the Archimedian axiom. Sinaceur suggests that we should similarly view Artin and Schreier as showing that 'the order relation on the field of real numbers cannot be dissociated from its topology', for we have just seen that the order is definable in terms of the field operations.[58] So: whereas Dedekind had 'reduced continuity to order [via Dedekind cuts], Artin and Schreier reduced order to calculation'.[59]

Most recent work on the model theory of the real numbers does not, however, work directly with the axiomatisation provided by Artin–Schreier and Tarski. Instead, it works with a generalisation known as *o-minimality*. Its origins lie in a remark from Tarski's 1931 paper, where he notes that the parameter-free definable subsets of the reals in the signature $\{0, 1, +\}$ are precisely finite unions of intervals of the following form, with rational or infinite endpoints:[60]

$$(a, b), \quad [a, b), \quad (a, b], \quad [a, b]$$

Tarski's idea seems to have been forgotten, until van den Dries's 1982 address on Tarski's problem of the decidability of the real field expanded by exponentiation.[61]

[54] For a more formal treatment of the axioms of an ordered field see e.g. Lang (2002: 449ff).

[55] See Tarski (1948) for the decidability result and Tarski (1967) for the completeness result. The methods are very similar in each case.

[56] See the introduction to Tarski (1931).

[57] Sinaceur (1991).

[58] Sinaceur (1991: 146). See also Sinaceur (1991: 29, 217, 223).

[59] Sinaceur (1994: 200). See also Sinaceur (1991: 28, 187, 1994: 194) for quotations from Hasse and Weil which Sinaceur uses to buttress her case for this view of the significance of Artin and Schreier.

[60] Tarski (1931: 233). As usual, we define $(a, b) = \{c \in \mathcal{R} : a < c < b\}$, and the closed bracket symbol indicates that we use '$\leq$' in lieu of '$<$'. Here, infinite endpoints are only allowed when we do not employ a closed bracket symbol. That is, $(-\infty, b]$ is allowed and is defined as $\{c \in \mathcal{R} : c \leq b\}$, but e.g. $[-\infty, b)$ is not allowed since this would incorrectly suggest that $-\infty$ is an element of the reals.

[61] van den Dries (1984). For the original problem, see Tarski (1948: 45).

In the associated paper, van den Dries noted that one could take something very much like Tarski's result as an axiom and deduce many of the known properties of the definable sets of reals from it. This notion was later generalised by Pillay and Steinhorn as follows:[62]

Definition 4.19: *Let $\mathcal{M}$ be an expansion of a dense linear order without endpoints. Then $\mathcal{M}$ is* o-minimal *iff every subset of $\mathcal{M}$ which is definable with parameters is a finite union of points and open intervals.*

Here, open intervals are of the form (a, b), and a dense linear order without endpoints is simply a linear order which does not have greatest or least elements and which satisfies the density condition: if $a < b$ then there is some c with $a < c < b$. The canonical example of an o-minimal structure is provided by Tarski's own work, namely the real field.[63] The other important example is due to Wilkie, who showed that the real field with exponentiation is also o-minimal.[64] Wilkie's result inaugurated much recent activity in this area. Nonetheless, Tarski's original question, about the decidability of the real field with exponentiation, is still open.[65]

While much present interest in o-minimality stems from its applications to other areas of mathematics,[66] van den Dries also indicates a more foundational motivation. Grothendieck once suggested that there should be a *tame topology*: a notion of topology which a priori excluded all of the 'pathological' functions from elementary analysis, such as space-filling continuous curves.[67] A basic result about o-minimal structures is that every definable function in them is both piecewise continuous and piecewise differential.[68] As such, van den Dries suggests that 'o-minimal structures provide an excellent framework for developing tame topology'.[69]

4.A Gödel's Completeness Theorem

In this appendix, we prove Gödel's Completeness Theorem 4.24, which we invoked both in §2.A and in our proof of Compactness in §4.1. Our proof of Gödel's result comes in a series of lemmas, the first of which is very easy.

[62] Pillay and Steinhorn (1984, 1986); van den Dries (1998) is a standard reference to, and overview of, o-minimality.

[63] This follows almost automatically from the 'quantifier elmination' results which one uses to establish the decidability and completeness results. See Marker (2002: 99) and van den Dries (1998: 37).

[64] Wilkie (1996).

[65] There are some conditional results. Macintyre and Wilkie (1996) show that it is decidable if a number-theoretic conjecture known as Schaunel's conjecture is true.

[66] For example, see Scanlon (2012).

[67] See Grothendieck (1997: §5).

[68] For piecewise continuity, see the monotonicity theorem in van den Dries (1998: 43); for piecewise differentiability see van den Dries (1998: 115).

[69] van den Dries (1998: vii). For further discussion, see Giaquinto (2015: §3.3) and Galebach (2016).

Proposition 4.20: *For any first-order theory T: T is consistent iff every finite subset of T is consistent.*

Proof. *Left-to-right* is trivial. *Right-to-left* holds because any deduction in first-order logic is only ever finitely long, and so a deduction of T's inconsistency would only use finitely many sentences from T. □

The next result is more interesting: it gives a method for keeping tight control over existential claims.

Lemma 4.21 (Henkin): *Let T be a consistent $\mathscr{L}$-theory. There is a signature $\mathscr{L}^* \supseteq \mathscr{L}$ with the same cardinality as the set of $\mathscr{L}$-formulas, and a consistent $\mathscr{L}^*$-theory $T^* \supseteq T$ with the* witness property, *namely: for every $\mathscr{L}^*$-formula $\varphi(x)$, there is some $\mathscr{L}^*$-constant symbol c such that $(\exists x\varphi(x) \rightarrow \varphi(c)) \in T^*$.*

Proof. Let κ be the cardinality of $\mathscr{L}$-formulas (see the end of §1.B for a brief discussion of cardinals). We define the signature $\mathscr{L}^*$ by expanding $\mathscr{L}$ with *new* constant symbols, c_α, for every $\alpha < \kappa$. We can enumerate all the $\mathscr{L}^*$-formulas with one free variable, x, by $\varphi_1, \ldots, \varphi_\alpha, \ldots$ for $\alpha < \kappa$. We now define a sequence of theories, by recursion up through κ:

$$T_0 = T$$

$$T_{\alpha+1} = T_\alpha \cup \{\exists x\varphi_{\alpha+1}(x) \rightarrow \varphi_{\alpha+1}(c_\beta)\}, \text{ where } c_\beta \text{ is the new constant with least index not appearing in any sentence in } T_\alpha \text{ or in } \varphi_{\alpha+1}(x)$$

$$T_\beta = \bigcup_{\alpha<\beta} T_\alpha \text{ for limit ordinals } \lambda$$

Recall that there are κ-many new constants; so, since κ is a cardinal, at any stage T_α with $\alpha < \kappa$ there are always some constants which have not yet been used. We now prove, by induction, that T_α is consistent for each $\alpha \leq \kappa$.

The consistency of $T_0 = T$ was assumed.

For induction, suppose T_α is consistent. Suppose, for reductio, that $T_{\alpha+1}$ is inconsistent. So $T_\alpha \vdash \neg(\exists x\varphi_{\alpha+1}(x) \rightarrow \varphi_{\alpha+1}(c_\beta))$ and hence $T_\alpha \vdash \exists x\varphi_{\alpha+1}(x)$ and $T_\alpha \vdash \neg\varphi_{\alpha+1}(c_\beta)$, by simple deductive manipulations. But c_β does not occur in any sentence in T_α, nor in $\varphi_{\alpha+1}(x)$. So we have $T_\alpha \vdash \forall x\neg\varphi_{\alpha+1}(x)$, and hence T_α is inconsistent, contrary to our assumption. So $T_{\alpha+1}$ is consistent after all.

Finally, let β be a limit ordinal, and suppose for induction that T_α is consistent for all $\alpha < \beta$. Then any finite subset of T_β occurs in some T_α and so is, by assumption, consistent. So T_β is consistent, by Proposition 4.20.

So $T_\kappa = T^*$ is consistent, and has the witness property by construction. □

Before stating our next result, we need a version of Zorn's Lemma. Against the background of ZF, Zorn's Lemma is equivalent to the Axiom of Choice (as stated in

§1.B), and so is always available in ZFC. Here is the version we use: a partial order of subsets of a given set has a maximal element when the union of any non-empty linearly ordered subset is in the partial order. More precisely, let $P \subseteq \wp(A)$, and suppose that for any non-empty $L \subseteq P$ satisfying $x \subseteq y$ or $y \subseteq x$ for all x, y in L, one has $\bigcup L$ in P; then Zorn's Lemma says that P has a maximal element m, i.e. if $m \subseteq x$ then $m = x$ for any $x \in P$.[70] Our next result uses Zorn's Lemma to turn consistent theories into *complete*, consistent theories. (As in §3.B, an $\mathcal{L}$-theory T is *complete* iff either $\varphi \in T$ or $\neg\varphi \in T$ for every $\mathcal{L}$-sentence φ.)

Lemma 4.22 (Lindenbaum): *For any consistent $\mathcal{L}$-theory T, there is some complete, consistent $\mathcal{L}$-theory $T^* \supseteq T$.*

Proof. Consider the set of all consistent $\mathcal{L}$-theories extending T:

$$C = \{S \text{ is an } \mathcal{L}\text{-theory} : T \subseteq S \text{ and } S \text{ is consistent}\}$$

We first argue that there is some maximal $T^* \in C$, i.e. a set T^* which is not a proper subset of any member of C. By Zorn's Lemma, it suffices to show that if L is a set of elements of C which are linearly ordered by $\subseteq$, then $\bigcup L$ is in C. But trivially $T \subseteq \bigcup L$; and every finite subset of $\bigcup L$ is consistent, since any finite subset of it must be included in some element of L, since its elements are linearly ordered by $\subseteq$, so that $\bigcup L$ itself is consistent (as in Proposition 4.20). By Zorn's Lemma, then, there is indeed some maximal $T^* \in C$.

It remains to show that T^* is complete. Suppose that $\varphi \notin T^*$. Then $T^* \cup \{\varphi\}$ is inconsistent, since T^* is maximal. So $T^* \cup \{\varphi\} \vdash \bot$, and hence $T^* \vdash \neg\varphi$. So $T^* \cup \{\neg\varphi\}$ is consistent and hence $\neg\varphi \in T^*$ because T^* is maximal. □

Finally, we show how to construct models for complete, consistent theories with the witness property:

Lemma 4.23: *Let T be a complete, consistent $\mathcal{L}$-theory with the witness property. Then T has a model $\mathcal{M}$ whose domain consists of equivalence classes of $\mathcal{L}$-constant symbols.*

Proof. Define an equivalence relation $\sim$ on the $\mathcal{L}$-constant symbols, by $c \sim d$ iff $T \vdash c = d$. We use $[c]$ to denote the equivalence classes under $\sim$, i.e. $[c] = \{d : c \sim d\}$. Let $\mathcal{M}$ be the $\mathcal{L}$-structure whose underlying domain M is the set of these $\sim$-equivalence classes, and which interprets the $\mathcal{L}$-symbols as follows:

[70] For more on Zorn's Lemma and its equivalence with the Axiom of Choice, see Hrbáček and Jech (1999: 142).

$$c^{\mathcal{M}} = [c]$$
$$R^{\mathcal{M}} = \{([c_1], \dots, [c_n]) : T \vdash R(c_1, \dots, c_n)\}$$
$$f^{\mathcal{M}} = \{([c_1], \dots, [c_n], [c_{n+1}]) : T \vdash f(c_1, \dots, c_n) = c_{n+1}\}$$

In this last line, we are describing the interpretation of the n-place function symbol f by means of its graph. Further, while the interpretation of the function and relation symbols is defined in terms of how the theory acts on the representatives from the equivalence classes, the definition of $\sim$ ensures that nothing depends on this choice of representatives. For instance, if R is a two-place relation and f is a one-place function, then we have:

$$T \vdash (c_1 = d_1 \wedge c_2 = d_2 \wedge R(c_1, c_2)) \rightarrow R(d_1, d_2)$$
$$T \vdash (c_1 = d_1 \wedge c_2 = d_2 \wedge f(c_1) = c_2) \rightarrow f(d_1) = d_2$$

Expressed in terms of equivalence classes and the structure $\mathcal{M}$, these imply that if $[c_1] = [d_1]$ and $[c_2] = [d_2]$, then $R^{\mathcal{M}}([c_1], [c_2])$ implies $R^{\mathcal{M}}([d_1], [d_2])$, and likewise $f^{\mathcal{M}}([c_1]) = [c_2]$ implies $f^{\mathcal{M}}([d_1]) = [d_2]$. Hence the interpretations of the relation and function symbols in $\mathcal{M}$ does not depend on the choice of the representatives from the equivalence classes.

We now show, by induction on complexity of $\mathscr{L}$-formulas $\varphi(\bar{x})$ with all free variables displayed, that for all tuples $\bar{c}$ of $\mathscr{L}$-constant symbols:

$$T \vdash \varphi(\bar{c}) \text{ iff } \mathcal{M} \vDash \varphi\left(\overline{[c]}\right)$$

For atomic formulas, the biconditional holds by the definition of $\mathcal{M}$. The induction step associated with conjunction follows easily by the induction hypothesis, as does the induction step for negation, since T is complete. For the induction step associated with the existential quantifier, first consider the right-to-left direction of the biconditional. That is, first suppose that $\mathcal{M} \vDash \exists x \varphi\left(\overline{[c]}, x\right)$. Then $\mathcal{M} \vDash \varphi\left(\overline{[c]}, [d]\right)$ for some $\mathscr{L}$-constant symbol d. With an appeal to the induction hypothesis, this finishes the right-to-left direction. For the left-to-right direction, suppose that $T \vdash \exists x \varphi(\bar{c}, x)$. Then since T has the witness property, $T \vdash \varphi(\bar{c}, d)$ for some $\mathscr{L}$-constant symbol d. And we are done, by invoking the induction hypothesis again. □

Putting all of these results together, we obtain Gödel's Completeness Theorem.

Theorem 4.24 (Gödel Completeness): *Let T be any first-order theory. If $T \vDash \varphi$ then $T \vdash \varphi$. In particular, if T is consistent then T has a model.*

Proof. Suppose that $T \vDash \varphi$ and, for reductio, that $T \nvdash \varphi$. Then $T \cup \{\neg\varphi\}$ is consistent. Apply Henkin's Lemma 4.21 to $T \cup \{\neg\varphi\}$, then Lindenbaum's Lemma 4.22

to that. The resulting theory is a complete, consistent $\mathcal{L}^*$-theory with the witness property. By Lemma 4.23, it has a model, whose $\mathcal{L}$-reduct satisfies $T \cup \{\neg\varphi\}$. Since we assumed $T \vDash \varphi$, this model also satisfies φ, which is a contradiction. □

4.B A model-theoretic proof of Compactness

In §4.1, we proved the Compactness Theorem 4.1 from Gödel's Completeness Theorem 4.24. That involves a detour through a *deductive* system, but the idea can be made more direct.[71] We start with a definition:

Definition 4.25: *Let T be any theory. Say that T is* satisfiable *iff T has a model. Say that T is* finitely-satisfiable *iff every finite subset of T has a model.*

In these terms, to prove the Compactness Theorem 4.1, we just need to show that finite-satisfiability entails satisfiability. Our strategy is basically to repeat most of the steps of §4.A, replacing 'consistent' with 'finitely-satisfiable'. In detail: we will reprove Lemmas 4.21–4.23, using only semantic notions; the Compactness Theorem then follows immediately:

Lemma (4.21†): *Let T be a finitely-satisfiable $\mathcal{L}$-theory. There is a signature $\mathcal{L}^* \supseteq \mathcal{L}$ with the same cardinality as the set of $\mathcal{L}$-formulas, and a finitely-satisfiable $\mathcal{L}^*$-theory $T^* \supseteq T$ with the witness property.*

Proof. We define T_α, for $\alpha \leq \kappa$, exactly as in Lemma 4.21. We must then prove that T_α is finitely satisfiable for each $\alpha \leq \kappa$. The only interesting case is successor ordinals. For induction, suppose T_α is finitely-satisfiable, and consider any finite $S_\alpha \subseteq T_\alpha$. Let $\mathcal{M}$ be any model of S_α. If $\mathcal{M} \nvDash \exists x \varphi_{\alpha+1}(x)$, then $\mathcal{M} \vDash \{\exists x \varphi_{\alpha+1}(x) \rightarrow \varphi_{\alpha+1}(c_\beta)\}$. Alternatively, if $\mathcal{M} \vDash \exists x \varphi_{\alpha+1}(x)$, then $\mathcal{M} \vDash \varphi_{\alpha+1}(a)$ for some $a \in M$. Since c_β does not occur in $\mathcal{M}$'s signature, form a model $\mathcal{M}^*$ by augmenting $\mathcal{M}$ so that $c_\beta^{\mathcal{M}^*} = a$. Now $\mathcal{M}^* \vDash \{\exists x \varphi_{\alpha+1}(x) \rightarrow \varphi_{\alpha+1}(c_\beta)\}$. Either way, we have a model of $S_\alpha \cup \{\exists x \varphi_{\alpha+1}(x) \rightarrow \varphi_{\alpha+1}(c_\beta)\}$. Since S_α was an arbitrary finite subset of T_α, $T_{\alpha+1}$ is finitely-satisfiable. □

Lemma (4.22†): *For any finitely-satisfiable $\mathcal{L}$-theory T, there is some complete, finitely-satisfiable $\mathcal{L}$-theory $T^* \supseteq T$.*

Proof. As in the proof of Lemma 4.22, Zorn's Lemma yields a maximal $\mathcal{L}$-theory T^* which extends S and is finitely-satisfiable. It remains to show that T^* is complete.

[71] Paseau (2010) surveys and discusses five proofs of compactness. The proof in this appendix is the first of those five.

Suppose, for reductio, that $\varphi \notin T^*$ and $\neg\varphi \notin T^*$. Since T^* is maximal, some finite $S_1 \subseteq T^*$ is such that $S_1 \cup \{\varphi\}$ is unsatisfiable, i.e. $S_1 \vDash \neg\varphi$, and similarly some finite $S_2 \subseteq T^*$ is such that $S_2 \vDash \neg\neg\varphi$. But then $S_1 \cup S_2$ is a finite unsatisfiable subset of T^*, contradicting the fact that T^* is finitely-satisfiable. □

Lemma (4.23†): *Let T be a complete, finitely-satisfiable $\mathscr{L}$-theory with the witness property. Then T has a model $\mathcal{M}$ whose underlying domain consists of equivalence classes of $\mathscr{L}$-constant symbols.*

Proof. Repeat the proof of Lemma 4.23, replacing $\vdash$ with $\vDash$ throughout. □

4.c The valuation function of §4.6

This appendix proves the existence of a valuation function, v, with the properties described in Theorem 4.12. In fact, we define v in terms of the negative of the natural logarithm. So, during the proof of Theorem 4.12, it may help to consider this graph:

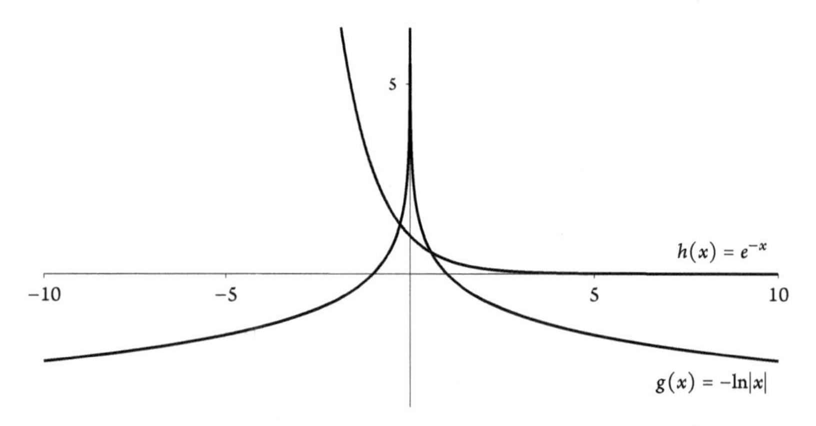

It depicts $g(x) = -\ln|x|$ for $x \neq 0$, and its inverse on positive values, $h(x) = e^{-x}$, which we use in later results. The asymptotic behaviour of g is representative of v. For example, small non-zero finite values of x yield large but finite values of $g(x)$, and infinitesimal values of x yield infinite values of $g(x)$.

Theorem (4.12): *Let the natural logarithm function,* ln, *be definable in $\mathcal{R}$, and let $\mathcal{R}^* \geq \mathcal{R}$. Then there is an ordered abelian group $\mathcal{G}$ and a surjective map $v : \mathcal{R}^* \longrightarrow (G \cup \{\infty\})$ satisfying the following properties:*

(1) $v(a) = \infty$ *iff* $a = 0$
(2) $v(ab) = v(a) + v(b)$
(3) $v(a+b) \geq \min(v(a), v(b))$
(4) $0 < v(a) < \infty$ *iff a is infinitesimal*

(5) $v(a) < 0$ *iff* a *is infinite*
(6) $v(a) = 0$ *iff* a *is finite, non-zero, and not infinitesimal*
(7) *If* $|a| \leq |b|$*, then* $v(a) \geq v(b)$

Proof. Let $\sim$ be an equivalence relation on $\mathcal{R}^*$ defined by:

$$a \sim b \text{ iff } a - b \text{ is finite}$$

Let $\mathcal{G}$ be the quotient group of $\mathcal{R}^*$ induced by this equivalence relation. If H denotes the set of finite elements of $\mathcal{R}^*$, then $a \sim b$ iff $a - b \in H$, so that $\mathcal{G}$ may be written $\mathcal{R}^*/H$. Thus the elements of $\mathcal{G}$ are the equivalence classes $[a] = \{b \in R^* : a \sim b\}$, where a ranges over R^*. Note that $\mathcal{G}$ has a natural order defined by $[a] < [b]$ iff $b - a$ is positive and infinite. Finally, note that the unit element of the group $\mathcal{G}$ is $[0]$, but we will write this as 0, as in the statement of the Theorem, since this is the traditional way of writing the units of groups whose operation is written additively. We now define our map v:

$$v(a) = \begin{cases} \infty & \text{if } a = 0 \\ [-\ln^* |a|] & \text{otherwise} \end{cases}$$

Here, ∞ is stipulated to be greater than all elements of $\mathcal{G}$ and to satisfy $a + \infty = \infty$. Clearly we have $v(a) = \infty$ iff $a = 0$. To see that $v(ab) = v(a) + v(b)$, not:

$$v(ab) = [-\ln^* |ab|] = [(-\ln^* |a|) + (-\ln^* |b|)] = v(a) + v(b)$$

In this equation, we appeal to the fact that $[x+y] = [x] + [y]$; this follows from the fact that the map sending x to $[x]$ is a homomorphism of groups. (This is because the projection of a group onto one of its quotients is always a homomorphism.)

Next, suppose that $0 < v(a) < \infty$. So $a \neq 0$ and $-\ln^* |a| > 0$ is not finite. Then $|a|$ is infinitesimal, and hence so is a. Now, suppose that $v(a) < 0$. Then $v\left(\frac{1}{a}\right) = -v(a) > 0$ and so $\frac{1}{a}$ is infinitesimal and hence a is infinite. Finally, suppose that $v(a) = 0$. Then $-\ln^* |a|$ is finite. Then a is finite, non-zero and not infinitesimal.

Suppose that $|a| \leq |b|$. Since the natural logarithm is non-increasing, $\ln^* |a| \leq \ln^* |b|$ and hence $-\ln^* |a| \geq -\ln^* |b|$. If the two quantities are finitely far from one another then $v(a) = v(b)$. Otherwise $-\ln^* |a| + \ln^* |b| > 0$ is infinite and hence $v(a) > v(b)$.

Finally, we show that $v(a + b) \geq \min(v(a), v(b))$. Without loss of generality we may suppose that $v(a) \geq v(b)$, so that $v(b) = \min(v(a), v(b))$. If $a = 0$ then $a + b = b$ and so we are done. Hence we may assume that $a \neq 0$. If $b = -a$ then $a+b = 0$ and so $v(a+b) = \infty$ and hence we are trivially done. Thus we may further assume that $b \neq -a$. Then by our assumptions that $a \neq 0$ and $b \neq -a$ we have that $\frac{a}{b}$

is finite non-zero and hence also $\frac{a+b}{b}$ is finite non-zero. So now $v(a+b) - v(b) = v\left(\frac{a+b}{b}\right) \geq 0$ and hence $v(a+b) \geq v(b) = \min(v(a), v(b))$. □

Moreover, as promised in §4.6, the value group is unique up to isomorphism:

Proposition 4.26: *Any two ordered abelian groups with the properties described in Theorem 4.12 are isomorphic.*

Proof. Suppose that $\mathcal{G}'$ is an ordered abelian group with the properties of Theorem 4.12, with a witnessing surjective map $v' : R^* \longrightarrow (G' \cup \{\infty\})$. It suffices to show that $\mathcal{G}$ and $\mathcal{G}'$ are isomorphic, where $\mathcal{G}$ is the group constructed in the proof of Theorem 4.12. Define a map $f : \mathcal{R}^* \longrightarrow G'$ by $f(a) = v'(e^{-a})$. Then f is a group homomorphism:

$$f(0) = v'(e^{-0}) = v'(1) = 0$$
$$f(a+b) = v'(e^{-(a+b)}) = v'(e^{-a} \cdot e^{-b}) = v'(e^{-a}) + v'(e^{-b}) = f(a) + f(b)$$

Also, f is surjective. For, suppose that $c \in G'$. Since v' is surjective, choose $b \in R^*$ such that $v'(b) = c$. Then $b \neq 0$ since $c \neq \infty$. Since $v'(b) = v'(-b)$, without loss of generality let $b > 0$. Let $a = -\ln^* b$, so that $e^{-a} = b$. Then $f(a) = v'(e^{-a}) = v'(b) = c$.

We now show that the *kernel of* f, also called $\ker(f) = \{a \in \mathcal{R}^* : f(a) = 0\}$, is precisely the finite elements, which we wrote as H in the proof of Theorem 4.12. For, suppose that a in $\mathcal{R}^*$ is finite. Then e^{-a} is also finite non-zero and not infinitesimal, so that $v'(e^{-a}) = 0$ by hypothesis on v'. Hence $f(a) = 0$. Conversely, suppose that a is infinite. If $a < 0$ then e^{-a} is infinite and thus $v'(e^{-a}) < 0$ by hypothesis on v', so that $f(a) \neq 0$; and if $a > 0$ then e^{-a} is infinitesimal and then $0 < v'(e^{-a}) < \infty$ by hypothesis on v', so that again $f(a) \neq 0$. Hence, indeed $\ker(f)$ is precisely the finite elements.

Now the group epimorphism $f : \mathcal{R}^* \longrightarrow \mathcal{G}'$ induces a group isomorphism $\tilde{f} : \mathcal{G} \longrightarrow \mathcal{G}'$ given by $\tilde{f}([a]) = f(a)$.[72] Further, this isomorphism $\tilde{f}$ respects the ordering. For, suppose that $[a] < [b]$. Then $b - a$ is positive and infinite, so that $e^{-(b-a)}$ is infinitesimal. Hence $0 < v'(e^{-(b-a)}) < \infty$ by hypothesis on v', so that $0 < f(b-a) < \infty$. Since f is a homomorphism, we have $0 < f(b) - f(a)$, i.e. $f(a) < f(b)$. An identical argument shows that $f(a) < f(b)$ implies $a < b$. Hence, $\mathcal{G}$ and $\mathcal{G}'$ are indeed isomorphic as ordered abelian groups. □

[72] This is sometimes called the *first isomorphism theorem* for groups. See any standard algebra text, e.g. Hungerford (1980: 43).

5

Sameness of structure and theory

In Chapters 1–4, we introduced structures and used them to explore a wide range of philosophical topics. In this chapter, we step back from these immediate *applications* of the notion of a structure, and instead consider the various ways in which two structures might be regarded as *the same*. From a semantic point of view, theories can be taken to be the sets of sentences true on a class of structures, so notions of sameness of structure also induce notions of sameness of theory.

Much of this chapter involves simply *setting out* various different notions of sameness of structure and sameness of theory. However, we also comment on their philosophical significance. Early on, we consider whether there is a better notion than *isomorphism* for characterising what it means for two mathematicians to have 'the same structure' in mind; we suggest that this is likely to be interest-relative (§§5.1–5.4). And in the later parts of this chapter, we consider three claims to the effect that interpreting one theory within another preserves some key property. In all three cases, however, we are sceptical that interpretability provides what has been demanded of it (§§5.6–5.8).

The notions of sameness of structure and theory which we introduce in this chapter also recur frequently in this book. To anticipate: in §7.11 we will look at how sameness of theory establishes an entanglement between arithmetic and syntax; in §10.3 we appeal to notions of sameness of theory to show how incompleteness infects 'internally categorical' theories; and in §14.7 we consider how a specific notion of 'sameness of theory' plays out in a disagreement about metaphysical priority.

5.1 Definitional equivalence

Structures, as defined in Definition 1.2, are relative to a specific signature. For example, the natural numbers with $<$, and the natural numbers with both $<$ and 0, are technically different structures. Invoking isomorphism-types cannot correct this, since isomorphisms are defined so that they only obtain between structures in the same signature (see Definition 2.1). But this jars with our informal ways of thinking and speaking about 'mathematical structures', like the natural numbers. When we develop arithmetic, we do not think much of introducing a defined notion, like '0 is the $<$-least natural number' or 'n divides m iff there is some x such that $n \times x = m$'; and there is certainly nothing in our practice which suggests that introducing these new expressions changes the topic of conversation.

The notion of *definitional equivalence* was designed to handle this. It was presented as a way to explicate the thought that two structures are mere 'notational variants'[1] of each other, or that they have the 'same import', or are 'various descriptions amount[ing] to the same thing'.[2] Roughly, two structures are definitionally equivalent iff they share the same domain and each is definable in the other.[3] More precisely:

Definition 5.1: *An* $\mathcal{L}$*-structure* $\mathcal{M}$ *and an* $\mathcal{L}^*$*-structure* $\mathcal{N}$ *are* definitionally equivalent *iff they have the same underlying domain and* $\mathcal{M}$*'s interpretation of the* $\mathcal{L}$*-constant, relation, and function symbols are definable in* $\mathcal{N}$*, and likewise* $\mathcal{N}$*'s interpretation of the* $\mathcal{L}^*$*-constant, relation, and function symbols are definable in* $\mathcal{M}$*.*

We outlined *definability* in Definition 1.8 of §1.13. To illustrate: the natural numbers with $<$ is definitionally equivalent to the natural numbers with $<$ and 0, because 0 can be defined in terms of $<$ as the $<$-least element. More interestingly, perhaps, we can develop the theory of lattices either by taking the $\leq$-relation as primitive, or by taking the meet $\cdot$ and join $+$ operations as primitives.[4]

5.2 Sameness of structure and ante rem structuralism

The idea of definitional equivalence has been used to discuss a *metaphysical* version of the question of when two 'structures' are identical.

As explained in §2.4, Shapiro's *ante rem structuralism* is a version of objects-platonism which holds that mathematics is about ante-structures. These are not the model-theorist's structures, of Definition 1.2, but instead are abstract entities consisting of places with intra-structural relations holding between them. Now, since his ante-structures are supposed to be *bona fide* objects, we can ask Shapiro: *When are ante-structures identical?*

We remarked in §2.4 that Shapiro's ante-structures are a bit like isomorphism types. And, if we think of ante-structures as being as fine-grained as isomorphism types, this would suggest that Shapiro should answer the question as follows:

For any structures $\mathcal{M}$ *and* $\mathcal{N}$*:* $ante(\mathcal{M}) = ante(\mathcal{N})$ *iff* $\mathcal{M} \cong \mathcal{N}$*.*

Here, 'ante($\mathcal{M}$)' is to be read as *the (most detailed) ante-structure which is instanti-*

[1] Corcoran (1980b: 232).

[2] Bouvère (1965b: 622), cf. Bouvère (1965a). She originally called definitional equivalence 'synonymy'.

[3] Hodges (1993: 59–63) and Visser (2006: §3).

[4] Givant and Halmos (2009: 43) and Corcoran (1980b: 233).

ated by $\mathcal{M}$.[5] Unfortunately, this answer will not do. Very sensibly, Shapiro wants to respect the fact mentioned in §5.1, that practising mathematicians do not really distinguish between the natural numbers with $<$, and the natural numbers with both $<$ and 0; so he wants to regard these as the *same* ante-structure. As such he needs to replace *isomorphism*, in the above, with some other relation between structures. Shapiro suggests that *definitional equivalence* 'is a good candidate for "sameness of structure"',[6] and this would yield the following:

For any structures $\mathcal{M}$ and $\mathcal{N}$: ante$(\mathcal{M})$ = ante$(\mathcal{N})$ iff $\mathcal{M}$ and $\mathcal{N}$ are isomorphic to definitionally equivalent structures (i.e. iff there exists some $\mathcal{M}^*$ such that $\mathcal{M} \cong \mathcal{M}^*$ and $\mathcal{M}^*$ is definitionally equivalent to $\mathcal{N}$).

This avoids the previous problem. However, there is a nuance here: when someone invokes definitional equivalence, we need to be told what *definability* amounts to. In mathematical contexts, we typically consider definability in *first-order logic.* However, we could consider definability in stronger logics. And our choice of logic will affect which ante-structures are identical, according to the proposed criterion.

To take a simple example: in first-order logic, the natural numbers in the signature $\{0, S\}$ are *not* definitionally equivalent to the natural numbers in the signature $\{0, S, +, \times\}$. However, these structures *are* definitionally equivalent in second-order logic.[7] Since the ante rem structuralist will probably want to regard these as the *same* ante-structure, she will likely look to second-order logic. But definitional equivalence in second-order logic is quite coarse-grained. For example: in second-order logic, the natural numbers in the signature $\{0, S\}$ are definitionally equivalent to an isomorphic copy of the *integers* in the same signature (we show this in §5.B). So, the proposed criterion would tell us that the natural number ante-structure is identical to the integer ante-structure. And that seems contrary to Shapiro's interests. For there *does* seem to be a change in topic when we move from the naturals to the integers: we seem to expand our domain of discourse, by including the *negative* whole numbers.

It is not immediately obvious, then, that there is a nice, mathematically definable equivalence relation which can be used to address the question, *When are ante-structures identical?* This is unfortunate for ante rem structuralism, since we doubt

[5] We need the caveat 'most detailed', since if $\mathcal{M}$ is a standard model of Peano Arithmetic, it will instantiate the natural number ante-structure (whatever that is), but it will also instantiate the cardinal-$\aleph_0$ ante-structure (i.e. the structure consisting of $\aleph_0$ positions with *no* intrastructural relations between them). On cardinal ante-structures, see Shapiro (1997: 115–23).

[6] Shapiro (1997: 91). See also Resnik (1981: 533–6, 1997: 205–9).

[7] The inequivalence in first-order logic holds because, in first-order logic, all the definable subsets of the natural numbers with zero and successor are finite or cofinite (see Marker 2002: 104). For the equivalence in second-order logic, see our remarks immediately after Definition 1.10 in §1.A (this only requires the Comprehension Schema, rather than e.g. the full semantics for second-order logic).

that they can countenance any *indeterminacy* of identity between ante-structures.[8] But it is worth explicitly noting that none of this is *only* a problem for ante rem structuralists. When asked whether two mathematicians are discussing 'the same structure', in some intuitive (and not very metaphysical) sense, those who are not ante rem structuralists should be happy to let the answer be context-dependent and possibly indeterminate.

5.3 Interpretability

We have just considered a *philosophical* use to which definitional equivalence might be put. In this section and the next, though, we will consider some of the *mathematical* or *logical* reasons for considering notions of equivalence between structures. Ultimately, the aim is to find a notion of 'sameness of structure' which states that the structures share all the relevant 'logical properties'. For the remainder of the chapter, we will assume for simplicity that we are working with *first-order* logic, although it will be obvious how to generalise the definitions to other logics.

For mathematical purposes, definitional equivalence is too restrictive to be of much use, and for two reasons.

First, definitionally equivalent structures must share the same domain. But there is no mathematical reason to expect that structures which we see as the same, in this or that respect, will share an underlying domain. This point is familiar from §2.2: mathematicians are often indifferent between isomorphic structures.

Second, definitional equivalence has no way to deal with the fact that we often form one mathematical structure from another by taking equivalence classes. For example, we form the rationals from the integers by considering pairs of integers (a_1, a_2) where $a_2 \neq 0$ under the equivalence relation $(a_1, a_2)E(b_1, b_2)$ iff $a_1 b_2 = b_1 a_2$. Likewise, we form the points of the real projective plane from triples of real numbers (not all of which are zero) by considering the equivalence relation E' of 'being on the same line through the origin', i.e.:

$$(a_1, a_2, a_3)E'(b_1, b_2, b_3) \text{ iff } (\exists \lambda \neq 0)(a_1 = \lambda b_1 \wedge a_2 = \lambda b_2 \wedge a_3 = \lambda b_3)$$

Obviously, E is definable in the integers, and E' is definable in the real numbers. But what we would like is a way to move from these structures, to a structure that has just *one* representative for each collection of equivalent elements.

To accomplish this, we must build up to the notion of a quotient structure. We start by recalling some notation. If E is an equivalence relation on a set X, then *the equivalence class of* a is the set of elements equivalent to a, i.e.:

[8] Shapiro (1997: 79–82, 92–3) experimented with versions of indeterminacy of identity for positions between structures, but changed his mind (2006a: 124, 128–31). For further discussion on the indeterminacy of identity here, see Evans (1978), Chihara (2004: 81–3), and MacBride (2005: 570–1).

$$[a]_E = \{b \in X : aEb\}$$

Often, X will be a set of n-tuples. In this case, we reuse the notation introduced in §2.B, writing for example $\bar{a}_1 = (a_{1,1}, \ldots, a_{1,n})$ and $\bar{a}_2 = (a_{2,1}, \ldots, a_{2,n})$. To illustrate this notation, we can rewrite the relation E' from above as follows:

$$\bar{a}_1 E' \bar{a}_2 \text{ iff } (\exists \lambda \neq 0)(a_{1,1} = \lambda a_{2,1} \wedge a_{1,2} = \lambda a_{2,2} \wedge a_{1,3} = \lambda a_{2,3})$$

With this notation in place, we can say that one structure is a *quotient structure* of another iff: the domain of the first is the collection of equivalence classes of some definable equivalence relation in the second, such that all of the structure of the first is definable in the second. More formally:

Definition 5.2: *Let $\mathcal{A}$ be an $\mathcal{L}$-structure and $\mathcal{B}$ be an $\mathcal{L}^*$-structure. We say that $\mathcal{B}$ is a* quotient structure *of $\mathcal{A}$ iff both of the following hold:*

(1) *the domain of $\mathcal{B}$ is the set $B = \{[a]_E : a \in X\}$ of equivalence classes of some $\mathcal{A}$-definable set $X \subseteq A^m$ under an $\mathcal{A}$-definable equivalence relation $E \subseteq X \times X$*

(2) *for each $\mathcal{L}^*$-constant symbol c, each n-place $\mathcal{L}^*$-relation symbol R, and each n-place $\mathcal{L}^*$-function symbol f, the following sets are $\mathcal{A}$-definable:*

$$\begin{aligned}
(c^{\mathcal{B}})^{-1} &= \{\bar{a} \in X : \mathcal{B} \models c = [\bar{a}]_E\} \\
(R^{\mathcal{B}})^{-1} &= \{(\bar{a}_1, \ldots, \bar{a}_n) \in X^n : \mathcal{B} \models R([\bar{a}_1]_E, \ldots, [\bar{a}_n]_E)\} \\
(f^{\mathcal{B}})^{-1} &= \{(\bar{a}_1, \ldots, \bar{a}_n, \bar{a}_{n+1}) \in X^{n+1} : \mathcal{B} \models f([\bar{a}_1]_E, \ldots, [\bar{a}_n]_E) = [\bar{a}_{n+1}]_E\}
\end{aligned}$$

To illustrate the definition, consider the example of the rationals from the previous paragraph. Since $\frac{a}{b} + \frac{c}{d} = \frac{u}{v}$ iff $adv + cbv = bdu$, the additive structure on the rationals is definable in the integers. Expressed in the notation of Definition 5.2, we say that $(+^{\mathbb{Q}})^{-1}$ is $\mathbb{Z}$-definable, because it may be written as:

$$\begin{aligned}
(+^{\mathbb{Q}})^{-1} &= \{(a, b, c, d, u, v) \in (\mathbb{Z}^2)^3 : \mathbb{Q} \models b, d, v \neq 0 \wedge \tfrac{a}{b} + \tfrac{c}{d} = \tfrac{u}{v}\} \\
&= \{(a, b, c, d, u, v) \in \mathbb{Z}^6 : \mathbb{Z} \models b, d, v \neq 0 \wedge adv + cbv = bdu\}
\end{aligned}$$

We then use this notion of a quotient structure to define the key notions of interpretability and mutual interpretability:

Definition 5.3: *Let $\mathcal{A}$ be an $\mathcal{L}$-structure and $\mathcal{B}$ be an $\mathcal{L}^*$-structure. We say that $\mathcal{A}$ is* interpretable *in $\mathcal{B}$ iff $\mathcal{A}$ is isomorphic to a quotient structure of $\mathcal{B}$. We say that $\mathcal{A}$ and $\mathcal{B}$ are* mutually interpretable *iff each is interpretable in the other.*

As a notion of sameness of structure, mutually interpretability avoids the two problems we raised for definitional equivalence at the start of this section. First, mutually interpretable structures need not share a common domain; second, mutual interpretability is compatible with constructing structures via equivalence relations. For instance, by what was said above, the rationals are interpretable in the integers. Additionally, by an important result of Julia Robinson, the integers are definable in the rationals. So, the integers and rationals are *mutually interpretable.*[9]

Later in this chapter, it will be useful to have some explicit terminology for some special kinds of interpretability. Suppose that $\mathcal{B}$ is isomorphic to a quotient structure of $\mathcal{A}$, where the relevant equivalence relation is just identity, i.e.:

$$\bar{a}_1 E \bar{a}_2 \text{ iff } a_{1,1} = a_{2,1} \wedge \ldots \wedge a_{1,n} = a_{2,n}$$

In this case, we say that $\mathcal{B}$ is interpretable in $\mathcal{A}$ with *identity interpreted absolutely.*[10] Further, if the equivalence relation E is an equivalence relation on n-tuples, we say that the interpretation is *n-dimensional.*[11] Hence, the simplest interpretations are one-dimensional interpretations where identity is interpreted absolutely; in such cases, no quotienting has taken place.

For all its virtues, mutual interpretability tends not to be used directly by mathematicians. The reason for this is that the isomorphisms that witness mutual interpretability need not be 'recognisable' by the structures in question, so that whether two structures are mutually interpretable is decided by facts of the ambient metatheory, rather than just by facts about the structures themselves. To illustrate the point, suppose that a geometric structure $\mathcal{G}$ and an algebraic structure $\mathcal{A}$ are mutually interpretable.[12] Then $\mathcal{G}$ defines an isomorphic copy $\mathcal{A}_1$ of the algebraic structure. We might hope to use $\mathcal{A}_1$ to study the original geometric structure, $\mathcal{G}$, and to do this, we might consider the isomorphic copy $\mathcal{G}_1$ of $\mathcal{G}$ defined in $\mathcal{A}_1$. Evidently, $\mathcal{G}_1$ is itself definable in $\mathcal{G}$, just by composing the two definitions. Unfortunately, though, there is no guarantee that there is a $\mathcal{G}$-definable isomorphism between $\mathcal{G}$ and $\mathcal{G}_1$. So if we obtain information about $\mathcal{G}_1$ by the algebraic means of $\mathcal{A}_1$, and we want to pull this information back to the original geometric structure $\mathcal{G}$, we may have to rely upon metatheoretic resources.

Fortunately, in practice the relevant isomorphisms tend to be definable. And this leads to a stronger notion of sameness of structures, namely biinterpretability.

---

[9] J. Robinson (1949: Theorem 3.1 pp.106–7).

[10] There is no standard notation for this. Marker (2002: 24–9) uses 'definably interpreted'.

[11] This terminology is from Hodges (1993: 212).

[12] A concrete example of this arises from the 'introduction of coordinates' to go back and forth between an algebraic field structure and a geometric point-line structure. For more, see Hodges (1993: 222–3 Example 1), Artin (1957: ch.2), and Hartshorne (2000: Theorem 21.1 p.137).

5.4 Biinterpretability

The notion of biinterpretability expands upon the idea of mutual interpretability, by insisting that the isomorphisms between the structures should, indeed, be definable. Here is the formal definition:[13]

Definition 5.4: *Let V be a class of $\mathscr{L}$-structures and let W be a class of $\mathscr{L}^*$-structures. We say that V and W are* biinterpretable *iff the following conditions all obtain:*

(1) *Every structure $\mathcal{M}$ from V uniformly defines a structure $I(\mathcal{M})$ from W which is a quotient structure of $\mathcal{M}$.*

(2) *Every structure $\mathcal{N}$ from W uniformly defines a structure $J(\mathcal{N})$ from V which is a quotient structure of $\mathcal{N}$.*

(3) *For every structure $\mathcal{M}$ from V there is a uniformly $\mathcal{M}$-definable bijection $g_{\mathcal{M}} : J(I(M)) \longrightarrow M$ that induces an isomorphism $g_{\mathcal{M}} : J(I(\mathcal{M})) \longrightarrow \mathcal{M}$.*

(4) *For every structure $\mathcal{N}$ from W there is a uniformly $\mathcal{N}$-definable bijection $h_{\mathcal{N}} : I(J(N)) \longrightarrow N$ that induces an isomorphism $\hbar_{\mathcal{N}} : I(J(\mathcal{N})) \longrightarrow \mathcal{N}$.*

We say that $\mathcal{M}$ and $\mathcal{N}$ are biinterpretable *iff the two classes $\{\mathcal{M}^* : \mathcal{M}^* \cong \mathcal{M}\}$ and $\{\mathcal{N}^* : \mathcal{N}^* \cong \mathcal{N}\}$ are biinterpretable.*

Here, the adjective 'uniformly' means that the formulas used in the definitions are the same for every case. So we could rephrase (1) as follows: there is a sequence of formulas such that, for any structure $\mathcal{M}$ from V, the first formula defines an equivalence relation whose equivalence classes form the domain of the structure $I(\mathcal{M})$ from W, and the other formulas define the interpretation of the constant, relation, and function symbols on this structure. Likewise, in (3), the function is 'uniformly definable' in that there is a single formula which, when applied to any structure $\mathcal{M}$ from V, defines a function from an appropriate Cartesian power of M back to M itself, such that the associated function $g_{\mathcal{M}} : J(I(M)) \longrightarrow M$ defined on the equivalence classes is a bijection.[14]

Biinterpretability has a central place in mathematical logic. To illustrate: the Slaman–Woodin conjecture predicts that the Turing degrees are biinterpretable with the standard model of PA_2. This is a major open problem in computability, and here is how Slaman describes its significance:

> The [Slaman–Woodin] Conjecture, if true, reduces all the logical questions that one could ask of [the Turing degrees] to the exact same questions about second order arithmetic. The structures would be logically identical, though presented in different first order languages.[15]

---

[13] Hodges (1993: 222) and Visser (2006: §3.3).

[14] In the formulas which witness uniform definability, parameters may be allowed from some parameter-free definable set, so long as any choice of parameter effects the task at hand.

[15] Slaman (2008: 99).

Within model theory itself, Pillay speaks of biinterpretability as being partially constitutive of the aims of *pure* model theory, where the relevant contrast is to *applications* of model theory to other areas of mathematics:

> Interpretability is a key (even characteristic) notion, and in a tautological sense the business of 'pure' model theory becomes the classification of first order theories up to biinterpretablity.[16]

Part of the idea behind Slaman's and Pillay's remarks is that biinterpretability preserves many properties that arise in mathematical logic. For instance, biinterpretability preserves *stability* (which we introduced in Definition 14.11 of Chapter 14) and biinterpretable structures have the same automorphism group.[17]

It may well be, then, that structures are biinterpretable iff they share all the features that matter for central purposes of mathematical logic. This is not to say, though, that biinterpretability is the once-and-for-all *correct* notion of 'sameness of structure'. Consider, again, the question from §5.2, of when two mathematicians are discussing 'the same structure', in some intuitive, informal sense. If one mathematician is talking about the natural numbers and the other is talking about the integers, then in certain contexts it will surely be right to say that they are talking about importantly *different* structures, since the former has a least element and the latter does not. This is so, even though these two structures will be biinterpretable (given the right signature).[18] And this again suggests that it may be purpose-relative, context-sensitive, or indeterminate, which features of a structure are relevant to the question of whether they are 'the same'.

5.5 From structures to theories

We now move from discussing 'sameness of structures' to 'sameness of theories'. In fact, this can be done very simply:

Definition 5.5: *Where T is an $\mathcal{L}$-theory and T^* is an $\mathcal{L}^*$-theory:*

(1) *T is* interpretable *in T^* iff every model $\mathcal{M}^*$ of T^* uniformly interprets a model $\mathcal{M}$ of T.*[19]

(2) *T and T^* are* mutually interpretable *iff each interprets the other.*

[16] Buss et al. (2001: 186).

[17] For the fact about automorphism groups, see Hodges (1993: Exercise 8 p.226).

[18] They are biinterpretable in a signature containing addition and multiplication, since these resources allow one to do all of the usual coding tricks deployed in e.g. the proofs of Gödelian incompleteness.

[19] The sense of 'uniformly' is, as above, that the same formulas are used each time. The restriction on parameters is similar to that mentioned in footnote 14: these formulas may be allowed to include parameters from a certain parameter-free definable class in the interpreting structure $\mathcal{M}^*$, so long as one stipulates that any choice of parameters from this class succeeds in effecting such a definition of a model of T (cf. Hájek and Pudlák 1998: 149; Visser 2006: §B.3; Hodges 1993: 215).

(3) *T and T^* are* biinterpretable *iff they are mutually interpretable, and there are isomorphisms which witness the correctness of the interpretations (i.e. iff the two classes $V = \{\mathcal{M} : \mathcal{M} \models T\}$ and $W = \{\mathcal{N} : \mathcal{N} \models T^*\}$ are biinterpretable in the sense of Definition 5.4.)*

(4) *T and T^* are* definitionally equivalent *iff they are biinterpretable and the domains do not change under the interpretations and the isomorphisms are given by the identity map.*[20]

This definition characterises interpretation (between theories) in terms of *definability*. However, when dealing with logics with a sound and complete proof-system, such as first-order logic, it is natural to characterise interpretation in terms of *provability*. Then key idea will be: *translations of theorems are theorems.*

At a first approximation, we say that a theory T is *interpretable* in a theory T^* iff the primitives of the interpreted theory T can be translated into formulas of the interpreting theory T^* so that the translation of every theorem of T is a theorem of T^*. However, to make this rigorous, we must say more about the relevant notion of translation. We do this in two steps.[21] The first step is to define the interpretation of a signature into a theory. For the sake of simplicity we focus on the case where the interpretations are one-dimensional and where identity is interpreted absolutely (see §5.3):[22]

Definition 5.6: *Let $\mathcal{L}$ be a signature and let T^* be an $\mathcal{L}^*$-theory. Then an* interpretation *I of $\mathcal{L}$ into T^* is given by the following data:*

(1) *an $\mathcal{L}^*$-formula $D^I(x)$ such that $T^* \vdash \exists x D^I(x)$*

(2) *for every $\mathcal{L}$-constant symbol c, an $\mathcal{L}^*$-formula $C^I(x)$ such that $T^* \vdash \exists! x \left(C^I(x) \wedge D^I(x)\right)$*

(3) *for every n-place $\mathcal{L}$-relation symbol R, an n-place $\mathcal{L}^*$-formula $R^I(\bar{x})$ such that $T^* \vdash \forall \bar{x} \left(R^I(\bar{x}) \rightarrow \bigwedge_{i=1}^n D^I(x_i)\right)$*

(4) *for every n-place $\mathcal{L}$-function symbol f, an n+1-place $\mathcal{L}^*$-formula $F^I(\bar{x}, y)$ such that both $T^* \vdash \forall \bar{x} \forall y \left(F^I(\bar{x}, y) \rightarrow \left(\bigwedge_{i=1}^n D^I(x_i) \wedge D^I(y)\right)\right)$, and $T^* \vdash \forall \bar{x} \left(\bigwedge_{i=1}^n D^I(x_i) \rightarrow \exists! y \left(F^I(\bar{x}, y) \wedge D^I(y)\right)\right)$.*

---

[20] For an equivalent characterisation of definitional equivalence in terms of 'having a common definitional extension', see Hodges (1993: 59–61).

[21] Here we follow Lindström (2003: 96–7), Hájek and Pudlák (1998: 148–9), and Visser (2006: §2.2).

[22] To cover m-dimensional interpretations, we must modify Definition 5.6 as follows. First, the unary formulas D^I and C^I would be replaced with m-place formulas; the n-place R^I would be replaced with $n \cdot m$-place formulas, and the $(n+1)$-place formula F^I would be replaced by an $(n+1) \cdot m$-place formula. Second, all the bound variables featured in Definition 5.6 would be replaced by bound m-tuples of variables. To allow identity to be interpreted non-absolutely, we must add to the data of Definition 5.6 a formula $E^I(\bar{x}_1, \bar{x}_2)$, which is T^*-provably an equivalence relation on $D^I(\bar{x})$. We would then modify (2)–(4) to ensure that the interpretations of the constant, relations, and function symbols respected this equivalence relation. Finally, we would modify Definition 5.6 so that $(x = y)^I$ is $E^I(x, y)$, and replace 'uniqueness' by 'uniqueness up to equivalence' in clause (2) and (4).

We define a map $\varphi \mapsto \varphi^I$ of $\mathscr{L}$-formulas to $\mathscr{L}^*$-formulas, in the same number of free variables, starting with atomic formulas:

$$(x = y)^I := x = y \qquad (x = c)^I := C^I(x)$$
$$(R(\bar{x}))^I := R^I(\bar{x}) \qquad (f(\bar{x}) = y)^I := F^I(\bar{x}, y)$$

and offering recursion clauses as follows:

$$(\neg\varphi)^I := \neg(\varphi^I)$$
$$(\varphi \wedge \psi)^I := (\varphi^I \wedge \psi^I) \qquad (\varphi \vee \psi)^I := (\varphi^I \vee \psi^I)$$
$$(\forall x \varphi(x))^I := (\forall x : D^I)\varphi^I(x) \qquad (\exists x \varphi(x))^I := (\exists x : D^I)\varphi^I(x)$$

The intuitive idea behind clause (1) is that D^I serves to pick out the domain of the interpretation. The remaining clauses then specify the means by which the interpreting theory is to interpret the constants, relation, and function symbols. So, the uniqueness claim is needed in (2) because each constant symbol must pick out a unique object, and similar remarks apply to (4).[23]

The connection between interpretability, as laid down in Definition 5.5(1), and the notion of an interpretation, as just defined, is quite straightforward. Given the soundness and completeness of first-order logic, to provide a *uniform interpretation* of one structure in another, as in Definition 5.5(1), just is to provide an interpretation of a signature in a theory, as in Definition 5.6. So, in place of Definition 5.5(1), we could instead have defined interpretability in the following, equivalent fashion:

Definition 5.7: *Suppose that T is an $\mathscr{L}$-theory and that T^* is an $\mathscr{L}^*$-theory. Then T is* interpretable *in T^* iff: there is an interpretation I of $\mathscr{L}$ into T^* such that, for all $\mathscr{L}$-sentences φ, if $T \vdash \varphi$ then $T^* \vdash \varphi^I$.*

In the case of Definition 5.7, however, our guiding idea of interpretation comes through much more clearly: *translations of theorems are theorems.*

To illustrate the notion, ZFC interprets PA by taking 'x is a finite ordinal' as the domain formula $D^I(x)$, translating '0' with '$\forall y\, y \notin x$', and translating the relation '$x < y$' with '$x \in y$'. The translations of PA's theorems are then all theorems of ZFC.

Whilst this translation preserves theoremhood, it does not preserve *non-theoremhood.* We can see this by invoking Gödelian considerations. (We review

[23] This definition does not say how to translate atomic formulas such as $f(g(x)) = y$ or $f(x) = h(x)$. But it is easy to handle such cases: $f(g(x)) = y$ is equivalent to $\exists z(g(x) = z \wedge f(z) = y)$, and the latter can be handled using the definition; similarly, $f(x) = h(x)$ is equivalent to $\exists z(f(x) = z \wedge h(x) = z)$. Formally, we are appealing to the fact that each atomic formula is equivalent to a formula in which the only atomic subformulas are those of the form $R(\bar{x})$ and $f(\bar{x}) = y$; see Hodges (1993: 58) for a proof of this fact.

some of the technicalities of this in §5.A.) Gödel showed how to develop a theory of syntax within a theory of arithmetic, such as PA.[24] This allows us, in particular, to arithmetise PA's consistency sentence, as an arithmetical formula Con(PA). Gödel's Second Incompleteness Theorem then shows that PA does not prove Con(PA). However, ZFC does prove the translation of Con(PA), since we can easily build a model of PA within ZFC.

If we want also to ensure that the translations of non-theorems are non-theorems, we need to consider a more restrictive notion of interpretability. Specifically: a theory T is said to be *faithfully interpretable* in a theory T^* iff T is interpretable in T^* both so that translations of theorems are theorems *and* so that translations of non-theorems are non-theorems. More precisely:

Definition 5.8: *Where T is an $\mathcal{L}$-theory and T^* is an $\mathcal{L}^*$-theory:*

1. *T is* faithfully interpretable *in T^* iff there is some interpretation I of $\mathcal{L}$ into T^* such that, for all $\mathcal{L}$-sentences φ: $T \vdash \varphi$ iff $T^* \vdash \varphi^I$.*
2. *T and T^* are* mutually faithfully interpretable *iff each faithfully interprets the other.*

The easiest way to produce examples of faithful interpretability is to note that biinterpretability implies faithful interpretability:

Proposition 5.9: *Biinterpretable theories are (mutually) faithfully interpretable.*

Proof. Suppose that T and T^* are biinterpretable, i.e. the classes $\{\mathcal{M} : \mathcal{M} \vDash T\}$ and $\{\mathcal{N} : \mathcal{N} \vDash T^*\}$ are biinterpretable in the sense of Definition 5.4. Suppose that $T^* \vdash \varphi^I$; we want to show that $T \vdash \varphi$. Let $\mathcal{N}$ be a model of T; we must show that it is a model of φ. Using clause (1) of Definition 5.4, $J(\mathcal{N})$ is a model of T^*, and so of φ^I. Since $J(\mathcal{N})$ is a model of φ^I, we have that $I(J(\mathcal{N}))$ is a model of φ. But by clause (3) of Definition 5.4, $\mathcal{N}$ and $I(J(\mathcal{N}))$ are isomorphic, so that $\mathcal{N}$ is a model of φ as well.[25] □

Using this result, we can easily chart the relative strengths of our several different notions of interpretation:

Proposition 5.10: *The following Hasse diagram represents the relationships between notions of interpretation, where a variety of interpretability entails all and only those*

---

[24] No theory has the status of *the* theory of syntax, in the way that PA is *the* theory of arithmetic. But for some suitable theories of syntax, see Leigh and Nicolai (2013: 619) and Hájek and Pudlák (1998: 151).

[25] The interpretation I is used here both in the context of a map from structures $\mathcal{M}$ to structures $I(\mathcal{N})$ and a map from formulas φ to formulas φ^I. This makes good sense, because the uniformity in the 'uniform definability' from the definition of biinterpretation (Definition 5.4) can be produced by the formulas from an interpretation of a signature in a theory in the sense of Definition 5.6.

varieties of interpretability which are connected to it by a downward path. (So: biinterpretable theories are always mutually faithfully interpretable, but not vice versa.)

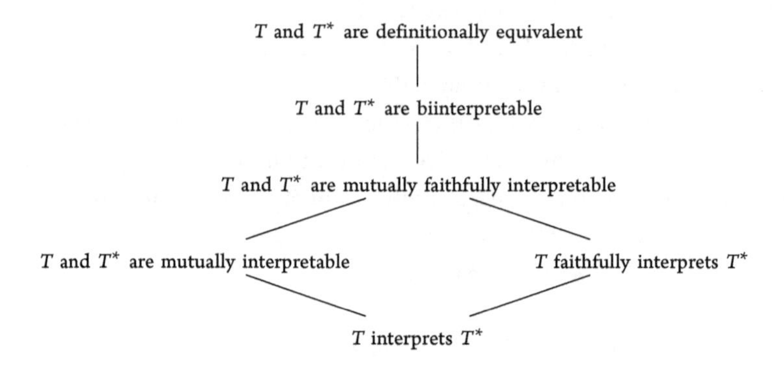

Proof sketch. Biinterpretability entails (mutual) faithful interpretability by Proposition 5.9. The other entailments are trivial. So it only remains to explain why the relevant entailments are strict.

Biinterpretation without definitional equivalence. Examples of such theories follow from a recent result of Visser and Friedman.[26]

Mutual faithful interpretation without biinterpretation. Consider the theory ZF_{fin}, which is obtained by taking the axioms of ZF and replacing the axiom of infinity with its *negation*. Enayat et al. show that PA and ZF_{fin} are mutually faithfully interpretable but *not* biinterpretable.[27]

Mutual interpretation without faithful interpretation. By Feferman's Theorem,[28] PA and PA + $\neg\text{Con}(\text{PA})$ are mutually interpretable. But neither faithfully interprets the other, since these theories have different Σ_1-consequences, and faithful interpretation requires sameness of Σ_1-consequences.[29]

Faithful interpretation without mutual interpretation. Consider PA and the theory SA which is the true theory of arithmetic in the signature $\{0, S\}$ (this was defined in §3.B). By Proposition 3.10, SA is complete. So PA faithfully interprets SA, via the 'identity' interpretation. But SA cannot interpret PA, since SA is both complete and computable but PA is not. □

The case of ZF_{fin}, which we just mentioned, is particularly interesting. Whilst ZF_{fin} is *not* biinterpretable with PA, Kaye and Wong show that we *can* obtain a biinterpretation, if we add an axiom to ZF_{fin} which states that every set is a subset of a transitive set.[30] Evidently then, we must be very careful when making claims like

[26] See Visser and H. M. Friedman (2014: §7).

[27] Enayat et al. (2011).

[28] Feferman (1960/1961: 76); see footnote 35 for more.

[29] See Lindström (2003: 106).

[30] Kaye and Wong (2007).

'finite set theory and number theory are "notational variants" of each other'. Whilst their standard *models* are biinterpretable, the *theories* only become biinterpretable by supplementing the axioms.

5.6 Interpretability and the transfer of truth

In the remainder of this chapter, we critically consider three possible philosophical roles for *interpretability*. We start by criticising the following thesis:

The Truth-Transfer Thesis. *If T^* interprets T and T^* is true, then T is true.*

This is a very natural idea. Moreover, it has been frequently deployed in the philosophy of mathematics by neo-logicists. Frege's Theorem states that PA_2 is interpretable in a formal theory known as HP.[31] The axioms of HP are just the Comprehension Schema instances and Hume's Principle, which is:

$$\forall X \forall Y (\#X = \#Y \leftrightarrow E(X, Y))$$

Here, '$E(X, Y)$' abbreviates the claim that there is a bijection between the Xs and the Y's, and '$\#X$' is read 'the number of Xs', where $\#$ is a function symbol from one-place second-order objects to first-order objects. Neo-logicists have insisted that HP has a particular epistemic status—something akin to a logical principle—so that it is knowable a priori. By combining Frege's Theorem with the Truth-Transfer Thesis, they have hoped to vindicate arithmetic, in the form of PA_2.[32]

Unfortunately for neo-logicists and others, there are at least two different kinds of counterexamples to the Truth-Transfer Thesis.

Counterexamples to the Truth-Transfer Thesis

The first kind of counterexample to the Truth-Transfer Thesis arises by considering theories T and sentences φ such that $T + \varphi$ and $T + \neg\varphi$ are *mutually* interpretable. Such sentences φ are known as *Orey sentences*. For instance, the continuum hypothesis and $V = L$ are Orey sentences against the background of ZFC, while the Axiom of Foundation is an Orey sentence against the background of ZFC *minus* Foundation.[33] The problem is simple: if interpreting $T + \varphi$ in a true theory T^* is sufficient for the truth of $T + \varphi$—as the Truth-Transfer Thesis insists—then in these cases, absurdly, both φ and $\neg\varphi$ are true.[34]

[31] See Wright (1983: ch.4) or Walsh (2012: §2.2).

[32] The paper Walsh (2014) discusses a more epistemic reading of the neo-logicist programme.

[33] Koellner (2009: 99) and Visser (2012: 413).

[34] See Walsh (2014: §3.3 pp.93ff.) for a related point in an epistemic context.

A second kind of counterexample relates to Feferman's Theorem. As mentioned in the proof of Proposition 5.10, this states that we can find theories, T, such that T and $T + \neg\text{Con}(T)$ are mutually interpretable.[35] For example, Feferman's Theorem applies to PA, PA_2, ZFC, and ZFC_2. Now, there is a long tradition of thinking that accepting a theory T also commits one to accepting $\text{Con}(T)$,[36] and to *rejecting* $\neg\text{Con}(T)$. And this leads Niebergall to write 'if there is *any* metatheorem presenting a problem for relative interpretability', then it is Feferman's Theorem.[37] (It is worth emphasising that Feferman-style counterexamples to the Truth-Transfer Thesis are *not* Orey-style counterexamples, simply because consistency statements are not Orey sentences. Indeed, the theory $\text{PA} + \text{Con}(\text{PA})$ is *not* interpretable in PA.[38])

Both kinds of counterexample illustrate that there are simply too *many* interpretations to sustain the Truth-Transfer Thesis. The only way to defend some version of the Thesis, then, will be to build in some restrictions on the nature of the interpretations that allow for truth-transfer.

A plausible idea would be to restrict the Thesis, saying only that truth is transferred between theories when we have *mutual* interpretability, or perhaps something stronger still. Unfortunately, that idea also fails. Our Orey-style and Feferman-style counterexamples all involved mutual interpretability.[39] Admittedly, there cannot be a Feferman-style counterexample for *faithful* interpretability.[40] But it is easy to find Orey-style counterexamples for *definitional equivalence*:

Proposition 5.11: *Let* $\psi(x)$ *be any formula in the signature of* PA *such that* $\text{PA} \vdash \exists x\psi(x)$ *and* $\text{PA} \vdash \exists x\neg\psi(x)$. *Let* F *be a new one-place predicate, and let* T *be the theory* $\text{PA} + [\forall x(\psi(x) \leftrightarrow F(x)) \vee \forall x(\psi(x) \leftrightarrow \neg F(x))]$. *Let* φ *be* $\exists x(\psi(x) \wedge F(x))$. *Then* $T + \varphi$ *and* $T + \neg\varphi$ *are definitionally equivalent.*

Proof. First note that $T + \varphi$ is equivalent to $\text{PA} + \forall x(\psi(x) \leftrightarrow F(x))$, while $T + \neg\varphi$ is equivalent to $\text{PA} + \forall x(\psi(x) \leftrightarrow \neg F(x))$. Then $T + \varphi$ interprets $T + \neg\varphi$ via $F^I(x) := \neg F(x)$, leaving all other vocabulary unchanged. Further, $T + \neg\varphi$ interprets $T + \varphi$ in the same way. Since the composition of the two interpretations is identity, this yields definitional equivalence. □

[35] See Feferman (1960/1961: 76), Lindström (2003: 104). The precise restrictions on T are that it satisfies the usual conditions of Gödel's first incompleteness theorem and that it is *essentially reflexive*, i.e. it proves the consistency of all its finite subtheories. The theories PA, PA_2, ZFC, and ZFC are all essentially reflexive. An example of a theory which is *not* essentially reflexive is the theory ACA_0 mentioned in footnote 50 below.

[36] For classical statements of this view, see the introduction to Dean (2015).

[37] Niebergall (2000: 44).

[38] See Lindström (2003: 98).

[39] And it is also worth noting that HP and PA_2 are mutually interpretable but *not* biinterpretable; see Walsh (2012: Corollary 24 p.1691, 2014: §6.3 pp.111–3).

[40] PA does not faithfully interpret $\text{PA} + \neg\text{Con}(\text{PA})$, nor vice versa (see the proof of Proposition 5.10).

Bridge principles

Defenders of the Truth-Transfer Thesis must find some other way to restrict it. To do this, we suggest that they think of interpretations as supplying 'bridge principles' between theories.[41] Then the new idea is that only interpretations with *true* bridge principles are guaranteed to preserve truth.

To implement this idea, let I be an interpretation of the $\mathscr{L}$-theory T in the $\mathscr{L}^*$-theory T^*. For readability, we assume that $\mathscr{L}$ is relational, that I is one-dimensional and interprets identity absolutely (see §5.3), and that the signatures $\mathscr{L}$ and $\mathscr{L}^*$ are disjoint. We also assume that T contains an axiom of the form $\forall x D(x)$, for a distinguished one-place predicate D. For instance, if T is a theory of numbers, then we can think of '$\forall x D(x)$' as meaning 'everything is a number'.

The following sentence essentially says that T's intended domain aligns with the formula D^I provided by the interpretation:

$$\forall x \left(D(x) \leftrightarrow D^I(x)\right) \qquad \textit{(bridge:D)}$$

Likewise, for each n-place relation symbol R from $\mathscr{L}$, the following sentence essentially says that the intended interpretation of the relation lines up with the formula R^I provided by the interpretation:

$$(\forall \bar{x} : D) \left(R(\bar{x}) \leftrightarrow R^I(\bar{x})\right) \qquad \textit{(bridge:R)}$$

We define *Bridge*(I) as the union of (*bridge:D*) and (*bridge:R*), as R varies over the relation symbols of $\mathscr{L}$. Note that Bridge(I) is a theory in the combined signature $\mathscr{L} \cup \mathscr{L}^*$, since in (*bridge:R*) the predicate R is from $\mathscr{L}$, while R^I is an $\mathscr{L}^*$-formula delivered by the interpretation. Using this notation, we can formulate:

The Bridged Truth-Transfer Thesis. *If I is an interpretation of T in T^*, and both T^* and Bridge(I) are true, then T is true.*

This Bridged Thesis does well in dealing with Orey-style counterexamples to the original Truth-Transfer Thesis. To see this, we note a preliminary result:

Proposition 5.12: *Suppose that I and J are both interpretations of the signature $\mathscr{L}$ in the $\mathscr{L}^*$-theory T^*. Then, for every $\mathscr{L}$-formula $\varphi(\bar{x})$:*

$$\textit{Bridge}(I) \cup \textit{Bridge}(J) \vdash (\forall \bar{x} : D) \left(\varphi^I(\bar{x}) \leftrightarrow \varphi^J(\bar{x})\right)$$

[41] We are using this in the same sense as used in the philosophy of science; see e.g. Nagel (1961). For standard objections to, and defence of, this notion of reduction, see Dizadji-Bahmani et al. (2010).

Proof. The proof is by induction on the complexity of formulas. For the base case of atomic $\mathscr{L}$-formula $R(\bar{x})$, assume that $D(x_k)$ holds for each $k \leq n$. Then:

$$(R(\bar{x}))^I \leftrightarrow R^I(\bar{x}) \leftrightarrow R(\bar{x}) \leftrightarrow R^J(\bar{x}) \leftrightarrow (R(\bar{x}))^J$$

The first biconditional follows from the definition of the interpretation I on atomics; the second follows from (*bridge:R*) applied to I; the third follows from (*bridge:R*) applied to J; and the fourth follows from the definition of the interpretation J on atomics.

For the inductive steps, consider e.g. the case where φ is the formula $\varphi_0 \wedge \varphi_1$, and suppose that we are given some $\bar{x}$ such that $D(x_k)$ holds for each $k \leq n$. Then:

$$\varphi(\bar{x})^I \leftrightarrow \left(\varphi_0^I(\bar{x}) \wedge \varphi_1^I(\bar{x})\right) \leftrightarrow \left(\varphi_0^J(\bar{x}) \wedge \varphi_1^J(\bar{x})\right) \leftrightarrow \varphi^J(\bar{x})$$

The first and last biconditionals follow from how interpretations are defined on conjunctions, while the middle biconditional follows from the inductive hypothesis. The other inductive steps are similar. □

So now let φ be an Orey sentence for T. Let I be an interpretation of the $\mathscr{L}$-theory $T+\varphi$ in T^*, and J be an interpretation of the $\mathscr{L}$-theory $T+\neg\varphi$ in T^*. Let $T^\cup$ be the theory $T^* \cup \text{Bridge}(I) \cup \text{Bridge}(J)$. Then $T^\cup \vdash \varphi^I \leftrightarrow \varphi^J$ by Proposition 5.12. And $T^\cup \vdash \varphi^I$, since T^* interprets $T + \varphi$ via I. Similarly, $T^\cup \vdash (\neg\varphi)^J$ i.e. $T^\cup \vdash \neg(\varphi^J)$, since T^* interprets $T + \neg\varphi$ via J. So $T^\cup$ proves all three of $\neg(\varphi^J)$, φ^I, and $\varphi^I \leftrightarrow \varphi^J$, and so is inconsistent. It follows that $T^\cup$ is *false*. So Orey sentences pose no threat to the Bridged Thesis: what such sentences show is that at least one of the two sets of bridge principles must be false.

To see how the Bridged Thesis deals with the Feferman-style counterexamples, let I be an intepretation of PA + $\neg$Con(PA) in PA. Then I lets us consider, within PA itself, a definable model of PA + $\neg$Con(PA). If we bought into the idea that our reasons for accepting PA were reasons for *rejecting* $\neg$Con(PA), then we will be able to recognise this (code of a) model as *non-standard*. And, as such, we have good reasons to reject Bridge(I); for Bridge(I) will tell us, in effect, that the non-standard model's order relation, $<^I$, aligns perfectly with *the* order relation, which is not something we can believe whilst also regarding the non-standard model *as* non-standard.

Difficulties concerning bridge principles

There is clearly much to say in praise of the Bridged Thesis. Unfortunately, even if it is *correct*, it is hard to see how it can ever be *applied*. The crucial issue concerns how we could ever be in a position to regard the bridge principles as true.

To see the problem, suppose we consider an interpretation of arithmetic in set theory. In that case, the bridge principles will consist in *identifications* of numbers with specific types of sets. For example, one bridge principle might entail that $2 = \{\emptyset, \{\emptyset\}\}$. But if we are thinking about set theory as providing a metatheory for arithmetic, then the Benacerraf-style considerations from §2.2 kick in: another equally acceptable bridge principle might entail that $2 = \{\{\emptyset\}\}$. It is not unusual to draw the conclusion that the truth-value of any specific set of bridge principles (in this case) is indeterminate. But then the Bridged Thesis cannot be applied.

The Benacerraf-style point arises for neo-logicists just as it does for set-theoretic reductionists. After all, HP pins down the truth condition of all sentences of the form $\#F = \#G$, but it says *nothing* about sentences of the form $2 = \#F$. And different bridge principles will give different verdicts here.

In response to this, a logicist might suggest that a particular set of bridge principles is *true by stipulation*. They might, for example, conceive of those chosen bridge principles as simple stipulations concerning how to define numerals in terms of '#'. But this move has certain costs. Logicists typically have quite ambitious goals: they may want to show how arithmetical reasoning is applicable to every domain of enquiry, and to show how knowledge of Hume's (quasi-logical) principles can 'settle the status of the arithmetical laws we already have, involving those arithmetical concepts we already grasp'.[42] Regarding PA_2 as true-by-stipulation seems incompatible with those goals.

Indeed, since many (if any) mutually inconsistent stipulations are possible, it seems that stipulated-to-be-true bridge principles can only provide us with a proof that PA_2 is *consistent* relative to HP. But that seems to achieve very little. Doubts about the consistency of PA_2 are not widely held. When they are voiced at all, they tend to arise from specific sceptical concerns such as whether there are actually infinitely many objects. But of course HP *also* entails the existence of infinitely many objects. So no one who was concerned about PA_2's consistency for this reason will be mollified by the observation that HP interprets PA_2.

5.7 Interpretability and arithmetical equivalence

We now consider a second, rather different, philosophical use of interpretability, which arises specifically within the philosophy of set theory. It is no surprise that interpretability plays a significant role within set theory. After all, we can readily use interpretability to express the way in which disparate mathematical theories are formalisable within set theory. But the particular case we wish to consider concerns the interpretability of set-theoretic axioms within set theory itself.

[42] Dummett (1996: 20). See Walsh (2014: §1) for references to these aims of logicism.

Set theorists have considered many different large cardinal axioms, determinacy axioms, and forcing axioms. Many of these axioms form a linear order under *provability*, but there are some recent examples which are incompatible with one another. It turns out, however, that even these are linear under *interpretability*, and that within each class of mutually interpretable theories there is a large cardinal axiom.[43] Writing of this hierarchy, Steel says: 'we know of only one road upwards, and large cardinals are its central markers.'[44]

Our question is simply this: *What attitude should we take towards mutually interpretable but incompatible set theories?*

One answer to this question is suggested by Koellner's use of the Guaspari–Lindström Theorem.[45] This entails that two extensions of ZFC by finitely many new axioms in the same signature as ZFC are mutually interpretable iff they prove exactly the same Π^0_1-sentences. Here, a Π^0_1-sentence is simply a sentence which begins with a block of universal quantifiers over natural numbers and all of whose other quantifiers are bounded. Invoking this theorem, Koellner then suggests a two-step strategy for choosing new axioms in set theory:

(a) 'for a given degree of [mutual] interpretability [...show] that the Π^0_1-consequences of the theories in the degree are true'; then
(b) choose a particular axiom '*from the degree*', on the basis of 'theoretical reasons' such as unity and simplicity.[46]

In this context, though, the Guaspari–Lindström Theorem is a double-edged sword. For, within the classes of mutual interpretability, there are theories which *differ* as regards the arithmetical sentences they prove. And we might well wonder why, in step (a), we select from rival axioms which *only* prove all the same Π^0_1-sentences, rather than from rival axioms which prove *all* the same arithmetical sentences.[47] The stricter alternative would correspond naturally to the idea that, whilst strong extensions of set theory may disagree with one another about the nature of sets, the theories that merit our consideration must agree with everything we know about our close friends, the natural numbers.[48]

---

[43] See theories T_2 and T_4 on Koellner (2009: 100), and remark (7) of Koellner (2009: 102).

[44] Feferman, H. M. Friedman, et al. (2000: 427).

[45] See Lindström (2003: 103, 115) and Koellner (2009: 98). The Guaspari–Lindström theorem says that for two such theories, T is interpretable in T^* iff every Π^0_1-sentence provable from T^* is also provable from T. This does not hold for all theories, but rather for essentially reflexive theories which additionally satisfy the usual conditions for Gödel's first incompleteness theorem. See footnote 35 for the definition of essentially reflexive.

[46] Koellner (2009: 99). See in particular the discussion in the numbered items (1)–(8) on Koellner (2009: 101–2) for the types of theoretical reasons which particularly interest Koellner.

[47] Koellner (2009: 98) suggests that the Π^0_1-sentences are 'the analogues of observational generalizations', in that they 'can be definitely refuted but never definitely verified'.

[48] Indeed, Koellner (2009: 98) suggests that 'for *any* arithmetical sentence φ, the choice between PA+φ and PA + $\neg\varphi$ is not one of mere expedience.' This raises the question of why we stop at claims about the natural numbers, and do not continue into claims about the reals. The operational reason for this exclusion is that strong extensions of set theory disagree precisely about the second-order structure of the reals.

There may, though, be a way around this. As Steel emphasises, the mutually interpretable extant extensions of set theory do, in fact, agree on *all* arithmetical consequences.[49] So we could, perhaps, continue to follow Koellner's project for invoking mutual interpretability, if we replace his invocation of Guaspari–Lindström in step (a) with a less formal, but initially plausible claim, that for all the natural cases in (current) set theory, two extensions of the usual set-theoretic axioms are mutually interpretable iff they agree on all the same arithmetical sentences.

However, this suggestion involves the use of the vague term 'natural', and there might be different understandings of what counts as 'natural' in this context. To show how this might cause problems, we will introduce the idea of a 'one-and-a-half-order' theory.

Where T is a first-order theory, let $T_{1.5}$ be the second-order theory which replaces T's schemas by axioms, but which only includes the *Predicative* Comprehension Schema of §1.11. So, for example, $\mathrm{PA}_{1.5}$ has the Induction Axiom from PA_2, but its comprehension principle only includes comprehension for first-order formulas (this theory is usually called ACA_0).[50] Equally, $\mathrm{ZFC}_{1.5}$ has the Separation and Replacement Axioms from ZFC_2, but restricted comprehension (this theory is normally called NBG).[51] It is easy to see that $T_{1.5}$ is a conservative extension of T for first-order formulas.[52] Hence, where we take T to be PA or ZFC itself, T and $T_{1.5}$ prove all the same arithmetical sentences.

Importantly, though, T and $T_{1.5}$ are not always mutually interpretable. This can arise because, in paradigmatic cases, $T_{1.5}$ is finitely axiomatisable, whereas T is not finitely axiomatisable (this happens for both PA and ZFC).[53] Now, this is not a violation of the Guaspari–Lindström Theorem, because $T_{1.5}$ is not an extension of its first-order counterpart T in the same signature as T.[54] Nonetheless, one-and-a-half-order theories provide 'natural' cases where *sameness of arithmetical consequences* diverges from mutual interpretability.

This suggests that Koellner's project for selecting axioms of set theory should not be understood in terms of mutual interpretability, but instead as follows:

(a′) for a given class of (strong) extensions of set theory, which have all the same

---

[49] Feferman, H. M. Friedman, et al. (2000: 427). This is because these interpretations are usually provided by inner models, which agree on all Σ^1_2-sentences by the Shoenfield Absoluteness Theorem (see Jech 2003: 490).

[50] See Simpson (2009: I.3 pp.6ff) for a formal definition of ACA_0. The origins of ACA_0 go back to Weyl; see Dean and Walsh (2017) for historical details.

[51] For references to original work of von Neumann, Gödel, and Bernays, see Mendelson (1997: ch.4).

[52] This is because any $\mathcal{M}$ of T can be expanded into a model $T_{1.5}$ by taking as the second-order objects all the subsets of $\mathcal{M}$ which are first-order definable over $\mathcal{M}$. In the terminology of Chapter 3, what we are appealing to here is that expansion-conservation implies consequence-conservation (see Definition 3.4 and Definition 3.3).

[53] See Simpson (2009: Lemma VIII.1.5 p.311) for a proof of the finite axiomatisability of ACA_0. See Mendelson (1997: Proposition 4.4) for the key step in the proof of the finite axiomatisability of NBG.

[54] And because $T_{1.5}$ is not essentially reflexive. See footnote 45.

arithmetical consequences as one another, show that they agree with everything we hold true on the natural numbers; and then

(b′) choose a particular axiom from that class, on the basis of 'theoretical reasons' such as unity and simplicity.

We think this would amount only to a minor revision of Koellner's project.

Nevertheless, if we were looking to Koellner's project for a reason to think that the notion of (mutual) *interpretability* tracks something of prior philosophical interest, then there is a serious problem here. To illustrate further, consider Steel's statement that 'what we are trying to maximize here is the *interpretative power* of our set theory.'[55] For theories to which the Guispari–Lindström theorem applies, interpretative power aligns with a natural prior notion of deductive strength: given such theories, T and T^*, the theorem implies that T^* interprets but is not interpretable in T iff the Π^0_1-consequences of T^* are a strict superset of those of T. But, as the case of $T_{1.5}$ and T shows, this does not hold for theories in general: $T_{1.5}$ interprets but is not always interpretable in T, and yet they have exactly the same deductive strength when it comes to their first-order consequences. Why, then, should we follow Steel in seeking to maximise interpretative power?

In some situations, an answer is obvious: if we are sure that T^* is consistent, then we can be sure of the consistency of any theories which T^* interprets. This seems to have been why Gauss (and others) attached such great significance to the interpretation of the complex numbers in the real numbers: they were certain that there was a model of the real numbers—some notion of quantity that was in good standing—and their interpretation then gave them confidence that there was a model of the complex numbers.[56] But in the particular case of set theory, nothing similar can be said. After all, it is not obvious that there is anything like an intended model of the wildly largest large cardinal axioms,[57] and it is surely *these* axioms of which we should be less certain.

5.8 Interpretability and transfer of proof

In §§5.6–5.7 we considered some rather ambitious philosophical programmes. We will end the chapter by considering a more parochial explanation of why mathematicians seek and discover interpretations: because *an interpretation allows us to*

---

[55] Feferman, H. M. Friedman, et al. (2000: 423).

[56] Ewald (1996: v.1 p.310) and Gauss (1863–1929: v.2 p.174). For Gauss's own subsequent reservations about the significance of the geometric interpretation of the complex numbers, see Schlesinger's summary (in Gauss 1863–1929: v.10 pt.2 p.56).

[57] Obviously we should mention the inner model programme, which (roughly stated) attempts to show that variations on Gödel's constructible hierarchy that admit large cardinals can be 'built from below by well-understood operations' (Jensen 1995: 402). Since Steel is an eminent inner model theorist, perhaps the thought behind his 'maximise interpretability strength' comment is that one ought to find inner models 'built from below' which satisfy theories of higher and higher interpretability strength.

transfer a proof from one setting into another. This is a very natural thought. For example, speaking of the mutual interpretability of PA and ZF_{fin} (see §5.5), Just and Weese write that 'one can thus think of the theorems provable in PA as precisely the theorems about hereditarily finite sets that are provable without employing the notion of an infinite set.'[58]

This idea has been especially emphasised for a certain kind of interpretation. Let us define an interpretation I of an $\mathscr{L}$-theory T *in itself* to be a *duality* iff T proves that $(\varphi^I)^I \leftrightarrow \varphi$. Using the terminology of §5.5, this condition guarantees that T faithfully interprets itself, i.e. that $T \vdash \varphi$ iff $T \vdash \varphi^I$.[59] When the duality I is clear from context, we call φ^I the *dual* of φ. Hence, when the interpretation is a duality, the provability of a theorem aligns with the provability of its dual.

The most famous example of a duality is from projective geometry, where the interpretation is given by swapping the words 'point' and 'line', and 'lies on the same line' and 'shares a common point'.[60] Another well-known example comes from category theory, where the interpretation is provided by switching the expression 'the domain of f is a' with 'the codomain of f is a' and vice-versa.[61] Speaking of how duality in category theory allows us to prove only results about limits and automatically get results about colimits, or to prove only results about products and automatically get results about coproducts, Simmons writes: 'by using the opposite category [...] we can make precise this left-right symmetry, and halve the work.'[62]

But in a recent paper, Detlefsen has urged caution as to this understanding of the significance of dualities.[63] He suggests that, whatever might be gained by 'halving the work' via the duality, something equally important might be lost by not developing the dual proof 'directly'; that is, by not building up the relevant concepts, motivating examples, lemmas, and other results which are preliminary to the dual proof. He focusses his discussion around projective duality, but the point seems to generalise to any duality (in our sense). And, in support of Detlefsen's concern, we note that, in category theory, it is still important to teach products and coproducts as separate constructions, and for students to learn how to work with both.

Detlefsen's point concerns dualities; but since dualities are a specific case of faithful interpretations, it is worth considering how to apply his point to faithful interpretations more generally. So: suppose we have a faithful interpretation I of $\mathscr{L}$-theory

---

[58] Just and Weese (1997: 54). But see our remarks about finite set theory and number theory after Proposition 5.9.

[59] The left-to-right direction follows from the definition of an interpretation. For the right-to-left direction, suppose that $T \vdash \varphi^I$; then by the definition of an interpretation, we have $T \vdash (\varphi^I)^I$; and hence $T \vdash \varphi$ by the provable equivalence of $(\varphi^I)^I \leftrightarrow \varphi$.

[60] See e.g. Veblen and Young (1965: ch.1) or Hartshorne (1967: ch.4).

[61] See e.g. Mac Lane (1998: §II.1) or Awodey (2010: §3.1).

[62] See Simmons (2011: 71). For sentiments to this effect about duality in projective geometry, see the many quotations and references in the introductory sections to Detlefsen (2014).

[63] Detlefsen (2014).

T in $\mathscr{L}$-theory T^*. Then given any $\mathscr{L}$-sentence φ, we have two ways to establish the T-derivability of φ:

(i) we can try to develop a T-proof of φ; or
(ii) we can try to develop a T^*-proof of φ^I.

The faithful interpretation, I, might be thought valuable precisely because it makes option (ii) available. But now the Detlefsen-like worry arises: leaning on I comes at the cost of not learning how to develop a T-proof of φ.

Here is a quick way to make the worry vivid. Given our interpretation I and some T^*-proof of φ^I, we know *that there is* a T-proof of φ. But we do not necessarily *have* such a proof. So even if we get two *theorems* for the price of one, we do not get two *proofs*.

In the general setting of faithful interpretations, we can bolster this point with a technical observation. If T^* interprets T via I and π is a T-proof of φ, then I provides an effective procedure for delivering a T^*-proof of φ^I: just apply I to the T-axioms in π, and then appeal to the fact that interpretations act compositionally on logical connectives and hence on logical inferences used in π. However, a *faithful interpretation* offers no such guarantee in the opposite direction. For, to say that T^* *faithfully* interprets T, is just to add the bare existential claim that if there is a T^*-proof π^* of φ^I then there is a T-proof π of φ. That bare existential provides no guarantee that there is any relation between the complexity of π^* and π, nor that there is any effective procedure for obtaining π from π^* to π.

Now, there are ways to blunt this objection. Most known cases of faithful interpretations result from combining interpretations in various ways, and so provide a guarantee in both directions. For instance, in the case of dualities, if π^* is a T-proof of φ^I, then applying I to π^* yields a T-proof of $(\varphi^I)^I$, and by combining this with a T-proof of $(\varphi^I)^I \leftrightarrow \varphi$ we get our T-proof π of φ. In the case of biinterpretations, if $\pi*$ is a T^*-proof of φ^I, applying J to π^* yields a T-proof of $(\varphi^I)^J$, and by combining this with a T-proof π of $(\varphi^I)^J \leftrightarrow \varphi$ we get a T-proof of φ.[64]

Moreover, in most known cases, these procedures for transforming π^* into π are not just *effective*, but tractable. Visser and Verbrugge introduce the following species of interpretation: an interpretation I of T into T^* is *feasible* if there is a polynomial p such that for all the axioms φ of T there is a proof π^* of φ^I from T^* with $|\pi^*| \leq p(|\varphi|)$. (Here, $|\cdot|$ is the length of the formula, where we measure length of proof by summing the lengths of the lines of the proof.)[65] It will follow from this that for all proofs π of φ from T there is a proof π^* of φ^I from T^* with $|\pi^*| \leq p(|\pi|)$. Further,

[64] Recall that in Proposition 5.9 we showed that biinterpretations are faithful. The proof of this Proposition shows that we have a T-proof of the equivalence $(\varphi^I)^J \leftrightarrow \varphi$, because isomorphism implies elementary equivalence.

[65] See Verbrugge (1993: 389). Note, though, that claims about length of proof are heavily dependent on the proof-system being employed, and that Verbrugge works with a Hilbert-style proof system.

Verbrugge notes that if the interpreted theory is finite, then the interpretation is automatically feasible,[66] and that:

> All in all it seems that the only examples of theories [which are interpretable but not feasibly so] are contrived theories obtained by fixed-point constructions [...] It would be nice to find a more natural counterexample.[67]

So, in 'natural' cases, it may be possible to insist that the value of a faithful interpretation I is as follows: given a T^*-proof of φ^I, we can obtain a T-proof of φ.

Nevertheless, three barriers to the utility of faithful interpretations remain. First: whilst one *can* obtain a T-proof of φ, we may well not *bother*. And if we do not bother, then we are still relying on the bare existential, doubling the number of theorems we have but not the number of proofs and so, perhaps, not our level of understanding. Second: actually obtaining the T-proof of φ will not be as simple as running some (feasible) algorithm on the T^*-proof φ^I. After all, mathematicians rarely offer fully formalised proofs—they provide informal, discursive proofs which omit tedious or routine steps—so that there can be serious work to do in obtaining a machine-readable T^*-proof (and turning the formal T-proof into something human-readable). But third, and most interesting: if π is obtained from π^* just by an effective translation, we might wonder whether π and π^* are genuinely different proofs.

In the specific case of projective duality, indeed, Detlefsen argues that they will *not* be genuinely different.[68] Recall that projective duality is obtained just by by swapping the words 'point' and 'line', and 'lies on the same line' and 'shares a common point'. Consequently, π is obtained from π^* just by trivially permuting the non-logical primitives. As such, Detlefsen denies that π and π^* can count as different proofs, in anything other than a purely syntactic sense. We entirely agree with Detlefsen in the case of projective duality, but we would urge caution before generalising this point to *any* faithful interpretation. After all, in the more general setting, there is no guarantee that π and π^* will differ in such a trivial way. Nonetheless, and crucially, there is no *guarantee* that a faithful interpretation always yields two *interestingly* different proofs for the price of one.

5.9 Conclusion

We have canvassed some of the most prominent notions of sameness of structure and sameness of theory that populate mathematical logic. One clear message is that there is a dizzying array of possibilities here. Even after the choice of logic has been

[66] Verbrugge (1993: 389).
[67] Verbrugge (1993: 401).
[68] Detlefsen (2014: §6).

made, and even after one has settled upon looking at structures as opposed to theories, and even after one has fixed whether equivalence relations are allowed to go proxy for identity and whether objects can be interpreted as n-tuples of objects and whether parameters are allowed, some fundamental decisions remain. To use an image: interpretability merely requires that my interpreting perspective can mimic certain aspects of your interpreted perspective; mutual interpretability requires that each of us can mimic the other; and stronger notions, like biinterpretability and faithful interpretability, also require (in different ways) that these copies be accurate and indeed verifiability so.

Put this way, it may well seem that these notions of interpretation *must* track something (or several things) of chief philosophical importance. But one of the main messages of this chapter is that it is remarkably difficult to defend anything beyond the most parochial versions of this idea.

The parochial point is just that interpretations of theories result in proofs from one area being translated into proofs in another area. There are obvious reasons why this can be useful, and especially so if the two areas initially seemed only to be distantly related. (It *was* important and useful to discover that geometry and algebra are closely related, and similarly for number theory and finite set theory.)

However, in the last three sections of this chapter we surveyed some more ambitious reasons for caring about interpretations. There, we found much less cause for optimism. Interpretability has no clear role to play in truth-transfer; no clear role to play in safeguarding arithmetical consequences; and it is not guaranteed to give you 'two proofs for the price of one', in anything but the most parochial way.

5.A Arithmetisation of syntax and incompleteness

In this chapter, we have mentioned arithmetisation of syntax, and Gödel's theorems. We sketch the main idea behind arithmetisation in this appendix, and state versions of Gödel's two incompleteness theorems.

Given a fixed stock of symbols, we can easily define a primitive recursive function which encodes every (finite) string of such symbols with a unique natural number. Here is a very simple approach. First, we enumerate the symbols in our fixed stock, $s_1, s_2, s_3, \ldots$. Then, when we encounter the string:

$$\sigma = s_{i_1} s_{i_2} s_{i_3} \ldots s_{i_n}$$

we encode σ with a number $\ulcorner\sigma\urcorner$, defined thus:

$$\ulcorner\sigma\urcorner = \pi_1^{i_1} \times \pi_2^{i_2} \times \pi_3^{i_3} \times \ldots \times \pi_n^{i_n}$$

where π_m is the mth prime number. By the Fundamental Theorem of Arithmetic—that every number has a unique prime factorisation—this coding function from strings to natural numbers is an injection. So every string of symbols from our fixed finite stock can be treated as a natural number without loss.

Coding of deductions proceeds similarly. We can think of a deduction simply as a sequence of sentences $\sigma_1, \ldots, \sigma_n, \sigma_{n+1}$, with σ_{n+1} the conclusion, and then simply encode that deduction with the number:

$$\pi_1^{\ulcorner \sigma_1 \urcorner} \times \pi_2^{\ulcorner \sigma_2 \urcorner} \times \ldots \times \pi_n^{\ulcorner \sigma_n \urcorner} \times \pi_{n+1}^{\ulcorner \sigma_{n+1} \urcorner}$$

Now, the coding function we just outlined uses exponentiation, which is not a primitive of PA. This poses no real obstacle, since exponentiation can be defined in PA. However, technical-cum-philosophical subtleties can arise if we want to carry out coding in theories which are *weaker* than PA. For example: Q is strong enough to capture any recursive function, in the sense that, for any recursive function $f : \omega \longrightarrow \omega$, there is a formula $\varphi(x, y)$ such that:[69]

$$f(n) = m \text{ iff } \mathrm{Q} \vdash \varphi(S^n(0), S^m(0)), \text{ and}$$
$$f(n) \neq m \text{ iff } \mathrm{Q} \vdash \neg\varphi(S^n(0), S^m(0))$$

Since our coding function is itself recursive, Q also captures our coding function with some formula $\varphi(x, y)$. However, Q is so weak that it can prove almost nothing *about* this formula. Consequently, it is potentially misleading to say that Q is sufficiently strong to *express* syntactic notions; rather, in the spirit of Bezboruah and Shepherdson, we might say that Q defines an *algebraic* notion, which only expresses syntactic notions when realised in stronger theories (or in certain models).[70] One such stronger theory would of course be PA, but theories between Q and PA might suffice. For example, the theory $I\Sigma_0 + exp$ consists of: the axioms of Q, a primitive exponentiation function symbol along with the basic recursive definition of exponentiation in terms of multiplication, and induction for Σ_0-formulas.[71] This (very weak) theory might be sufficiently strong that it can reasonably be said to *express* syntactic notions.

In any case, we can now state the two incompleteness theorems. We start with a version of the first incompleteness theorem:[72]

Theorem 5.13: *No consistent, computably enumerable theory which interprets Robinson's Q is arithmetically complete.*

[69] See e.g. Hájek and Pudlák (1998: 155ff) and Rautenberg (2010: 210ff).

[70] See Bezboruah and Shepherdson (1976: 504).

[71] I.e. formulas containing no unbounded quantifiers; see Hájek and Pudlák (1998: 37ff).

[72] For Gödel's original paper, see Gödel (1986: 144ff); this version employs strengthenings due to Rosser (1936) and Craig (1953). Modern presentations of the theorem can be found in any number of places, including Enderton (2001: 236) and Rautenberg (2010: 252).

Here, we say that T is *computably enumerable* iff there is an algorithm which sequentially enumerates all of T's members. So, when T is a computably enumerable theory, the algorithm outputs a list of the sentences in T, and every sentence in T will eventually be listed at some point. We say that T is *arithmetically incomplete* iff there is some φ in Q's signature, such that $\varphi^I \notin T$ and $\neg\varphi^I \notin T$ (where I is T's interpretation of Q).

Here is a version of Gödel's Second Incompleteness Theorem:[73]

Theorem 5.14: *No consistent, computably enumerable theory T which interprets* Q *proves* $\mathrm{Con}(T)$.

Here, $\mathrm{Con}(T)$ is T's interpretation of Q's arithmetisation of the claim that no contradiction can be deduced from T's axioms, i.e. something of the form $\forall x \neg \mathrm{Prf}_T(x, \ulcorner\bot\urcorner)$.

5.B Definitional equivalence in second-order logic

We now show that definitional equivalence in second-order logic is more coarse-grained than in first-order logic, as mentioned in §5.2.

Proposition 5.15: *Let $\mathcal{N}$ be the natural numbers in the signature consisting just of zero and successor. Let $\mathcal{Z}$ be the integers in the same signature. An isomorphic copy of $\mathcal{Z}$ is second-order definitionally equivalent to $\mathcal{N}$.*

Proof. As discussed after Definition 1.10 in §1.A, we can define $<$, $+$, and $\times$ in $\mathcal{N}$ using second-order resources. The same holds for $\mathcal{Z}$. So we can employ these notions in both models and in their isomorphic copies. Since the result only concerns an isomorphic copy of $\mathcal{Z}$, we can assume (by a Push-Through Construction) that the domain and operations on $\mathcal{Z}$ are computable subsets of the natural numbers.

Let $\mathcal{M}$ be an isomorphic copy of $\mathcal{Z}$ with with the same underlying domain as $\mathcal{N}$ and with an isomorphism $h : \mathcal{M} \longrightarrow \mathcal{Z}$ given by $h(2k) = k$ and $h(2k+1) = -(k+1)$. That is, $\mathcal{M}$ is the result of pushing through h^{-1} from $\mathcal{Z}$ onto the natural numbers.

Now all $\mathcal{M}$'s structure is trivially definable in $\mathcal{N}$, since all $\mathcal{M}$'s structure is the result of pushing through a computable map. So it suffices to show that $\mathcal{N}$'s structure is definable in $\mathcal{M}$. Since these two models have the same interpretation of 0, we only need to show how to define $\mathcal{N}$'s successor in $\mathcal{M}$. For this we note that the following claims are equivalent:

[73] For a proof and a more precise statement, see e.g. Hájek and Pudlák (1998: Theorem 2.21 p.164).

(1) $\mathcal{N} \models n = m + 1$

(2) $\mathcal{Z} \models (h(n) + h(m) = -1 \wedge h(n) < h(m)) \vee (h(n) + h(m) = 0 \wedge h(m) < 0 < h(n))$

(3) $\mathcal{M} \models (n + m = -1 \wedge n < m) \vee (n + m = 0 \wedge m < 0 < n)$

We check each claim in turn.

(2) $\Leftrightarrow$ *(3).* Trivial, because h is an isomorphism

(1) $\Rightarrow$ *(2).* Suppose that (1) holds. If m is even in $\mathcal{N}$ then $m = 2k$ and $n = 2k+1$ for some natural number k, and then by definition of h we have that $h(m) = h(2k) = k$ and $h(n) = h(2k+1) = -(k+1)$, so $\mathcal{Z} \models h(n) + h(m) = -1 \wedge h(n) < h(m)$. If m is odd in $\mathcal{N}$ then $m = 2k+1$ and $n = 2k+2$ for some natural number k, and similarly we have that $h(m) = h(2k+1) = -(k+1) < 0$ and $h(n) = h(2k+2) = k+1 > 0$. Thus $\mathcal{Z} \models h(n) + h(m) = 0 \wedge h(m) < 0 < h(n)$ again.

(2) $\Rightarrow$ *(1).* Suppose first that $\mathcal{Z} \models h(n) + h(m) = -1 \wedge h(n) < h(m)$. If m were odd, say $m = 2k+1$ for some natural number k, then $h(m) = h(2k+1) = -(k+1)$, and so $h(n) = k$, contradicting that $h(n) < h(m)$. Hence m must rather be even, say $m = 2k$ for some natural number k. Then $h(m) = h(2k) = k$ and so $h(n) = -(k+1)$ and $n = 2k+1$ and so indeed $\mathcal{N} \models n = m + 1$. Supposing alternatively that $\mathcal{Z} \models h(n) + h(m) = 0 \wedge h(m) < 0 < h(n)$, one can conclude by a similar argument that $\mathcal{N} \models n = m + 1$. □

Proposition 5.16: *Let $\mathcal{Z}$ and $\mathcal{N}$ be as in Proposition 5.15. No isomorphic copy of $\mathcal{Z}$ is first-order definitionally equivalent to $\mathcal{N}$.*

Proof. For reductio, suppose there is an isomorphic copy $\mathcal{M}$ of $\mathcal{Z}$ such that $\mathcal{M}$ is first-order definitionally equivalent to $\mathcal{N}$. Then the successor function of $\mathcal{M}$, which we write as $S^{\mathcal{M}}$, is $\mathcal{N}$-definable. So we have an $\mathcal{N}$-definable function $\sigma : N \longrightarrow N$ such that $\sigma(n) = m$ iff $\mathcal{M} \models S(n) = m$.

Since $S^{\mathcal{M}}$ is a bijection on $\mathcal{M}$, clearly σ is a bijection. So σ^{-1} is an $\mathcal{N}$-definable bijection. For readability, we write π in place of σ^{-1}, noting that $\pi(n) = m$ iff $\mathcal{M} \models n = S(m)$.

Now for each natural number k, consider the following formula, where we employ the numeral-notation from equation (*numerals*) of §1.13:

$$\varphi_k(x, y) := \bigvee_{0 \leq i < k} \left(S^i(x) = y \vee y = S^i(x)\right)$$

On $\mathcal{N}$, this defines the notion of $|x - y| \leq k$, where $|\cdot|$ denotes the ordinary absolute value function on the natural number.

We now argue that there is a natural number K such that $|\sigma(n) - n| \leq K$ for all natural numbers n. For reductio, suppose otherwise, i.e. that for all K there is a natural number n such that $|\sigma(n) - n| > K$. Where c is some new constant, consider the theory T which consists of all the sentences true on $\mathcal{N}$ in the signature $\{0, S\}$, together with all the axioms of the form $\neg\varphi_K(\sigma(c), c)$ as K ranges over the natural numbers. (In this, we are viewing σ as an abbreviation for the formula defining it in $\mathcal{N}$.) By the Compactness Theorem 4.1, T has a model. Using the notion of a 'chain' from Proposition 3.10, the interpretation of c must be on a different chain then the interpretation of $\sigma(c)$. And this is a contradiction. For σ cannot now be defined by a quantifier-free formula; but elementary results concerning $\mathcal{N}$ show that any $\mathcal{N}$-definable set *can* be defined by a quantifier-free formula.[74]

Exactly similarly, there is a natural number K such that $|n - \pi(n)| \leq K$ for all natural numbers n. By taking the maximums of the two constants, there is a natural number K such that for all n both $|\sigma(n) - n| \leq K$ and $|n - \pi(n)| \leq K$. Using this bound K, two notions become $\mathcal{N}$-definable (here, $<$ just stands for the ordinary ordering on the natural numbers, as given in the metatheory):

$$\sigma(n) > n \quad \text{iff} \quad K + n \geq \sigma(n) > n \quad \text{iff} \quad \mathcal{N} \vDash \bigvee_{i=1}^{K} \sigma(n) = S^i(n)$$

$$n > \pi(n) \quad \text{iff} \quad K + \pi(n) \geq n > \pi(n) \quad \text{iff} \quad \mathcal{N} \vDash \bigvee_{i=1}^{K} n = S^i(\pi(n))$$

As such, these two sets are $\mathcal{N}$-definable:

$$X = \{n \in N : \sigma(n) > n\} \qquad Y = \{n \in N : n > \pi(n)\}$$

Note that σ is an injection from X to Y. For, suppose that n is in X, i.e. $\sigma(n) > n$; then $\sigma(n) > \pi(\sigma(n))$, so that $\sigma(n)$ is in Y. Similarly, π is an injection from Y to X. Since the two functions are inverses of one another, σ is a bijection from X to Y, and π is a bijection from Y to X.

We now appeal to a second elementary fact about $\mathcal{N}$: any $\mathcal{N}$-definable subset of N is either finite or *cofinite*, where 'cofinite' means 'has finite complement'.[75]

Suppose first that X is cofinite. Then Y is also cofinite, since it is the image of an injection σ, and so $X \cap Y$ is infinite and hence cofinite. Choose the least n_0 greater than every element in $\omega \setminus (X \cap Y)$; then $\sigma(n) > n > \pi(n)$ for all $n \geq n_0$. Now, define the following π-version of the numerals by recursion in the metatheory:

$$\pi^0(x) := x \qquad \pi^{n+1}(x) := \pi(\pi^n(x))$$

Note that if $\pi(x) = x$ for some x then by definition of π we would have $\mathcal{M} \vDash x = S(x)$, a contradiction. From this and the fact that π is an injection, the map $n \mapsto \pi^n(x)$ is an injection when x is fixed. Choose n_1 such that $\pi^n(0) \geq n_0$ for all $n \geq n_1$. Then $\sigma(\pi^n(0)) > \pi^n(0) > \pi(\pi^n(0))$, i.e. $\pi^{n-1}(0) > \pi^n(0) > \pi^{n+1}(0)$, for all $n \geq n_1$. So we have an infinite decreasing sequence of natural numbers, a contradiction.

[74] See Marker (2002: Corollary 3.1.6 p.75).

[75] See Marker (2002: 104).

Suppose, instead, that X is finite. Then Y is also finite, and so $X \cup Y$ is finite. Choose the least n_0 greater than every element in $X \cup Y$; then $\sigma(n) \leq n \leq \pi(n)$ for all $n \geq n_0$. Now, define the following σ-version of the numerals by recursion in the metatheory:

$$\sigma^0(x) := x \qquad \sigma^{n+1}(x) := \sigma(\sigma^n(x))$$

As above, we can choose n_1 such that $\sigma(\sigma^n(0)) \leq \sigma^n(0) \leq \pi(\sigma^n(0))$, i.e. $\sigma^{n+1}(0) \leq \sigma^n(0) \leq \sigma^{n-1}(0)$ for all $n \geq n_1$. Since the map $n \mapsto \sigma^n(0)$ is an injection, the inequalities are strict. So again we have an infinite decreasing sequence of natural numbers, a contradiction. $\square$

B

Categoricity

Introduction to Part B

A theory is said to be *categorical* iff all of its models are isomorphic. Part B focuses on the technical and philosophical issues around categoricity.

We start by outlining a certain attitude to model theory, which we call *modelism*. The modelist idea is that structure-talk, as used informally by mathematicians, is to be understood in terms of isomorphism, in the model theorist's sense. For example, modelists will want to explicate talk of 'the natural numbers' in terms a particular isomorphism type. As such, modelists face an important doxological question: *How can we pick out particular isomorphism types?* This question will occupy us for much of Part B; so we spend some time in Chapter 6 explaining the nature of (various versions of) this question, and in particular what it means to say that it is a *doxological* question. Chapter 6 therefore introduces not just one but two neologisms: *modelism* and *doxology*. We apologise for the nomenclature. But the ideas are important enough to name, and we will use these names throughout Part B.

In Chapter 7, we focus on modelists who want to pin down the isomorphism type of the natural numbers. That aim immediately runs into two technical barriers: the Compactness Theorem 4.1, which we encountered back in §4.1, and the Löwenheim–Skolem Theorem 7.2, which we prove in §7.A. These results show that no first-order theory with an infinite model can be categorical; all such theories have non-standard models. In Chapter 4, non-standard models were our friends, for they allowed us to introduce and reason with infinitesimals. Here, they seem more scary: they threaten to leave modelists unable to explain how (if at all) humans can talk about 'the natural numbers'.

The Compactness and Löwenheim–Skolem theorems encapsulate the expressive limitations of first-order logics. Other logics, such as second-order logic with its full semantics, are not so expressively limited. Indeed, Dedekind's Categoricity Theorem 7.3 tells us that all full models of PA_2 are isomorphic. However, it is a subtle philosophical question, whether one is entitled to invoke the *full* semantics for second-order logic. In particular, we argue that the full semantics is out of reach for any *moderate* modelist (i.e. any modelist who embraces the moderate naturalism which we first discussed in §2.3, which rejects any appeal to 'mathematical intuition' or anything similar). Moreover, by generalising the problems concerning full second-order logic, we show that moderate modelists cannot, by their own lights, pin down 'the natural numbers'.

In Chapter 8, we switch from numbers to sets. Again, no first-order set theory can hope to get anywhere near categoricity, but Zermelo famously proved the *quasi*-categoricity of second-order set theory ZFC_2; i.e. that all full models of ZFC_2 are isomorphic, 'so far as they go'. As in Chapter 7, we face the question of who is *entitled* to invoke full second-order logic, and that question is as subtle as before. However, the *quasi*-categorical nature of Zermelo's Theorem gives rise to some specific questions concerning the *aims* of axiomatic set theories, which we explore. Moreover, given the status of Zermelo's Theorem in the philosophy of set theory, we offer a stand-alone proof of the result in §8.A. In §§8.B–8.C, we also prove a similar quasi-categoricity result for Scott–Potter set theory, a theory which axiomatises the idea of an arbitrary stage of the iterative hierarchy.

The overarching moral of Chapters 7–8, however, is that *moderate* modelists cannot explain how they could hope to pin down any particular isomorphism type, and so cannot deliver on their goal of explicating structure-talk in terms of isomorphism types. This observation can lead to a kind of *model-theoretical scepticism*; that is, a moderate modelist might think that model theory has shown to us that we simply *cannot* pick out the 'the natural numbers'. But in Chapter 9, we present two *transcendental* arguments which show that this line of thought is incoherent.

The simple conclusion is that one cannot be a moderate modelist. But this still leaves us with a choice between abandoning moderation and abandoning modelism. In Chapters 10–12, we speculatively outline a way to save moderation by abandoning modelism. The rough idea is to do *metamathematics without semantics*, by working deductively in higher-order logics. In Chapter 10, we discuss the internal categoricity of arithmetic. In Chapter 11, we discuss internal categoricity for pure set theories. We emphasise the promise of such results, stressing that they may provide a non-semantic way to draw the boundary between algebraic and univocal theories. Finally, in Chapter 12, we explore how internal categoricity might allow us to make certain claims about *mathematical truth*. Along the way, we outline an internalist attitude towards model theory *itself*, and use this to illuminate the cryptic conclusions of Putnam's famous 'Models and Reality'.

Chapters 6–12 are driven by questions about our ability to pin down mathematical entities and to articulate mathematical concepts. Chapter 13 is driven by similar questions about our ability to pin down the semantic frameworks of our languages. It transpires that there are not just non-standard models, but non-standard ways of doing model theory *itself*. In more detail: whilst we normally outline a two-valued semantics which makes sentences True or False in a model, the inference rules for first-order logic (or faithful Henkin second-order logic) are compatible with a *four*-valued semantics; or an $\aleph_0$-valued semantics; or what-have-you. This gives rise to perhaps the 'deepest' level of indeterminacy questions: *How can humans pin down the semantic framework for their languages?* This question is asked much less frequently than the questions we raised in earlier chapters about e.g. the natural numbers or sets. But there is no good reason for this; and the dialectic surrounding these questions is always the same.

Readers who only want to dip into particular topics of Part B can consult the following diagram of dependencies, whilst referring to the table of contents:

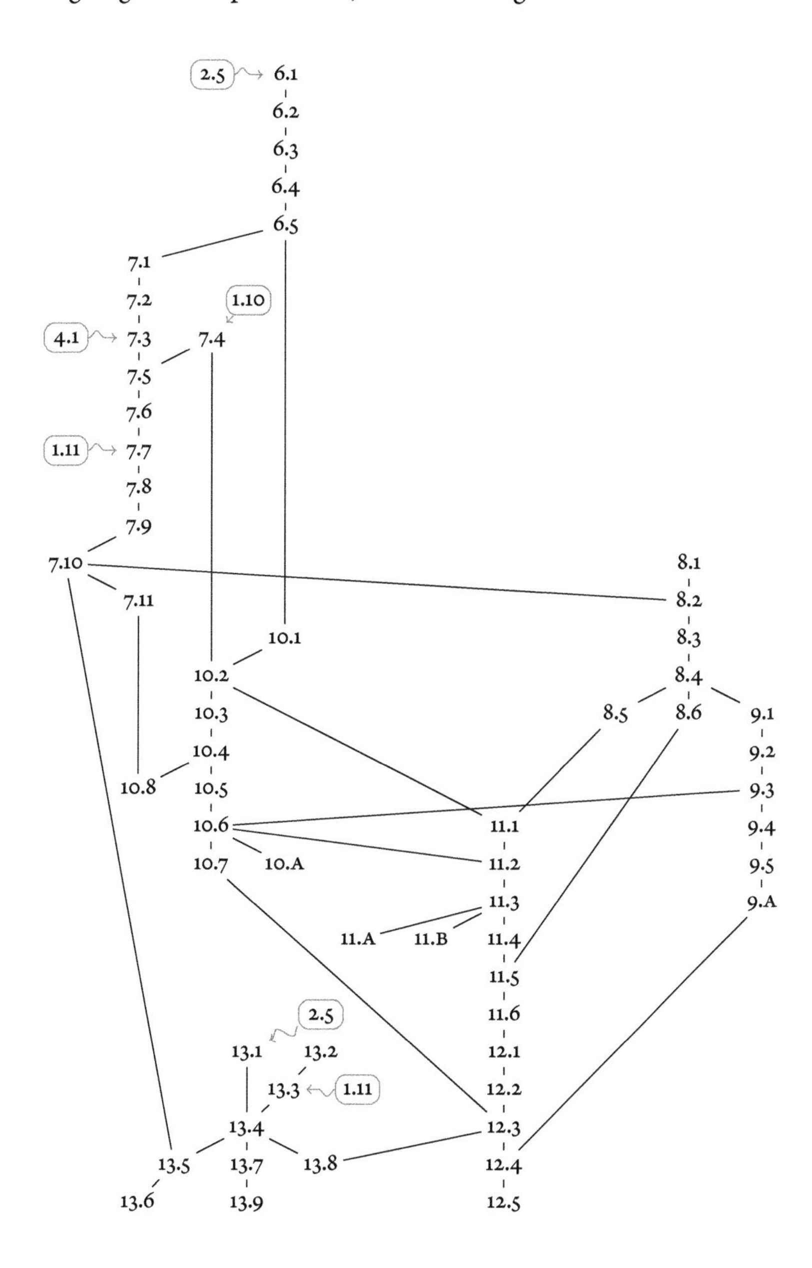

6
Modelism and mathematical doxology

As mentioned in the introduction to this part, *modelism* and mathematical *doxology* form central themes in Part B. So we begin with a brief chapter which explains what modelism is and outlines a doxological challenge facing it.

6.1 Towards modelism

To motivate the move towards modelism, we will review the morals of Chapter 2.

In §2.3, we introduced a philosopher whom we called the *moderate objects-platonist*. Her *objects-platonism* amounts to treating mathematical entities as abstract entities, which are not of our own creation. Her *moderation* amounts to embracing a certain kind of naturalism. So she does not think that mathematical intuition, or quasi-perception, or anything similar, relates her to these entities.

In §§2.2–2.4, we used the Push-Through Construction to argue that moderate objects-platonists must accept that mathematical language is radically referentially indeterminate. We pointed out that, for plenty of mathematical theories—specifically what we called *algebraic* theories in Chapter 2—this is of no concern at all. The problem—if it is one—arises only for *univocal* theories, such as arithmetic, for which one might think that there is such a thing as 'the intended model'.

As noted in §2.5, however, radical indeterminacy of mathematical language may not even be a problem for *non*-algebraic theories. Many philosophers and mathematicians are comfortable with the idea that our arithmetical terms, such as '27', do not pick out any particular entity; they insist only that our *entire theory* of arithmetic picks out a particular 'structure', in some informal sense.

Now, one way to approach this idea is by explicating 'structures', in the informal sense, in terms of $\mathcal{L}$-structures, in the model-theorist's sense. We first considered this in §2.2, suggesting that we might explicate informal talk of 'structures' in terms of isomorphism types (i.e. equivalence classes of isomorphic $\mathcal{L}$-structures). This is typical of an attitude towards model theory which we call *modelism*. The word is our own.[1] We apologise for its ugliness. But we could not find any word in the existing literature with exactly the right connotations. Here is a slightly more fleshed-out presentation of the attitude we call *modelism*.

[1] In fact, the word 'modelism' does have an existing use, covering the practice of building miniature models of aircraft, trains and submarines (for example). We doubt this will lead to confusion.

The modelist manifesto. *Mathematicians frequently engage in structure-talk, and model theory provides precise tools for explicating the notion of 'structure' they invoke. Specifically, mathematical structure should be understood in terms of* ISOMORPHISM, *in the model theorist's sense.*[2]

This is only a rough, initial statement of modelism; as we will soon see, there are many ways to sharpen it. But we should emphasise that, in today's broadly structuralist climate, modelism might seem entirely *innocuous*. After all, modelism can apparently be motivated with a simple, rhetorical question:

What better way could there be to make sense of informal structure-talk, than via model theory, the branch of mathematics whose explicit subject matter is structures?

Perhaps its seemingly innocent nature explains why modelism has never before been named in its own right. Nevertheless, one of the main themes of Part B is that modelism is not innocuous at all.

6.2 Objects-modelism

The main purpose of this chapter, though, is to get clear on various versions of modelism. The easiest version to describe is *objects-modelism*. It holds that mathematicians who speak informally of 'structures' are talking about *bona fide* objects. Really, this just amounts to combining modelism with objects-platonism. So, objects-modelists think that talk about 'the natural number structure' really *does* pick out a particular abstract entity, and that this entity is well explicated via a specific isomorphism type from model theory.

At its most blunt, this explication can be effected by *identifying* informal 'structures' with isomorphism types. According to this blunt version of objects-modelism, then, '*the* natural number structure' *is* just a particular isomorphism type, which is a specific, *bona fide*, abstract object. And on this blunt approach, model theory directly supplies us with the formal tools to explore mathematical ontology.

Shapiro's ante rem structuralism is a slightly less blunt version of objects-modelism. As explained in §2.4, ante rem structuralism holds that mathematics is about *ante-structures*. These are construed as *bona fide* abstract entities, consisting of places with certain intra-structural relations holding between the places. They

[2] In §5.2, we suggested that definitional-equivalence might be better than isomorphism, for certain explicatory purposes. This sort of point could be repeated frequently throughout Part B, but repeatedly mentioning it would be tiresome and would not affect the dialectic. So we simply focus on isomorphism in what follows.

are something like platonic forms, which can be multiply instantiated. So ante-structures are not exactly the model-theorist's $\mathscr{L}$-structures, and nor are they isomorphism types. Rather, as suggested in §5.2, they are something like *abstractions* from isomorphism types. Still, according to ante rem structuralists, model theory still supplies us with the formal tools for investigating our mathematical ontology. Indeed, when Shapiro describes his own theory of ante-structures as 'an axiomatization of the central framework of model theory.'[3] In our terms, then, Shapiro is an exemplar objects-modelist.

6.3 Doxology, objectual version

We have considered two versions of objects-modelism. There may be others. But here is a question which arises for *any* variety of objects-modelism:

The Modelist's Doxological Challenge. *How can we pick out particular isomorphism types?*

We consider some possible *answers* to this Challenge in Chapters 7 and 8. In this chapter, we just want to get clear on the *kind* of question that is being asked. Specifically, we want to emphasise that it is *doxological* rather than epistemological.

Epistemology is an important concern in contemporary philosophy of mathematics. An easy way to see this is to glance at the vast literature surrounding the 'access problem', as first formulated by Benacerraf, and punchily refined by Field:

> Benacerraf's challenge [...] is to provide an account of the mechanisms that explain how our beliefs about these remote entities [i.e. mathematical entities as construed by objects-platonists] can so well reflect the facts about them.[4]

Model theory is unlikely to be of much help in answering such epistemological questions. For example, consider the question of how we can know about the natural numbers, or (in Field's terms) how our beliefs about the natural numbers can so well reflect the facts about them. Model theory is officially implemented within set theory; and any epistemological concerns about numbers are likely to be eclipsed by epistemological concerns about sets.

Still, within the philosophy of mathematics, *doxology* is just as important as epistemology. Or rather, it should be. Unfortunately, doxology is sometimes sidelined, perhaps in part because 'doxology' is not really a word.[5] We have adopted the word from McGee, who introduces it as follows:

[3] Shapiro (1997: 93); he calls this 'structure theory'.

[4] Field (1989: 26), discussing Benacerraf (1973). For a recent exploration of the problem, with a wide discussion of the literature, see Clarke-Doane (2016).

[5] In fact, the word 'doxology' does have an existing use, covering a specific kind of theological praise. As in footnote 1, the risk of confusion seems low.

[Benacerraf's access] problem is sometimes posed as a problem in mathematical epistemology: How can we know anything about mathematical objects, since we don't have any causal contact with them? But to put it as a problem in epistemology is misleading. The problem is really a puzzle in mathematical doxology: Never mind knowledge, how can we even have mathematical beliefs? Mathematical beliefs are beliefs about mathematical objects. To have beliefs about mathematical objects, we have to refer to them; to refer to them, we have to pick them out; and there doesn't appear to be anything we can do to pick out the referents of mathematical terms.[6]

The doxological problem that McGee has formulated should sound familiar: it is exactly the problem which we raised against moderate objects-platonism in Chapter 2, and summarised in §6.1, when we asked how we can determinately refer to the natural number 27 (for example).

The Modelist's Doxological Challenge raises a similar problem for objects-modelists. In asking them how we can pick out particular isomorphism types, we are asking them to provide an account of how we are able to refer to any of the entities which populate their mathematical ontology. For 'blunt' objects-modelists, this is immediately obvious, for they *identify* 'structures' with isomorphism types. But the point is essentially the same for ante rem structuralists: they need to explain how we can refer to any particular ante-structure.

6.4 Concepts-modelism

So far, we have considered objects-modelism and a doxological challenge which arises for it. However, modelism need not be sung in a particularly ontological key. Instead, a modelist might think that model theory supplies us with a toolkit for understanding mathematical *concepts*.

In saying this, we are opening an enormous can of worms. So, before they have all escaped, we should issue some caveats. We do not know of any great account of mathematical concepts: of what they are, of how exactly they work, or of how they relate to what other branches of philosophy call 'concepts'. We are using the terminology of 'mathematical concepts' rather naïvely, and maybe ill-advisedly. But we are doing so, because it allows us to highlight an important tendency within the philosophy of mathematics.

Here is a naïve way to bring that tendency into focus. We have repeatedly observed that mathematicians talk of 'structures'. The objects-modelist wants to reify these 'structures', but it is not obvious that we must follow her. We might have some specific metaphysical qualms about the entities that she postulates. Alternatively, we might just think that reification will not be very illuminating. Nonetheless, we

[6] McGee (1993: 103).

might well want to continue to use structure-talk within mathematics. For, on this line of thought, many mathematical concepts are *structural*, and so structure-talk allows us to explore mathematical concepts.

It might help to consider a specific example. Suppose we speak about 'an ω-sequence', or describe something as being 'an ω-sequence'. In some sense of the word 'concept', we are clearly *deploying* a central and structural mathematical concept: the concept *ω-sequence*. Moreover, according to a fairly natural line of thought, we can probe the full richness of that concept by first considering a class of isomorphic $\mathscr{L}$-structures, and then considering the properties of those structure that are invariant under isomorphism. But if we had wanted to investigate a *different* concept—e.g. the linear order concept exemplified by the linear order $\mathbb{N} + \mathbb{Z}$ consisting of the natural numbers followed by a copy of the integers—we would have needed to work with a *different* isomorphism type.

This rather rough line of thought culminates in *concepts-modelism*. The crucial idea is this: mathematical concepts can be as fine-grained as isomorphism types; certain key mathematical concepts are this fine-grained; and such fine-grained mathematical concepts are well explicated via isomorphism types.[7]

For clarity: according to the concepts-modelist, concepts *can be* as fine-grained as isomorphism type, but they *need* not be so fine-grained. Plenty of non-isomorphic structures are linear orders, and so the concept *linear order* is less fine-grained than an isomorphism type. The crucial point, for the concepts-modelist, is that there are plenty of fine-grained concepts, and that these *are* as fine-grained as isomorphism types. Such concepts will presumably include: *ω-sequence, real line, complex plane,* and perhaps *initial segment of the iterative hierarchy of (pure) sets.*

For further clarity: given a fine-grained mathematical concept, the concepts-modelist will employ some isomorphism type as an ersatz for that concept; but she need not think that *every* isomorphism type determines a mathematical concept. To see why she might doubt this, consider Isaacson's view that there are only *countably* many mathematical concepts. This follows from his view that (a) concepts 'involve the element of understanding inherently',[8] and that (b) we can only hope to articulate countably many concepts.[9] Some concepts-modelists might agree with Isaacson; others might disagree; and we will not pick sides at the outset. The essential point, for the concepts-modelist, is just that every fine-grained mathematical concept is well-explicated via an isomorphism type.

With these clarifications out of the way, we should present some actual *examples* of concepts-modelists. First, we consider Martin. He is interested in the question

[7] Cf. Benacerraf (1984: 479–80) on 'the complexity problem'.

[8] Isaacson (1994: 127).

[9] 'But if structures are given by concepts, and concepts are the kind of thing that are in our minds, there cannot be sufficiently many concepts to characterize these continuum many structures.' Isaacson (1994: 129; cf. 2011: 38).

of whether one can rule out non-isomorphic instances of the concept *set*,[10] with the specific aim of honing in on a concept that *is* as fine-grained as an isomorphism type. We shall discuss the prospects of his view in detail in §8.5 and §11.B. For present purposes, our point is simply that Martin's ambitions make him count as a concepts-modelist.

Hellman's *modal structuralism* is also a version of concepts-modelism.[11] Unlike ante rem structuralists, such as Shapiro, modal structuralists do not postulate any particular mathematical entities. Consequently, they are not objects-platonists, and so they are not objects-modelists either. However, modal structuralists do believe that mathematical claims have determinate, truth-evaluable content. That content is cashed out via necessitated conditionals. For example, modal structuralists recast each arithmetical sentence, φ, as follows:

Necessarily: if S is an ω-sequence, then φ holds in S

In so doing, the modal structuralist has (speaking naïvely) deployed the mathematical concept *ω-sequence*. And so we can reasonably ask the modal structuralist how she understands this phrase. There are many things that a model-structuralist might say at this point. But, as we show in §7.11, Hellman himself explicates the concept *ω-sequence* via a specific isomorphism type. (In effect, he asks us to consider what *possible systems* could belong to a particular isomorphism-type.) And this makes him, too, an exemplar concepts-modelist.

6.5 Doxology, conceptual version

We have given some sense of what concepts-modelism looks like. We now want to consider how concepts-modelism is affected by the Modelist's Doxological Challenge: *How can we pick out particular isomorphism types?*

In §6.3, we outlined McGee's shift from the epistemological to the doxological. In particular, we considered McGee's question: 'Never mind knowledge, how can we even have mathematical beliefs?'[12] McGee himself went on to gloss this doxological question in ontological terms: he asked how we could have beliefs *about* particular

[10] 'If so then the important question is: Can one rule out non-isomorphic instances of the concept (in my sense) of set of x's? Can one rule out non-isomorphic instances of the general iterative concept of set? [¶] As I have said elsewhere, I think that the answer to these questions is yes.' Martin (2005: 365; see also 2001: 5–8, 2005: 366).

[11] Hellman (1989, 1996, 2001, 2005) introduces modal structuralism. See Shapiro (1997: 88–9, 228–9) for critical discussion. As Hellman (1989: viii, 7–9) notes, Putnam (1967) seems to have been the first philosopher to outline modal structuralism. However, unlike Hellman, Putnam did *not* advocate modal structuralism as 'a *new* school in the foundations of mathematics'. Instead, Putnam suggested that (what is now called) modal structuralism was just one of 'many different "equivalent descriptions"' that one might adopt (1967: 19–20; but cf. 1975: 72).

[12] McGee (1993: 103).

mathematical objects. But the central, doxological question—*how can we even have mathematical beliefs?*—can be presented in conceptual, rather than objectual, terms.

To illustrate, consider the epistemological question of how we could know that there actually are any ω-sequences. The very attempt to *ask* this epistemological question presupposes that we possess the concept *ω-sequence*: if we lacked that concept, then we could not even pose the question. In that sense, even before we turn to the epistemological question, there are certain prior *doxological* questions. These might include:

- How can we *acquire* the concept?
- How do our utterances manage to *express* the concept?
- How can we have thoughts which *involve* the concept?

We can ask these (and similar) questions about an individual human, or about how distinct humans can (for example) acquire the *same* concept. And we can ask similar questions about any mathematical concept.

With this in mind, let us revisit the Modelist's Doxological Challenge: *How can we pick out particular isomorphism types?* The objects-modelist takes this as a challenge to explain how she can *refer* to certain abstract object. However, the concepts-modelist can make equally good sense of the Challenge. Since she explicates fine-grained mathematical concepts via isomorphism types, she should regard talk of 'picking out particular isomorphism types' as shorthand for a series of doxological questions. In particular, consider the specific Challenge of how we pick out the isomorphism type of an ω-sequence. For the concepts-modelist, this amounts to the series of bullet-pointed questions above: How can we acquire, possess, express, or deploy the concept *ω-sequence,* where that concept is thought of as being at least as fine-grained as an isomorphism type, so that picking out a *different* isomorphism type would amount to possessing a *different* concept?

Having posed these questions, we should repeat our earlier caveats about concepts. We do not have a developed theory of mathematical concepts, and so in particular we do not have a theory of concept-acquisition, concept-expression, etc. However, if it is legitimate to speak of mathematical concepts at all, then it is legitimate to ask questions about concept-acquisition, expression, possession, and so forth, and to ask for the development of such a theory.

This much is quite clear, if we consider some *non*-mathematical concepts. It is surely legitimate to ask similar doxological questions concerning the concept of *redness,* such as: How do humans acquire the concept? How is it expressed? To what extent is it shared between humans? And so forth. Our answers to these questions might begin by pointing to the fact that most people's eyes have cones with certain peak sensitivities, and then explaining how this shared physiological detection mechanism is crucial to our (collective) ability to acquire, possess, and express the concept of *redness*.

In the case of *redness*, though, we can point both to red things and also to organs which are good at detecting red things. In the case of mathematical concepts, such as the concept *ω-sequence*, it is doubtful that there is anything similar to point to. And, to anticipate a little, this is what makes doxological questions in the philosophy of mathematics look deep and difficult.

But we are getting ahead of ourselves. In this chapter, we have formulated two broad families of *modelism*: objectual and conceptual. We have also formulated the Modelist's Doxological Challenge: *How can we pick out particular structures?* And we have emphasised that this is *doxological* rather than epistemological. For objects-modelists, the Challenge is to explain how we can refer to the objects which populate her mathematical ontology. For concepts-modelists, the Challenge is to explain how we are able to acquire, possess, express, or deploy concepts which are thought of as being at least as fine-grained as isomorphism types. And, though the terminology of 'modelism' and 'doxology' is new, we hope this chapter has made clear that the ideas involved are not new at all. They tap into deep questions in the philosophy of mathematics.

7
Categoricity and the natural numbers

In Chapter 6, we introduced both objects-modelism and concepts-modelism, and introduced the ensuing Modelist's Doxological Challenge. In this chapter, we show that modelists who are also moderate naturalists (in the sense of §2.3) cannot answer this Challenge. Throughout the chapter, we focus on arithmetic.

The Modelist's Doxological Challenge, recall, is to explain how we can pick out particular isomorphism types. In the setting of arithmetic, the specific challenge is to explain how we can pick out the standard isomorphism type of the natural numbers. We start by arguing that moderate modelists must hope to answer the doxological challenge 'by description'. However, we then use model theory to show that the relevant isomorphism type *cannot* be picked out by any *first-order* theory.

The modelist will probably respond to this problem by invoking some logic which is *stronger* than first-order logic. There are plenty of candidate logics and, in general, there may be nothing wrong with invoking them. However, the Modelist's Doxological Challenge can also be pitched at the level of selecting a logic, for these stronger logics must be characterised using mathematical concepts. And the *moderate* modelist will be unable to answer this version of the challenge.

Before diving in, though, we should comment on a phrase that we will use frequently. In Chapter 6, we formulated the Modelist's Doxological Challenge in a deliberately loose way, in terms of 'picking out particular structures'. Using such deliberately loose vocabulary allows us to discuss certain common problems that arise for objects-modelists and concepts-modelists, in a suitably general way. However, it is always possible to turn our loose phrasing into something more specific, as more specific versions of modelism are advanced. So, whilst we often speak loosely for readability, we unpack matters more slowly when greater care is required.

7.1 Moderate modelism

In §2.3, we introduced a zeitgeisty brand of naturalism which, following Putnam, we described as 'moderate'. The moderate rejects all appeals to a special faculty of *mathematical intuition*, or anything similar, within her philosophy of mathematics. We say that a *moderate modelist* is a modelist—of any variety—who embraces this moderation. So, in particular, moderate modelists refuse to invoke mathematical intuition when they attempt to answer the Challenge: *How do we pick out particular isomorphism types?*

Having spurned intuition, we suggest that moderate modelists must embark on a programme of 'doxology by description'. For—loosely but powerfully—how *else* could a moderate modelist claim to 'pick out' a particular isomorphism type, but by attempting to describe it? This, however, is one of those moments where it *is* worth unpacking the loose question, and considering how it applies specifically to objects-modelists and concepts-modelists.

The Challenge facing objects-modelists is to explain how we can *refer* to particular isomorphism types. They regard such isomorphism types as abstract objects of a special sort. Now, an *im*moderate objects-modelist could invoke a special faculty of quasi-perception, which enables her to quasi-perceive particular isomorphism types and hence, perhaps, refer to them directly. But *moderate* modelists can offer no such reply to the Challenge. The problem is exactly as in §2.3. Following Putnam, moderates must deny that any 'neural process [...] could be described as the perception of' the isomorphism type common to all ω-sequences.[1] It would be more plausible for a moderate modelist to suggest that we come to refer to specific isomorphism types by learning some mathematical theory. But then, following Dummett, she cannot allow that the theory merely gestures in roughly the right direction, with the slack picked up by some latent faculty of intuition (as it were):[2] moderates cannot countenance any appeal to intuition, latent or otherwise. Rather, the moderate must insists that the theories *themselves* precisely pin down the isomorphism types. Exactly as in §2.3, then, *moderate* objects-modelists cannot have 'reference by acquaintance' to isomorphism types; they must make do with 'reference by description'.

The Challenge facing concepts-modelist, on the other hand, is to explain how we can acquire, possess, or deploy concepts which she thinks are as fine-grained as isomorphism-types. An *im*moderate concepts-modelist could simply answer the Challenge by invoking a special faculty of mathematical intuition which straightforwardly furnishes her with the concept *ω-sequence* (for example). But moderate concepts-modelists can make no such claim. Their moderation rules out the idea that we possess these concepts innately (for example), or acquire them via flash of rational insight, or anything similar. Once again, then, it is hard to see how we could possibly *acquire* such concepts, except by articulating them with some theory. Again, then, concepts-modelists are committed to 'doxology by description'.

To summarise: throughout this chapter and indeed Part B, we shall assume that *moderate* modelists—of any sort—must try to answer the Modelist's Doxological Challenge by laying down some formal theory.

---

[1] Putnam (1980: 471).

[2] Dummett (1963: 189–90).

7.2 Aspirations to Categoricity

And here is how that just *might* be done. We say that a theory is *categorical* iff all of its models are isomorphic. Now, if we think of theories as picking out their models, then a categorical theory will pick out a single isomorphism type. So this gives modelists—particularly moderate modelists—a reason to aim for categorical theories. For, if T is categorical, then objects-modelists can maintain that presenting T allows us to refer to some particular 'structure', and concepts-modelists can maintain that presenting T allows us to express some specific, fine-grained, mathematical concept. In short, it seems like the Doxological Challenge if the modelist can find a categorical theory.

A second reason to aim for categoricity is brought out by an elementary result:[3]

Proposition 7.1: *If T is categorical, then T is complete.*

We essentially proved this result in §2.5, in our remarks 'concerning (i)', and we leave it to the reader to fill in the details. Now, suppose that we present a *supervaluational* semantics for the sentences in T's language, following the template of §2.5, and so we offer these clauses:

(a) φ is *true* iff $T \vDash \varphi$
(b) φ is *false* iff $T \vDash \neg\varphi$
(c) φ is *indeterminate* otherwise

By Proposition 7.1, if T is categorical, then every sentence in T's language has a determinate truth value on the supervaluational semantics.

7.3 Categoricity within first-order model theory

There are, then, at least two reasons to *aim* for categorical theories. Unfortunately, elementary model theory imposes two immediate barriers to *finding* any.

The Compactness Theorem 4.1 supplies the first barrier. As we saw in §4.1, there are models of PA which contain non-standard elements; that is, elements which look infinitely large (see Definition 4.2). Such non-standard models are never isomorphic to the standard model. However, they can be elementarily equivalent to the standard model, i.e. they can make all the same first-order sentences true (see Definition 2.4). So, no amount of first-order theorising can rule out these non-standard models.

[3] Indeed, the search for categorical theories has often been motivated by the desire to secure the determinacy of truth-value; see e.g. Kreisel (1967: 147–52) and Weston (1976: 286). Corcoran (1980a, 1981) and Awodey and Reck (2002: 18) note that it is not always clear whether the late-19th and early-20th century practitioners of categoricity (like Dedekind and Veblen) were motivated by categoricity *per se*, or by the completeness of theories which it delivers.

The Löwenheim–Skolem Theorem supplies the second barrier to categoricity.[4] We shall first state the result, and then explain it below. (We reserve its proof for §7.A. The notation $\preceq$, for elementary extensions, was set out in Definition 4.3, and see the end of §1.B for a brief discussion of cardinals.)

Theorem 7.2 (Löwenheim–Skolem): *Let $\mathcal{M}$ be an infinite $\mathcal{L}$-structure:*

(1) *For any $A \subseteq M$, there is an $\mathcal{L}$-structure $\mathcal{H}$ such that $\mathcal{H} \preceq \mathcal{M}$, and $A \subseteq H$, and $|H| \leq \max(|A|, |\mathcal{L}|, \aleph_0)$*
(2) *For any cardinal $\kappa \geq \max(|M|, \mathcal{L})$, there is an $\mathcal{L}$-structure $\mathcal{N}$ such that $\mathcal{M} \preceq \mathcal{N}$ and $|N| = \kappa$*

The theorem comes in two parts. Part (1) states that, if we start with a structure $\mathcal{M}$ and a subset A of $\mathcal{M}$'s domain, then we can build a model $\mathcal{H}$ which contains A but which is an elementary substructure of $\mathcal{M}$. This is sometimes called the Skolem *Hull* Construction, since $\mathcal{H}$ is a hull built around A. Equally, it is sometimes called the *Downward* Löwenheim–Skolem Theorem, since we are making a *smaller* model from $\mathcal{M}$. The simplest case of this is where $\mathcal{L}$ is countable and where $A = \varnothing$, so that the hull $\mathcal{H}$ is itself countable. Part (2) is sometimes called the *Upward* Löwenheim–Skolem Theorem since, starting with an infinite model $\mathcal{M}$, we can build arbitrarily *large* elementary extensions of $\mathcal{M}$.[5]

Together, the two parts of the Löwenheim–Skolem Theorem show that every countable first-order theory with an infinite model has a model of every infinite size. But, more generally, since isomorphic models must be of the same size, *no first-order theory with an infinite model is categorical.* This is significant, since the first-order theories which we might have wanted to be categorical—arithmetic, analysis, and set theory—all have infinite models.

In brief: categoricity is out of reach for the first-order theories we most care about. And this puts an enormous blockade on the *only* route by which moderate modelists might hope to answer their Doxological Challenge.

7.4 Dedekind's Categoricity Theorem

For those seeking a categorical theory of arithmetic, the most obvious thought is to move from first-order logic to full second-order logic (for the technical differences, see §1.10). After all, Dedekind famously proved the categoricity of second-order

[4] Having linked Compactness and the Löwenheim–Skolem Theorem, it is worth mentioning that Lindström's Theorem provides a precise sense in which Compactness and Löwenheim–Skolem-type-results exhaustively *characterise* first-order logic (Lindström 1969; Väänänen 2012a).

[5] On the history of this theorem, see Mancosu et al. (2009: §4 pp.352ff) and Badesa (2004).

Peano Arithmetic, PA_2, whose axioms are given in Definition 1.10. Here is a proof:[6]

Theorem 7.3 (Dedekind): *All full models of* PA_2 *are isomorphic.*

Proof. Let $\mathcal{M}$ be a full model of PA_2. It suffices to show that $\mathcal{M}$ is isomorphic to the standard model $\mathcal{N}$ of PA_2. Using our notation for numerals from (*numerals*) of §1.13, consider the set of all elements 'finitely far' from $\mathcal{M}$'s 'zero' element, i.e.:

$$C = \{a \in M : \text{for some natural number } n,\ \mathcal{M} \vDash a = S^n(0)\}$$

Note that the phrase 'for some natural number n' here occurs in the *metalanguage*. Now, $\mathcal{M}$ satisfies the Induction Axiom; and since $\mathcal{M}$ is a *full* model of second-order logic, C falls within the 'range' of that Axiom's universal quantifier, i.e.:

$$\mathcal{M} \vDash [C(0) \wedge \forall y\,(C(y) \rightarrow C(S(y)))] \rightarrow \forall y C(y)$$

The antecedent of this conditional is satisfied: clearly $0^{\mathcal{M}} = a$ is such that $\mathcal{M} \vDash a = S^0(0)$, and if $\mathcal{M} \vDash a = S^n(0)$ then, applying successor to both sides, we have $\mathcal{M} \vDash S(a) = S^{n+1}(0)$. And, since the antecedent of the conditional is satisfied, we have that $M \subseteq C$. Since also $C \subseteq M$ by definition, we have that $C = M$.

Using this fact, we define an isomorphism $h : \mathcal{N} \longrightarrow \mathcal{M}$ from the standard model $\mathcal{N}$ of PA_2 to $\mathcal{M}$ as follows: $h(n) = a$ iff $\mathcal{M} \vDash a = S^n(0)$. Since $C = M$, the map h is a surjection. To see that it is an injection, suppose that $h(n) = h(m)$; then $\mathcal{M} \vDash S^n(0) = S^m(0)$ and, since PA_2 contains an axiom asserting that the successor operation S is injective, we have that $n = m$. Now, by definition $h(0) = 0^{\mathcal{M}}$ since $\mathcal{M} \vDash S^0(0) = 0$. And h preserves the successor operation, since if $h(n) = a$ then $\mathcal{M} \vDash a = S^n(0)$, so that $\mathcal{M} \vDash S(a) = S^{n+1}(0)$ and so $h(n+1) = S^{\mathcal{M}}(a) = S^{\mathcal{M}}(h(n))$. □

7.5 Metatheory of full second-order logic

Before we discuss the philosophical significance of Dedekind's Theorem 7.3, it is worth making three technical observations which follow from it, concerning the differences between first-order logic and full second-order logic. First:

(i) The Compactness Theorem 4.1 fails for full second-order logic.

To see this, as in §4.1, consider the theory $T^* = PA_2 \cup \{S^n(0) < c : n \geq 0\}$. Exactly as in §4.1, every *finite* subset of T^* has a full model. But T^* itself has no full model. For if $\mathcal{M} \vDash T^*$, then $c^{\mathcal{M}} \notin C$—where C is defined as in the proof of Theorem 7.3—contradicting the fact that $C = M$. Next,

[6] For Dedekind's original proof, see Dedekind (1888: ¶132, 1930–32: v.3 p.376) and Ewald (1996: v.2 p.821). For contemporary proofs in a different style, see Shapiro (1991: 82–3) and Enderton (2001: 287).

(ii) The Upward Löwenheim–Skolem Theorem 7.2(2) fails for full second-order logic.

After all, every full model of PA_2 is isomorphic to the standard *countable* model. Now, in fact, Dedekind's Theorem 7.3 can be tweaked to yield the categoricity of second-order real analysis.[7] Consequently, every full model of second-order real analysis has cardinality continuum. And so also:

(iii) The Downward Löwenheim–Skolem Theorem 7.2(1) fails for full second-order logic.

In short, the meta-theorems about first-order logic and full second-order logic are very different. These differences once led Kalmár to quip:

> One can say humorously, while first order reasonings are convenient for proving true mathematical theorems, second order reasonings are convenient for proving false metamathematical theorems. Of course, instead of calling them false we honour them by calling them second order meta-theorems.[8]

Mostowski was willing to spoil a good joke, and replied:

> [they] are neither misleading nor false. They just require stronger assumptions than many metamathematical theorems concerning systems based on first order logic.[9]

Here is the point, shorn of all trace of humour. Considered just as pieces of pure mathematics, results like Dedekind's Theorem 7.3 are utterly unimpeachable. But, considered as pieces of pure mathematics, their philosophical significance remains up for grabs. Indeed, this illustrates a rather good rule of thumb: it is impossible to extract philosophical juice from a piece of pure mathematics, without invoking some philosophical thesis. Or, as George put it, 'A mathematical result shines only when illuminated by other views.'[10]

7.6 Attitudes towards full second-order logic

Modelists, however, *do* supply some 'other views' with which to illuminate Dedekind's Categoricity Theorem 7.3. In particular, modelists explicate 'structure', informally construed, via isomorphism types. So, they will take Dedekind's Theorem to establish the following conditional:

If we can appeal to full second-order logic, then we can pin down the specific isomorphism type of the natural numbers, and so can pick out THE *natural number structure (informally construed).*

[7] See e.g. Shapiro (1991: 84).
[8] In Mostowski (1967: 104).
[9] Mostowski (1967: 107).
[10] George (1985: 87); he was specifically focussing on the Löwenheim–Skolem Theorem.

However, at least *four* distinct attitudes can be taken towards this conditional.

The Algebraic Attitude. Imagine a character who denies that there is any such thing as *the* natural number structure (informally construed). Instead, she thinks of arithmetic as algebraic, in the sense used in §2.2, that the theory has no one intended model. Note that this character can quite happily be a modelist *herself*. For she is happy to employ model-theoretic notions, in explicating the idea of mathematical structure (informally construed). Note also that she admits the perfect mathematical rigour of Dedekind's Theorem. Her point is simply that the *indeterminacy* in the natural number structure is mirrored by an *indeterminacy* in second-order quantification, so that second-order logic, too, is inevitably algebraic in character. In short: she performs a *modus tollens* on the preceding conditional.[11]

The Logic-First Attitude. The next attitude instead involves a *modus ponens*. More specifically, it begins with the idea that our grasp on full second-order logic is unproblematic. As such, by the preceding conditional, it insists that we can pin down the natural numbers (up to isomorphism).[12]

The Infer-to-Stronger Logic Attitude. One might think that mathematical practice itself dictates that we are able to pin down the uniquely intended natural number structure (informally construed). If we treat this structure (informally construed) as an isomorphism type, then we can use the Compactness Theorem 4.1 or the Löwenheim–Skolem Theorem 7.2 to argue that we must have access to resources beyond first-order logic. Indeed, for those who are not too worried about the Algebraic Attitude, this pattern of argumentation may look like nothing more than an inference to the best explanation: an *abduction* on the preceding conditional.[13]

The Holistic Attitude. The previous two attitudes seem to presuppose a clean separation between full second-order logic and mathematics. However, one might well think that the boundary between the two is rather artificial.[14] So the final choice is to embrace a more holistic attitude, according to which our grasp of (for example, full second-order) logic and our grasp of determinate mathematical structures (in the informal sense) come together, with each helping to illuminate the other.

On any of the last three attitudes, finding a categorical theory is likely to be treated as a hallmark of *success* in the project of axiomatising arithmetic.[15] Whether that is correct or not, our point here is fairly simple. By themselves, categoricity results are just pieces of pure mathematics. If they are to be deployed in philosophical discussion, we first need some bridging principle which connects informal structure-talk with the technical notions supplied by model theory. But even

---

[11] See Shapiro (2012: 308) and cf. Mostowski (1967: 107) and Read (1997: 92).

[12] See Read (1997: 89).

[13] This is part of Shapiro's approach to second-order logic (1991: xii–xiv, 100, 207, 217–8, 2012: 306), and Shapiro cites Church (1956: 326fn.535) as a precedent.

[14] As suggested by Shapiro (2012: 311–22) and Väänänen (2012b).

[15] Cf. Read (1997: 92) and Meadows (2013: 525–7, 536–40) but also Corcoran (1980a: 203–5).

the modelist's bridging principle—to explicate structure via isomorphism types—is compatible with many different attitudes concerning the philosophical significance of the categoricity result.

7.7 Moderate modelism and full second-order logic

We have just outlined four different attitudes that modelists might take towards Dedekind's Theorem 7.3. In §7.1, however, we specifically focussed on *moderate* modelism. So we now want to consider which of these attitudes are available to *moderate* modelists. In particular, we want to show that moderate modelists *cannot* legitimately invoke the Logic-First Attitude, and why this matters.

Recall from §§7.1–7.2 that moderate modelists must try to answer their Doxological Challenge by laying down a *categorical* theory. In particular, then, they are likely to attempt to answer the Doxological Challenge by appealing to Dedekind's Categoricity Theorem for PA_2, and then invoking the Logic-First Attitude towards full second-order logic.

The 'second-order' component of 'full second-order logic' is wholly unobjectionable. No one can prevent mathematicians from speaking a certain way, or from formalising their theories using any symbolism they like. The qualifying expression 'full', however, is more delicate. As outlined in §1.10, this describes a particular semantics for second-order logic: one in which the second-order quantifiers essentially range over the entire powerset of the first-order domain of the structure. We can instead supply a faithful *Henkin* semantics for second-order logic, as in §1.11. And we can obtain a Löwenheim–Skolem result for faithful Henkin semantics, just by making some minor tweaks to our original proof of Theorem 7.2.[16] So, to be quite explicit, *Dedekind's Theorem 7.3 fails if we replace 'full' with 'Henkin'*.

Such observations have been made before. Repeatedly.[17] But here is Putnam's brief statement of why this raises a problem for moderate modelists:

> [...] the 'intended' interpretation of the second-order formalism is not fixed by the use of the formalism (the formalism itself admits so-called 'Henkin models' [...]), and [so] it becomes necessary to attribute to the mind special powers of 'grasping second-order notions'.[18]

---

[16] For more, see e.g. Shapiro (1991: 70–6, 92–5), Manzano (1996: ch.6), and Enderton (2001: §§4.3–4.4). An alternative way to raise essentially the same point is as follows: if we formalize the semantics for 'full' second-order logic and its constituent notions (like powerset) in a *first-order* set theory, then the Löwenheim–Skolem result applies once again at the level of the metatheory.

[17] See e.g. Weston (1976: 288), Parsons (1990b: 14–17, 2008: 394–6), Field (1994: 308n.1, 2001: 319, 321–2, 338–9, 352–3), Shapiro (2012: 273–5), Väänänen (2012b: 120), Meadows (2013: 535–42), and Button (2013: 28).

[18] Putnam (1980: 481).

This is another instance of Putnam's just-more-theory manoeuvre, which we first discussed in the context of his permutation argument (see §2.3). Moreover, we think it is successful, as wielded specifically against *moderate modelists*. But we must spell this out carefully.

Using the Löwenheim–Skolem Theorem 7.2, we can present our modelist with various alternatives for what the theory PA_2 picks out: 'the standard model' (i.e. one particular isomorphism type), or some 'non-standard model' (i.e. some other isomorphism type). In trying to spell out why PA_2 picks out the former, our moderate modelist appeals to Dedekind's Theorem 7.3. In order for that Theorem to do the job she wants it to, she must have ruled out the Henkin semantics for second-order logic; indeed, in the vocabulary of §2.3, she must have shown that full models are *preferable* to Henkin models. But the distinction between full models and Henkin models essentially invokes abstract *mathematical* concepts. And, so the worry goes, the distinction between full and Henkin models is *just more theory*, and hence up for reinterpretation.

This, however, is another moment when it pays to unpack the problem in two slightly different ways, depending upon whether we are dealing with moderate *concepts*-modelists or moderate *objects*-modelists.

For moderate concepts-modelists, the issue is straightforward. They need to explain how creatures like us are able to possess a particular mathematical concept which (they claim) is as fine-grained as an isomorphism type. To answer this, they tried to invoke full second-order logic. But this only pushes back the problem: we must ask them how creatures like us are able to possess the particular mathematical concepts invoked in the *semantics* for full second-order logic. Indeed, when the aim is to *secure* a grasp of certain abstract mathematical concepts, invoking Dedekind's Theorem 7.3 is simply *question-begging*, since the use of the full semantics simply *assumes* precisely what was at issue, namely, an unproblematic grasp of various abstract mathematical concepts.

For moderate objects-modelists, the issue is only slightly more complicated; we just need to rephrase all of the points about *grasping concepts* in terms of *reference*. The moderate objects-modelist is looking for a mechanism that allows her to refer to the isomorphism type of the natural numbers. If she tries to invoke Dedekind's Categoricity Theorem 7.3, then she needs to explain that she is working with the *full* semantics for second-order logic. As such, she needs to refer to the *full* powerset of the underlying domain which, if everything has gone successfully, is (isomorphic to) the natural numbers. But the full powerset of this domain has the same cardinality as the real numbers. Moreover, via familiar coding mechanisms, we can view each real number as a certain subset of natural numbers. So, to explain how she can pick out the natural numbers (up to isomorphism), it seems she must *first* explain how she can pick out the real numbers (up to isomorphism). Her problems have only worsened: whatever problems arise in referring to the naturals will pale in comparison to the problems which arise in referring to the reals.

Either way, then, in appealing to full second-order logic, the moderate modelist is simply out of the frying pan, and into another frying pan. And, if anything, the second frying pan is slightly larger and hotter than the first.

7.8 Clarifications

The argument we just gave is central to Part B. So we will pause to consider several (unsuccessful) ways in which a moderate modelist might seek to undermine it. We imagine three responses from the moderate modelist:

I grant that there is a certain 'circularity' in attempting to SECURE *certain mathematical resources by* INVOKING *them. And yes, this 'circularity' occurs when I explain the differences between faithful Henkin semantics and full semantics. But this 'circularity' is not vicious; it is merely a benign instance of holism.*

This response misunderstands the problem facing moderates. Crudely put, moderates need to provide us with a way to break into the circle of mathematical concepts. As we discussed in §7.1, moderates cannot simply insist that they 'just can' refer to the isomorphism type of an ω-sequence, or insist that 'as a matter of brute fact' their words 'just do' express the concept *ω-sequence* (rather than some non-standard concept). Well then, by the arguments of the preceding section, moderates cannot insist that they 'just do' employ second-order logic with its *full* semantics (rather than some Henkin semantics).

At this point, the moderate modelist may change tack:

My grasp of full second-order quantifiers is simply LOGICAL, *or purely* COMBINATORIAL, *and so is not up for reinterpretation via Henkin semantics.*[19]

This merely labels the problem, without solving it. Whatever honorific they give these notions—logical, combinatorial, or something else—the question remains how moderate modelists can lay claim to them.

In desperation, the moderate modelist might retreat further:

Well, all I know is that SOMETHING *has gone wrong in your argument, because we do in fact manage to pin down the isomorphism type of an ω-sequence!*

---

[19] Some people read Shapiro (1991) this way. However, as remarked in footnote 14, above, Shapiro does not think that second-order logic can be used in the present context to secure the categoricity of arithmetic.

There is a well-known interpretation of second- and higher-order logic in terms of plural logic (see Boolos 1984; Oliver and Smiley 2013). It is sometimes suggested that this perspective automatically rules out any Henkin-like semantics and so requires a full semantics. For representative quotations, and excellent criticisms of this approach, see Florio and Linnebo (2016).

We will say much more about this in Chapter 9. But even if this is right, it does not let the moderate modelist off the hook. To be very clear: our argument does not aim to establish that *we* cannot pin down this isomorphism type, or that *we* do not have a grasp of full second-order logic. Our argument establishes only that the *moderate modelist* cannot think we do. So, we agree that '*something* has gone wrong here'; but *what* has 'gone wrong' is just moderate modelism *itself*.

The short point is that the moderate modelist has no answer (yet) to her Doxological Challenge. She seems to have no way to explain how either full second-order logic, or arithmetic, could be anything other than *algebraic* theories. Indeed, despite her aim of pinning down *the* natural number structure, she is being dragged towards the Algebraic Attitude of §7.6.

7.9 Moderation and compactness

The problem outlined in §§7.7–7.8 is rooted in the fact, mentioned in §7.5, that full second-order logic is not compact.[20] To show this, we will start with a simple observation:

(i) No theory of arithmetic is categorical in any compact logic.

The argument to this effect is exactly as in §4.1. Let T be any theory of arithmetic. Then define a theory T^* by adding infinitely many new sentences to T which state, in effect, that c is an element such that $0 < c$, that $1 < c$, that $2 < c$, etc. If the logic in question is compact, then T^* has a model; but any model of T^* contains a non-standard element; so T is not categorical.

Consequently, anyone who wants to provide a categorical theory of arithmetic must use a non-compact logic. But, speaking very crudely, non-compact logics are hard to get to grips with. This follows from a second simple observation:

(ii) If a logic has a sound and complete proof system, whose proofs are always finitely long, then that logic is compact.

The argument for this is exactly as in our proof of the Compactness Theorem 4.1 for first-order logic, from the soundness and completeness of its proof system. Suppose that every finite subset of T has a model. Since our proof system is *sound*, there is no sentence φ such that some finite subset of T proves both φ and $\neg\varphi$. Since our proofs are *finite*, T itself does not prove both φ and $\neg\varphi$. Since our proof system is *complete*, we do not have both $T \vDash \varphi$ and $T \vDash \neg\varphi$. So T has a model, since otherwise we vacuously have $T \vDash \psi$, for any sentence ψ.

Combining our two observations, we arrive at the following:

[20] Many thanks to Catrin Campbell-Moore for suggesting that we make this point in terms of *compactness*, rather than Gödelian incompleteness.

(iii) Any logic which allows for a categorical theory of arithmetic lacks a sound and complete finitary proof system.

Our understanding of what consequence amounts to according to any such logic must, then, come from a specification of the formal semantics for that logic. But the specification of a formal semantics invariably looks like *just more mathematical theory*. As such, it will always be at least as hard for moderate modelists to explain how they grasp the intended semantics for the logic in question, as it is for them to explain how to pin down the 'standard model' of arithmetic in the first place.

7.10 Weaker logics which deliver categoricity

The preceding argument is very abstract, because it considers arbitrary logics. Indeed, it is essentially an informal argument in *abstract model theory*. To make the point more concrete, we outline seven logics which are weaker than full second-order logic, but within which categorical theories of arithmetic can be given.

We can easily obtain a categorical theory of arithmetic if we:

(a) Employ a fragment of second-order logic equipped with one-place second-order relation-variables $X, Y, Z, \ldots$ but no second-order quantifiers, with the semantics organised so that $\varphi(X)$ is true iff every subset of the first-order domain satisfies φ.

It is sometimes suggested that this logic is significant, because it accommodates the idea that arithmetical induction is *totally open-ended*. The contrast is as follows: the induction schema used in PA only gives us induction for formulas in the signature $\{0, 1, +, \times\}$. The hope is that (a), by contrast, allows us to consider induction in any *possible extension* of our language, thereby reflecting induction's open-endedness.[21]

Now, it is easy to see that we can provide a categorical theory of arithmetic using the logic sketched in (a): just take the usual second-order theory, PA_2, and delete the '$\forall X$' in front of its Induction Axiom.[22] Dedekind's Theorem 7.3 then goes through just as before. But that this theory is categorical is neither surprising nor very interesting.[23] For this logic is obviously just a notational variant for the fragment of full second-order logic in which formulas begin with at most one higher-order universal quantifier and contain no further higher-order quantifiers. In this context, this notational variant is surely no more philosophically significant

[21] McGee (1997: 56ff), Lavine (1994: 224–40, 1999), and Parsons (2008: 262–93) have all invoked (a) in defence of philosophical arguments based upon categoricity results. However, they all did so in the context of the *internal* categoricity results of Chapters 10–11. The applicability of induction to formulas in any signature whatsoever had been stressed previously by Feferman: see footnote 43 in Chapter 4.

[22] The technical point here goes back to Corcoran (1980a: 192–3), and is also discussed by Shapiro (1991: 247–8).

[23] As noted by Field (2001: 354), Walmsley (2002: 253), Pedersen and Rossberg (2010: 333–4), and Shapiro (2012: 309–10).

than the fact that, in propositional logic, we can omit the outermost pairs of brackets in a sentence without risk of ambiguity. Simply put: our grasp on the idea of *totally* open-ended induction is exactly as precarious as our grasp on *full* second-order quantification.

Other intermediate logics have been considered. The following three augmentations of first-order logic also suffice for categorical theories of arithmetic:[24]

(b) Treat '0' and 'S' as logical constants, i.e. as having fixed interpretations.
(c) Add a new quantifier which expresses that there are finitely many φs.
(d) Introduce a single one-place predicate, whose fixed interpretation in an arbitrary model of arithmetic is given by the numbers finitely far from (the interpretation of) zero.

Given (b), PA itself is trivially categorical, for it has *exactly* one model. Given (c), we obtain categoricity by supplementing PA with the axiom 'for any x, there are only finitely many entities less than x', since this claim this would be false of all non-standard numbers. And given (d), we obtain categoricity by supplementing PA with the axiom 'everything is finitely far from zero' (compare our definition of C in the proof of Dedekind's Theorem 7.3). But, as Read notes, if we attempt to secure categoricity by invoking any of these logics, then we simply shift 'the problem from the identification of postulates characterizing [the natural numbers] categorically [...] into the semantics and model theory of the logic used to state the postulates'.[25] This is the just-more-theory manoeuvre all over again.

A marginally more interesting approach is to:

(e) Add Härtig's two-place quantifier, $\mathsf{H}xy\,(\varphi(x), \psi(y))$, which expresses that there are exactly as many φs as ψs.[26]

Given (e), we obtain a categoricity by supplementing PA with an axiom stating 'if there are exactly as many entities less than x as there are entities less than y, then $x = y$', for this would be false of certain non-standard numbers. But, again, to grasp the (intended) semantics of Härtig's quantifier, we need to grasp the behaviour of cardinality *in general*, which again presupposes within the semantics precisely the notions that we were seeking to secure by providing a categorical theory. Similarly, we might:

(f) Allow sentences containing (countably) infinitely long conjunctions and disjunctions.[27]

We then obtain categoricity by supplementing PA with the countable disjunction: 'everything is either 0, or $S(0)$, or $S(S(0))$, or, ...'. But to grasp this proposal, we need to grasp the meaning of the ellipsis; and that looks exactly like the original

[24] Read (1997: 89–92) discusses (b), (c) and (e). Field (1980: ch.9, 2001: 320, 338–40) defends (c).
[25] Read (1997: 91).
[26] We discuss Härtig's quantifier more in Chapter 16.
[27] In terminology we will introduce in §15.4, this is the logic $L_{\omega_1\omega}$.

challenge of grasping the natural number sequence. (Indeed, using our notation (*numerals*) of §1.13, and the notation for infinitary connectives which we will define in §15.4, we would typically write this axiom as $\forall x \bigvee_{n<\omega} x = S^n(0)$, thereby explicitly invoking a grasp of the natural numbers in the metatheory.)

Mathematically, the most interesting alternative is to:

(g) Insist that the arithmetical function symbols + and ×, even if not logical constants, must always stand for *computable* functions.

Given (g), PA itself becomes categorical, as a consequence of Tennenbaum's Theorem that all computable models of (first-order) PA are isomorphic.[28] However, this option again faces an obvious challenge. To make sense of the notion of *computability*, we need to make sense of the idea of a process specified with an arbitrary but *finite* number of instructions, which can run for any arbitrary but *finite* number of steps. In short, the notion of *computability* seems to presuppose precisely the arithmetical notions it was supposed to vouchsafe. Again: this is the just-more-theory manoeuvre.[29]

Perhaps further logics could be advanced, within which we could provide a categorical theory of arithmetic. But the general problem facing moderate modelists should now be clear. They need to explain how we grasp certain mathematical concepts. They must answer by invoking some categoricity theorem. But to prove categoricity, they must spell out the semantics of their chosen logic (given that the logic must be non-compact, and hence has no sound and complete finitary proof system). In so doing, they will invoke precisely the kinds of mathematical concepts that were at issue in the first place, and which they were hoping to secure by *appeal to* a categoricity theorem.

In short: the moderate modelist's attempts to go beyond first-order logic invariably amount to *just more mathematical theory*. With this, *moderate modelism is dead*.

7.11 Application to specific kinds of moderate modelism

Triumphant as this sounds, we should probably pause to remember that no one has ever *called* themselves a 'moderate modelist'. So, to bring out the significance of the death of moderate modelism, we will explain what becomes of the modelist positions which we outlined in Chapter 6.

As we noted in §6.2, Shapiro is an objects-modelist par excellence. He believes in ante-structures, which are something like abstractions from isomorphism types. His version of the Modelist's Doxological Challenge is to explain how we are able

---

[28] Tennenbaum (1959); for proofs, see Kaye (2011) and Ash and J. Knight (2000: 59).

[29] For more on this appeal to Tennenbaum's Theorem, see McCarty (1987: 561–3), Dean (2002, 2014), Halbach and Horsten (2005), Quinon (2010), Button and P. Smith (2012), and Horsten (2012).

to refer to particular ante-structures. As one might expect, Shapiro explicitly concedes that there would be a serious problem here, *if* we were limited to first-order logic. But Shapiro denies that our resources are limited in this way, and answers the Challenge by explicitly invoking full second-order logic and Dedekind's Categoricity Theorem 7.3.[30] Our observation is simple: Shapiro cannot do so whilst remaining a *moderate*.

We also explained in §6.4 that certain modal structuralists, like Hellman, are concepts-modelists. To recall, modal structuralists unpack each arithmetical sentence, φ, along the following lines:

Necessarily: if S is an ω-sequence, then φ holds in S

We must ask the modal structuralist how to understand the phrase 'S is an ω-sequence'. And, just like Shapiro, at this point Hellman himself explicitly explains the use of this phrase via full second-order logic and Dedekind's Categoricity Theorem 7.3.[31] Once again, our point is just this: Hellman cannot do so whilst remaining a *moderate*.

These observations are, perhaps, to be expected. To end the chapter, though, we shall show how certain versions of *formalism* can run into an unanswerable Doxological Challenge. This will take longer to explain. But it is time well spent, since formalism is sometimes (mistakenly) thought to provide an *escape* from the kinds of problems that we have discussed in this chapter.[32]

Consider a kind of formalist who wants to emphasise the 'formal' part of 'formalism'. More specifically, she conceives of mathematics in terms of *formal proofs from formal theories*. Just as we asked a modal structuralist to unpack the phrase 'S is an ω-sequence', we should ask the formalist to unpack the phrase 'π is a formal proof from a formal theory'. In some (perhaps naïve) sense of the word 'concept', it involves *deploying* a specific concept, and so we should ask: *How do we pin down the concept of a formal proof from a formal theory?*

To make matters tangible, suppose that our formalist wants (at some point) to talk about formal proofs from the theory PA. Since there are infinitely many formal PA-proofs, they do not exist in any concrete form. So, in speaking of 'the formal PA-proofs', our formalist character must be speaking of an abstract type. Well then: *How do we pin down that type?*

An *im*moderate formalist might answer that some (limited) faculty of mathematical intuition allows her to grasp the notions which are invoked in specifying this abstract type. (In particular, she will probably invoke concepts like *recursively specifiable* and *arbitrary but finite*.)[33] But *moderate* formalists will have to tell a differ-

[30] See references in earlier footnotes, and Shapiro (1997: 133).

[31] Hellman (1989: 18ff, 1996: 105ff, 2001: 188ff, 2005: 552ff).

[32] See for example Klenk (1976: 485–7).

[33] Hilbert's own brand of formalism was rather *immoderate*, since he maintained that intuition supplied our concept *finite*. For discussion, see e.g. Detlefsen (1986: 16–22) and Potter (2000: 228–32).

ent story. Indeed, moderate formalists will surely have to say that we pin down the type 'by description' (cf. §7.1). So, in particular, they might *axiomatise* the theory of PA-provability, by fully regimenting the contents of §1.A and §1.C within a theory of *syntax*, *S*, whose 'objects' of *S* are *strings* (i.e. formal sentences and formal proofs.)[34]

Now, no particular theory has the status of *the* theory of syntax, in the way that Peano Arithmetic has the status of *the* theory of arithmetic. Nonetheless, ever since Gödel taught us how to arithmetise syntax, we have known how to treat syntax arithmetically, within PA. (We mentioned Gödel's arithmetisation of syntax in §§5.5 and 5.A.) As such, we can innocuously assume that some suitable extension of PA will *interpret* *S* (in the sense of 'interprets' defined in Definition 5.7).[35] Call that extension *T*.

A problem now looms into view. We know that *T* has non-standard models (or non-standard Henkin models, if *T* is second-order). Invoking interpretability, there are therefore non-standard models of the formalist's theory of syntax, *S*. And these give rise to what we might call *non-standard formal* PA-*proofs*.

To bring this out, consider Con(PA), the arithmetised sentence which intuitively states that PA is consistent. Unpacked slightly, this sentence has the form $\forall x \neg \mathrm{Prf}_{\mathrm{PA}}(x, \ulcorner \bot \urcorner)$, telling us that no natural number codes a formal PA-proof of some canonical contradiction. But since PA + ¬Con(PA) is consistent, the Completeness Theorem 4.24 entails that it has a non-standard model, $\mathcal{M}$, with a non-standard element c such that $\mathcal{M} \vDash \mathrm{Prf}_{\mathrm{PA}}(c, \ulcorner \bot \urcorner)$. Since *T* interprets *S*, there are non-standard models of *S*. And these will contain a non-standard 'sequence of sentences' which constitute a 'non-standard proof' of the *inconsistency* of PA.

And so we have a problem. *If* it is a doxologically open question, which notion of 'formal PA-proof' our formalist picks out using her theory *S*, *then* it is a doxologically open question, whether PA is consistent. And that seems absurd.

(We should perhaps emphasise, though, that this issue is doxological rather than epistemological. The worry is not that we might some day *discover* a concrete PA-proof of absurdity. The worry is even more mind-boggling: it is that we might have somehow acquired a *non-standard* concept of *formal proof*, according to which PA counts as 'inconsistent'.)

To rule this out, of course, our formalist character will need to tell us why the 'non-standard proofs' in the non-standard model of S are not *really* proofs. In some sense, the problem is clear enough: they are infinitary objects, corresponding to non-standard numbers, rather than finitary ones. But, in order for her to say this, she must pin down the concept *finite*, or, equivalently, the concept ω-*sequence*.[36]

---

[34] See Quine (1946: 105) and Corcoran et al. (1974).

[35] Note that we have discussed interpretation of *theories*, rather than of *structures*, since we are here tackling a version of formalism.

[36] Cf. Weir's (2010: 240) remark that 'many abbreviatory concrete realizations of IPA depend on prior grasp of finitary arithmetic'.

She needs, it seems, to pin down the standard models of the arithmetical theory T which interprets her theory S. But now she has run slap-bang into the problems facing concepts-modelism.

Some versions of formalism, then, are untenable, for just the reasons that moderate concepts-modelism is untenable. But all of the caveats of §7.8 apply to this point. In particular: it does not follow that *no* version of formalism is tenable. For a start, formalists might reject moderation. Equally, formalists might deny that *formal* proofs are especially important, and so duck the entire problem. Finally, formalists might reject concepts-modelism, and deny that the concepts she is trying to articulate in S are as fine-grained as a model-theorist's isomorphism types. Maybe there are other problems waiting down the line for these alternative versions of formalism. But our point is simple: formalism *alone* offers no guarantee of escaping the doxological problems discussed in this chapter.

7.12 Two simple problems for modelists

At its heart, this chapter contains only two simple ideas.

First: moderate modelists must attempt to pin down an isomorphism type by providing a *description* of that isomorphism type. But the description cannot be presented in first-order logic. For first-order logic is too weak to allow for a categorical theory of arithmetic. More generally, no compact logic allows for a categorical theory of arithmetic. And so no logic with a finitary deductive system allows for a categorical theory of arithmetic.

Second: logics which are strong enough to provide categorical theories of arithmetic must be articulated semantically. And the semantic concepts involved in specifying such logics are just as mathematical as those required in grasping arithmetic. Consequently, it is at least as difficult for moderates to explain how we can pin down those concepts, as it is for them to explain how we can pick out the isomorphism type of an ω-sequence.

In sum: moderate modelism is untenable. It is brought down by the Modelist's Doxological Challenge. And the significance of this point is straightforward: either we must abandon moderation, and embrace the idea that we have something like a faculty of mathematical intuition; or we must abandon any version of modelism. That is quite some choice to make.

7.A Proof of the Löwenheim–Skolem Theorem

In this appendix, we prove the Löwenheim–Skolem Theorem 7.2. We start with a helpful lemma. Recall that the notation $\preceq$ for elementary extensions was set out in

Definition 4.3, and that the notion of a substructure was set out in Definition 3.7.

Lemma 7.4 (Tarski–Vaught Test): *Let $\mathcal{M}$ and $\mathcal{N}$ be $\mathcal{L}$-structures, with $\mathcal{M}$ a substructure of $\mathcal{N}$. Then the following are equivalent:*

(1) $\mathcal{M} \preceq \mathcal{N}$

(2) *for any $\mathcal{L}$-formula $\varphi(\bar{v}, x)$ and any $\bar{a}$ from M, if $\mathcal{N} \vDash \exists x\varphi(\bar{a}, x)$ then there is $b \in M$ such that $\mathcal{N} \vDash \varphi(\bar{a}, b)$*

Proof. (1) ⇒ (2). Suppose $\mathcal{N} \vDash \exists x\varphi(\bar{a}, x)$, with $\bar{a}$ from M. Then by elementarity, $\mathcal{M} \vDash \exists x\varphi(\bar{a}, x)$. Choose a witness b from M with $\mathcal{M} \vDash \varphi(\bar{a}, b)$. By elementarity again, $\mathcal{N} \vDash \varphi(\bar{a}, b)$.

(2) ⇒ (1). We aim to show that $\mathcal{M} \vDash \varphi(\bar{a})$ iff $\mathcal{N} \vDash \varphi(\bar{a})$ for all $\bar{a}$ from M and all $\mathcal{L}$-formulas φ. This is by induction on complexity. For atomic formulas, the proof is just like Theorem 2.3 (1) ⇒ (2), replacing the isomorphism with the identity map. In particular, we first prove that $t^{\mathcal{M}}(\bar{a}) = t^{\mathcal{N}}(\bar{a})$ for every term and every $\bar{a}$ from M, and the rest is easy. For conjunctions and negations, the proof is just like Theorem 2.3 (2) ⇒ (3). Finally, for existentials:

$$\begin{aligned}\mathcal{M} \vDash \exists x\varphi(\bar{a}, x) &\text{ iff } \mathcal{M} \vDash \varphi(\bar{a}, b) \text{ for some } b \in M\\ &\text{ iff } \mathcal{N} \vDash \varphi(\bar{a}, b) \text{ for some } b \in M\\ &\text{ iff } \mathcal{N} \vDash \exists x\varphi(\bar{a}, x)\end{aligned}$$

The second biconditional holds by the induction hypothesis, while the non-trivial direction of the third biconditional holds by (2). □

Armed with this, we can prove the Löwenheim–Skolem Theorem.

Theorem (Löwenheim–Skolem Theorem 7.2): *Let $\mathcal{M}$ be an infinite $\mathcal{L}$-structure:*

(1) *For any $A \subseteq M$, there is an $\mathcal{L}$-structure $\mathcal{H}$ such that $\mathcal{H} \preceq \mathcal{M}$, and $A \subseteq H$, and $|H| \leq \max(|A|, |\mathcal{L}|, \aleph_0)$.*

(2) *For any cardinal $\kappa \geq \max(|M|, \mathcal{L})$, there is an $\mathcal{L}$-structure $\mathcal{N}$ such that $\mathcal{M} \preceq \mathcal{N}$ and $|N| = \kappa$*

Proof. (1). The strategy is to begin with A, and then sequentially add witnesses for existentially quantified claims in $\mathcal{M}$. Having done this infinitely often, we can lean upon the fact that any formula of first-order logic is only finitely long, and invoke the Tarski-Vaught test (Lemma 7.4). Here is the detail. Where $\lhd$ is a well-ordering on M (given by the Axiom of Choice as laid down in §1.B), we define:

$$H_0 = A$$
$$H_{m+1} = H_m \cup \{b \in M : \mathcal{M} \vDash \varphi(\bar{a}, b) \text{ for some } \mathscr{L}\text{-formula } \varphi \text{ and some } \bar{a} \in H_m^n, \text{ where for all } d \in M, \text{ if } \mathcal{M} \vDash \varphi(\bar{a}, d) \text{ then } b \trianglelefteq d\}$$
$$H = \bigcup_{m<\omega} H_m$$

It is clear that $|H| \leq \max(|A|, |\mathscr{L}|, \aleph_0)$, since we only added $\max(|A|, |\mathscr{L}|, \aleph_0)$-many new elements at each stage of the construction and there are only ω-many stages. We now define an $\mathscr{L}$-structure $\mathcal{H}$, with domain H, with the following clauses for each $\mathscr{L}$-constant symbol c, each n-place $\mathscr{L}$-relation symbol R, and each n-place $\mathscr{L}$-function symbol f:

$$c^{\mathcal{H}} = c^{\mathcal{M}}$$
$$R^{\mathcal{H}} = R^{\mathcal{M}} \cap H^n$$
$$f^{\mathcal{H}} = f^{\mathcal{M}}|_{H^n}$$

We should note why $f^{\mathcal{H}} : H^n \longrightarrow H$. If $\bar{a} \in H^n$, then both $\bar{a} \in H_m^n$ for some (least) m and $\mathcal{M} \vDash \exists x\, f(\bar{a}) = x$. Since this x is unique and indeed identical to $f^{\mathcal{H}}(\bar{a}) = f^{\mathcal{M}}(\bar{a})$, we have that $f^{\mathcal{H}}(\bar{a}) \in H_{m+1}$, by construction.

To show that $\mathcal{H} \preceq \mathcal{M}$, we invoke the Tarski–Vaught Lemma 7.4. Let $\varphi(\bar{v}, x)$ be an $\mathscr{L}$-formula, let $\bar{a} \in H^n$, and suppose $\mathcal{M} \vDash \exists x \varphi(\bar{a}, x)$. As before, there is some (least) m such that $\bar{a} \in H_m^n$, and so some $b \in H_{m+1} \subseteq H$ such that $\mathcal{M} \vDash \varphi(\bar{a}, b)$. Hence $\mathcal{H} \preceq \mathcal{M}$.

(2). Let $c_1, \ldots, c_\alpha, \ldots$ be a sequence of κ-many constant symbols, none of which are in $\mathscr{L}(M)$, and consider the theory

$$T := \{\varphi \text{ is an } \mathscr{L}(\mathcal{M})\text{-sentence} : \mathcal{M}^\circ \vDash \varphi\} \cup \{c_\alpha \neq c_\beta : \alpha < \beta < \kappa\}$$

(Both $\mathscr{L}(M)$ and $\mathcal{M}^\circ$ are defined in Definition 1.5.) Let T_0 be any finite sub-theory of T. Since $\mathcal{M}$ is infinite, T_0 has a model which results from interpreting the finitely many new constants c_α as standing for distinct elements in $\mathcal{M}^\circ$. So by the Compactness Theorem 4.1, T has a model $\mathcal{N}$ of size at least κ. By (1), we can assume its size is *exactly* κ, and by a Push-Through Construction, we obtain that $\mathcal{M} \preceq \mathcal{N}$. □

8

Categoricity and the sets

In Chapter 7, we focussed on arithmetic. We now turn our attention to set theory.

From the outset of this book, we have approached model theory set-theoretically. So, in considering the model theory of *set theory itself*, we can consider the extent to which a model of set theory aligns with our set-theoretic metatheory. This leads to some subtle differences—both technical and philosophical—between discussions of the 'intended interpretation' of arithmetic and of set theory. Nonetheless, many of the philosophical upshots are similar to those of Chapter 7.

As in Chapter 7, we will start by discussing the barrier to any kind of categoricity imposed by Compactness and the Löwenheim–Skolem Theorem. We then present Zermelo's *Quasi*-Categoricity Theorem for ZFC_2 and survey philosophical reactions to it, noting the parallels between arithmetic and set theory as we go.

The chapter ends with some lengthy technical appendices. In §8.A we prove Zermelo's Theorem. We provide the result because, although it is often invoked by philosophers of mathematics, we know of no reasonably self-contained treatment of it. In §8.B we show how to build up a more minimal set theory, due to Scott and Potter, and in §8.C we prove a quasi-categoricity theorem for this minimal theory.

8.1 Transitive models and inaccessibles

Throughout this chapter, we assume that all models are in the signature of set theory, i.e. the signature whose only primitive is $\in$. We start with a brief technical preamble concerning models of set theory. (Readers who are familiar with the technicalities should feel free to skim this section.) The notion of a *transitive* model is key to our discussion. Intuitively, a transitive model interprets set membership *correctly*, so far as it interprets it at all. More precisely:

Definition 8.1: *A set is* transitive *iff every element of it is also a subset of it. A model* $\mathcal{M}$ *is* transitive *iff both (i) its underlying domain,* M*, is a transitive set and (ii) if* a, b *are elements of* M*, then:* $a \in b$ *iff* $\mathcal{M} \vDash a \in b$.[1]

[1] Other texts describe what we are calling transitive models simply as a pair $(M, \in)$, or a pair $(M, \in_{\restriction M})$, where M is a transitive set. See Kunen (1980: 112, 141) and Jech (2003: 163).

Perhaps the most important transitive sets are those which constitute the cumulative hierarchy:[2]

Definition 8.2: *The* cumulative hierarchy *comprises the following sets, defined by transfinite recursion:*

$$V_0 = \varnothing \qquad V_{\alpha+1} = \wp(V_\alpha) \qquad V_\alpha = \bigcup_{\beta<\alpha} V_\beta \textit{ if } \alpha \textit{ is a limit ordinal}$$

It is easy to prove by simultaneous induction on α that: (i) each of these sets is transitive; that (ii) if $\alpha \leq \beta$ then $V_\alpha \subseteq V_\beta$; and that (iii) the members of V_α which are ordinals are precisely the ordinals less than α.

The Axiom of Foundation implies that every set is a member of some element of the cumulative hierarchy.[3] This allows us to define a *rank function* from sets to ordinals as follows: $\text{rank}(x) = \beta$ iff β is the least ordinal such that $x \in V_{\beta+1}$. So this is a map from sets to ordinals which tracks when a set 'first enters' the hierarchy.[4]

For reasons we explain below, some particularly important stages in the hierarchy are associated with inaccessible cardinals.[5] These are defined as follows:

Definition 8.3: *A cardinal κ is* regular *iff there is no $\alpha < \kappa$ and function $f : \alpha \longrightarrow \kappa$ whose image is unbounded in κ. A cardinal κ is* inaccessible *iff: $\kappa > \omega$, and κ is regular, and if $\lambda < \kappa$ then $|\wp(\lambda)| < \kappa$.*

The basic idea behind regular cardinals is that they cannot be 'approached from below'. An equivalent characterisation is: κ is regular iff the union of $< \kappa$ many sets, each of cardinality $< \kappa$, is itself of cardinality $< \kappa$.[6] Since cardinal exponentiation may be defined by $2^\lambda = |\wp(\lambda)|$, the additional thought behind inaccessible cardinals is just that they are closed under cardinal exponentiation.

The foundational interest of inaccessible cardinals is indicated by this result, which we prove in §8.A:

Theorem 8.4: *$\mathcal{A} \vDash \text{ZFC}_2$ iff $\mathcal{A} \cong V_\kappa$ for some inaccessible κ.*

---

[2] See end of §1.B for a brief review of notation for, and elementary results about, ordinals.

[3] In fact, against the other axioms of ZFC, Foundation is *equivalent* to the claim that every set is a member of some element of the cumulative hierarchy. See Kunen (1980: 101).

[4] For more on the rank function, see Kunen (1980: 104) and Jech (2003: 68).

[5] Sometimes what we are calling 'inaccessible cardinals' are called 'strongly inaccessible cardinals'. The relevant contrast is to 'weakly inaccessible cardinals' which are defined to be infinite regular cardinals κ which satisfy the weaker condition that if $\lambda < \kappa$ then $\lambda^+ < \kappa$, where λ^+ denotes the least cardinal greater than λ. We will not need the notion of 'weakly inaccessible cardinal', and so simply shorten 'strongly inaccessible cardinal' to 'inaccessible cardinal'.

[6] See Jech (2003: 32).

The second-order theory ZFC_2 is set out in Definition 1.12, and the semantics here are the *full* semantics for second-order logic, rather than Henkin semantics (for the differences, see §§1.10–1.11). So Theorem 8.4 states that, up to isomorphism, the full models of ZFC_2 are exactly the inaccessible stages of the cumulative hierarchy. Moreover, since ZFC_2 trivially entails the ordinary first-order theory ZFC, these are also very natural models of ZFC.

With these technical points behind us, we turn to some of the philosophical issues concerning models of set theory.

8.2 Models of first-order set theory

As in Chapter 7, we begin by considering barriers to the very idea of producing anything like a categorical set theory. The classic contemporary reference for this is Putnam's famous 'Models and Reality'.[7] In this, Putnam criticises a kind of moderate objects-platonism about set theory. Very briefly put: using model theory, he shows that there are unintended models of ZFC, and then challenges moderate objects-platonists to explain how to can rule them out.

Refining the challenge

Putnam's criticism of moderate objects-platonism is, though, an instance of something slightly more general, namely: it is a set-theoretical version of the Modelist's Doxological Challenge. The Challenge for objects-modelists is exactly as stated above: they must explain how we can refer to particular models of set theory. But a similar Challenge arises for concepts-modelists: they must explain why, for example, some models of ZFC are very poor explications of our concept *uncountable set*, or *ordinal*.

Having reframed Putnam's argument as a Doxological Challenge, we can use model theory to give the Challenge bite. As noted in §7.3, the Löwenheim–Skolem Theorem 7.2 shows that every countable first-order theory with an infinite model has models of every infinite size. In particular, then, ZFC, has models of every infinite cardinality. And this is sufficient to run an argument just like that of Chapter 7 against the modelist, but focussing on set theory rather than arithmetic.

Equally, we can use the Compactness Theorem 4.1 to generate non-standard models of any first-order set theory. In particular: in §1.B we defined the set-theoretic operation s and the sets $\varnothing$ and ω. Let T be our favourite first-order set theory, presumably some consistent extension of ZFC. Let c be a new constant, and define a new theory:

---

[7] Putnam (1980).

$$T^* = T \cup \{c \in \omega\} \cup \{s^n(\varnothing) \in c : n < \omega\}$$

As in §4.1, every finite subset of T^* has a model; so T^* itself has a model. But the interpretation of c makes this model a poor explication of our concept *ordinal*: in effect, there are infinitely many things that the model thinks are members of c, but the model also thinks that c is a finite von Neumann ordinal.

Putnam himself, though, attempted to use slightly different model-theoretic results than those we just mentioned. He attempted to apply the Skolem Hull Construction, Theorem 7.2(1), to the set-theoretic universe *itself*.[8] This generates certain complications, which we explore for the rest of this section.

Gödelian considerations

The first complication is clear:[9] we *cannot* apply the Skolem Hull Construction to the set-theoretic universe itself. That result only applies to the notion of a *structure* as formally defined in Definition 1.2, where the underlying domain of the structure is a *set*. And the set-theoretic paradoxes dictate that there is no set of all sets. So the set-theoretic universe should not be thought of as a *model* of set theory, in the sense of Definition 1.2, and Theorem 7.2(1) cannot be directly applied to it.

An obvious response to this difficulty would be to augment the theory ZFC, so that it entails the existence of a (set-sized) model of ZFC.[10] We might do this by adding 'there is an inaccessible cardinal' to ZFC. By the comments in §8.1, this would be like adding to ZFC the claim 'ZFC has a very natural model'. We could then simply relativise Putnam's argument to the initial segment of the cumulative hierarchy, up to the first inaccessible, which is a model of ZFC by Theorem 8.4.

Since set-theorists often invoke inaccessible cardinals themselves, the move just suggested on Putnam's behalf does not seem *ad hoc*. Nevertheless, as Bays has noted, it raises a dilemma for Putnam.[11] Where we use φ to abbreviate the claim 'there is an inaccessible cardinal', Bays' dilemma can be put as follows:

Rejection. *A modelist who accepts* ZFC *has not yet incurred any commitment to* ZFC + φ*. So she can simply reject* φ*, and deny that there are any unintended models of* ZFC.

[8] Putnam (1980). Putnam offered a further, more complicated, argument against a moderate objects-platonist who is considering whether to accept the axiom of constructibility. For more on that argument, see Putnam (1980: 466–70), Velleman (1998), Bays (2001, 2007), Bellotti (2005), and Button (2011). Scowcroft (2012) rectifies some technical errors in Button (2011).

[9] Here we follow Bays (2001: 335–6, 2007: 119–23) and Velleman (1998). Bays explains the problem with the Skolem Hull Construction as follows: If it were legitimate to apply that Construction to the set-theoretic universe itself, then one could prove in ZFC that there was a model of ZFC. Then, by the Completeness Theorem, one could prove Con(ZFC) in ZFC. So, by Gödel's Second Incompleteness Theorem, ZFC would be inconsistent.

[10] See Bays (2001: 338, 2007: 123–124) and Bellotti (2005: 396).

[11] Bays (2001: 340, 2007: 122). For a more general version of the dilemma, see Button (2011: 322–3).

Nonchalance. *A modelist who accepts* ZFC $+ \varphi$ *can react with nonchalance to the (mere) existence of unintended models of* ZFC*. She would only care about the existence of unintended models of* ZFC $+ \varphi$*, and Putnam has not (yet) shown that there are any such models.*

Both strategies seem to offer good ways for side-stepping the doxological problem of dealing with unintended models of one's favourite set theory.

To cut off the *Rejection* horn, Putnam must weaken the auxiliary hypothesis, φ, so that someone who accepts ZFC has no real option but to *accept* ZFC $+ \varphi$. One natural way to do this is to take φ to be 'ZFC is consistent', abbreviated Con(ZFC); after all, as mentioned in §5.6, there is a long tradition of thinking that acceptance of a theory T entails acceptance of Con(T).[12] Moreover, the hypothesis Con(ZFC) will serve the purpose of providing an unintended model of set theory. After all, the Completeness Theorem 4.24 shows that there is a countable model of ZFC.

In this case, though, *Nonchalance* looks very plausible. By Gödel's Second Incompleteness Theorem, we know that a model of T need not be a model of Con(T). So a moderate modelist who accepts T + Con(T) can briskly dismiss models of T which are not also models of Con(T);.

But matters now get complicated, since the preceding notion of *acceptance* iterates. The modelist who accepts T + Con(T) because she accepts T is probably also committed to Con(T + Con(T)). And with acceptance of this further theory, Putnam can again produce an unintended model of T + Con(T).

We are now set to embark on an infinite sequence of iterations. At any given stage β of this process, the modelist will attempt to opt for *Nonchalance*, saying that she would only care about the existence of unintended models of T_β + Con(T_β). Putnam will then point out that her acceptance of *this* theory, call it $T_{\beta+1}$, commits her to Con($T_{\beta+1}$), and he will then appeal to the Completeness Theorem to obtain an unintended model of $T_{\beta+1}$.

When an infinite regress arises, it is often hard to assess who wins. But the moderate modelist is definitely the loser in this case.[13] Either the moderate modelist has a *final* set theory which she accepts and thinks is consistent, or she admits that she *cannot* provide such a set theory. In the first case, she must accept that her theory has unintended, countable models. In the second case, she must also admit that she cannot 'pin down the sets'; for, as a moderate, she must accept that the only possible route for pinning down the sets is through providing theories (see §7.1), and she has just accepted that no particular theory is up to the job. So the moderate modelist is damned either way.

However, two points have emerged during this discussion. First: we have abandoned Putnam's attempt to use the Skolem Hull Construction 7.2(1) to turn a given

[12] For references, see footnote 36 of Chapter 5.

[13] For more on this regress, see Bays (2001: 126–7) and Button (2011: 329–33).

model of set theory into a countable model. Instead, we have presented a countable model by applying the Completeness Theorem 4.24 to a first-order *theory*. Second: once we focus on theories rather than models, nothing much depends upon the fact that we are considering *set* theory rather than any other theory. In sum: the case of unintended models of set theory no longer seems very *distinctive*.

Transitive models and Skolem's Paradox

We now consider a second issue, which is more specific to models of set theory.

When we discussed unintended models of arithmetic in §7.10, we considered the idea that the intended model of arithmetic should have *computable* addition and multiplication functions (see option (g) of §7.10 and our comments on Tennenbaum's Theorem). We might want to consider a similar idea in the case of set theory. In particular, it is reasonably common to insist that the intended model of set theory should be *transitive*, in the sense of Definition 8.1.[14] And, if it is legitimate to insist upon transitivity, this will undercut any attempt to generate a countable model using simply the Skolem Hull Construction 7.2(1) or the Completeness Theorem, since neither Theorem is guaranteed to produce transitive models.

Towards the end of this subsection, we shall consider whether it *is* legitimate to insist on transitivity. First, we note a basic result in the model theory of set theory which can restore transitivity. To state the result, we need some definitions.

Definition 8.5: *Let $\mathcal{A}$ be a structure. Then $\mathcal{A}$ is* extensional *iff it satisfies the Extensionality axiom. And $\mathcal{A}$ is* well-founded *iff there are no infinite descending $\in^{\mathcal{A}}$-chains in $\mathcal{A}$, i.e., there is no sequence a_n from A indexed by natural numbers $n \geq 0$ (in the metatheory) such that $\mathcal{A} \vDash a_{n+1} \in a_n$ for all $n \geq 0$.*

Using this, terminology we can state Mostowski's Collapse Lemma (we leave its proof to §8.A):[15]

Lemma 8.6 (Mostowksi Collapse): *Any well-founded, extensional structure is isomorphic to a transitive structure.*

The following simple Corollary illustrates the scope and limits of Mostowski's Lemma. (The notations $\preceq$ and $\equiv$ are set out in Definitions 4.3 and 2.4):[16]

---

[14] This suggestion was made by Tarski (quoted in Skolem 1970: 638); see also Benacerraf (1985: 101–4) and Wright (1985: 118).

[15] See Mostowski (1969: 20–1), Kunen (1980: 106), and Jech (2003: 69). There is also a more general version of the Mostowski's Collapse Lemma which applies to classes, and in these renditions of the Lemma, there will be a further condition that the interpretation of the membership relation is 'set-like'.

[16] See Mostowski (1969: Theorem III.3.8 p.43), McIntosh (1979: 321–2), and Button (2011: 344–6).

Corollary 8.7: *Let $\mathcal{A}$ be a transitive model of ZF. Then there is a countable, transitive model $\mathcal{B}$ such that $\mathcal{A} \equiv \mathcal{B}$.*

Proof. Let $\mathcal{A}$ be a transitive model $\mathcal{A}$ of ZF. Applying the Skolem Hull Construction 7.2(1), obtain a countable substructure $\mathcal{H} \preceq \mathcal{A}$. Since $\mathcal{A}$ is transitive, by Definition 8.1 the membership relation on $\mathcal{A}$ is just the usual membership relation, and so it is has no infinite descending membership chains. And since $\mathcal{H} \preceq \mathcal{A}$, we also have that $\mathcal{H}$ is well-founded. Further, $\mathcal{H}$ is extensional since $\mathcal{H} \vDash$ ZF and hence $\mathcal{H}$ models the Extensionality Axiom. So Lemma 8.6 applies, and $\mathcal{H}$ is isomorphic to a transitive structure $\mathcal{B}$. Since $\mathcal{B} \cong \mathcal{H} \preceq \mathcal{A}$, we have that $\mathcal{A} \equiv \mathcal{B}$. □

The model generated by this Corollary has some nice features. In particular: if $\mathcal{B}$ is a countable, transitive model of ZF, then every element of B is a countable set. Now, because ZF proves 'there is an uncountable set', we have:

$$\mathcal{B} \vDash \exists x \neg \exists y (y \text{ is an enumeration of } x)$$

Here, 'y is an enumeration of x' is an informal abbreviation for the usual set-theoretic rendering of the idea that y is a surjective function from ω to x. Hence, for some a in $\mathcal{B}$'s domain:

$$\mathcal{B} \vDash \neg \exists y (y \text{ is an enumeration of } a) \qquad (sko{:}big)$$

But because $\mathcal{B}$ is both countable and transitive, every member of $\mathcal{B}$'s domain is countable. So in particular:

$$\exists y (y \text{ is an enumeration of } a) \qquad (sko{:}small)$$

This is perhaps the sharpest version of what is called *Skolem's Paradox*. It is called a 'paradox' because of the apparent tension between (*sko:big*) and (*sko:small*). But the tension can be readily explained and dissolved. There is an enumeration of a. However, this enumeration lies outside $\mathcal{B}$'s domain, and so beyond the range of the quantifier '$\exists y$' as interpreted *within* $\mathcal{B}$.[17] In a slogan: to be uncountable-according-to-$\mathcal{B}$ is not to be uncountable *simpliciter*. Paradox dissolved.

We shall, then, set aside Skolem's Paradox, and simply return to the use of Corollary 8.7 in attacking moderate modelism about set theory.

First, note that Corollary 8.7 has exactly the same drawbacks as the use of the Skolem Hull Construction 7.2(1): it operates on a given *model* to generate a new model. As we noted when discussing Bays' dilemma, the set-theoretic universe is

[17] Because $\mathcal{A} \vDash$ ZF is countable and transitive, 'being an enumeration of' is absolute in the sense of Definition 8.11. So, it is the relativity of the quantifier, and not the relativity of the notion of an enumeration, that does the work in this version of Skolem's Paradox. For more on the mathematics of Skolem's Paradox, see Bays (2014: §2).

not a *model*, since there is no set of all sets. So whatever problems arose when considering the Skolem Hull Construction 7.2(1) also arise when using Corollary 8.7.

Second, the countable transitive model $\mathcal{B}$ obtained by Corollary 8.7 is *not* a set of the form V_α, in the sense of Definition 8.2.[18] But, if it makes sense to insist that models of set theory must be *transitive*—which is to say that they must interpret membership in the same way as the metatheory—it is only a small further step to insist that a model of set theory must be a stage of the iterative hierarchy.

All of this suggests the following: *if* it is legitimate to insist that the intended interpretation must make membership transitive, *then* it is very hard to attack moderate modelism about set theory via results from model theory. However, we must emphasise the conditional nature of this point. In §7.10, we argued that moderate modelists will succumb to the just-more-theory manoeuvre if they try to insist that the intended model of arithmetic should be *computable* (for example). A similar fate will befall moderate modelists who insist that models of set theory should be *transitive*.

Ultimately, to insist on transitivity amounts to insisting that we are *really* picking out set-membership, and not something which merely looks and quacks like set-membership from a first-order perspective. Given the sheer abstractness of sets, we can—as usual—ask the moderate modelist how her set-theoretic language succeeds in picking out the *actual* set-membership relation. And, just as before, model theory can be used to apply pressure. At the start of this section, we showed how to generate non-standard models of any consistent first-order set theory using the Compactness Theorem 4.1. Such models are not well-founded, and so they are not transitive; but nothing expressible in first-order logic rules them out.

The upshot of all of this is as follows. The best attacks on moderate modelism about set theory involve first using the just-more-theory manoeuvre to block (e.g.) appeals to transitivity, and then appealing to the same kinds of elementary model-theoretic results as we invoked in Chapter 7. So, at this point, there is very little about the attack that is specific to *set theory*. Much as in the case of arithmetic, the point is *merely* this: first-order theories with infinite models cannot be categorical, and attempts to go beyond first-order logic inevitably look like just more mathematical theorising.

8.3 Zermelo's Quasi-Categoricity Theorem

In the setting of arithmetic in the last chapter, we considered Dedekind's Categoricity Theorem 7.3 for the second-order theory PA_2. Similarly, in the case of set theory,

[18] To see this, note that $\mathcal{B} \neq V_\alpha$ for any $\alpha > \omega$, since these are all *uncountable*; and $\mathcal{B} \neq V_\alpha$ for any $\alpha \leq \omega$, since none of these model the Axiom of Infinity. For more, see the proof of Lemma 8.16.

we shall consider Zermelo's Quasi-Categoricity Theorem for the second-order theory ZFC_2. Roughly put, Zermelo's Theorem states that any two models of ZFC_2 are either isomorphic, or one is isomorphic to an initial segment of the other. But we must state this result precisely.

As in Definition 8.2, the cumulative hierarchy is defined by transfinite recursion. Transfinite recursion can be carried out in any model of ZFC. So, where $\mathcal{A}$ is a model of ZFC, we can define the entity which $\mathcal{A}$ thinks is the cumulative hierarchy. Slightly more precisely, we define a sequence of elements $V_a^{\mathcal{A}}$, where a ranges over the elements of A which $\mathcal{A}$ thinks are ordinals. (Note: this somewhat cute talk, about what a model 'thinks', abbreviates something perfectly well defined. When we say, for example, that $\mathcal{A}$ 'thinks that a is an ordinal', we mean that a is an element of A which satisfies the formalisation of the predicate 'v is an ordinal' in $\mathcal{A}$.)

For each model $\mathcal{A}$ of ZFC, we can now naturally treat $V_a^{\mathcal{A}}$ as a set-theoretic structure itself. Intuitively, the structure is given just by looking inside $\mathcal{A}$ for what is in $\mathcal{A}$'s cumulative hierarchy up to level a, and then restricting $\mathcal{A}$'s membership relation to these entities. More precisely, $V_a^{\mathcal{A}}$'s domain is $A_a = \{b \in A : \mathcal{A} \vDash b \in V_a\}$ and the membership symbol is interpreted by $\{(d, e) \in A_a \times A_a : \mathcal{A} \vDash d \in e\}$. To ease readability, we simply denote this structure by $V_a^{\mathcal{A}}$ as well.

We can now state Zermelo's Theorem, leaving its proof to §8.A:[19]

Theorem 8.8 (Zermelo's Quasi-Categoricity Theorem): *Let $\mathcal{A}$ and $\mathcal{B}$ be full models of ZFC_2. Then exactly one of the following obtains:*

(1) $\mathcal{A} \cong \mathcal{B}$
(2) $\mathcal{A} \cong V_a^{\mathcal{B}}$, *for some a which $\mathcal{B}$ thinks is an inaccessible cardinal*
(3) $\mathcal{B} \cong V_a^{\mathcal{A}}$, *for some a which $\mathcal{A}$ thinks is an inaccessible cardinal*

8.4 Attitudes towards full second-order logic: redux

In §7.6, we noted that Dedekind's Categoricity Theorem 7.3 for PA_2 is unimpeachable, as a bit of pure mathematics, but that several different basic philosophical attitudes can be adopted in response to it: the Algebraic, Logic-First, Infer-to-Stronger

---

[19] This essentially rolls together Zermelo's First and Second Isomorphism Theorems (1930: §4), with two differences. First: Zermelo's results were formulated so as to allow for *urelements*, whereas we have considered only *pure* sets. (We explain the use of urelements slightly more in Chapter 11.) Second: we have added to clauses (2) and (3) the claim that the larger model 'thinks a is an inaccessible cardinal'; this extension is useful in §8.5.

Zermelo was interested in this Theorem since it allows us to conceive of the models of ZFC_2 as a linear hierarchy. (This is also a consequence of Theorem 8.4, from which we deduce Theorem 8.8 in §8.A.) Zermelo thought that this gave a compelling answer to the set-theoretic paradoxes, since what is a second-order object at one level becomes a first-level object at a higher level. This is at least one natural way to read the complex passage in Zermelo about the 'ultrafinite antinomies' (see Zermelo 1930: 47, 2010: 429; Ewald 1996: v.2 p.1233).

Logic, and Holistic Attitudes. Exactly the same point applies to Zermelo's Theorem 8.8, and there are representatives of each of these four attitudes.

The Logic-First Attitude. In 1963, Cohen announced his proof of the independence of the continuum hypothesis, CH, from first-order ZFC.[20] An explosion of set-theoretic independence results soon followed. Against this background, Kreisel drew attention to Zermelo's Theorem 8.8, arguing that CH has a determinate truth value, despite its independence from ZFC. His reasoning was as follows: since CH only concerns sets of low rank, it is either true in all full models of ZFC_2 or false in all full models of ZFC_2, and hence either true or false *simpliciter* (and not indeterminate) on the supervaluational semantics explained in §2.5 and §7.3.[21] Kreisel, then, assumed that he *had* unproblematic access to full second-order logic. His was the Logic-First Attitude.

The Algebraic Attitude. Like Kreisel, and indeed contemporaneously, Mostowski was interested in the philosophical implications of set-theoretic independence results. Mostowski, however, drew a rather different conclusion than Kreisel:

> [...] the incompleteness of set-theory [...] is comparable [...] to the incompleteness of group theory or of similar algebraic theories. These theories are incomplete because we formulated their axioms with the intention that they admit many non-isomorphic models. In [the] case of set-theory we did not have this intention but the results are just the same.[22]

In short, Mostowski argued that the independence results concerning ZFC had shown that set theory itself is *algebraic*, in the sense of §2.2, that it has no intended model. So, in contrast with Kreisel, Mostowski adopted a similarly algebraic attitude towards second-order logic; an attitude that was inevitable, given his claim that second-order logic 'is a part of set theory'.[23] Hamkins has more recently espoused the same attitude: on the basis that '[s]et theory appears to have discovered an entire cosmos of set-theoretic universes', he holds that Zermelo's Theorem 8.8 simply reveals that the idea of 'full' second-order quantification is exactly as indeterminate as the sets themselves.[24] This is precisely the Algebraic Attitude.

The Infer-to-Stronger Logic Attitude. In fact, immediately after expressing his view that set theory is algebraic, Mostowski voiced a potential concern:

> [...] if there are a multitude of set-theories then none of them can claim the central place in mathematics. Only their common part could claim such a position; but it is debatable

[20] Cohen (1963, 1964). That is, ZFC $\nvdash$ CH and ZFC $\nvdash \neg$CH, where CH is the statement that there are no cardinals strictly between $\aleph_0$ and $2^{\aleph_0}$.

[21] See Kreisel (1967: 150). Kreisel later revealed slightly more of his attitude towards this result: 'CH *is* decided by the full (second order) axioms of Zermelo; by the above this is already something although we don't know which way [CH is decided ...]. Our *present* analysis of Zermelo's axioms, that is the first order schemata in the usual language of set theory, is not sufficient to decide CH. Put succinctly: not the notion of set, but our analysis (present knowledge) of this notion is at fault' (1971: 196).

[22] Mostowski (1967: 94).

[23] Mostowski (1967: 107).

[24] Hamkins (2012: 418, 427–8).

whether this common part will contain all the axioms needed for a reduction of mathematics to set-theory.[25]

This suggests why independence in set theory might be more philosophically troubling than independence in more obviously algebraic theories (such as group theory): it might threaten the frequently stated idea that all of mathematics can be represented within the sets.[26] And so one might contrapose Mostowski's reasoning: If the image of set theory as foundational is thought to be sufficiently central to our conception of mathematics, and if additionally that image is genuinely threatened by regarding set theory as algebraic, then we might be able to mount an 'inference to the best explanation', that we can grasp full second-order logic. That is the Infer-To-Stronger Logic Attitude.

The Holistic Attitude. The final attitude is simply that there is a false dichotomy between set theory and higher-order logic. This is precisely the attitude recently expressed by Shapiro:

> [...] second-order logical consequence *is* intimately bound up with set theory [...]. But that does not disqualify second-order logic from logical and foundational studies. Mathematics and logic are a seamless whole, and it is impossible to draw a sharp boundary between them.[27]

All told, then, all four philosophical reactions to a (quasi-)categoricity result are on display. Which are genuinely *available* to you, though, will depend on your starting philosophical assumptions.

In particular, and exactly as in Chapter 7, the moderate modelist about set theory will want to embrace something other than the Algebraic Attitude. She will want to do so, since she can only hope to answer the Modelist's Doxological Challenge, in the specific case of set theory, by supplying a (quasi-)categorical theory (see §§7.1–7.2). So she will want to appeal to Zermelo's Theorem 8.8 to answer the Challenge. And to do that, she must insist that she has an unproblematic grasp of full second-order logic. But, exactly as in §7.7, her *moderation* will prevent her from being able to say why *full* models of ZFC_2 are preferable to *Henkin* models of ZFC_2. As such, she will be unable to make anything positive out of Zermelo's Theorem 8.8.

At this point, the moderate modelist might well protest. In response, we would re-run all of the arguments of Chapter 7, occasionally replacing the phrase 'arithmetic' with 'set theory'. Flogging this dead horse would, though, be exhausting and

---

[25] Mostowski (1967: 94–5) and G. H. Moore (1982: 4).

[26] There is obviously much more one could say about Mostowski's concern that the common part may be too meagre for reductive purposes. For example: the notion of 'reduction' which Mostowski has in mind would be most naturally explicated via the notion of *interpretability* from Chapter 5. However, Koellner (2009: 99) points out that *both* ZFC + CH and ZFC + $\neg$CH are reducible to ZFC in this sense. Hence, the 'core' might well be less meagre than Mostowski had feared.

[27] Shapiro (2012: 312).

unilluminating. So we shall pass over the corpse of moderate modelism without further ado.

8.5 Axiomatising the iterative process

Once we have set aside moderate modelism, though, two philosophical reactions that *are* distinctive to the case of set theory can emerge. These reactions are due to Martin and Isaacson, and they raise the interesting question: *What does set theory axiomatise?* We discuss Martin here, and Isaacson in the next section.

According to the *iterative conception of set*, sets are 'constructed' iteratively in a transfinite process of set formation, much as is suggested by the cumulative hierarchy of Definition 8.2. Martin explicitly holds that ZFC 'should be thought of as an attempt to axiomatise the iterative concept'.[28] In this light, Martin views Zermelo's Quasi-Categoricity Theorem 8.8 as showing that any two implementations of the iterative conception of set will end up generating isomorphic structures (as far as they go). As a result, and as in the discussion of the Logic-First Attitude in §8.4, any two such implementations will agree on the truth-value of 'low level' statements such as CH.[29]

Martin is a modelist about the iterative conception of sets. So, for familiar reasons, Martin cannot appeal to Zermelo's Theorem 8.8 whilst remaining a moderate. However, for those who are prepared to abandon moderate modelism, nothing obviously prohibits the appeal to full second-order logic in general and Zermelo's Theorem 8.8 in particular.

The more interesting (because more specific) issue for Martin is that, in considering the iterative conception of set, one might object to the use of second-order logic *even in* the object language. As Reinhardt once put it, anyone who uses second-order logic just seems to have 'forgotten' to add a level of sets:

> [...] our idea of set comes from the cumulative hierarchy, so if you are going to add a layer at the top it looks like you just forgot to finish the hierarchy.[30]

Given Martin's invocation of the iterative conception of set, we can reformulate Reinhardt's complaint as follows. On the one hand, the idea of the iterative conception is that properties at one stage are transformed into sets at a subsequent stage. On the other hand, the use of a second-order theory seems to involve postulating a fixed class of properties that are never transformed into sets. This gives rise to a clear tension.

---

[28] Martin (1970: 112).

[29] Martin (2001, 2015). See §11.B for further discussion of Martin's views.

[30] Reinhardt (1974: 196), cf. Burgess (1985: 546). Note that Reinhardt raised this concern, not in the context of Zermelo's Theorem, but in the context of his own use of higher-order logic to extend the standard set-theoretic axioms via reflection principles.

In response, Martin might maintain that ZFC_2 simply aims to axiomatise an *arbitrary stage* in the iterative process. This would involve replying to Reinhardt as follows: *we did not forget to finish the hierarchy; it is just that the second-order entities in one model of* ZFC_2 *will become first-order entities in a model of* ZFC_2 *which occurs later in the process.* The basic idea is reasonable, but ZFC_2 itself is poorly suited to the task of axiomatising an *arbitrary* stage in the process of set formation. After all, plenty of typical stages, like $V_{\omega+2}$, $V_{\omega+3}$, $V_{\omega+3}$, ..., are not models of ZFC_2, for they do not even satisfy the Power Set axiom.

For this reason, we suggest that Martin should move away from ZFC_2 and consider an alternative axiomatic set theory, namely second-order *Scott–Potter* level theory, SP_2. This theory is neither as well known as ZFC_2 nor as well known as it should be, so we must start by setting it down. Our axiomatisation is essentially due to Potter, who built on work by Scott.[31] However, their theories are first-order, and so we have tweaked the axiomatisation slightly (we explain the tweaks at the end of §8.B; we defined the notation $(\exists x : X)$ and $(\forall x : X)$ in §1.9).

Definition 8.9: *We define three formulas:*[32]

$$\begin{aligned} A(x,y) &:= \forall v(v \in x \leftrightarrow (\exists u \in y)(v \in u \lor v \subseteq u)) \\ H(x) &:= (\forall z \in x)A(z, x \cap z) \\ L(x) &:= (\exists y : H)A(x,y) \end{aligned}$$

The theory SP_2 *then consists of just two axioms:*
Extensionality. $\forall x \forall y(\forall z[z \in x \leftrightarrow z \in y] \rightarrow x = y)$
Levelling. $\forall X[\exists x \forall z(z \in x \leftrightarrow X(z)) \leftrightarrow (\exists y : L)(\forall z : X)z \in y]$

We read $A(x,y)$ as *x is the accumulation of y*, $H(x)$ as *x is a history*, and $L(x)$ as *x is a level.* The distinctive axiom, Levelling, can then be glossed as follows: *A property determines a set iff all of its instances are members of some level.* Stated like this, SP_2 offers an extremely minimal axiomatisation of the *very idea* of the iterative conception of sets. In particular, and unlike ZFC_2, it makes no comment on 'how far' the iterative process goes: there is no axiom of infinity, axiom of powersets, or anything similar, in SP_2.

That said, simply staring at the definition of $L(x)$ gives little clue as to why SP_2 has much to do with the iterative conception of sets. In §8.B, we run through some

[31] Scott (1974) and Potter (2004). This approach improves on the Boolos–Schoenfield stage axioms, since that uses two primitive notions, for set-membership and rank (Shoenfield 1967; Boolos 1971). Those stage axioms were partly anticipated by Scott (1960) himself. Potter (1993: 183) credits John Derrick with the trick which enables the deletion of a rank-primitive.

[32] *Note:* the use of '$\cap$' in defining H is for readability and to highlight a conceptual connection; but '$u \in x \cap z$' can simply be read as '$u \in x \land u \in z$' and can easily be eliminated from the definition.

deductive theorems of SP_2 which provide a better sense of how $L(x)$ behaves; but here we will simply state a key result (which we prove in §8.c):[33]

Theorem 8.10 (SP_2 Quasi-Categoricity): $\mathcal{A} \vDash SP_2$ *iff* $\mathcal{A} \cong V_\alpha$ *for some ordinal* $\alpha > 0$

Indeed, the stages of the cumulative hierarchy of Definition 8.2 are precisely what SP_2 calls 'levels' (up to isomorphism; see Proposition 8.32(2)). So, unlike ZFC_2, *the theory* SP_2 *axiomatises the very idea of an arbitrary stage in the iterative process of set-formation.*

As such, SP_2 seems better suited than ZFC_2 to Martin's philosophical purposes. Moreover, assuming the legitimacy of full second-order logic, SP_2 can still be used in the manner suggested by Martin (and Kreisel). For, whilst there are models of SP_2 which are too small to comment on CH, all of the models of SP_2 that allow the iterative process to go on for sufficiently long will agree about CH. For instance: all models isomorphic to V_α for some $\alpha \geq \omega + 1$ decide CH;[34] and if we specifically want to restrict our attention to these models, we can simply augment SP_2 with a sentence stating that there is an $\omega{+}1^{\text{th}}$-level. So Martin could happily invoke SP_2, rather than ZFC_2, to argue that CH has a determinate truth value.

8.6 Isaacson and incomplete structure

We now consider Isaacson's rather different attitude to Zermelo's Theorem 8.8. One of Isaacson's basic ideas is that any consistent theory describes a 'structure'. However, he also thinks that the *categoricity* of a theory determines 'whether that structure is general or particular'.[35] Now, Isaacson's 'particular structures' are essentially what we have been calling mathematical structures, informally construed. So, in our terminology, Isaacson is suggesting that the existence of a consistent, categorical theory is necessary and sufficient for the existence of a mathematical structure, informally construed.

Zermelo's Theorem 8.8, though, is a mere *quasi*-categoricity result. It states that any two (full) set hierarchies agree 'as far as they go', but it allows that one of them may outstrip the other. So, if we had hoped to use Zermelo's Theorem to explain

[33] Potter (2004) does not explicitly state or prove this result, but the availability of the result is implicit in the entire approach to axiomatising set theory in this way. The result is explicitly mentioned (though not proved) by Incurvati (2010: 130). Tait (1998: 474–5) describes a different theory with the same property, and suggests that it is 'the natural system with which to begin'. Finally, it ought be mentioned that Uzquiano (1999: §5) proves a categoricity result for V_α where $\alpha > \omega$ is a limit. Thanks to Luca Incurvati, Michael Potter, and Dana Scott for discussions about this.

[34] In this setting, we can formulate the continuum hypothesis as follows: if $X \subseteq 2^{\aleph_0}$ and there is a second-order injection $f : \mathbb{N} \longrightarrow X$, then either there is a second-order injection $g : X \longrightarrow \mathbb{N}$ or there is a second-order injection $g : 2^{\aleph_0} \longrightarrow X$. (Thanks to Hugh Woodin for suggesting this formulation over something more cumbersome.)

[35] Isaacson (2011: 32).

how we grasp *the* set hierarchy—in the sense of a mathematical structure which is unique-up-to-isomorphism—then we have a problem. Isaacson himself is aware of this, and suggests the following response:

> [...] what is undecided in virtue of this degree of non-categoricity is genuinely undecided, in the same way that the fifth postulate of Euclid's geometry is genuinely undecided by the axioms. This includes GCH (or some version that survives the refutation of CH), and the existence of large cardinals [...].[36]

Hence, on Isaacson's view, ZFC_2 does not aim to describe an arbitrary stage in an absolutely infinite sequence (as it did in our discussion of Martin). Rather, it aims to describe a 'particular structure' which is somehow 'incomplete'.

To see why Isaacson might say this, note that some expansions of ZFC_2 are fully categorical. For instance, for any natural number $n \geq 0$, the theory ZFC_2 plus 'there are exactly n inaccessibles' is categorical, given full second-order logic.[37] But for anyone who thinks of ZFC_2 in terms of the iterative conception of set, these categorical theories seems to fall short of the mark. After all, a theory stating 'there are exactly n inaccessibles' seems only to characterise some *initial segment* of the iterative hierarchy.

More generally, no *fully* categorical extension of ZFC_2 seems adequate to handle the purportedly *all-encompassing* nature of set theory. For, where $\mathcal{M}$ is a model of T, its domain, M, must omit some entities. For example, M's Russell set—the set of all non-self-membered sets in M, as characterised in the model theory—is not a member of M. So we know that we could have formed a more encompassing structure with a more encompassing domain than M. For this reason, it seems that a *quasi*-categoricity result is the most we *ought* to hope for from our set theory; that mere quasi-categoricity is somehow *inevitable*.[38]

Nevertheless, none of this forces us to follow Isaacson in holding that our quasi-categorical set theory describes a *single* but *incomplete* structure. We can equally hold that ZFC_2 is an inevitably *incomplete* axiomatisation, which applies to many *different* (complete) structures; namely, as suggested by Martin, to the different sequential stages of set-formation. And to our ears, at least, this is somewhat easier to understand than the idea of an incomplete entity.

---

[36] Isaacson (2011: 53; see also 4, 50). GCH is the generalised continuum hypothesis. This asserts that the cardinality of the powerset $P(\kappa)$ of an infinite cardinal κ is as small as possible, namely it is the next infinite cardinal beyond κ.

[37] To see this, suppose that $\mathcal{A}$ and $\mathcal{B}$ model ZFC_2, plus this additional axiom for some fixed $n \geq 0$. By Zermelo's Quasi-Categoricity Theorem 8.8, there are three options to consider.

Option (2) says that $\mathcal{A} \cong V_a^{\mathcal{B}}$, for some a which $\mathcal{B}$ thinks is an inaccessible cardinal. But $\mathcal{A}$ and hence its isomorphic copy $V_a^{\mathcal{B}}$ think that there are exactly n inaccessibles $a_0, \ldots, a_{n-1}$, which all lie below the inaccessible a in $\mathcal{B}$. But then $\mathcal{B}$ will think that there are at least $n + 1$ inaccessibles, a contradiction. This rules out option (2). Similar considerations, with the roles of $\mathcal{A}$ and $\mathcal{B}$ reversed, rule out option (3). So option (1) must obtain, i.e. $\mathcal{A} \cong \mathcal{B}$, as required.

[38] In Chapter 11 we discuss full categoricity results for second-order set theories (including McGee 1997). But crucially, those are *internal* categoricity results, in a sense we explain in Chapters 10–11.

8.A Zermelo Quasi-Categoricity

In this appendix, we prove Zermelo's Quasi-Categoricity Theorem 8.8. Our proof outline follows Kanamori, but his proof occurs in an advanced set theory text, and so is only a few paragraphs long.[39] Given the importance attached to Zermelo's Theorem, we want to offer a detailed and reasonably self-contained proof.

We begin with a proof of the Mostowski Collapse Lemma 8.6. We used this to present Skolem's Paradox in §8.2, and we will need it to prove Zermelo's Theorem.

Lemma (Mostowski Collapse Lemma 8.6): *Any well-founded, extensional structure is isomorphic to a transitive structure.*

Proof. Let $\mathcal{A}$ be well-founded and extensional (see Definition 8.5). We first define a notion of rank relative to $\mathcal{A}$:

$$\text{rank}^{\mathcal{A}}(a) = \begin{cases} 0 & \text{if } \mathcal{A} \vDash \forall x\, x \notin a \\ \sup\left\{\text{rank}^{\mathcal{A}}(b) + 1 : \mathcal{A} \vDash b \in a\right\} & \text{otherwise} \end{cases}$$

This is well-defined, since $\mathcal{A}$ is well-founded and extensional. By transfinite recursion on $\text{rank}^{\mathcal{A}}$, we define a *collapse function, h*, such that $\text{rank}^{\mathcal{A}}(x) = \text{rank}(h(x))$:

$$h(a) = \begin{cases} \varnothing & \text{if rank}^{\mathcal{A}}(a) = 0 \\ \{h(b) : \mathcal{A} \vDash b \in a\} & \text{otherwise} \end{cases}$$

We now use h to define a model $\mathcal{B}$, the *Mostowski collapse* of $\mathcal{A}$:

$$B = \{h(x) : x \in A\}$$
$$\in^{\mathcal{B}} = \{(x,y) \in B \times B : x \in y\}$$

To see that $\mathcal{B}$ is transitive, suppose $b \in a \in B$. So for some $a' \in A$, we have $a = h(a') = \{h(b') : \mathcal{A} \vDash b' \in a'\}$. So $b = h(b')$, for some b' such that $\mathcal{A} \vDash b' \in a'$, so that $b \in B$.

It remains to prove that $h : \mathcal{A} \longrightarrow \mathcal{B}$ is an isomorphism. We prove by induction that h is an injection. Suppose for induction that whenever $\max(\text{rank}^{\mathcal{A}}(x), \text{rank}^{\mathcal{A}}(y)) < \gamma$ and $h(x) = h(y)$, we have $x = y$. Now suppose $\max(\text{rank}^{\mathcal{A}}(a), \text{rank}^{\mathcal{A}}(b)) = \gamma$ and $h(a) = h(b)$; we show that $a = b$ by showing inclusions in both directions inside $\mathcal{A}$. If $\mathcal{A} \vDash d \in a$, then $h(d) \in h(a) = h(b) = \{h(e) : \mathcal{A} \vDash e \in b\}$. So there is some e such that $\mathcal{A} \vDash e \in b$ and $h(d) = h(e)$, and since $\max(\text{rank}^{\mathcal{A}}(d), \text{rank}^{\mathcal{A}}(e)) < \gamma$, the induction hypothesis yields that $d = e$; hence $\mathcal{A} \vDash d \in b$. Similarly, if $\mathcal{A} \vDash e \in b$, then $\mathcal{A} \vDash e \in a$. So $a = b$, since $\mathcal{A}$

[39] Kanamori (2003: 18–19).

is extensional. So h is an injection; and since h is obviously a surjection, it is a bijection. Finally, h preserves the structure of $\in^{\mathcal{A}}$ by construction. For, on the one hand, if $\mathcal{A} \models b \in a$ then by definition of the function h, we have $h(b) \in h(a)$. On the other hand, if $h(b) \in h(a)$, then by definition of $h(a)$, we have $h(b) = h(b')$ for some $b' \in A$ satisfying $\mathcal{A} \models b' \in a$. Then $b = b'$ by the injectivity of h, so that $\mathcal{A} \models b \in a$. □

Now we move to results more directly concerned with the proof of Zermelo's Quasi-Categoricity Theorem 8.8.

We start with an important definition, which concerns the alignment between a model of the set-theoretic axioms, and our set-theoretic metatheory.

Definition 8.11: *Suppose that $\mathcal{A}$ is a model and $\varphi(\bar{x})$ is a formula. Then $\varphi(\bar{x})$ is* absolute *for $\mathcal{A}$ iff for all $\bar{a}$ from A we have: $\varphi(\bar{a})$ iff $\mathcal{A} \models \varphi(\bar{a})$.*

The left-side of the biconditional, '$\varphi(\bar{a})$', is a claim made in the metatheory itself. So the absoluteness of a formula in a structure indicates agreement between the structure and the metatheory. Unsurprisingly, absoluteness is a demanding condition. In these terms, Skolem's Paradox, discussed in §8.2, shows that 'x is uncountable' is not absolute for all transitive models.

The most widely applicable sufficient condition for a formula to be absolute is a syntactic condition, given by the following definition:

Definition 8.12: *The* Δ_0-formulas *are the smallest set of formulas in the signature of set theory which contain the quantifier-free formulas, are closed under the propositional connectives, and are such that if $\varphi(x)$ is Δ_0, then so are both $(\exists x \in y)\varphi(x)$ and $(\forall x \in y)\varphi(x)$.*

The quantifiers in the last part of this definition are called *bounded quantifiers*. So: the Δ_0-formulas are the formulas whose quantifiers (if any) are all bounded.

The Δ_0-formulas are not absolute for all structures. However, an easy induction on complexity shows that they are absolute for transitive models:[40]

Proposition 8.13: *If $\mathcal{A}$ is a transitive model, then every Δ_0-formula is absolute for $\mathcal{A}$.*

This Proposition is surprisingly applicable. For instance, there are Δ_0-formulas which define notions like ordered-pair, subset, cross-product, ordinal, being a function, being the domain of a function, and being the range of a function.[41]

[40] See Kunen (1980: 118–19) and Jech (2003: 163–4).

[41] See Kunen (1980: 119ff) and Jech (2003: 164–5). The only one of these which perhaps deserves comment is being an ordinal. The official definition of an ordinal is 'a transitive set *well-ordered* by mem-

We now start the march to Zermelo's Theorem, beginning with some elementary results concerning inaccessibles (as defined in Definition 8.3):

Lemma 8.14: *Let κ be an inaccessible cardinal:*

(1) *If $\gamma < \kappa$, then $|V_\gamma| < \kappa$.*
(2) *If $a \in V_\kappa$, then $|a| < \kappa$.*
(3) *If $a \subseteq V_\kappa$ and $|a| < \kappa$, then $a \in V_\kappa$.*

Proof. (1). This is a simple transfinite induction. For $\gamma = 0$ this follows immediately from the fact that $V_0 = \varnothing$. For $\gamma = \beta + 1$, this follows from κ's inaccessibility, which implies that if $|V_\beta| < \kappa$ then $|V_\gamma| = |V_{\beta+1}| = |\wp(V_\beta)| < \kappa$. For γ a limit, this follows from κ's regularity, since the union of $< \kappa$ many sets of cardinality has cardinality $< \kappa$ for regular κ.

(2). Suppose $a \in V_\kappa$. Since κ is a limit, there is $\gamma < \kappa$ such that $a \in V_\gamma$, and so by transitivity $a \subseteq V_\gamma$. So $|a| \leq |V_\gamma| < \kappa$, by (1).

(3). Suppose $a \subseteq V_\kappa$ and $|a| < \kappa$. Since $a \subseteq V_\kappa$ and κ is limit, $\text{rank}(b) < \kappa$ for any $b \in a$. Choose a bijection $F : |a| \longrightarrow a$ and compose it with the rank function. Since κ is regular and $|a| < \kappa$, the composition $\text{rank} \circ F : |a| \longrightarrow \kappa$ has bounded range. So $\text{rank} : a \longrightarrow \kappa$ also has bounded range, say by $\gamma < \kappa$. Then $\text{rank}(a) \leq \gamma + 1 < \kappa$, and so $a \in V_\kappa$. □

The proof of Zermelo's Theorem 8.8 takes several stages. Throughout, we use the expression '$\mathcal{A}$ thinks that...', as introduced in §8.3. We also use the full semantics for second-order logic throughout. We start with a result which effectively entails the right-to-left direction of Theorem 8.4, along with some additional information about the interpretation of power set and cardinality:

Theorem 8.15: *Let κ be an inaccessible cardinal:*

(1) $V_\kappa \vDash \text{ZFC}_2$.
(2) *If $a \in V_\kappa$, then $\wp^{V_\kappa}(a) = \wp(a)$, where $\wp^{V_\kappa}(a)$ is V_κ's interpretation of the powerset of a.*
(3) *If $a, b \in V_\kappa$, then $|a| = |b|$ iff $V_\kappa \vDash |a| = |b|$. Hence, a in V_κ is a cardinal iff V_κ thinks that a is a cardinal.*

Proof. For (1), we must verify each of the axioms of ZFC_2 in turn (see §1.B for the axioms). When we come to the Power Set Axiom, we will additionally verify (2). After this, we will verify (3).

Extensionality. This follows from the transitivity of V_κ. Suppose that V_κ thinks that a and b have exactly the same members, i.e. that $c \in a$ iff $c \in b$, for all $c \in V_\kappa$; we

bership'. But, on transitive models, this is equivalent by the Axiom of Foundation to being a transitive set *linearly* ordered by membership, which is obviously a Δ_0-condition.

must show that this holds for *all* c. Since $a, b \in V_\kappa$ and V_κ is transitive, both a and b are subsets of V_κ. So suppose $c \in a$; then since $a \subseteq V_\kappa$, we have $c \in V_\kappa$, and so by hypothesis $c \in b$. Similarly, if $c \in b$ then $c \in a$. So $a = b$.

Separation. Suppose $F \subseteq V_\kappa$ and $a \in V_\kappa$. Then $|a| < \kappa$ by Lemma 8.14(2). Since $a \cap F \subseteq a \subseteq V_\kappa$ by transitivity, and $|a \cap F| \leq |a| < \kappa$, we have $a \cap F \in V_\kappa$ by Lemma 8.14(3).

Since we have Separation, by the remarks in §1.B, we only need to verify one direction of the biconditionals from the official versions of Pairing and Union.

Pairing. Suppose that $a, b \in V_\kappa$. Since κ is a limit, there is $\gamma < \kappa$ such that $a, b \in V_\gamma$. Then $\{a, b\} \in \wp(V_\gamma) = V_{\gamma+1}$. Since $\gamma + 1 < \kappa$, it follows that $\{a, b\} \in V_\kappa$. Finally, V_κ thinks that $c = \{a, b\}$ contains both a and b, since the relation $a \in c \wedge b \in c$ is Δ_0 and thus absolute.

Union. Suppose that $a \in V_\kappa$. Since κ is a limit, there is $\gamma < \kappa$ such that $a \in V_\gamma$. Now suppose that $c \in \bigcup a$. Then $c \in b \in a$ for some set b. Since $a \in V_\gamma$ and V_γ is transitive, $b \in a \subseteq V_\gamma$, and so $b \in V_\gamma$. Similarly, $c \in b \subseteq V_\gamma$ so that $c \in V_\gamma$. Hence $\bigcup a \subseteq V_\gamma$, and so $\bigcup a \in V_{\gamma+1}$ and thus $\bigcup a \in V_\kappa$. Finally, note that V_κ thinks that $d = \bigcup a$ contains all the members of the members of a, since the condition $(\forall b \in a)(\forall c \in b) c \in d$ is Δ_0 and thus absolute.

Power Set. Suppose that $a \in V_\kappa$. There is $\gamma < \kappa$ such that $a \in V_\gamma$. Suppose that $b \in \wp(a)$. Then $b \subseteq a \subseteq V_\gamma$, by the transitivity of V_γ, so that $b \in V_{\gamma+1}$. So $\wp(a) \subseteq V_{\gamma+1}$ and so $\wp(a) \in V_{\gamma+2}$ and thus $\wp(a) \in V_\kappa$. Further, V_κ thinks that $d = \wp(a)$ contains all and only the subsets of a. For, first suppose that $b \in V_\kappa$ and V_κ thinks that $b \subseteq a$; then by the absoluteness of subsethood, in fact $b \subseteq a$, so that $b \in \wp(a) = d$, which is again absolute, so that V_κ thinks that $b \in d$. Conversely, suppose that $b \in V_\kappa$ and V_κ thinks that $b \in d$. Then by absoluteness, $b \in d$ and since $d = \wp(a)$ one has that $b \subseteq a$; then by absoluteness of subsethood, one has that V_κ thinks that $b \subseteq a$. This both shows that Power Set holds in V_κ and that $d = \wp^{V_\kappa}(a) = \wp(a)$, thereby establishing (2).

Infinity. The empty set has a Δ_0-definition, via $d = \varnothing$ iff $(\forall z \in d) z \neq z$. The successor operation $s(x) = y$ also has a Δ_0-definition, via

$$x \in y \wedge (\forall z \in x) z \in y \wedge (\forall z \in y)(z \in x \vee z = x)$$

Note that $\varnothing$ is in V_κ and that V_κ is closed under successor. So, as ω is in V_κ and ω contains $\varnothing$ and is closed under successor, by absoluteness V_κ thinks this too.

Foundation. Suppose that $a \in V_\kappa$ and V_κ thinks that a is not empty. By absoluteness of being empty (see above), indeed $a \neq \varnothing$. So by Foundation in the metatheory, there is $b \in a$ such that $b \cap a = \varnothing$. By the transitivity of V_κ, we have $b \in V_\kappa$, and since the relation $b \cap a = \varnothing$ is Δ_0-definable, V_κ indeed thinks that $b \cap a = \varnothing$.

Replacement. Suppose that $G \subseteq V_\kappa \times V_\kappa$, that $a \in V_\kappa$, and that G restricted to members of a is functional. By Lemma 8.14(2), $|a| < \kappa$. Equally, where $b \subseteq V_\kappa$ is the image of a under G, the assumption of G's functionality yields $|b| \leq |a| < \kappa$; so Lemma 8.14(3) again yields $b \in V_\kappa$.

Choice. Suppose that $a \in V_\kappa$. By what we have established so far, the set $a \times a$ is in V_κ and is identical to what V_κ thinks the product of a with itself is. Then $\wp(a \times a)$ is in V_κ by part (2) of this result (established above). Appealing to Choice in the metatheory, choose some R in $\wp(a \times a)$ that well-orders a. It is easy to check that V_κ thinks that R also well-orders a.

This completes parts (1) and (2). For part (3), suppose that $a, b \in V_\kappa$ are such that $|a| = |b|$, so that there is a bijection $f : a \longrightarrow b$. Since functions are identified with their graphs, $f \subseteq a \times b$ and so f is also a member of V_κ by Lemma 8.14(3). Since being a bijection is absolute, $V_\kappa \vDash |a| = |b|$. The proof of the converse is similar. □

Next, we show a converse to the previous theorem.

Lemma 8.16: *If $V_\kappa \vDash$ ZFC$_2$, then κ is an inaccessible cardinal.*

Proof. First, note that κ is a cardinal. Otherwise there would be a bijection $F : \beta \longrightarrow \kappa$ with $\beta < \kappa$, and since $\beta \in V_\kappa$, an application of Replacement inside V_κ would imply that $\kappa \in V_\kappa$.

Second, note that $\kappa > \omega$. To see this, choose an element a of V_κ which V_κ thinks is an infinite ordinal. (This is possible, since $V_\kappa \vDash$ ZFC$_2$.) By the absoluteness of being an ordinal, a is indeed an ordinal. Further by the absoluteness of the empty set and the successor operation, a is indeed infinite. So κ is infinite too, since $a \in V_\kappa$.

Third, note that κ is regular. For consider any map $F : \beta \longrightarrow \kappa$ for some $\beta < \kappa$, and let a be F's image. Then $a \in V_\kappa$, by applying Replacement inside V_κ. Since κ is a limit, there is $\gamma < \kappa$ such that $a \in V_\gamma$, so that $a \subseteq \gamma$. So F's range is bounded.

Finally, we must show that $\lambda < \kappa$ implies $|\wp(\lambda)| < \kappa$. To begin, note that for any set $a \in V_\kappa$ we have $\wp^{V_\kappa}(a) = \wp(a)$. That $\wp^{V_\kappa}(a) \subseteq \wp(a)$ holds by transitivity and the absoluteness of subset, exactly as in Theorem 8.15(2). To show that $\wp(a) \subseteq \wp^{V_\kappa}(a)$, suppose $X \subseteq a$; so, by transitivity, $X \subseteq a \subseteq V_\kappa$. Then by Separation and the absoluteness of set-theoretic intersection, we have that $X = (a \cap X) \in V_\kappa$. Hence indeed $\wp^{V_\kappa}(a) = \wp(a)$ for elements $a \in V_\kappa$.

Now, since V_κ satisfies Choice, V_κ thinks $\wp^{V_\kappa}(\lambda) = \wp(\lambda)$ is in a bijection with some ordinal. But then, by the absoluteness of being a bijection and being an ordinal, $\wp(\lambda)$ is indeed in a bijection with some ordinal in V_κ, i.e. some ordinal $< \kappa$, as required. □

So far in this section, we have considered models of the form V_κ. To complete the proof of Zermelo's Theorem, we must work with *arbitrary* models of ZFC$_2$. We start on this in Lemma 8.18. First, we need a preliminary result. The following

Lemma is a *deductive consequence* of ZFC_2, and our argument for it is entirely proof-theoretic. (The advantage of such an argument is that we know it will hold in *any* model of ZFC_2.)

Lemma 8.17: ZFC_2 *proves: if F is non-empty, then* $\exists v\,(F(v) \wedge (\forall x \in v)\neg F(x))$.

Proof. Suppose a is such that $F(a)$. Choose a transitive set t such that $a \subseteq t$. By Separation, the set $t \cap F$ exists.

If $t \cap F$ is empty, then a is the desired witness. After all, any $x \in a$ with $F(x)$ would also satisfy $x \in a \subseteq t$ and hence be in the empty $t \cap F$.

If $t \cap F$ is not empty, then by Foundation, there is b in $t \cap F$ such that $b \cap (t \cap F) = \varnothing$. Since t is transitive and $b \in t$, we have $b \subseteq t$. Now $b \in (t \cap F)$ implies $F(b)$. So b is the desired witness, since if there were $x \in b$ with $F(x)$ then we would have $x \in b \subseteq t$ and so x would be in the empty set $b \cap (t \cap F)$. □

We are now on the verge of completing Theorem 8.4:

Lemma 8.18: *If* $\mathcal{A} \vDash \mathsf{ZFC}_2$, *then* $\mathcal{A} \cong V_\alpha$ *for some ordinal* α.

Proof. Evidently $\mathcal{A}$ is extensional. To see that $\mathcal{A}$ is well-founded, suppose for reductio that there is an infinite descending $\in^{\mathcal{A}}$-chain a_n in $\mathcal{A}$, so that $\mathcal{A} \vDash a_{n+1} \in a_n$ for all $n \geq 0$; since we are working with the full semantics, $F = \{a_n : n \geq 0\}$ is a second-order entity in $\mathcal{A}$ and so Lemma 8.17 applies, from which we immediately obtain a contradiction. Hence, by the Mostowski Collapse Lemma 8.6, $\mathcal{A}$ is isomorphic to a transitive structure $\mathcal{B}$. It now suffices to show that $\mathcal{B} = V_\alpha$ for some ordinal α.

We first show that $\wp^{\mathcal{B}}(a) = \wp(a)$ for any $a \in B$. That $\wp^{\mathcal{B}}(a) \subseteq \wp(a)$ holds by transitivity and the absoluteness of subset, exactly as in the proof of Theorem 8.15(2). To see that $\wp(a) \subseteq \wp^{\mathcal{B}}(a)$, suppose that $X \subseteq a$. Then X determines a second-order entity in $\mathcal{B}$ and by Separation, $X \cap a = X$ exists as a first-order entity in $\mathcal{B}$. So indeed $\wp^{\mathcal{B}}(a) = \wp(a)$. Using this identity, an easy induction on γ in $\mathcal{B}$ shows that $V_\gamma^{\mathcal{B}} = V_\gamma$.

Where α is the least ordinal not in the underlying domain B of $\mathcal{B}$, we now claim that $B = V_\alpha$. For the left-to-right direction, suppose that $a \in B$. Then since $\mathcal{B} \vDash \mathsf{ZFC}_2$, we have that $a \in V_\gamma^{\mathcal{B}} = V_\gamma$ for some $\gamma < \alpha$, so that $a \in V_\alpha$. Conversely, suppose that $a \in V_\alpha$. Since $\mathcal{B} \vDash \mathsf{ZFC}_2$, it follows that α is a limit, so that $a \in V_\gamma$ for some $\gamma < \alpha$, and hence $a \in V_\gamma = V_\gamma^{\mathcal{B}} \subseteq B$. □

Putting these results together, we obtain what we wanted:

Theorem (8.4): $\mathcal{A} \vDash \mathsf{ZFC}_2$ *iff* $\mathcal{A} \cong V_\kappa$ *for some inaccessible* κ.

Proof. Left-to-right. Combine Lemmas 8.18 and 8.16. *Right-to-left.* Combine Theorem 8.15 with Theorem 2.3. □

Theorem (Zermelo's Quasi-Categoricity Theorem 8.8): *Let $\mathcal{A}$ and $\mathcal{B}$ be full models of* ZFC_2. *Then exactly one of the following obtains:*

(1) $\mathcal{A} \cong \mathcal{B}$
(2) $\mathcal{A} \cong V_a^{\mathcal{B}}$, *for some a which $\mathcal{B}$ thinks is an inaccessible cardinal.*
(3) $\mathcal{B} \cong V_a^{\mathcal{A}}$, *for some a which $\mathcal{A}$ thinks is an inaccessible cardinal.*

Proof. By Theorem 8.4, there are inaccessibles κ and λ such that $\mathcal{A} \cong V_\kappa$ and $\mathcal{B} \cong V_\lambda$. Then there are there three cases to consider.

Case 1: $\kappa = \lambda$. Then $\mathcal{A} \cong \mathcal{B}$, by the transitivity of isomorphism, and (1) obtains.

Case 2: $\kappa < \lambda$. In this case, $\kappa \in V_\lambda$. By the absoluteness of being an ordinal, V_λ thinks that κ is an ordinal. Moreover, since κ is regular, there is no $\gamma < \kappa$ and function $f : \gamma \longrightarrow \kappa$ with unbounded range. Since the property of being such a function is absolute, there is also no such function satisfying that property in V_λ; so indeed V_λ thinks that κ is regular. Finally, since κ is inaccessible, $|\wp(\gamma)| < \kappa$ for any $\gamma < \kappa$; so by Theorem 8.15(2), $|\wp^{V_\lambda}(\gamma)| = |\wp(\gamma)| < \kappa$ and thus by Theorem 8.15(3), V_λ thinks that the powerset of γ has cardinality $< \kappa$. So indeed V_λ thinks that κ is inaccessible. Then, by pushing through the two isomorphisms $\mathcal{A} \cong V_\kappa$ and $V_\lambda \cong \mathcal{B}$, we are done.

Case 3: $\lambda < \kappa$. Exactly as above, establishing that (3) obtains. □

8.B Elementary Scott–Potter foundations

In our discussion of Martin in §8.5, we introduced second-order Scott–Potter theory, SP_2. In §8.C, we shall prove that SP_2 is quasi-categorical. In this appendix, though, we will present some elementary deductive consequences of SP_2. *From now until the end of this appendix, we work deductively in* SP_2.

Our presentation largely follows Potter's 2004-book, and we provide more exact references to it as we proceed. But we provide these deductions ourselves for three reasons. First, including these deductions allows us to provide a *self-contained* proof the SP_2 Quasi-Categoricity Theorem 8.10, just as we presented a self-contained proof of Zermelo's Quasi-Categoricity Theorem 8.8. Second, as we explain at the end of this appendix, our formulation of SP_2 differs slightly from Potter's. Third and most important, though, this sequence of results gives some 'intuitive' content to the definitions of A, H and L, which can otherwise look rather magical.

We begin with a little notation. Given a set y, if there is a set x such that $A(x, y)$, then x is unique by Extensionality. So when $A(x, y)$, we abbreviate:

$$x = acc(y) = \{v : (\exists u \in y)(v \in u \vee v \subseteq u)\}$$

We call this operation *accumulation*, and the intuitive idea is that it 'fleshes out' a set. We will later see that histories are initial sequences of levels. Levels themselves are just the accumulations of all *lower* levels. Finally, and crucially, levels are well-ordered by membership. These comments, we hope, give an intuitive sense of how SP_2 works. But now we must get our hands dirty.

We begin by proving some elementary facts about *accumulation*:

Proposition 8.19: *For all a, b such that $acc(a)$ and $acc(b)$ exist:*

(1) $a \subseteq acc(a)$
(2) *if $a \subseteq b$, then $acc(a) \subseteq acc(b)$*
(3) *if $b \in a$, then $b \subseteq acc(a)$.*

Proof. (1). Suppose that $x \in a$. Then trivially there is $x \in a$ such that $x \in x \vee x \subseteq x$, so $x \in acc(a)$.

(2). Suppose that $a \subseteq b$, and suppose that $x \in acc(a)$. Then there is $c \in a$ such that $x \in c \vee x \subseteq c$. Since $a \subseteq b$, we have $c \in b$ and so $x \in acc(b)$.

(3). Suppose that $b \in a$ and $c \in b$. Then clearly there is some $b \in a$ such that $c \in b \vee c \subseteq b$, so that $c \in acc(a)$. □

Another simple fact is that, intuitively, all of a set's subsets exist:

Proposition 8.20 (Aussonderung): *for all X and all a, there is a set $\{v \in a : X(v)\}$*

Proof. Fix X and a. Since a is a set, there is a level ℓ such that $a \subseteq \ell$, by Levelling. By Comprehension, there is a property G such that $\forall v(G(v) \leftrightarrow (v \in a \wedge X(v))$. Since all of G's instances are members of ℓ, G determines the set $\{v \in a : X(v)\}$. □

We say that a set a is *supertransitive* iff for all $c \subseteq b \in a$ we have $c \in a$. We now prove some elementary facts about levels and supertransitivity:[42]

Proposition 8.21: *For any level ℓ and history h such that $\ell = acc(h)$:*

(1) *if $\ell_0 \in h$, then $\ell_0 = acc(\ell_0 \cap h)$ and $\ell_0 \cap h$ is a history and ℓ_0 is a level*
(2) *if $a \subseteq h$ is non-empty, then $(\exists \ell_0 \in a)(\forall \ell_1 \in a)\ell_1 \notin \ell_0$*
(3) *if $x \in \ell$, then there is a level $\ell_0 \in h \cap \ell$ such that $x \subseteq \ell_0$*
(4) *ℓ is both transitive and supertransitive*

Proof. (1). Let $h_0 = \ell_0 \cap h$, which exists by Aussonderung. Since $\ell_0 \in h$, we have that $\ell_0 = acc(\ell_0 \cap h) = acc(h_0)$. Suppose $x \in h_0$. Since $h_0 \subseteq h$, we have $x \cap h_0 \subseteq$

[42] For (1) see Potter (2004: Proposition 3.4.1 p.41); for (2) see Potter (2004: Lemmas 3.6.2–4 pp.43–44). For transitivity in (4) see Potter (2004: Proposition 3.6.5 p.44).

$x \cap h$. By Aussonderung, $acc(h_0) = acc(\ell_0 \cap h) \subseteq acc(h)$ exists. Since $x \in h_0$, by Proposition 8.19(3) we have $x \subseteq acc(h_0) = \ell_0$. Then $x \cap h \subseteq \ell_0 \cap h = h_0$ and so $x \cap h \subseteq x \cap h_0$. Hence $x \cap h_0 = x \cap h$. Then since $x \in h$, we have that $x = acc(x \cap h) = acc(x \cap h_0)$, so that h_0 is a history.

(2). For reductio, suppose there is a nonempty $a \subseteq h$ with $(\forall \ell_0 \in a)(\exists \ell_1 \in a)\ell_1 \in \ell_0$. Since a is nonempty, there is some $\ell_1 \in a \subseteq h$ and, by (1), ℓ_1 is a level. So $b = \{x \in \ell_1 : (\forall \ell_0 \in a)x \in \ell_0\} = \{x : (\forall \ell_0 \in a)x \in \ell_0\}$ is a set by Aussonderung.

Suppose $c \subseteq b$ and fix any level $\ell_0 \in a$. By our reductio assumption, there is $\ell_1 \in a \subseteq h$ such that $\ell_1 \in \ell_0$. Then $c \subseteq b \subseteq \ell_1$ and so there is $\ell_1 \in \ell_0 \cap h$ such that $c \in \ell_1 \vee c \subseteq \ell_1$. Then by (1), $c \in acc(\ell_0 \cap h) = \ell_0$. Since $\ell_0 \in a$ was arbitrary, $c \in b$.

Now, by Aussonderung, in particular $c = \{x \in b : x \notin x\}$ exists and $c \in b$. So by definition of c we have $c \notin c$ iff $c \in c$, a contradiction.

(3). Fix $x \in \ell = acc(h)$; by Aussonderung the set $a = \{\ell_0 \in h : x \in \ell_0 \vee x \subseteq \ell_0\}$ exists and is clearly nonempty. By (2), there is some $\ell_0 \in a$ such that $(\forall \ell_1 \in a)\ell_1 \notin \ell_0$. First, note that $\ell_0 \in h$, so that ℓ_0 is a level by (1). Moreover, since $\ell_0 \in h \subseteq acc(h) = \ell$, $\ell_0 \in h \cap \ell$. It remains to show that $x \subseteq \ell_0$. Since $\ell_0 \in a$, we have $x \in \ell_0 \vee x \subseteq \ell_0$; so it suffices to show that $x \notin \ell_0$. For reductio, suppose $x \in \ell_0$. Since $x \in \ell_0 = acc(\ell_0 \cap h)$ by (1), there is $\ell_1 \in \ell_0 \cap h$ with $x \in \ell_1 \vee x \subseteq \ell_1$. So $\ell_1 \in a$ by definition of a, and so $\ell_1 \notin \ell_0$ by the earlier fact that $(\forall \ell_1 \in a)\ell_1 \notin \ell_0$, contradicting the fact that $\ell_1 \in \ell_0$.

(4). Assume $x \in \ell$. By (3), fix $\ell_0 \in h \cap \ell$ with $x \subseteq \ell_0$. For transitivity, suppose $y \in x$, so that $y \in \ell_0$. Then there is $b \in h$ with $y \in b \vee y \subseteq b$, namely $b = \ell_0$. Hence $y \in acc(h) = \ell$. For supertransitivity, suppose $y \subseteq x$. Then $y \subseteq x \subseteq \ell_0$ and so there is $b \in h$ with $y \in b \vee y \subseteq b$, namely (again) $b = \ell_0$. Hence we have $y \in acc(h) = \ell$. □

In what follows, we often appeal to the (super)transitivity of levels, i.e. Proposition 8.21(4). Indeed, we can now show that accumulations *always* exist:[43]

Proposition 8.22: *acc(a) is a set for any a*

Proof. Fix any set a. By Levelling, choose a level ℓ such that $a \subseteq \ell$. So $acc(a) \subseteq acc(\ell)$ by Proposition 8.19(2), provided these accumulations exist. By Aussonderung, it suffices to show that $acc(\ell) \subseteq \ell$. Suppose that $x \in acc(\ell)$. Then there is $b \in \ell$ such that $x \in b \vee x \subseteq b$. Then if $x \in b$ then $x \in \ell$ by transitivity of ℓ; and if $x \subseteq b$ then $x \in \ell$ by supertransitivity of ℓ; and $x \in \ell$ either way. □

Combining Proposition 8.22 with Proposition 8.21(1), we see that every member of a history is a level. Moreover, we can now show that every level is the accumulation of the levels which are its members:[44]

---

[43] Potter (2004: Propositions 3.6.6–7 p.44).

[44] See Potter (2004: Proposition 3.6.8 p.45).

Proposition 8.23: *Let ℓ be any level. Then where $h = \{\ell_0 \in \ell : L(\ell_0)\}$,*

(1) *h is a history*
(2) *$\ell = acc(h)$*

Proof. (1). Let $\ell_0 \in h$; we must show that $acc(\ell_0 \cap h) = \ell_0$. Suppose $x \in acc(\ell_0 \cap h)$, i.e. there is $\ell_1 \in \ell_0 \cap h$ with $x \in \ell_1 \lor x \subseteq \ell_1$. Then $x \subseteq \ell_1$ by transitivity of ℓ_1, and so $x \in \ell_0$ by supertransitivity of ℓ_0. Conversely, suppose that $x \in \ell_0$. Since ℓ_0 is a level, by Proposition 8.21(3) there is a level $\ell_1 \in \ell_0$ such that $x \subseteq \ell_1$; since $\ell_0 \in h$ we have $\ell_0 \in \ell$ and so by transitivity $\ell_1 \in \ell$ so that $\ell_1 \in h$. Hence $x \subseteq \ell_1 \in \ell_0 \cap h$ and so $x \in acc(\ell_0 \cap h)$.

(2). Since ℓ is a level, there is a history h_0 such that $\ell = acc(h_0)$; we must show $acc(h_0) = acc(h)$. Suppose that $x \in acc(h_0)$; then there is a level $\ell_0 \in h_0$ such that $x \in \ell_0 \lor x \subseteq \ell_0$. And since $h_0 \subseteq acc(h_0) = \ell$, we have $\ell_0 \in \ell$, so that $x \in acc(h)$. Conversely, suppose $x \in acc(h)$. Then there is a level $\ell_0 \in \ell$ such that $x \in \ell_0 \lor x \subseteq \ell_0$. Since ℓ_0 is transitive, we must have $x \subseteq \ell_0$; and since ℓ is supertransitive we have $x \in \ell$. □

We can now prove that the levels are well-ordered by membership:[45]

Proposition 8.24:

(1) $\forall X((\exists \ell : L)X(\ell) \rightarrow$
$\qquad (\exists \ell_0 : L)[X(\ell_0) \land (\forall \ell_1 : L)(\ell_1 \in \ell_0 \rightarrow \neg X(\ell_1))])$
(2) $(\forall \ell : L)(\forall \ell_0 : L)(\ell \in \ell_0 \lor \ell = \ell_0 \lor \ell_0 \in \ell)$

Proof. (1). Suppose that ℓ is a level with $X(\ell)$. Let $a = \{\ell_1 \in \ell : X(\ell_1) \land L(\ell_1)\}$ which exists by Levelling. If $a = \varnothing$ then ℓ itself is our witness. Otherwise, let $h = \{\ell_1 \in \ell : L(\ell_1)\}$, and note that h is a history and $\ell = acc(h)$ by Proposition 8.23. Since $a \subseteq h$ is nonempty, by Proposition 8.21(2) we have a level $\ell_0 \in a$ such that $(\forall \ell_1 \in a)\ell_1 \notin \ell_0$, so that $(\forall \ell_1 : L)(\ell_1 \in \ell_0 \rightarrow \neg X(\ell_1)$ by definition of a.

(2). Suppose not, for reductio. We define the following abbreviations:

$$\Upsilon(\ell_0, \ell_1) := \ell_0 \in \ell_1 \lor \ell_0 = \ell_1 \lor \ell_1 \in \ell_0$$
$$\Phi(\ell) := \exists \ell_1 \neg \Upsilon(\ell, \ell_1)$$

By our reductio assumption, there is a level ℓ such that $\Phi(\ell)$. So by (1), there is a level ℓ_0 such that $\Phi(\ell_0)$ but for all levels $\ell_1 \in \ell_0$ we have $\neg\Phi(\ell_0)$. Since $\Phi(\ell_0)$, there is some ℓ_1 such that $\neg(\ell_0, \ell_1)$; and by (1) again we can assume that for all levels $\ell_2 \in \ell_1$ we have $\Upsilon(\ell_0, \ell_2)$. We shall argue that $\ell_0 = \ell_1$, which contradicts the claim $\neg\Upsilon(\ell_0, \ell_1)$.

[45] See Potter (2004: Proposition 3.6.10–11 p.45)

Suppose $x \in \ell_0$; we show that $x \in \ell_1$. By Proposition 8.21(3), there is $\ell_2 \in \ell_0$ with $x \subseteq \ell_2$. Since $\ell_2 \in \ell_0$, we have $\neg\Phi(\ell_2)$ and so $\ell_2 \in \ell_1 \vee \ell_2 = \ell_1 \vee \ell_1 \in \ell_2$. In the first case we have $x \in \ell_1$ by supertransitivity of ℓ_1. The other two cases lead us to contradict $\neg\Upsilon(\ell_0, \ell_1)$. For, in the second case we would have $\ell_1 \in \ell_0$, while in the second case we would have $\ell_1 \in \ell_2 \in \ell_0$ and thus $\ell_1 \in \ell_0$ by transitivity of ℓ_0.

Conversely suppose that $x \in \ell_1$; we show that $x \in \ell_0$. By Proposition 8.21(3), there is $\ell_2 \in \ell_1$ with $x \subseteq \ell_2$. Since $\ell_2 \in \ell_1$, we have $\Upsilon(\ell_0, \ell_2)$, i.e. $\ell_0 \in \ell_2 \vee \ell_0 = \ell_2 \vee \ell_2 \in \ell_0$. In the third case we have $x \in \ell_0$ by the supertransitivity of ℓ_0. The other two cases again lead us to contradict $\neg\Upsilon(\ell_0, \ell_1)$. Indeed, in the second case we have $\ell_0 \in \ell_1$, while in the third case we have $\ell_0 \in \ell_2 \in \ell_1$ and so $\ell_0 \in \ell_1$ by transitivity of ℓ_1. □

From this result, and the consequence of Levelling that every set is a subset of some level, within SP_2 we can now introduce a term, $\text{ord}(x)$, to be read *the order of x*, for the $\in$-least level ℓ such that $x \subseteq \ell$.[46] These results explain ord's behaviour:[47]

Proposition 8.25: $\text{ord}(a)$ *exists for any a. Moreover:*

(1) *if* $\exists x X(x)$, *then there is some* ord-*minimal set a such that* $X(a)$, *i.e.* $(\forall v : X)(\text{ord}(a) = \text{ord}(v) \vee \text{ord}(a) \in \text{ord}(v))$.
(2) $a \notin \text{ord}(a)$
(3) *if* $a \in b$, *then* $\text{ord}(a) \in \text{ord}(b)$
(4) *for any level* ℓ: $\text{ord}(a) = \ell \leftrightarrow [a \subseteq \ell \wedge (\forall x \in \ell)(\exists v \in a)x \subseteq \text{ord}(v) \in \ell]$

Proof. The initial claim, and (1), are immediate from Proposition 8.24.

(2). For reductio, suppose that $a \in \text{ord}(a)$. By Proposition 8.23(2), $\text{ord}(a) = acc(\{\ell_0 \in \text{ord}(a) : L(\ell_0)\})$. So there is a level $\ell_0 \in \text{ord}(a)$ such that $a \in \ell_0 \vee a \subseteq \ell_0$, and by ℓ_0's transitivity we have $a \subseteq \ell_0$, contradicting that $\ell_0 \in \text{ord}(a)$.

(3). For reductio, suppose that $a \in b$ but $\text{ord}(a) \notin \text{ord}(b)$. By Proposition 8.24(2), either $\text{ord}(b) \in \text{ord}(a)$ or $\text{ord}(b) = \text{ord}(a)$. So $\text{ord}(b) \subseteq \text{ord}(a)$ by the transitivity of $\text{ord}(a)$, which together with $a \in b$ gives that $a \in \text{ord}(a)$. But this contradicts (2).

(4). For left-to-right, suppose $\text{ord}(a) = \ell$. Then by definition $a \subseteq \ell$. Now fix $x \in \ell$. By Proposition 8.21(3) there is a level $\ell_0 \in \ell$ with $x \subseteq \ell_0$, so that either $\text{ord}(x) \in \ell_0$ or $\text{ord}(x) = \ell_0$ by Proposition 8.24(2). Suppose for reductio that $(\forall v \in a)\text{ord}(v) \in \ell_0$; then $(\forall v \in a)v \in \ell_0$ by supertransitivity of ℓ_0, and so $a \subseteq \ell_0 \in \ell$, contradicting the fact that $\text{ord}(a) = \ell$. So there is some $v \in a$ with $\text{ord}(v) \notin \ell_0$; and now by Proposition 8.24(2) and transitivity of levels, $\ell_0 \subseteq \text{ord}(v)$.

For right-to-left, suppose $\text{ord}(a) \neq \ell$. By Proposition 8.24(2), either $\text{ord}(a) \in \ell$ or $\ell \in \text{ord}(a)$. If $\ell \in \text{ord}(a)$ then $a \nsubseteq \ell$. If $\text{ord}(a) \in \ell$ then $(\forall v \in a)\text{ord}(v) \in \text{ord}(a)$

[46] Potter (2004: 48). Potter writes $V(a)$, but we want to avoid overloading the letter 'V'.

[47] See Potter (2004: Exercise 1 p.49).

by (3), and so $\neg(\exists v \in a)\text{ord}(a) \subseteq \text{ord}(v)$ by supertransitivity of levels. So we have $\text{ord}(a) \in \ell$ and $\neg(\exists v \in a)(\text{ord}(a) \subseteq \text{ord}(v) \in \ell)$. □

We will use these deductive results in proving the quasi-categoricity of SP_2 in the next appendix, and also revisit them in Chapter 11. But, to complete this appendix, we explain how our axiomatisation of SP_2 differs from Potter's.

First: Potter allows for urelements. We have set these aside, not because we object to them, but because the statement of the quasi-categoricity result would require setting them aside anyway.

Second: Potter's system is first-order, whereas SP_2 is a second-order system.

Third: Potter's axioms directly employ the term-forming operator, $acc(y)$. At the outset, he cannot *assume* that $acc(y)$ exists for any y, so he must allow that (in principle) '$acc(y)$' may be an empty term.[48] To avoid this complexity, we define the formula $A(x, y)$ explicitly, and then introduce $acc(y) = x$ as an abbreviation, to be used when a set x exists such that $A(x, y)$.[49]

Fourth: Potter does not use the axiom which we called Levelling. Instead, he offers an axiom scheme of separation, whose second-order version is:[50]

SP-Separation. $\forall X(\forall v : Lev)\exists x \forall z(z \in x \leftrightarrow [x \in v \wedge X(z)])$

And he also has the axiom:[51]

SP-Foundation. Every set is a subset of some level.

We use Levelling, since it allows for the pithy gloss: *A property determines a set iff all of its instances are members of some level.* But it is easy to see that Levelling is deductively equivalent to SP-Separation + SP-Foundation, thanks to the Comprehension Schema of second-order logic.

8.c Scott–Potter Quasi-Categoricity

Armed with our deductive consequences of SP_2, in this appendix we prove that SP_2 categorically axiomatises the idea of an arbitrary level of the cumulative hierarchy. *From here on, we work in our set-theoretic metatheory.* Our target result is:

Theorem (SP_2 Quasi-Categoricity 8.10): $\mathcal{A} \vDash \text{SP}_2$ *iff* $\mathcal{A} \cong V_\alpha$ *for some ordinal* $\alpha > 0$.

To get started, recall that our definitions of A, H and L are offered in a set-theoretic object-language (see Definition 8.9), as is our defined operator *acc* from §8.B. Since

48 Potter (2004: 24, 41).

49 Potter (2004: 42) himself suggests this idea.

50 Potter (2004: 42).

51 Potter (2004: 41).

our metatheory is *also* set-theoretic, we can use exactly the same definitions *in the metatheory*. So, where a is a set according to the metatheory—i.e. it is an element of some V_α as defined in Definition 8.2—we can ask in the metatheory whether some set b is such that $A(b, a)$ and, if so, which set. More generally, we can ask under what circumstances the *metatheoretic* notions of *accumulation, history* and *level* coincide with the notions defined *within* a model of SP_2. Here are some answers to these questions:

Proposition 8.26: *For any ordinal $\alpha \geq 0$:*

(1) *if $\beta < \alpha$, then $V_\beta \in V_\alpha$*

(2) $V_\alpha = acc\{V_\gamma : \gamma < \alpha\}$

(3) $\{V_\gamma : \gamma < \alpha\}$ *is a history*

(4) V_α *is a level*

Proof. *(1).* The proof is by induction on α. For $\alpha = 0$ it is trivial. For $\alpha = \alpha_0 + 1$, suppose that $\beta < \alpha$. Then $\beta \leq \alpha_0$. If $\beta < \alpha_0$ then by induction hypothesis $V_\beta \in V_{\alpha_0}$ and so $V_\beta \in V_{\alpha_0} \subseteq V_\alpha$. If $\beta = \alpha_0$ then trivially $V_\beta = V_{\alpha_0} \in V_\alpha$. For α limit, if $\beta < \alpha$ then $\beta < \alpha_0 < \alpha$ for some α_0, and so by induction hypothesis $V_\beta \in V_{\alpha_0} \subseteq V_\alpha$.

(2). $V_\alpha = \{x : (\exists \gamma < \alpha)(x \in V_\gamma \vee x \subseteq V_\gamma)\} = acc\{V_\gamma : \gamma < \alpha\}$.

(3). If $\delta < \alpha$, then $acc(V_\delta \cap \{V_\gamma : \gamma < \alpha\}) = acc\{V_\gamma : \gamma < \delta\} = V_\delta$. The first equality follows by (1) and the second equality follows by (2).

(4). Immediate from (3) and (2). □

To prove the right-to-left direction of the SP_2 Quasi-Categoricity Theorem 8.10, it will now suffice to show that V_α is not *just* a level, but that V_β also *thinks* that V_α is a level, whenever $\beta > \alpha$. To obtain this result, we begin by introducing an extension of the idea of a transitive structure (cf. Definition 8.1). So a structure $\mathcal{M}$ is *supertransitive* iff: both $\mathcal{M}$ is transitive and also M is supertransitive in the sense of §8.B, i.e. if $c \subseteq b \in M$, then $c \in M$. It is quite easy to show:

Proposition 8.27: *For any ordinal $\alpha \geq 0$:*

(1) V_α *is supertransitive.*

(2) *if $\beta < \alpha$, then the history $\{V_\gamma : \gamma < \beta\}$ is in V_α.*

Proof. Both proofs are by induction on α.

(1). For $\alpha = 0$ this is trivial. For $\alpha = \alpha_0 + 1$, suppose that $b \in V_\alpha$ and $c \subseteq b$. Then $c \subseteq b \subseteq V_{\alpha_0}$ and so $c \in V_\alpha$. For α limit, suppose that $b \in V_\alpha$ and $c \subseteq b$. Then since α is limit, $b \in V_\gamma$ for some $\gamma < \alpha$. Then by induction hypothesis V_γ is supertransitive and so c is in V_γ and thus also in V_α.

(2). For $\alpha = 0$ this is trivial. For $\alpha = \alpha_0 + 1$, suppose that $\beta < \alpha$. Then $\beta \leq \alpha_0$. If $\beta < \alpha_0$ then by induction hypothesis $\{V_\gamma : \gamma < \beta\}$ is in $V_{\alpha_0} \subseteq V_\alpha$. If $\beta = \alpha_0$ then it

suffices to note that $\{V_\gamma : \gamma < \alpha_0\} \subseteq V_{\alpha_0}$; but this follows from Proposition 8.26(1). For α a limit, if $\beta < \alpha$ then $\beta < \alpha_0 < \alpha$ for some α_0, and so by induction hypothesis the history $\{V_\gamma : \gamma < \beta\}$ is in $V_{\alpha_0} \subseteq V_\alpha$. □

Just as Δ_0-formulas are absolute for transitive structures (see Proposition 8.13), there is a simple class of formulas which are absolute for super-transitive structures. Let the *super-Δ_0-formulas* be the smallest class of formulas extending the Δ_0-formulas such that if $\varphi(x)$ is super-Δ_0, then so are $(\exists x \subseteq y)\varphi(x)$ and $(\forall x \subseteq y)\varphi(x)$. Then, in parallel to Proposition 8.13, we have:

Proposition 8.28: *If $\mathcal{A}$ is a supertransitive structure, then every super-Δ_0-formula is absolute for $\mathcal{A}$.*

Proof. Suppose the result holds for $\varphi(x)$; it suffices to show it holds for $(\exists x \subseteq y)\varphi(x)$. Suppose that b is in A and that $(\exists x \subseteq b)\varphi(x)$. Since $\mathcal{A}$ is supertransitive, there is a in A such that $\varphi(a)$. Then since the result holds for $\varphi(x)$, we have $\mathcal{A} \models \varphi(a)$. Since subset is a Δ_0-notion, we have $\mathcal{A} \models a \subseteq b \wedge \varphi(a)$, so trivially $\mathcal{A} \models (\exists x \subseteq b)\varphi(x)$. The other direction is similar. □

We can use this to show that *acc*, H and L are all *absolute* within the stages of the iterative hierarchy:

Proposition 8.29: *For any ordinal α:*

(1) *if $a \in V_\alpha$, then $acc(a) \in V_\alpha$ and $acc(a) = acc^{V_\alpha}(a)$*
(2) *if $h \in V_\alpha$, then h is a history iff V_α thinks that h is a history*
(3) *if $\ell \in V_\alpha$, then ℓ is a level iff V_α thinks that ℓ is a level.*

Proof. (1). Let $a \in V_\alpha$. Then $a \in V_{\beta+1}$ for some $\beta < \alpha$. Then $a \subseteq V_\beta$. Suppose c is in $acc(a)$. Then $c \in b \vee c \subseteq b$ for some $b \in a$. Then $b \in V_\beta$ and $b \subseteq V_\beta$ by transitivity of V_β. If $c \in b$ then since $b \subseteq V_\beta$ we have $c \in V_\beta$. If $c \subseteq b$ then since $b \in V_\beta$ we have also $c \in V_\beta$ by supertransitivity of V_β. Hence $acc(a)$ is a subset of V_β and so $acc(a) \in V_{\beta+1} \subseteq V_\alpha$, which is the first part of (1).

To see that $acc(a) = acc^{V_\alpha}(a)$, simply note that the definition of $acc(a)$ is super-Δ_0 and so absolute. In particular, the condition that $acc(x) = w$ is equivalent to $acc(x) \subseteq w$ and $w \subseteq acc(x)$; but the condition $acc(x) \subseteq w$ is super-Δ_0, since it can be written as $(\forall y \in x)(\forall z \in y)z \in w \wedge (\forall y \in x)(\forall z \subseteq y)z \in w$, and the condition $w \subseteq acc(x)$ is Δ_0, since it can be written as $(\forall z \in w)(\exists y \in x)(z \in y \vee z \subseteq y)$.

(2). The definition of a history is clearly super-Δ_0.

(3). Since $h \subseteq acc(h)$, one has that ℓ is a level iff there is a history $h \subseteq \ell$ such that $\ell = acc(h)$, which is then super-Δ_0. □

We are now in a position to prove both halves of the SP_2 Quasi-Categoricity Theorem 8.10. We begin with the right-to-left half:

Proposition 8.30: *$V_\alpha \vDash SP_2$, for all ordinals $\alpha > 0$.*

Proof. Extensionality is trivial. To prove Levelling, fix $X \subseteq V_\alpha$. First suppose that V_α thinks that there is some level ℓ such that $(\forall z : X)z \in \ell$. Then $X \in V_\alpha$ by the supertransitivity of V_α. Next, suppose $X \in V_\alpha$; then there is some $\beta < \alpha$ such that $X \subseteq V_\beta \in V_\alpha$, and V_α thinks that V_β is a level by Proposition 8.29 and Proposition 8.26. □

We next prove the left-to-right half of the SP_2 Quasi-Categoricity:

Proposition 8.31: *If $\mathcal{M} \vDash SP_2$, then $\mathcal{M} \cong V_\alpha$ for some ordinal $\alpha > 0$.*

Proof. The model $\mathcal{M}$ is extensional by definition. Also, $\mathcal{M}$ is well-founded: by Proposition 8.24(1), there is no infinite-descending $\in^{\mathcal{M}}$-chain of $\mathcal{M}$-levels, and so by Proposition 8.25(1) there is no infinite descending $\in^{\mathcal{M}}$-chain. So, by Mostowski's Collapse Lemma 8.6, $\mathcal{M}$ is isomorphic to a transitive structure $\mathcal{N}$, which itself is a model of SP_2.

We claim $\mathcal{N}$ is supertransitive. For, suppose that $c \subseteq b \in N$. By Levelling, there is an $\ell \in N$ such that $\mathcal{N}$ thinks both that ℓ is a level and that $b \subseteq \ell$. By the transitivity of $\mathcal{N}$, indeed $b \subseteq \ell$ and hence $c \subseteq \ell$. So $c \in N$ by Levelling again.

The entities which $\mathcal{N}$ thinks are its levels are well-ordered, by Proposition 8.24. So let these be enumerated by ℓ_β for ordinals $\beta < \alpha$. We show, by induction on $\beta < \alpha$, that $\ell_\beta = V_\beta$.

Suppose, for induction, that $\ell_\gamma = V_\gamma$ for all $\gamma < \beta$. By Lemma 8.26, $h = \{V_\gamma : \gamma < \beta\}$ is a history. Since $V_\gamma = \ell_\gamma$, we have that $h \subseteq \ell_\beta$, so that $h \in N$ by supertransitivity of $\mathcal{N}$. Moreover, $\mathcal{N}$ thinks that h is a history, by the same argument as Proposition 8.29(2). We now claim that $acc(h) = \ell_\beta$. For, suppose that $x \in acc(h)$; then $x \in V_\gamma \vee x \subseteq V_\gamma$ for some $\gamma < \beta$, and so $x \in \ell_\gamma \vee x \subseteq \ell_\gamma$; and since this is Δ_0, the structure $\mathcal{N}$ thinks this too; so in fact $x \in \ell_\beta$, by Proposition 8.23(2), and so $acc(h) \subseteq \ell_\beta$. The converse inclusion is similar. So by Lemma 8.26(2), we have $V_\beta = acc(h) = \ell_\beta$. This completes the induction. So, $\ell_\beta = V_\beta$ for all $\beta < \alpha$.

Finally, we claim that $N = V_\alpha$. For the inclusion $N \subseteq V_\alpha$, note that if $x \in N$ then by Levelling there is $\beta < \alpha$ such that $x \subseteq \ell_\beta = V_\beta$, and then $x \in V_{\beta+1} \subseteq V_\alpha$. For the inclusion $V_\alpha \subseteq N$, observe that if $x \in V_\alpha$ then $x \subseteq V_\beta = \ell_\beta$ for some $\beta < \alpha$, and so $x \in N$ since $\mathcal{N}$ is supertransitive. □

This completes our proof of Theorem 8.10. To finish this appendix, we establish some quick *deductive* consequences of ZFC_2:

Proposition 8.32: ZFC_2 *deductively proves:*

(1) SP_2
(2) *the levels are precisely the* V_α
(3) $\text{ord}(a) = V_{\text{rank}(a)}$ *for all a.*

Proof. (1). Our proof that each V_α is a level from Proposition 8.26(4) carries over word-for-word into a proof in a deductive system for ZFC_2. From this, we can easily prove Levelling, working deductively in ZFC_2. First suppose a second-order X is such that $\exists x \forall z(z \in x \leftrightarrow X(z))$. Since we are working in ZFC_2, choose α such that $x \subseteq V_\alpha$. Then since $y = V_\alpha$ is a level, we have $(\exists y : L)(\forall z : X)z \in y$. Conversely, suppose that $(\exists y : L)(\forall z : X)z \in y$. Since the level y is by definition a set, this means that all instances of X are in y, so $\exists x \forall z(z \in x \leftrightarrow X(z))$ by Separation in ZFC_2.

Having established (1), for the remainder of the proof we may appeal to all the results established in §8.B.

(2). We argue by induction, using the fact that membership well-orders the levels. By Proposition 8.26(4), it suffices to show that every level is of the form V_α for some α. Suppose it is true for all levels $\ell' \in \ell$ that there is $\alpha_{\ell'}$ such that $\ell' = V_{\alpha_{\ell'}}$. Let $\alpha = \sup_{\ell' \in \ell}(\alpha_{\ell'} + 1)$. We show that $V_\alpha = \ell$. Proposition 8.23(2) tells us that $\ell = acc(\{\ell' \in \ell : L(\ell')\}$. Suppose that $x \in \ell$. Then there is $\ell' \in \ell$ such that $x \in \ell' \vee x \subseteq \ell'$. By the transitivity of ℓ' (see Proposition 8.21(4)) we have $x \subseteq \ell' = V_{\alpha_{\ell'}}$, and so since $\alpha_{\ell'} < \alpha$ we have $x \in V_\alpha$. Conversely, suppose that $x \in V_\alpha$. Then there is $\beta < \alpha$ such that $x \in V_{\beta+1}$. Then $\beta < \alpha_{\ell'} + 1$ for some $\ell' \in \ell$. Then $\beta + 1 \leq \alpha_{\ell'} + 1$ and so $\beta \leq \alpha_{\ell'}$ and hence $x \in V_{\alpha_{\ell'}} = \ell'$. So there is $\ell' \in \ell$ such that $x \in \ell' \vee x \subseteq \ell'$, and so $x \in \ell$.

(3). This follows from (2) and the definitions of $\text{ord}(a)$ and $\text{rank}(a)$. □

9

Transcendental arguments against model-theoretical scepticism

In Chapters 7 and 8, we explained the demise of moderate modelism. The moderate objects-modelist cannot explain how creatures like us can pin down the natural numbers. The moderate concepts-modelist cannot explain how creatures like us can possess the concept *ω-sequence*.

There is an obvious potential 'sceptical solution' to these concerns: accept that model theory shows that we *cannot* pin down the natural numbers, or determinately grasp the concept *ω-sequence*, or whatever. We call this position *model-theoretical scepticism*. In this chapter, we argue that model-theoretical scepticism is *incoherent*.

9.1 Model-theoretical scepticism

We start by spelling out the idea of model-theoretical scepticism.

Outside of the philosophy of mathematics, there are various classic sceptical challenges. An example is the question: *Are you in a brain in a vat?* No one, we hope, thinks the answer to this question is: *Yes*. No one, we hope, actually *is* a brain-in-vat sceptic. But the question poses a potentially interesting challenge. It serves as a stimulus to reflect on our relations to the external world.[1]

Within the philosophy of mathematics, there are also various sceptical challenges. Indeed, the Modelist's Doxological Challenge supplies the beginnings of one. A specific instance of that Challenge is HOW *can we pick out the natural number structure?*, where that structure is thought of as some specific isomorphism type. Having finished Chapters 7 and 8, we might well be inclined to ask the more panicked question: CAN *we even pick out the natural number structure?*

This formulation of a sceptical challenge continues to use the loose terminology of 'picking out a structure', introduced in Chapter 6. It is worth pausing to spell out both objectual and conceptual versions of the challenge. Construed objectually, the worry is that we never determinately *refer* to the standard model of arithmetic (a particular isomorphism type). For example, we might worry that when we talk

---

[1] For discussion of why we *should* bother engaging with brain-in-vat scepticism, see Putnam (1994: 284–5) and Button (2016a: §1).

about '*the* natural number structure', we end up referring indeterminately and indifferently to all the (non-isomorphic) models of first-order PA. Construed conceptually, the worry is that we do not possess a mathematical concept whose content is as specific as an isomorphism type. For example, we might worry that our concept *ω-sequence* is really only so determinate that it applies to (the order type of) *any* model of PA.

Chapters 7 and 8 are likely to generate such sceptical worries. And we say that a *model-theoretical sceptic* is someone who, on the basis of model-theoretical arguments, embraces the sceptical conclusion that we *cannot* refer determinately to the natural number structure, or that our concept *ω-sequence* is deeply indeterminate.

As in the case of the brain-in-vat sceptic, the challenges posed by our model-theoretical sceptic should at least prompt us to reflect on our relations to mathematical entities or concepts. But the analogy between brain-in-vat scepticism and model-theoretical scepticism is worth pursuing further. Our moderate modelist wants to pin down the isomorphism type of an ω-sequence. We have insisted that there is some difficulty in her doing so, related to her inability to offer a categorical axiomatisation of that notion. But, impressed by the analogy with brain-in-vat scepticism, she might reply as follows:

You never really managed to demonstrate that I cannot pin down the isomorphism type of an ω-sequence by appealing to categoricity. All you really managed to show is that I cannot demonstrate, to the satisfaction of some kind of radical sceptic, how I grasp the semantics for a sufficiently rich logic, such for as full second-order logic. But my inability to answer the model-theoretical sceptic is no more embarrassing than our general inability to answer the brain-in-vat sceptic. And we should not allow our philosophy to be entirely dictated by sceptics!

There are two points in this imagined speech which are perfectly correct. First, the moderate modelist has, indeed, failed to answer the model-theoretical sceptic. Second, it is true that we should not allow our philosophising to be entirely dictated by sceptics. But neither point lets her off the hook. The reason for this is simple: the moderate modelist *herself* invited the sceptic to the table.

Here is the point in more detail (and it may help to re-read §7.8). Being modelists, our moderate modelists want to pin down the isomorphism type of an ω-sequence. Given this, it is surely a fair question to ask *how* they can do that, i.e. to raise the Modelist's Doxological Challenge. Being moderates, they impose certain restrictions on what an acceptable answer to that question might look like. They shun appeal to intuition, to innate knowledge, or to brute facts. But then the question—which was fair to ask—becomes unanswerable for them. And, at the risk of stating the obvious: to point out that you find a question unanswerable is not to answer it; nor, if your own background assumptions render it unanswerable, does it indicate that the question itself was somehow intrinsically unfair.

At this point, though, an option suggests itself: *moderate modelists might simply embrace model-theoretical scepticism*. Indeed, that option seems like the *inevitable* upshot of moderate modelism. Moderate modelists can neither supply any means by which we could pin down (for example) *the* natural number structure, nor can they shrug and say 'we just *do* pin it down', without compromising their moderation (see §7.1 and §7.8). They must embrace the sceptical conclusion.

And this gives rise to an interesting *disanalogy* with brain-in-vat scepticism. Whilst no one sane thinks that we *are* envatted brains, some philosophers have been happy to accept model-theoretical scepticism. Here, for example, is Skolem:

> [...] the affirmation of the existence of uncountable sets can only be considered a play on words; this unprovable absolute is merely a fiction. The true significance of Löwenheim's Theorem is precisely this critique of the unprovable absolute. In brief: this critique does not reduce the higher infinities of elementary set theory to absurdity; it reduces them to non-objects.[2]

Model-theoretic considerations have here pushed Skolem towards a rejection of the existence of uncountable sets. So Skolem is model-theoretical sceptic.[3]

More recently, Hamkins has suggested that there is no such thing as *the* standard model of arithmetic.

> My long-term expectation is that technical developments will eventually arise that provide a forcing analogue for arithmetic, allowing us to modify diverse models of arithmetic in a fundamental and flexible way, just as we now modify models of set theory by forcing, and this development will challenge our confidence in the uniqueness of the natural number structure, just as set-theoretic forcing has challenged our confidence in a unique absolute set-theoretic universe.[4]

To be sure, Hamkins' suggestion is not merely based on existing results in model theory, such as the Löwenheim–Skolem Theorem 7.2, but on a prediction concerning model theory's future development. Nonetheless, his suggestion is that model-theoretic results can and should lead us to adopt a thorough relativism about 'the' concepts *natural number* and *finiteness*.

For all its potential appeal, we think that such model-theoretical scepticism is essentially *incoherent*. To show this, we present two *transcendental arguments* against model-theoretical scepticism.

---

[2] Skolem (1941: 468). Thanks to Chloé de Canson for discussions on translating Skolem's French.

[3] At least, in this passage; see Benacerraf (1985) and George (1985) for a discussion of Skolem's changing views over time. For a list of authors with somewhat similar views, see Bays (2014: §3.2).

[4] Hamkins (2012: 428); cf. also Hamkins and Yang (forthcoming: §6). Bays (2014: §3.3) also sees this as connected with Skolem's model-theoretical scepticism, though in a slightly different way.

9.2 Moorean versus transcendental arguments

Before presenting the arguments, we should explain what we mean by calling them *transcendental.* The purpose of a transcendental argument against a sceptic is to show that the scepticism in question is self-undermining. And the best way to illustrate the idea of such an argument is by comparing transcendental responses with *Moorean* responses to scepticism.[5]

Again, we start with the (perhaps more familiar) challenge raised by the brain-in-vat sceptic: *Are you a brain in a vat?* Moore's response to the sceptic is: *No, since I have hands, and mere brains-in-vats have no hands.* One way to understand Moore's response is as follows: the sceptic will use various principles to make Moore worry that he is a (handless) brain-in-a-vat, but Moore is simply more confident that he has hands than he is of those principles.

We might consider a similarly *Moorean* response against model-theoretical scepticism. Suppose the model-theoretical sceptic poses the challenge: *Do we ever pin down the natural numbers?* A Moorean response would be to say that the model-theoretical sceptic uses several model-theoretical assumptions in an attempt to make us worry that we fail to do so, but that our ability to do so is more secure than those assumptions.

We do not want to comment on the efficacy (or otherwise) of this Moorean response. We mention it, only to emphasise the contrast between this response, and a *transcendental* argument against the very same model-theoretical sceptic. Whilst there are certain similarities between various kinds of responses to the sceptic, the key difference lies in what the arguments aim to show. Moorean arguments aim to show that we are more *certain* of anti-scepticism than of scepticism. By contrast, a transcendental argument aims to show that scepticism cannot coherently be maintained. In our particular case, we want to show that no one can intelligibly use model-theory to motivate a sceptical position. Indeed, playfully put, our aim is to turn the game played by the model-theoretical sceptic against itself.[6]

9.3 The Metaresources Transcendental Argument

With our aims outlined, we now present our first transcendental argument. We call it the *Metaresources Transcendental Argument,* and it is essentially due to Bays.

[5] See G. E. Moore (1939). This section draws on Button (2013: 124–7, 2016a: §§2–3). There is an extensive literature on Moorean responses to brain-in-vat scepticism; this is just one possible reading.

[6] The classic reference is Kant's (1787: B276) 'refutation of idealism', which aims to conclude that 'the game played by the idealist is turned against itself'. This is how the notion of a *transcendental argument* has been used by English-speaking philosophers since Strawson (1959, 1966).

The argument

In her attempt to raise sceptical problems, the model-theoretical sceptic is happy to employ the full power of first-order model theory. In particular, as in Chapter 7, she may invoke the Compactness Theorem 4.1 or the Löwenheim–Skolem Theorem 7.2, in order to show us that any first-order theory of arithmetic has plenty of unintended models. These model-theoretic results make it hard to answer questions like: *How do we pin down the concept of finiteness?*

However, as Bays points out, 'the notions of finitude and recursion are needed to describe first-order model theory'. After all, first-order formulas 'can be of arbitrary *finite* length, but they cannot be infinite'. Similarly, first-order satisfaction is defined via recursion. We would add that first-order deductions can be of arbitrary finite length, but they cannot be infinite. Thus, the very use of first-order model theory itself *presupposes*, through and through, that we understand the concept *finite*. So if the model-theoretical sceptic employs first-order model theory in an attempt to argue that we lack the concept *finite*, she saws off the branch on which she sits.[7]

First and foremost, this anti-sceptical argument targets those who want to use model theory to argue that we have no firm grasp of the concept *finite*. But the same anti-sceptical argument can also be used against those who would use model theory to argue that we cannot pick out the natural numbers (thought of as *objects*).[8] Roughly: in using model theory to explain *why* we fail to pick out the natural numbers, she must invoke certain (conceptual) resources which would then *allow* us to pin them down after all.

So rephrased, the anti-sceptical argument is something like a *modus tollens* of our discussion of formalism in §7.11. But we should make this explicit. So. Suppose the sceptic first notes that there are many models of PA, and then asks us how we can refer to any one of these (up to isomorphism), hoping to embarrass us into admitting that we cannot (whilst remaining moderates). In response, we note that the sceptic's challenge is posed in terms of the formal theory PA, perhaps closed under provability. To understand the challenge at all, then, we must understand what that theory is. As such, we need to possess the general notion of an arbitrary but finitely long formula, and the general notion of an arbitrary but finitely long proof. All of this requires possession of the general concept of *finiteness*.[9] But in that case, if we fully understand the sceptical challenge: *How do you single out a particular model of* PA?, we are already equipped with the resources to single out the standard model:

---

[7] See Bays (2001: 345). Barton (2016: 206) gives a similar argument against Hamkins's multiversism.

[8] There is, of course, a potential gap between insisting that we securely grasp mathematical *concepts*, and insisting that we can refer determinately to certain mathematical *objects*. After all, as noted in §6.4, one might simply not *countenance* mathematical objects, perhaps citing metaphysical qualms or doubting that reification is very illuminating. Our point here is entirely restricted to model-theoretically inspired worries about referring to mathematical objects.

[9] As noted in footnote 24 of §7.11, known formalisations of theories of syntax interpret weak fragments of arithmetic like Robinson's Q.

It is the unique model (up to isomorphism) whose elements are all FINITELY FAR *from the initial element.* Indeed, insofar as we understand the sceptical challenge, we can straightforwardly also understand the proof of Dedekind's Categoricity Theorem 7.3 for PA_2, in a logic whose semantics allow the second-order quantifiers to range over the set:

$$C = \left\{(S^n(0))^{\mathcal{M}} \in M : \text{for some natural number } n\right\}$$

After all, C is precisely the subset of elements of M which are 'finitely far' from the initial element; and the categoricity of PA_2 only requires comprehension of this C.

The point of all this is the following. In order to understand why model theory is supposed to push us towards sceptical concerns, we must possess certain model-theoretic concepts. However, possession of those model-theoretic concepts enables us to brush aside the sceptical concerns. Accordingly: insofar as we can understand the sceptical challenge, we can dismiss it.

Connections with just-more-theory

As just noted, there is a tight link between the Metaresources Transcendental Argument, and our proof of Dedekind's Categoricity Theorem 7.3. In Chapter 7, however, we noted that any logic which is strong enough to provide a categorical theory of arithmetic is not compact. As such, the model-theoretical sceptic will insist that invoking any such logic inevitably amount to *just more theory*. So we must carefully explain how the just-more-theory manoeuvre plays out here.

The just-more-theory manoeuvre raises deep problems for certain *specific* philosophical positions. Specifically, it shows that moderate modelists cannot hope to regard the proof of Theorem 7.3 *as* a proof of categoricity. If *they* attempt to invoke the set C outlined above, then *their* attempt to do so will indeed succumb to the just-more-theory manoeuvre.

But this point is *localised* to moderate modelism. As we explained in §7.8, moderates need to explain how we can break into the circle of mathematical concepts that allow us to regard (for example) the proof of Theorem 7.3 *as* a proof of categoricity. Crucially, it is incompatible with moderation to insist that we 'just can' refer to the isomorphism type of an ω-sequence, or that 'as a matter of brute fact' our words 'just do' express the concept *ω-sequence* (rather than some non-standard concept), or that we 'just do' use second-order logic with its full semantics (rather than the Henkin semantics). These are serious problems for moderate modelists, and they will not go away.

It certainly does not follow, though, that *no one* can regard the proof of Theorem 7.3 *as* a proof of categoricity. In the simplest case, an *im*moderate modelist could simply insist that she 'just does' use second-order logic with full semantics. That might be objectionable on other grounds; but they would, indeed, be *other* grounds than those supplied by the just-more-theory manoeuvre itself. More intriguingly—though we must wait until §12.3 to elaborate on this idea—if the problem arises only for moderate *modelists*, then a moderate character who rejects modelism might also be able to follow the proof of Theorem 7.3. But the simple point here is that there are positions which can present the Metaresources Transcendental Argument as a simple and straightforward *ad hominem* against the model-theoretical sceptic, without themselves feeling any sting from the just-more-theory manoeuvre.

We can push this point further. Consider someone who, having read through Chapters 7 and 8, and finally §9.1, has come to embrace model-theoretical scepticism. So she wants to wield model theory to convince herself (for example) that she does not possess the concept of *finiteness*. She then encounters the Metaresources Transcendental Argument: it is pointed out to her that, in wielding model theory, she is wielding precisely the concept of *finiteness* which she claims not to grasp determinately. At this point, how the just-more-theory manoeuvre affects her *opponents* is irrelevant: she has boxed herself into an incoherent position.

The point, in brief, is just that the sceptical challenge undermines itself. But the correct conclusion to draw from this is *not* that moderate modelism is 'alright after all', and can somehow avoid falling into model-theoretical scepticism. Since moderate modelism was led inexorably *to* an incoherent sceptical position, the correct conclusion is rather that there is something *deeply* wrong, not just with model-theoretical scepticism, but with moderate modelism.[10]

Limited conclusions

The Metaresources Transcendental Argument successfully highlights a defect with model-theoretical scepticism. But we should emphasise that it does not do much more than that, and mention some intrinsic *limitations* on the Argument.

Again, we start by considering our classic sceptical challenge from outside the philosophy of mathematics: *Are you in a brain in a vat?* There is a strong transcendental argument against anyone who would wish to claim that we are (eternally) envatted brains.[11] Nonetheless, showing that we are *not* (eternally) envatted brains shows very little about what the world *is* like. Equally, it goes almost no way towards explaining how creatures like us can come to form beliefs about the world around

[10] For more, see Button (2016a). Button argues that this is a sensible reading of Putnam's own use of the model-theoretic arguments: as a (slightly enigmatic) reductio of a brand of moderate objects-platonism. Of course, there are other ways to read Putnam. Nevertheless, we should emphasise that Putnam *never* embraced model-theoretical scepticism himself (see e.g. Putnam (1977: 488, 495, 1980: 482, 1981: 53–4)). Here, then, we part company somewhat with Bays, who both explicitly presents Putnam as the *target* of his Metaresources Transcendental Argument (2001: 345), and suggests that Putnam holds 'that there's no fact of the matter as to what the terms and predicates of our language refer to' (2014: §3.4).

[11] Specifically, Putnam's (1981: ch.1) brain-in-vat argument; for defence and references to the literature, see Button (2013: chs.12–14, 2016a).

us, or come to know things about that world. In short: a transcendental argument in this context leaves all of empirical psychology standing and, with it, our ability to ask philosophical questions inspired by empirical psychology.

The situation is similar for our Metaresources Transcendental Argument. The Argument defeats a certain kind of model-theoretical scepticism. But the Argument offers very little help in deciding whether to accept or reject some proposed axiom of set theory, for example. Equally, the Argument goes almost no way at all towards explaining how creatures like us can come to hold mathematical beliefs, or acquire mathematical knowledge. Indeed, the Metaresources Transcendental Argument leaves standing all the empirical work that is directed at the question of how humans acquire mathematical concepts (either individually, as they mature; or as a species, as they evolved).

Switching to another limitation of our Metaresources Transcendental Argument, we must emphasise that it fails even to *connect* with certain sceptical challenges. At the most extreme: the Argument has no sway at all against someone who thinks that all of mathematics—model theory included—is a vast comedy of errors. If that kind of opponent can be dealt with at all, she must be dealt with by other means.

The preceding comments are generic, in the sense that they apply not just to the Metaresources Transcendental Argument, but to any transcendental argument against model-theoretical scepticism. However, the Metaresources Transcendental Argument also has some very specific limitations. In particular, the argument turns entirely on the fact that the concept of *finiteness* is required by the very idea of studying countable first-order theories. We can imagine a model-theoretical sceptic who raises no issues concerning the concept of *finiteness*, but limits their concerns to issues surrounding *uncountability*. Nothing in the Metaresources Transcendental Argument could speak to those kinds of concerns.

Such are the limitations on the Metaresources Transcendental Argument. Within its limitations, though, we endorse the Argument. It beautifully undermines a certain kind of model-theoretical scepticism.

9.4 The Disquotational Transcendental Argument

We will now consider a second transcendental argument: the *Disquotational Transcendental Argument*. It is best introduced by example.

Finiteness

Having read Chapter 7, you may be wondering whether your arithmetical vocabulary picks out a non-standard isomorphism-type. To ensure that you understand the sceptical concern, you remind yourself what a non-standard model of arithmetic is. Looking to Definition 4.2, you see that it is a model containing elements which are 'infinitely far' from that model's initial element, i.e. a model $\mathcal{M}$ containing an element a such that $a \neq (S^n(0))^{\mathcal{M}}$ for each natural number n. But, of course, this depends upon some kind of calibration to the natural numbers of the *metatheory*. Now, if we had already pinned down *the* standard model, and could make reference to it in our metatheoretical definition of non-standardness, there would be no problem. But the sceptical concern is precisely that we *cannot* pin down the standard model.[12]

Still, where $\mathcal{N}$ is a model of set theory, say, we can usefully distinguish between entities which $\mathcal{N}$ thinks are finite, and entities which $\mathcal{N}$ thinks are not finite (to use the terminology introduced in §8.3). For example: within set theory, we can formally define the notion of a 'finite von Neumann ordinal', and can then consider the elements which satisfy 'finite von Neumann ordinal' within some model of set theory. Following this template, we say that an entity a is finite$^{\mathcal{N}}$ iff $\mathcal{N}$ thinks that a is finite. Now the sceptical worry which you are trying to entertain has two distinct components. The sceptic is asking you to take seriously that you do not have access to the standard model of arithmetic, but instead access to some non-standard model, $\mathcal{N}$. So your worry is:

(1) My word 'finite' applies to the finite$^{\mathcal{N}}$ things

But in order to make this a *sceptical* challenge, the sceptic wants to remind you that $\mathcal{N}$ is, indeed, non-standard with respect to its natural numbers. Hence you are supposed simultaneously to entertain:

(2) Everything I call a 'natural number' is finite$^{\mathcal{N}}$, but some of the things I call 'natural numbers' are not finite

We maintain that *it is impossible to take both worries seriously, simultaneously.* The impossibility emerges when we ask what the sentence (2) *expresses* if worry (1) obtains. By assumption, the word 'finite', as it occurs as the end of (2), must apply only to the finite$^{\mathcal{N}}$ things. In which case, (2) does not succeed in expressing *that* your word 'natural number' applies to some non-finite entities. But that was precisely a part of the sceptical worry you were supposed to be entertaining. So it seems you cannot even *understand* the sceptical worry you were supposed to be entertaining.

From this, you should draw a disjunctive conclusion. *Either* you cannot understand the sceptical worry, *or*, since you can understand it, the sceptical worry is not realised. Hence, as for the Metaresources Transcendental Argument: insofar as we understand the sceptical challenge, we have the resources to answer it.

(Once again, we have presented the anti-sceptical argument in terms which are neutral between objectual and conceptual readings. But we can offer more specific versions if required. For objectual versions, we regard (1) as saying that my word 'finite' is *true of* (or *refers to*) the finite$^{\mathcal{N}}$ things. For conceptual versions, we regard

[12] Cf. Parsons' (1990b: 31–2, 2008: 279–80) reading of Dummett (1963: 191–3).

(1) as saying that my word 'finite' *expresses* the concept *finiteness*$^{\mathcal{N}}$. Making similar adjustments to (2), we arrive at the same tension.)

Countable Skolem hulls

Like the Metaresources Transcendental Argument, our first version of the Disquotational Transcendental Argument concerned *finiteness*. But, as we now show, the Disquotational Argument can be run using richer mathematical concepts.[13]

Suppose you are trying to take *skolemite* concerns very seriously. In particular, having thought long and hard about Mostowski-collapsing a Skolem-hull, as in Corollary 8.7, you might come to worry that your set-theoretic vocabulary just picks out a countable, transitive model, $\mathcal{H}$, of ZFC. In this situation, you would only ever talk about countable sets. Nonetheless, it will be false for you to say something like 'every set is countable' since, after all, $\mathcal{H}$ is a model of ZFC, and 'there are uncountable sets' is a *theorem* of ZFC. So, with the same notation as before, you are supposed to worry that you only speak about some non-standard model, $\mathcal{H}$, i.e.:

(3) My word 'countable' applies to the countable$^{\mathcal{H}}$ sets

But you must also be alive to the fact that $\mathcal{H}$ is *non-standard*, i.e.:

(4) Some things I call 'sets' are not countable$^{\mathcal{H}}$, but all of the things I call 'sets' are countable

Just as before, it is impossible to take both worries seriously, simultaneously. If (3) obtains, then the word 'countable', as it occurs as the end of (4), must by assumption apply only to the countable$^{\mathcal{H}}$ things. In which case, (4) does not succeed in expressing *that* your word 'set' applies only to countable sets. But that was precisely a part of the sceptical worry that you were supposed to be entertaining. And so, inevitably: insofar as we understand the sceptical challenge, we have the resources to answer it.

A metaphor

We have illustrated one argumentative strategy in two different ways. The strategy itself essentially comes down to the following. To express the sceptical worry, we have no other option than to use our own language. But if the sceptical worry obtains, then disquotation in our own language fails. This means that we fail to *express* the sceptical worry in our own language. But then we have no way to express it at all. This is why we called this the *Disquotational Transcendental Argument*.[14]

---

[13] The argument which follows is essentially a recasting of Button (2016a: §3). This argument, in turn, is hugely in debt to Tymoczko (1989: 287–9) and A. W. Moore (2001: 165–7, 2011: §3).

[14] Note that this argument does not require you to endorse the view that the concept of truth is *exhausted* by disquotational platitudes. (Indeed, one of us explicitly denies this; see Button 2014.) The point

It might help, though, to offer a more metaphorical explanation of the Disquotational Transcendental Argument. (*Caution: this is a metaphor!*) The model-theoretical sceptic wants to use model theory to convince you that there is no way for you to attain a viewpoint from outside the non-standard model that you are 'trapped' inside. (Her repeated use of the just-more-theory manoeuvre should convince you that this really is her aim.) But then it is hardly surprising that we encounter deep difficulties in the very enterprise of making sense of the sceptical worry itself. For, if you can understand the idea that yours is a '*non*-standard model', then you must be standing outside of the non-standard model, looking in. And then you are not truly *trapped* inside it.

At the risk of repetition: that was a hand-waving metaphor. We hope it provides an intuitive sense of the Disquotational Transcendental Argument. But if you find the metaphor unhelpful, set it aside. The crucial point is that no one can hold both (1) and (2) together, or both (3) and (4) together.

Limitations on the argument

At the further risk of repetition, we should be clear on the limitations of the Disquotational Transcendental Argument.

Unlike the Metaresources Transcendental Argument, the Disquotational Transcendental Argument has some capacity for handling concepts like *uncountability*. That point came through in our second instance of the Disquotational Transcendental Argument. Nonetheless, like the Metaresources Transcendental Argument, the Disquotational Transcendental Argument is, first and foremost, an *ad hominem* against the model-theoretical sceptic.

The *ad hominem* nature of the argument, again, limits what one can conclude from it. For example: the second instance of the Disquotational Transcendental Argument cannot deliver a conclusive verdict that there *are* any uncountable sets. To see why: imagine someone who believes that every set is countable, not because they worry that we are 'trapped in' a Skolem hull satisfying ZFC, but because they simply dislike the very idea of a power set of an infinite set. Such a *countabilist* would effectively believe that *uncountable* means *proper-class-sized*. We cannot brush countabilism aside with a transcendental argument; instead, we must try to convince them that the (unrestricted) Power Set axiom is acceptable after all. At the risk of dividing labour too far: that looks like a project in the philosophy of set theory, rather than a project in the philosophy of model theory. At the very least: it is not a project in the *transcendental* philosophy of model theory, which has been our present concern.

---

is just this: whatever the status of the concept of truth in general, the individual *semantic clauses* one can specify for one's own language in one's own language always look like instances of disquotational platitudes.

We also want to be clear that this insistence on disquotation does *not* provide a cheap demonstration that our mathematical vocabulary is completely *determinate*. This point has been made clearly by Field. As he notes, the word 'bald' applies to all and only the bald things, even though the word 'bald' has *vague* (and hence indeterminate) conditions of application. This is so, because the conditions of application of 'bald' are exactly as vague in the object language as in the metalanguage.[15]

We must admit, then, that it is *in principle coherent* to think that the word 'countable' applies to all and only the countable things, whilst also insisting that the word 'countable' has indeterminate application conditions. We should emphasise, however, that we have not yet seen any reason to think that 'countable' *does* have indeterminate application conditions. And we especially want to emphasise that *the motivations underpinning model-theoretical scepticism cannot be used to turn this in-principle possibility into a legitimately live concern.*[16] We can see this by considering an important *dis*analogy between 'bald' and 'countable'. We can coherently say that our word 'bald' is as likely to apply to (exactly) the people with fewer than n hairs as to (exactly) the people with fewer than $n + 1$ hairs. Indeed, saying this might help us to see why the word 'bald' actually has vague conditions of application. But suppose the model-theoretical sceptic attempted to do something similar in the case of 'countable'. So, she might try to say that our word 'uncountable' is as likely to apply to (exactly) the uncountable$^{\mathcal{H}}$ sets as to the uncountable sets, for some countable Skolem hull $\mathcal{H}$. However, the reason that $\mathcal{H}$ is even of interest to the sceptic is that *every uncountable*$^{\mathcal{H}}$ *set is countable.* So, to say that our word 'uncountable' might pick out the uncountable$^{\mathcal{H}}$ sets is like saying that our word 'bald' might pick out plenty of non-bald people: no matter how precise or vague the word 'bald' is, it is a *bad* claim. Once again, model-theoretical scepticism undermines itself.

9.5 Ineffable sceptical concerns

Both of our transcendental arguments have a simple conclusion: insofar as we understand the sceptical challenge, we have the resources to answer it. Put like this, the sceptic has an obvious comeback. Rather than resisting the conditional, she might embrace it, and go on to insist that the *real* worry she wants to pose is precisely that her sceptical challenge is *ineffable*. One might think of her as pushing towards the following, desperate, thought: *Our ability to pin down mathematical structure is so poor that we cannot even express how poor it is!*

Now, we cannot offer a knockdown objection against such ineffable scepticism. But this is scarcely *our* fault. There is something deeply peculiar about trying to talk

[15] Field (1994: 398–9).

[16] See Button (2016a: §3).

someone out of worries which they admit they cannot put into words. All we can do really do is emphasise the peculiarity.

Here is a worry that might look similar to the ineffable sceptic's concern: *I am so stupid that I cannot even appreciate how stupid I am.* To some extent, this worry is probably true of all of us. But, at the very least, there had better not be a transcendental argument which the deeply stupid can wheel out to show that they are not stupid after all. Fortunately, there is no such an argument; and the stupidity-worry is dissimilar from model-theoretical scepticism. The worry, *I cannot appreciate the full depths of my own stupidity*, is not ineffable at all. We just 'effed' it.

Here is another worry that might look similar: *I currently lack the mathematical resources to express a whole variety of interesting mathematics.* Again, this may well be true: none of the vocabulary of set theory, model theory, and category theory was around in the year 1787; and no one knows what mathematical vocabulary will be around in the year 2387. So again, at the very least, there had better not be a transcendental argument which we can wheel out to show that there will never be any genuinely new ways of approaching mathematics. Fortunately, there is no such argument, and the future-oriented worry is fundamentally dissimilar from model-theoretical scepticism. The thought that our language will expand is not ineffable at all. We just 'effed' it.

No; the ineffable sceptic—i.e. the model-theoretical sceptic who has embraced an ineffable variety of scepticism—is pushing towards much *murkier* worries than this. Moreover, the *murkiness* of her worries is cannot be dispelled even in principle; for, if it could be dispelled, then the worries would be effable after all. In short, if we are even to engage with the ineffable sceptic's concerns, the best we can hope to do is grapple with an *apt metaphor*.

To this end, the ineffable sceptic might tell us that what we (mere humans) think of as uncountable is *really countable from the God's Eye Point of View* (or whatever).[17] But if this speech is to succeed in expressing what it is *meant* to express, it must be offered from the God's Eye Point of View itself. Equally: if it is to succeed in expressing what it is meant to express, then the God's Eye Point of View must never be our own. At which point, again, one might be forgiven for thinking that no coherent worry has even been *gestured* at, let alone put into words.

Our sceptic might, then, reach for less divine imagery. Instead of invoking the God's Eye Point of View, she might consider different (but equally human) mathematical communities 'living in different models of set theory'. The thought, perhaps, is that we can imagine a community of mathematicians who inhabit a model which looks to us like a countable transitive model of ZFC; and we might pity their benighted state; and then suddenly it might dawn on us: *maybe someone pities our*

[17] This is a phrase Putnam repeatedly used to characterise his opponents (1981: 49, 1982: 38, 1983: x, xviii); and our rebuttal is essentially Putnam's.

community, in just the way we pity those beneath us![18]

To repeat: this is a mere *metaphor* which the sceptic has trotted out in order to push us towards a self-consciously *ineffable* concern. Consequently, we are honestly not sure that it is even a good idea to *engage* with the metaphor, since to engage with it may be simply to give ineffable worries air-time which they do not merit. Wisely or unwisely, though, we will engage with the metaphor; and we will meet it with a pugilistic question: *Why should any position within the philosophy of mathematics acknowledge a non-trivial notion of 'the inhabitants of a model'?*

We should be clear on the question we are asking. Obviously, we have no wish to impugn the mathematical fruitfulness of certain heurisitic modes of speaking. For example, when we do forcing in set theory, it can be illuminating to speak of 'moving from' one set-theoretic universe to another. But it should go without saying that, in such cases, *we are not really moving.*

Perhaps the sceptic wants us to think of 'inhabitants of a model' along the lines of denizens of a (philosopher's) possible world. In that case, the sceptic would be suggesting the following: there is a possible world, w, where the only (pure) sets in w form a countable transitive model; we in the actual world pity the denizens of w for their inability to access to Cantor's paradise; and then we dimly form a concern, which we attempt to express by saying: *perhaps we have also been shut out from paradise!* But for this thought to have *any* bite, we would already have to have bought into a very particular and peculiar way of conceiving of the interaction between mathematical objects and possible worlds. Not many philosophical positions, after all, are forced to entertain the possibility that *there could have been fewer pure sets than there actually are.*

The preceding paragraph put the metaphor in (strained) objectual terms. Interestingly, is doubtful that *this* metaphor can even be recast in conceptual terms. To describe two people as 'inhabiting different models' would be to suggest that they possess different mathematical concepts despite—if this is to be a sceptical worry—acting in essentially the same way with regard to all mathematical matters. Again, in order for this to have *any* bite, we would need to have bought into a very particular and peculiar way of conceiving of mathematical concepts.

We might keep struggling with metaphors for a while longer, but we hope the point is clear. We cannot offer a knockdown refutation of these ineffable sceptical concerns, but it is doubtful that anyone is obliged to take them seriously. If model-theoretical scepticism is not incoherent, it is at least genuinely incomprehensible.

---

[18] Cf. the way in which certain critics of Putnam's brain-in-vat argument suggest that Putnam defeats brain-in-vat scepticism only to open himself up to an ineffable scepticism. Thus Wright (1992: 93) says: 'the real spectre to be exorcised concerns the idea of a thought *standing behind* our thought that we are not brains-in-a-vat, in just the way that our thought that they *are* mere brains-in-a-vat would stand behind the thought—could they indeed thinking anything—of actual brains-in-a-vat that "We are not brains-in-a-vat".' For critical discussion see Button (2013: 137ff).

9.A Application: the (non-)absoluteness of truth

The main aim of this chapter is fulfilled: we have used transcendental arguments to explain why model-theoretical scepticism is a disaster. This philosophical-cum-technical appendix can be safely omitted, at least on a first reading, since it does not affect any of the preceding discussion. However, we include it in this chapter because it relates the Metaresources Transcendental Argument of §9.3 to the question of whether arithmetical *truth* is absolute.[19]

Non-standard 'sentences' and free-floating truth

Most of what we have said about *truth* in this book has concerned truth-in-a-model. More precisely, we have considered (various notions of) satisfaction, symbolised with '$\vDash$', and defined rigorously in Chapter 1. In this section, we adopt a different perspective. Instead of considering truth-in-a-model, we consider models *of* formal theories of truth. That is, we consider a theory containing a one-place truth-predicate, *Tr*, which is intuitively to be read as '... is true'. This is to be applied to elements which are thought of as *sentences*, which we pick out using another one-place predicate, *Sent*, which is intuitively to be read as '... is a sentence'.

Since we are reasoning about sentences, a formal theory of truth could be presented via a theory of syntax. However, it will be easier to use a theory of arithmetic; for we already noted in §7.11 that arithmetic can interpret syntax. Indeed, for reasons outlined in §5.A, very weak theories of arithmetic can be used as the base for a formal theory of truth. We just need to augment them with a truth-predicate, *Tr*.[20] However, we also need to add principles governing how *Tr* behaves. And at this point, we encounter Tarski's famous impossibility result (this uses the notion of Gödel-numbering, $\ulcorner\sigma\urcorner$, which we sketched in §5.A):

Theorem 9.1 (Tarski): *Let T be a satisfiable $\mathscr{L}$-theory which interprets Q. Then T has no truth-predicate; i.e. there is no $\mathscr{L}$-formula $\varphi(v)$ with the property that $T \vDash \sigma \leftrightarrow \varphi(\ulcorner\sigma\urcorner)$ for each $\mathscr{L}$-sentence σ.*

Still, nothing prevents us from adding a *new* predicate to our language, which behaves as a truth predicate for the sentences of the *old* language. In particular, where T is an $\mathscr{L}$-theory with $Tr \notin \mathscr{L}$, we define an $\mathscr{L} \cup \{Tr\}$-theory by:

[19] We are using the term 'absolute' in the sense of Definition 8.11; that is, we are asking whether there is a reasonable class of models whose interpretation of a truth-predicate must agree with the truth-predicate of the metalanguage. However, this discussion will be intelligible without any particular grasp on the material from §8.A.

[20] If T is an $\mathscr{L}$-theory which interprets Q, then we can regard *Sent* as a one-place $\mathscr{L}$-formula with the following property: for any (standard) natural number n, $T \vDash Sent(S^n(0))$ iff $n = \ulcorner\sigma\urcorner$ for some $\mathscr{L}$-sentence σ.

$$T^{Tr} = T \cup \{\forall x(Tr(x) \rightarrow Sent(x))\} \cup \{\sigma \leftrightarrow Tr(\ulcorner\sigma\urcorner) : \sigma \text{ is an } \mathscr{L}\text{-sentence}\}$$

The idea is that T^{Tr} is a minimally adequate theory of truth for the $\mathscr{L}$-sentences. Moreover, T^{Tr} is satisfiable if T is:

Lemma 9.2: *Let T be a consistent $\mathscr{L}$-theory which interprets Q, with $Tr \notin \mathscr{L}$. Then T^{Tr} is consistent.*

Proof. Suppose $\mathcal{M} \vDash T$. Define a subset of M by:

$$U = \{a \in M : \mathcal{M} \vDash \ulcorner\sigma\urcorner = a \text{ and } \mathcal{M} \vDash \sigma, \text{ for some } \mathscr{L}\text{-sentence } \sigma\}$$

Regard $(\mathcal{M}, U)$ as an $\mathscr{L} \cup \{Tr\}$-structure, which augments $\mathcal{M}$ just by taking U to be the interpretation of the predicate Tr. Then $(\mathcal{M}, U) \vDash T^{Tr}$. □

However, Tarski's Theorem 9.1 entails that this 'minimally adequate' theory of truth has some rather surprising models. To see why, we must first take note of the following beautiful theorem of Svenonius:

Theorem 9.3 (Svenonius)**:** *Let $\mathcal{M}$ be an $\mathscr{L}$-structure, let R be an n-place predicate not in $\mathscr{L}$ (for some $n \geq 1$), and let $X \subseteq M^n$ be undefinable. Where $(\mathcal{M}, X)$ is the $\mathscr{L} \cup \{R\}$-structure formed by letting R pick out X, there is an elementary extension $(\mathcal{N}, Y)$ of $(\mathcal{M}, X)$ with an automorphism $h : \mathcal{N} \longrightarrow \mathcal{N}$ such that $h(Y) \neq Y$.*

An automorphism is just an isomorphism from a structure to itself. The proof of Svenonius' Theorem uses types, which we do not develop until Chapter 14.[21] But this Theorem has a simple Corollary (whose significance we explain below):[22]

Corollary 9.4: *Let T be a satisfiable $\mathscr{L}$-theory interpreting Q, with $Tr \notin \mathscr{L}$. Then there is an $\mathscr{L}$-structure $\mathcal{N}$, with $V_1 \neq V_2$, such that both $(\mathcal{N}, V_1)$ and $(\mathcal{N}, V_2)$ satisfy T^{Tr}.*

Proof. By Lemma 9.2, T^{Tr} has a model, $(\mathcal{M}, U)$. By Tarski's Indefinability Theorem 9.1, U is not definable in $\mathcal{M}$. So, by Svenonius' Theorem, there is some $(\mathcal{N}, V) \succeq (\mathcal{M}, U)$ and an automorphism h on $\mathcal{N}$ such that $V \neq h(V)$. The result follows by letting $V_1 = V$ and $V_2 = h(V)$. □

This Corollary describes a very strange situation. Let a be an element of N which is a member of only *one* of V_1 and V_2. Within $\mathcal{N}$, the arithmetical properties of a are completely fixed: after all, $\mathcal{N} \vDash \varphi(a)$ or $\mathcal{N} \vDash \neg\varphi(a)$, for each $\mathscr{L}$-formula $\varphi(v)$.

[21] For a proof of Svenonius' Theorem, see Poizat (2000: 184).

[22] See Krajewski (1974: Lemma 3) and Hamkins and Yang (forthcoming: Theorem 2).

But this does *not* suffice to pin down *a*'s *semantic* properties, in the following sense: we can *either* regard a as true, or regard a as false, depending on whether we interpret the truth-predicate, *Tr*, via V_1 or V_2. Otherwise put: semantic properties seem to float completely free from any other properties. This is a little mind-boggling.

In fact, in order for this to happen, this peculiar element a must be *non-standard,* in exactly the sense of Definition 4.2:[23]

Proposition 9.5: *Let $T, \mathscr{L}, \mathcal{N}, V_1$ and V_2 be as in Corollary 9.4. Then any element of N which is a member of only one of V_1 and V_2 is non-standard.*

Proof. Without loss of generality, suppose $a \in V_1$ is standard. So $(\mathcal{N}, V_1) \vDash Tr(a)$ and hence $(\mathcal{N}, V_1) \vDash Sent(a)$. Given the properties of *Sent*, this implies that there is some $\mathscr{L}$-sentence σ such that $a = \ulcorner\sigma\urcorner$. Since $(\mathcal{N}, V_1) \vDash Tr(\ulcorner\sigma\urcorner) \leftrightarrow \sigma$, we have $\mathcal{N} \vDash \sigma$. Since also $(\mathcal{N}, V_2) \vDash Tr(\ulcorner\sigma\urcorner) \leftrightarrow \sigma$, we have $(\mathcal{N}, V_2) \vDash Tr(\ulcorner\sigma\urcorner)$, so that $\ulcorner\sigma\urcorner = a \in V_2$. □

We can combine Corollary 9.4 and Proposition 9.5 in one slogan: *Truth is absolute for standard entities and non-absolute for non-standard entities.*[24]

This observation allows us to stop boggling at Corollary 9.4. Since a is non-standard, there is no $\mathscr{L}$-sentence σ such that $a = \ulcorner\sigma\urcorner$. As such, we can never use a within the context of one of the 'disquotation' sentences, $\sigma \leftrightarrow Tr(\ulcorner\sigma\urcorner)$, which we find in T^{Tr}. So: a might be 'true according to $(\mathcal{N}, V_1)$', in the sense that $(\mathcal{N}, V_1) \vDash Tr(a)$, and 'false according to $(\mathcal{N}, V_2)$', in the sense that $(\mathcal{N}, V_2) \vDash \neg Tr(a)$; but a does not code an $\mathscr{L}$*-sentence* with a free-floating truth value.

In this light, indeed, Corollary 9.4 is not even very surprising. We have repeatedly seen that first-order theories have 'pathological' models. Corollary 9.4 just presents us with a very particular 'pathological' model for a formal theory of truth. Moreover, its particular 'pathology' can be traced specifically to the fact that it contains an arithmetically non-standard entity. Otherwise put: $\mathcal{N}$'s 'natural numbers' are not isomorphic to *the* natural numbers. And as usual: If we can understand *that* fact, then we can dismiss the problem supposedly raised by $\mathcal{N}$, that semantic properties can float free from all other properties. That, indeed, is just an application of our Metaresources Transcendental Argument.

Truth and object

As mentioned in the footnotes, Hamkins and Yang have recently discussed Corollary 9.4. Their primary aim is to show that simply pinning down 'a natural number structure' is not, in general, sufficient to pin down mathematical truth. This

[23] Cf. Hamkins and Yang (forthcoming: 27).

[24] See Hamkins and Yang (forthcoming: 3–4).

point is almost immediate from Corollary 9.4 and Proposition 9.5. If we want to pin down the extension of arithmetical truth just by pinning down particular mathematical objects, then we must have (somehow) already picked out the *standard* natural numbers. Otherwise, the non-absoluteness of truth for non-standard entities will rear its head: there will be rival possible extensions of truth, in the guise of V_1 and V_2.

Such a result may well pose further problems for moderate modelism (say), since the moderate modelist has repeatedly *failed* to pin down *the* standard natural numbers. But, to be very clear, this is no reason to think that the non-absoluteness of truth for non-standard entities raises any problems for *us*, who have rejected moderate modelism. And this last point is worth pursuing.

Hamkins and Yang raise Corollary 9.4 en route to a more elaborate result, which they phrase as follows:

> Every consistent extension of ZFC has two models $\mathcal{M}_1$ and $\mathcal{M}_2$, which agree on the natural numbers and on the structure $(\mathbb{N}, +, \cdot, 0, 1, <)^{\mathcal{M}_1} = (\mathbb{N}, +, \cdot, 0, 1, <)^{\mathcal{M}_2}$, but which disagree [on] their theories of arithmetic truth, in the sense that there is in $\mathcal{M}_1$ and $\mathcal{M}_2$ an arithmetic sentence σ, such that $\mathcal{M}_1$ thinks σ is true, but $\mathcal{M}_2$ thinks it is false.[25]

The route from our Corollary 9.4 to Hamkins and Yang's Theorem involves treating $\mathcal{N}$ with a bit of extra refinement. The essential idea is to start with some non-standard model $\mathcal{M}_1$ of the relevant extension of ZFC, and then to treat $\mathcal{N}$ as the natural numbers according to $\mathcal{M}_1$. Our pesky element a, from Corollary 9.4, is then precisely Hamkins and Yang's element σ. Hamkins and Yang then build $\mathcal{M}_2$ around σ in such a way that '$\mathcal{M}_1$ thinks σ is true, but $\mathcal{M}_2$ thinks it is false'. (This turn of phrase should be understood as in §8.3.)

As pure mathematics, this is impeccable. However, we want to take issue with Hamkins and Yang's informal claim that σ is an *arithmetic sentence*. This is quite misleading. As in Lemma 9.5, σ must be a *non-standard* element. Granted, according to both $\mathcal{M}_1$ and $\mathcal{M}_2$, σ satisfies the predicate *Sent*. In that sense, one might gloss the technical result as saying: *there is an object, common to $\mathcal{M}_1$ and $\mathcal{M}_2$, which each model thinks is an arithmetic sentence.* But, in the ambient model theory within which Hamkins and Yang's result is stated and proved, we can and must insist that σ is *not* (a code for) an arithmetic sentence, for sentences are finite. As such, Hamkins and Yang should more accurately end the statement of their result as follows:

> ...but which disagree [on] their theories of arithmetic truth, in the sense that there is in $\mathcal{M}_1$ and $\mathcal{M}_2$ an entity a such that $\mathcal{M}_1 \vDash Tr(a)$ and $\mathcal{M}_2 \vDash \neg Tr(a)$.[26]

[25] Hamkins and Yang (forthcoming: Theorem 1). For consistency of formatting with the rest of this book, we have used '$\mathcal{M}$' where Hamkins and Yang have 'M'.

[26] In fact, Hamkins and Yang use a full-blown *satisfaction* relation, rather than a mere truth-predicate. In these terms, the statement would end: ...an entity a such that $\mathcal{M}_1 \vDash (\mathbb{N} \vDash a)$ and $\mathcal{M}_2 \vDash (\mathbb{N} \nvDash a)$.

However, once the result has been stated in this way, it provides no reason whatsoever to worry that we might somehow agree on the natural numbers but disagree on which arithmetical *sentences* are true. The essential point here concerns the Metaresources Transcendental Argument once again. Insofar as we understand the result—i.e. insofar as we speak the language in which the result is stated and proved—we can see that it says *nothing about sentences.*

Arguing with ostriches

To round off our discussion of the (non-)absoluteness of truth, we will explain how our application of the Metaresources Transcendental Argument connects with another philosophical point discussed by Hamkins and Yang.

Hamkins and Yang imagine that someone might respond to their considerations as follows: An 'inhabitant' of $\mathcal{M}_1$ might insist that they cannot make any sense of the alternative model $\mathcal{M}_2$, rejecting 'transcendental knowledge of the objects in $\mathcal{M}_2$ as things-in-themselves outside of the ontology of their universe $\mathcal{M}_1$'. This 'inhabitant' would essentially refuse to acknowledge the existence of $\mathcal{M}_2$, and so would refuse to see any reason to think that pinning down some particular mathematical objects is insufficient to pin down the truths about them. Hamkins and Yang respond to this protesting 'inhabitant' with a rhetorical question: 'do we have the burden of arguing with ostriches?' They continue:

> [...] we have described how it could be that another universe has the same natural numbers and natural number structure as exists in the universe $\mathcal{M}_1$, but different arithmetic truths, whether or not people living only in $\mathcal{M}_1$ can appreciate it.[27]

We mention all this, since we want to distance ourselves from any ostriches. Our use of the Metaresources Transcendental Argument provides a related but different criticism of the philosophical significance of Corollary 9.4 (and its mathematically more sophisticated variants).

Our point is as follows. Insofar as we can make sense of the imagery of 'inhabiting' a model, anyone who can understand the contents of Hamkins and Yang's paper can see that they, reading the paper, *neither* 'inhabit' $\mathcal{M}_1$ *nor* $\mathcal{M}_2$ themselves. For, in exactly the same spirit as a reader can follow the results in Hamkins and Yang's paper, they can see that the natural number substructure common to both models is non-standard, as is σ, so that it does not code a sentence. That, again, is just the point of the Metaresources Transcendental Argument.

[27] Hamkins and Yang (forthcoming: 28).

10

Internal categoricity and the natural numbers

Since we are halfway through Part B, it might help to take stock.

In Chapter 6, we introduced *modelism*. This is the view that mathematical structure, informally construed, should be explicated in terms of a model-theorist's isomorphism types. In Chapters 7–9, we showed that this is incompatible with *moderation*. This is the (naturalist) view that we must reject all appeal to 'mathematical intuition', or anything similar, in the philosophy of mathematics. Sadly, moderate modelists cannot explain how creatures like us could possibly pin down structures up to isomorphism. So moderate modelism must be rejected.

However, we have not yet considered whether we should reject the *moderation* or the *modelism*.[1] Moderation is just a *zeitgeisty* form of naturalism, which stems from a nebulous but broadly anti-rationalist conception of the kinds of creatures we are. As such, any serious exploration of alternatives to moderation would take us far away from model theory as such.

We *can*, however, outline a rival to modelism by using some metamathematical tools which bear a family-resemblance to standard model theory. This rival, which we call *internalism*, aims to be compatible with moderation. We will explore internalism over the next three chapters, first focussing on arithmetic, then turning to set theory in Chapter 11, and finally turning to internalism about model theory itself in Chapter 12.

We should emphasise right now, though, that we are not *advocating* internalism, any more than we were advocating modelism. Rather, we are presenting internalism in a speculative spirit. It is a fascinating position, worthy of attention, and we want to develop it as best we can.

The very idea of internalism is hugely in debt, in various ways, to Putnam, Parsons, and McGee. We shall mention these authors several times in the next two chapters, but such mentions will fail to convey the extent of our debt to them.

Internalism relies heavily on deduction. In this chapter and the next, the deductive system in question is always the system of second-order natural deduction laid down in §1.c (though in this chapter we do not need Choice). So, when we say

[1] In fact, this is a point where we are inclined to part ways: Sean is inclined to reject the *moderation*; Tim is inclined to reject the *modelism*. But, at the time of writing, these are just inclinations, rather than settled positions.

something is *second-order deducible*, or use the single turnstile $\vdash$, we always have in mind that specific deductive system. When we speak of 'pure second-order logic', we mean second-order logic in the empty signature, i.e. formulas which contain only connectives, various types of quantifiers and associated variables, and the identity sign. We frequently use the notation introduced in §1.9, where $(\exists x : \Xi)\psi$ abbreviates $\exists x(\Xi(x) \wedge \psi)$ and $(\forall x : \Xi)\psi$ abbreviates $\forall x(\Xi(x) \rightarrow \psi)$.

10.1 Metamathematics without semantics

We begin by introducing the *very idea* of our internalist rival to modelism. The technical material in this section and the next owes much to Väänänen and Wang.

From infinity to arithmetic

A set is *Dedekind-infinite* if there is some function from the set to itself which is injective but not surjective. This notion can be transcribed into the formalism of pure second-order logic in the obvious way, just by writing out the notions of functionality, injectivity, and non-surjectivity by hand:

$$\begin{aligned} \exists F \exists R(\exists z : F)[(\forall x : F)(\exists! y : F)R(x,y) \wedge \\ (\forall x, x', y : F)((R(x,y) \wedge R(x',y)) \rightarrow x = x') \wedge \\ (\forall x : F)\neg R(x,z)] \end{aligned}$$

Here, R is the graph of the injective function, whose non-surjectivity as a function from F to F is witnessed by the element z of F. Since this expresses that there is a property, F, with Dedekind-infinitely many instances, we dub this sentence DI_2.

From DI_2, we can deduce something which looks *very arithmetical*.[2] We can explain what, by running through something very much like Dedekind's own reasoning.[3] Fix F, R and z witnessing DI_2, and define a formula with one free one-place relation-variable, X, which intuitively says that z is in X and that X is closed under the R-relation:

$$\Gamma(X) := X(z) \wedge (\forall x : X)\forall y(R(x,y) \rightarrow X(y))$$

By second-order logic's Comprehension Schema, there is a property N such that:

$$\forall v(N(v) \leftrightarrow \forall X[\Gamma(X) \rightarrow X(v)])$$

---

[2] Indeed, 10.1 shows that DI_2 interprets PA_2 with parameters. See Chapter 5 footnote 19.

[3] Dedekind (1888). For a recent treatment which keeps close to Dedekind's original, see Potter (2004: 88–92). The deductive version of the argument is in Väänänen and Wang (2015: Theorem 2).

In effect, N is the 'minimal closure' of z under R. Moreover, by Comprehension again, there is an S such that:

$$\forall x \forall y(S(x,y) \leftrightarrow [R(x,y) \wedge N(x) \wedge N(y)])$$

In effect, S is the restriction of R to N. It is now easy but tedious to confirm the following (see §10.B; we label each conjunct for future reference):[4]

$$N(z) \wedge (\forall x : N)(\exists! y : N) S(x) = y \wedge \qquad (pa{:}res)$$
$$(\forall x : N) S(x) \neq z \wedge \qquad (pa{:}q1)$$
$$(\forall x, y : N)(S(x) = S(y) \rightarrow x = y) \wedge \qquad (pa{:}q2)$$
$$\forall X[(X(z) \wedge (\forall x : N)[X(x) \rightarrow X(S(x))]) \rightarrow (\forall x : N)X(x)] \qquad (pa{:}ind)$$

We abbreviate this conjunction as $\text{PA}(NzS)$, where 'N' reminds us of 'natural number', where 'z' reminds us of 'zero', and where 'S' reminds us of 'successor'. Additionally, in what follows we will frequently need to write '$\exists N \exists z \exists S$', and for space and readability we write '$\exists NzS$' (and similarly for universal quantification). In these terms, then, we have shown:[5]

Proposition 10.1: $\text{DI}_2 \vdash \exists NzS\, \text{PA}(NzS)$

Now, if we glance at Definition 1.10, which lays down the theory PA_2, it is clear that $\text{PA}(NzS)$ essentially re-axiomatises PA_2 relative to N, z, and S: for (*pa:res*) relativises our attention to N, then (*pa:q1*) and (*pa:q2*) go proxy for (Q1) and (Q2), while (*pa:ind*) stands in for PA_2's Induction Axiom.[6] And that is what we meant, when we earlier said that we can deduce something which looks *very arithmetical* from DI_2. In particular, we might consider glossing the formal sentence of pure second-order, $\exists NzS\, \text{PA}(NzS)$, using the *informal* claim:

(1) there is an arithmetical structure

So glossed, we would say that DI_2 *proves the existence of an arithmetical structure.*

---

[4] The use of functional notation here, and throughout the next three chapters, is abbreviatory and formally justified by (*pa:res*). For instance, in (*pa:ind*), $X(S(x))$ abbreviates $\forall y(S(x,y) \rightarrow X(y))$, and $S(x) = S(y)$ abbreviates $\forall v(S(x,v) \leftrightarrow S(y,v))$. If we preferred, we could obtain *bona fide* second-order functions from such relations, using either Choice or a comprehension principle for functions. For example, using Comprehension there is a relation S^* such that $\forall x \forall y(S^*(x,y) \leftrightarrow [S(x,y) \vee \neg N(x)])$, and we can then obtain a genuine *function* by Choice. However, we would prefer not to muddy the waters by using Choice.

[5] This is Väänänen and Wang (2015: Theorem 2).

[6] But note that $\text{PA}(NzS)$ *omits* the infinitely many Comprehension Schema instances which we build in to the axiomatisation of PA_2. So, in this context, the instances of the Comprehension Schema are simply relegated to the ambient deductive system.

The idea of an arithmetical internal-structure

However, regarding (1) as a good gloss of $\exists NzS\, \mathrm{PA}(NzS)$ is the key first step away from modelism and towards internalism.

Since (1) is an informal sentence concerning 'structure', modelists will want to explicate it via model theory. Their explication will involve three components. First, there is a particular object, an $\mathscr{L}$-structure, $\mathcal{N}$. This object has a domain, N, which contains a distinguished element, 0, and upon which we have some function $S : N \longrightarrow N$. Second, there is the theory which is true in this $\mathscr{L}$-structure. This is PA_2, a set of second-order sentences. Third, there is the relation, $\vDash$, between the $\mathscr{L}$-structure and the theory. This is a recursively defined, *bona fide* language–object relation. So modelists explicate (1) as follows: *there is some* $\mathcal{N}$ *such that* $\mathcal{N} \vDash \mathrm{PA}_2$.[7]

Our *internalist*, by contrast, explicates (1) as $\exists NzS\, \mathrm{PA}(NzS)$. This explication is very different. In speaking informally of 'an arithmetical structure', here, the internalist is *not* aiming to draw attention to some specific object which stands in a language–object satisfaction relation to some theory. She is *not* engaging in semantic ascent. She is simply saying something in a second-order object language, along the following lines: *some property (a second-order entity), some (first-order) object, and some function (a second-order entity) collectively behave arithmetically.*

Prima facie, either explication of (1) is entirely reasonable. But we must never confuse the two. To guard against such confusion, when we discuss what model-theorists call 'a model', we use the phrase 'an $\mathscr{L}$-structure'. When we discuss what internalists calls 'a model' (i.e. some entities of various types interacting with one another), we use the phrase 'an internal-structure'. To confuse $\mathscr{L}$-structures with internal-structures is not just a mistake; *it is literally a type confusion.* To repeat: $\mathscr{L}$-structures are first-order objects, presented within some model theory, whereas to talk of 'an internal-structure' is to gesture at three entities of different logical types.

To further unpack the (typed) differences between $\mathscr{L}$-structures and internal-structures, imagine we develop claim (1) a little, and say:

(2) there is an arithmetical structure, and φ holds in it

As before, modelists take 'an arithmetical structure' to be some $\mathscr{L}$-structure, $\mathcal{N}$. They then take φ to be a sentence which is mentioned (rather than used). And they take the notion of 'holding' to be the language–object satisfaction relation. So, they explicate (2) by saying: *there is some* $\mathcal{N}$ *such that* $\mathcal{N} \vDash \mathrm{PA}_2$ *and* $\mathcal{N} \vDash \varphi$.

As before, internalists regard (2) as drawing attention to some property, object and function which collectively behave arithmetically. Since this explication makes no mention of any language–object relation, internalists must explicate the claim 'φ holds in it' as a further constraint on the property, object and function which collectively behave arithmetically. So, internalists explicate (2) by saying:

[7] Depending on context, modelists may want to add an additional level of abstraction, by lifting all of this up to the level of isomorphism types.

$\exists NzS\,[\mathrm{PA}(NzS) \wedge \varphi(NzS)]$. That is, they regard φ as a formula containing free variables N, z, and S, which are then bound by the same quantifiers that bind the free variables in $\mathrm{PA}(NzS)$.

In this chapter, we will encounter several other informal claims about 'mathematical structure' which modelists and internalists will explicate differently. We will say more about them as we go; but we hope that the general pattern is becoming clear. Model theory involves metalinguistic ascent: treating theories and their models as (syntactically first-order) objects of study, and considering a *bona fide* language–object satisfaction relation which holds between them. Modelists embrace this as a device for thinking about 'mathematical structure', but internalists shun that way of thinking. Instead, the their idea is as follows:

The internalist manifesto. *For philosophical purposes, the metamathematics of second-order theories should not involve semantic ascent. Instead, it should be undertaken within the logical framework of very theories under investigation. Our slogan is:* METAMATHEMATICS WITHOUT SEMANTICS!

Over the next three chapters, we want to see just how far this idea can be pushed.

Before the pushing begins, we should probably explain why we have dedicated three chapters of this book to a somewhat *anti*-model-theoretical position. First, then: the sheer similarity of topic suggests that internalism ought be discussed next to modelism. Second: in spite of the (typed) differences, there are clear technical parallels between ordinary model-theoretic methods and the deductive approach recommended by internalists. Third, and relatedly: what set theorists call *class-models* are easily regarded as internal-structures.[8] And finally, as we show in Chapter 12, we can consider internalism about model theory *itself*, and so take a new perspective *on* model theory.

10.2 The internal categoricity of arithmetic

In this chapter, though, we set our sights a little lower, and simply consider internalism about arithmetic. So, the internalist's hope is to explicate informal talk of 'the natural numbers' in terms of arithmetical internal-structures. The key step towards realising this hope is via a result which we might gloss as follows: *all arithmetical internal-structures are isomorphic.*

[8] But depending on the set theory, one might not allow quantification over classes. For more on class models, see Jech (2003: 161–2). McGee presents his internal categoricity result for set theory (see §11.A) in terms of class-models, whilst describing the use of class-models as 'merely figurative, second-order logic in wolf's clothing' (1997: 56). Indeed, as we will reconstruct McGee's result, his internal-structures are class-models in the sense of Kelly–Morse set theory (with urelements).

Our first task, though, is to state that result precisely. Following the discussion in §10.1, the phrase 'internal-structure' needs to be treated with care here. So: just as internalists explicate (1) via $\exists NzS\,\text{PA}(NzS)$, they should explicate:

(3) for any arithmetical internal-structure, ...
via: $\forall NzS(\text{PA}(NzS) \rightarrow \ldots)$

Equally, they should explicate:

(4) for any two arithmetical internal-structures, ...
via: $\forall N_1z_1S_1N_2z_2S_2\,([\text{PA}(N_1z_1S_1) \wedge \text{PA}(N_2z_2S_2)] \rightarrow \ldots)$

To state our target result precisely, then, we just need a way to say 'there is an internal-isomorphism between $\text{PA}(N_1z_1S_1)$ and $\text{PA}(N_2z_2S_2)$'. But note: we have said 'internal-isomorphism' here, rather than 'isomorphism', and for good reason. Internal-structures are not $\mathcal{L}$-structures; indeed, as we explained in §10.1, it is literally a *type confusion* to identify them. Similarly, it would literally be a type confusion to expect to find an isomorphism, in the model-theorist's sense, between $\text{PA}(N_1z_1S_1)$ and $\text{PA}(N_2z_2S_2)$. After all, $\text{PA}(N_1z_1S_1)$ and $\text{PA}(N_2z_2S_2)$ are simply two open-formulas of pure second-order logic, containing a one-place relation-variable, a first-order variable, and a one-place function-variable.

Guided by the idea of *metamathematics without semantics*, however, it is quite easy to define the idea of an internal-isomorphism between $\text{PA}(N_1z_1S_1)$ and $\text{PA}(N_2z_2S_2)$. Here is the definition (we label the conjuncts for future reference):

$$\begin{aligned}
\text{IsoN}_{1\triangleright 2}(R) := {} & \forall v \forall y\,(R(v,y) \rightarrow [N_1(v) \wedge N_2(y)]) \wedge && (\textit{in:1})\\
& (\forall v : N_1)\exists! y R(v,y) \wedge && (\textit{in:2})\\
& (\forall y : N_2)\exists! v R(v,y) \wedge && (\textit{in:3})\\
& R(z_1,z_2) \wedge \forall v \forall y\,(R(v,y) \rightarrow R(S_1(v),S_2(y))) && (\textit{in:4})
\end{aligned}$$

Roughly, conjunct (*in:1*) says that R maps N_1 to N_2, (*in:2*) says that R is functional, (*in:3*) says that R is a bijection, and (*in:4*) says that R 'preserves arithmetical structure'. Assembling all of this, the following sentence of pure second-order logic roughly says 'all arithmetical internal-structures are internally-isomorphic':

$$\forall N_1z_1S_1N_2z_2S_2\,([\text{PA}(N_1z_1S_1) \wedge \text{PA}(N_2z_2S_2)] \rightarrow \exists R\,\text{IsoN}_{1\triangleright 2}(R))$$

Moreover, this sentence is a *deductive theorem* of second-order logic. So we have:[9]

Theorem 10.2 (Internal Categoricity of PA):

$$\vdash \forall N_1z_1S_1N_2z_2S_2\,([\text{PA}(N_1z_1S_1) \wedge \text{PA}(N_2z_2S_2)] \rightarrow \exists R\,\text{IsoN}_{1\triangleright 2}(R))$$

[9] This is Väänänen and Wang (2015: Theorem 1).

The proof is in §10.B, but it essentially consists in 'internalising' one version of a proof of Dedekind's Theorem 7.3, in roughly the same way that, in §10.1, we 'internalised' Dedekind's proof that an infinite system yields an arithmetical structure.

This suggests that we should compare Theorem 10.2 with Dedekind's Theorem 7.3. Speaking *very* loosely, both theorems say 'all models of arithmetic are isomorphic'. But speaking *that* loosely is unwise. Dedekind's Theorem 7.3 concerns $\mathcal{L}$-structures and full second-order satisfaction, and is proved in a model-theoretic metatheory which treats PA_2 as an object-theory. Theorem 10.2 involves no semantic ascent, no mention of sentences, and no semantic notions. It just amounts to a *deduction* of a single sentence of pure second-order logic.[10]

In what follows, we call Theorem 10.2 an *internal* categoricity result, and contrast this with Dedekind's *external* categoricity result.[11] These labels are appropriately suggestive. Crudely: an external result involves standing back from a theory's object language and considering its semantics in some model-theoretic metalanguage. By contrast, an internal result is proved deductively and *within* the object language. It concerns internal-structure, and is of chief concern to internalists.

10.3 Limits on what internal categoricity could show

The remainder of this chapter focusses on the philosophical significance of internal categoricity. We start, though, by emphasising some things that internal categoricity results *cannot possibly* show. The limitations here follow straightforwardly from the deductive nature of internal categoricity.

No pinning down $\mathcal{L}$-structures

The first limit is this: *no internal categoricity result can show that a theory pins down a unique $\mathcal{L}$-structure in the model-theorist's sense (even up-to-isomorphism).*

The reason for this is straightforward. Internal categoricity theorems are deductions within pure second-order logic. They involve no semantic notions. To suggest that an internal categoricity theorem could directly allow us to pin down an $\mathcal{L}$-structure would just be a type confusion again.

Of course, just as one *can* engage in semantic reflection about theories, so one *can* engage in semantic reflection about internal categoricity theorems. In particular, given the *full* semantics for second-order logic, Theorem 10.2 essentially amounts to

[10] However, it is worth highlighting that there is an important sense in which Theorem 10.2 requires a strong deductive second-order logic. In particular: as outlined §1.11, we are working with an Impredicative Comprehension Schema; Theorem 10.2 fails if we replace the Impredicative with the *Predicative* Comprehension Schema (see §10.C).

[11] Walmsley (2002: 249–51) seems to have been the first author to use the phrase 'internal categoricity'. The phrase 'relative categoricity' is also used here, e.g. Walsh and Ebels-Duggan 2015.

a statement of external categoricity, i.e. to Dedekind's Theorem. However, external categoricity vanishes with the*Henkin* semantics; for although all internal-structures are alike within a *single* Henkin interpretation, they need not be alike across *different* Henkin interpretations.[12] Unsurprisingly, then, treating internal categoricity via semantic ascent simply takes us on a long detour back to the issues of Chapter 7.

In short, if internal categoricity results are to show us anything new, they must be approached *as internal* categoricity results: as deductive theorems of pure second-order logic. And that is how we treat them in what follows.

No avoiding Gödelian incompleteness

The second limit is this: *internally categorical theories are incomplete.*

It is sometimes said that Gödelian incompleteness does not affect second-order theories. That claim is imprecise. Of course, if we invoke the full semantics for second-order logic, then the external categoricity of PA_2 yields its completeness (for all this, see §7.2). But the internalist aims to do metamathematics without semantics and eschews external categoricity results in favour of deductively proved internal categoricity results. And both versions of Gödel's incompleteness theorems (see §5.A) apply to deductive second-order theories.

To bring this out, we will use '*Num*' as a canonical (one-place) number-predicate, '0' as canonical (first-order) constant, and '*Succ*' as a canonical successor-function-symbol (formally, *Succ* will be a two-place relation symbol). Then we define PA_{int} to be this second-order sentence:

$$
\begin{aligned}
&Num(0) \land (\forall x : Num)(\exists! y : Num)Succ(x) = y \land \\
&(\forall x : Num)Succ(x) \neq 0 \land \\
&(\forall x, y : Num)(Succ(x) = Succ(y) \rightarrow x = y) \land \\
&\forall X[(X(0) \land (\forall x : Num)[X(x) \rightarrow X(Succ(x))]) \rightarrow (\forall x : Num)X(x)]
\end{aligned}
$$

Plainly, PA_{int} is just the theory obtained by taking the formula PA(NzS) from (*pa:res*)–(*pa:ind*) and replacing its free variables with our new canonical vocabulary. The name PA_{int} stands for *Peano Arithmetic, internalised.*

Now, PA_{int} is surely consistent.[13] Additionally, PA_{int} interprets Q, by the remarks in the last paragraph of §1.A. And, PA_{int} is obviously computably enumerable.[14] So, by Gödel's First Incompleteness Theorem 5.13, PA_{int} is arithmetically incomplete. And, by Gödel's Second Incompleteness Theorem 5.14, PA_{int} does not prove Con(PA_{int}).

---

[12] See Lavine (1999: 64) and Väänänen and Wang (2015: 99).

[13] Indeed, PA_{int} is consistent if DI_2 is consistent with our deductive second-order system (see §10.1).

[14] It is also computably enumerable when one explicitly includes in it the axioms of the background deductive system for the second-order logic—including the infinitely many instances of the Comprehension Schema, as in footnote 6—and closes the theory under provability.

This reinforces the earlier point, that internal categoricity provides no guide concerning semantics. Suppose, for reductio, that within PA_{int} we could define 'standard' semantic notions concerning internal-structures. Presumably, PA_{int} would trivially prove that some internal-structure—comprising *Num*, 0 and *Succ*—satisfies PA_{int}. Assuming that PA_{int} proves a version of the Soundness Theorem—as we would demand from a 'standard' semantics—then PA_{int} proves $\text{Con}(\text{PA}_{\text{int}})$, contradicting Gödel's Second Incompleteness Theorem 5.14. So, when we say that Theorem 10.2 tells us nothing about semantics, this is not just a quibble about what to call a 'semantics'. There are rock-solid impossibilities here.

No 'pinning down' internal-structure

These two observations pack some heavy punches concerning the potential significance of Theorem 10.2. To see why, consider this naïve reaction to the Theorem:

(5) Theorem 10.2 shows that PA_{int} pins down the natural numbers up to internal-isomorphism

This naïve reaction is both understandable and extremely appealing. We spent most of Chapters 6–8 considering the modelist's ambition of pinning down the natural numbers up to isomorphism (in the model-theorist's sense). In this chapter, we are presenting internalism as an alternative to modelism. But if the modelist-turned-internalist can insist on (5), she will be able to retain many of her former ambitions.

Sadly, (5) is too opaque to be of any use. As we repeatedly emphasised in our discussion of modelism in Chapters 6–8, talk of 'pinning down' is inevitably quite loose, and needs to be made more precise. But no way of making it more precise seems very conducive to internalism.

Suppose that the internalist unpacks (5) along these lines: PA_{int} *is true of, and only of, a rather limited range of entities*. Then she has invoked some language–object 'true of' relation, and engaged in semantic ascent after all. In so doing, she given up on her aim to treat Theorem 10.2 as an *internal* categoricity result. (And if she insists that the phrase 'true of' does *not* signal semantic ascent, then we would ask her to use some less misleading phrase.)

Suppose instead, then, that the internalist unpacks (5) along these different lines: PA_{int} *articulates our arithmetical concepts as fully as possible*. This may avoid semantic ascent, but it just seems wrong. Since $\text{Con}(\text{PA}_{\text{int}})$ is independent from PA_{int}, the theory $\text{PA}_{\text{int}} \wedge \text{Con}(\text{PA}_{\text{int}})$ outlines our arithmetical concepts more precisely than PA_{int} itself, for whatever exactly 'articulating our arithmetical concepts' amounts to, adding $\text{Con}(\text{PA}_{\text{int}})$ surely adds some detail.

For these reasons, we think that internalists should just abandon (5), and look for alternative ways to gloss the significance of internal categoricity.

The difficulties surrounding (5) point, in fact, towards a disconcerting dilemma.

On the one hand, suppose we think of 'internal-isomorphisms between internal-structures' *as if* they were isomorphisms between $\mathscr{L}$-structures. Then it is obvious why we might *care* about them. However, to treat them in this way involves semantic ascent, contrary to the aims of internalism. On the other hand, if we do not think of 'internal-isomorphisms between internal-structures' in that way, it is not at all obvious why we should *care* about them.

To avoid this dilemma, we think that internalists should switch their focus from Theorem 10.2 to a corollary which we will now introduce.

10.4 The intolerance of arithmetic

In §2.1, we saw that isomorphism entails elementary equivalence (this was Corollary 2.5). It turns out that *internal*-isomorphism entails something similar. To explain what, we need a short definition.

Roughly, we say that a formula's quantifiers are Ξ-restricted iff all its first-order and second-order quantifiers are restricted to Ξ. Formally, we employ the notation introduced in §1.9, where $(\exists x : \Xi)\psi$ abbreviates $\exists x(\Xi(x) \land \psi)$ and $(\forall x : \Xi)\psi$ abbreviates $\forall x(\Xi(x) \to \psi)$. We also introduce some similar second-order abbreviations. So, $(\exists X^n : \Xi^n)\psi$ abbreviates $\exists X^n(\forall \bar{v}[X^n(\bar{v}) \to \bigwedge_{i=1}^{n} \Xi(v_i)] \land \psi)$ and $(\forall X^n : \Xi^n)\psi$ abbreviates $\forall X^n(\forall \bar{v}[X^n(\bar{v}) \to \bigwedge_{i=1}^{n} \Xi(v_i)] \to \psi)$. Intuitively, $(\exists X^n : \Xi)$ draws attention to an n-place relation over the property Ξ. We then say that φ's *quantifiers are* Ξ*-restricted* iff every first-order quantified expression in φ is of the form $(\forall x : \Xi)$ or $(\exists x : \Xi)$, and every second-order quantified expression is of the form $(\forall X^n : \Xi)$ or $(\exists X^n : \Xi)$. In these terms, Theorem 10.2 entails the following (the proof is in §10.B):

Theorem 10.3 (Intolerance of PA): *For any formula $\varphi(NzS)$ whose quantifiers are N-restricted and whose free variables are all displayed:*

$$\vdash \forall NzS(\text{PA}(NzS) \to \varphi(NzS)) \lor \forall NzS(\text{PA}(NzS) \to \neg\varphi(NzS))$$

Given the discussion of §§10.1–10.2, the internalist will gloss this result as follows: *either φ holds in every arithmetical internal-structure, or $\neg\varphi$ holds in every arithmetical internal-structure.* This is why we call the result an *intolerance* theorem: no object-language deviation between internal-structures is tolerated.

10.5 A canonical theory

The first consequence import of the Intolerance Theorem 10.3 is that it allows us to introduce a *canonical* internalised theory of arithmetic.

Suppose we consider two arithmetical internal-structures. So, we have two candidate number properties N_1 and N_2, two candidate initial elements z_1 and z_2, and two candidate successor functions S_1 and S_2, such that $\text{PA}(N_1 z_1 S_1)$ and $\text{PA}(N_2 z_2 S_2)$. Given intolerance, exactly the same arithmetical claims must hold or fail in both internal-structures, modulo subscripts. So, for arithmetical purposes, there is simply no need to *bother* with the subscripts.[15] As such, we can all simply agree to use PA_{int}: the theory introduced in §10.3, with its canonical vocabulary of '*Num*', '0' and '*Succ*'.[16]

It is worth noting that the use of PA_{int} can be be justified by anyone who accepts both (a) our deductive system of second-order logic and (b) that there is *some* property with infinitely many instances. By Proposition 10.1, anyone who accepts (a) and (b) must accept that *some* arithmetical internal-structure exists; and then the Intolerance Theorem 10.3 shows that we can reason about all arithmetical internal-structures simultaneously using the canonical theory PA_{int}.

Note, though, that PA_{int} records only what matters *for arithmetic*; it is silent on all other matters. To illustrate the point: in our earlier example, perhaps the candidate initial elements, z_1 and z_2, were distinct from each other; maybe z_1 is Julius Caesar and z_2 is the abstract entity Goodness. But to ask whether 0 itself is Caesar, or Goodness, or something else, is just to misunderstand the *point* of employing the canonical theory, PA_{int}.[17] To repeat: PA_{int} is presented just as a canonical theory which provides a convenient common language for reasoning about all arithmetical internal-structures simultaneously.

10.6 The algebraic / univocal distinction

The ability to introduce such a canonical theory, off the back of the Intolerance Theorem, is already significant. But internalists can push this further, and use intolerance to explicate the distinction between algebraic and univocal theories.

Demarcating (non-)algebraic theories without modelism

In §2.2, we introduced a rough-and-ready distinction between algebraic and univocal theories. Algebraic theories are multiply-applicable (usually by design), and paradigm examples include theories governing rings, fields, topologies, and categories. By contrast, univocal theories aim to 'describ[e] a certain definite mathe-

[15] This relates to Parsons' own intended ambitions for the use of internal categoricity; see §10.A.

[16] Fans of Carnap may find it helpful to regard the Carnap-sentence $\exists NzS\, \text{PA}(NzS) \rightarrow \text{PA}_{\text{int}}$ as PA_{int}'s *meaning-postulate* (see footnote 7 of Chapter 3).

[17] In that respect, it realises a certain kind of 'structuralist' insight. It also connects with Parsons' claim that a sentence like '$2 = \{\{\emptyset\}\}$' has no absolute truth value, but depends upon context (2008: 103, 77).

matical domain'[18] or to 'specify *one particular interpretation*', speaking loosely.[19]

Modelists can provide a simple explication of this distinction. They explicate the idea of a 'particular interpretation' as an isomorphism type (or something similar). They then say that univocal theories aspire to pin down an isomorphism type (or similar), whereas algebraic theories have no such aspiration.[20]

Having abandoned modelism, however, internalists must explicate the distinction between algebraic and univocal theories in some other way (if at all). Moreover, given their desire to avoid semantic ascent, this is no easy task. There is a risk that the internalist will be unable to see any real contrast between, for example, group theory and arithmetic. Fortunately, the Intolerance Theorem 10.3 generates the required contrast.

Guided by the theory PA_{int}, let us introduce an internalised version of group theory, GT_{int}, which axiomatises group theory relative to Gr, e, and $\circ$:

$$\begin{aligned}\text{GT}_{\text{int}} := {} & Gr(e) \land (\forall x, y : Gr) Gr(x \circ y) \land \\ & (\forall x : Gr)\, x \circ e = e \circ x = x \land \\ & (\forall x, y, z : Gr)\, (x \circ y) \circ z = x \circ (y \circ z) \land \\ & (\forall x : Gr)(\exists y : Gr)\, x \circ y = y \circ x = e\end{aligned}$$

Now, in the course of doing group theory we will want to consider *multiple* groups. To take a simple example, we might say: if group A has only one element and group B has more than one element, then any function from B to A is a homomorphism.

Internalists can formalise this as follows. Let $\text{GT}_{\text{int}[\text{A}]}$ be the theory which results by subscripting every instance of 'Gr', 'e' and '$\circ$' in GT_{int} with 'A'. Similarly, where φ is a formula in the signature $\{Gr, e, \circ\}$, let $\varphi_{\text{int}[\text{A}]}$ be the formula which results by subscripting every instance of 'Gr', 'e' and '$\circ$' in φ with 'A'. Define $\text{GT}_{\text{int}[\text{B}]}$ and $\varphi_{\text{int}[\text{B}]}$ similarly. Then, where φ is $(\forall x : Gr)x = e$, to say that group A has one element but group B has more than one element is simply to assert:

$$\text{GT}_{\text{int}[\text{A}]} \land \varphi_{\text{int}[\text{A}]} \land \text{GT}_{\text{int}[\text{B}]} \land \neg\varphi_{\text{int}[\text{B}]}$$

And it is easy to prove deductively from this that any second-order relation which is functional with domain Gr_{B} and range Gr_{A} is an internal-homomorphism.

But suppose that we attempt something similar in the case of arithmetic, and try to consider two arithmetical structures which differ over some arithmetically-significant feature. Anyone who views arithmetic as univocal will immediately balk at this idea. And internalists can claim that balking is the *right* reaction. For, on

[18] Grzegorczyk (1962: 39).

[19] Kline (1980: 273).

[20] See, for example, Meadows (2013: 540).

the internalist's explication, to consider two arithmetical internal-structures which differ over some arithmetically-significant feature, is to assert:

$$\mathrm{PA}_{\mathrm{int}[\mathrm{A}]} \land \varphi_{\mathrm{int}[\mathrm{A}]} \land \mathrm{PA}_{\mathrm{int}[\mathrm{B}]} \land \neg\varphi_{\mathrm{int}[\mathrm{B}]}$$

for some φ in $\mathrm{PA}_{\mathrm{int}}$'s signature whose quantifiers are all *Num*-restricted. But, by Theorem 10.3, that is to assert something *deductively inconsistent*.

The short point here is that $\mathrm{PA}_{\mathrm{int}}$, unlike $\mathrm{GT}_{\mathrm{int}}$, is intolerant. And the internalist can now say: *the intuitive algebraic / univocal distinction should be formally explicated by the tolerant / intolerant distinction.*

Intolerance as a challenge for algebraic views of arithmetic

So far in this section, we have tacitly assumed that arithmetic *should* be regarded as univocal. Presumably this is the 'default view'.[21] However, in §7.6, we noted the possibility of an Algebraic Attitude towards arithmetic. And proponents of this Algebraic Attitude will deny that there is any great difference between considering two different groups, and considering two different natural number structures.

To be absolutely clear: the intolerance of $\mathrm{PA}_{\mathrm{int}}$ does *not* show that the Algebraic Attitude is deductively inconsistent. (*Of course* it does not: the Algebraic Attitude is a philosophical viewpoint, not a mathematical conjecture.) In what follows, our point is just that $\mathrm{PA}_{\mathrm{int}}$'s intolerance constrains the possible ways in which someone might try to *defend* the Algebraic Attitude.

One possible way to defend the Algebraic Attitude is to invoke (faithful) Henkin semantics. In particular, if φ is deductively independent from $\mathrm{PA}_{\mathrm{int}}$, then there are Henkin-structures $\mathcal{A}$ and $\mathcal{B}$ such that $\mathcal{A} \models \mathrm{PA}_{\mathrm{int}} \land \varphi$ and $\mathcal{B} \models \mathrm{PA}_{\mathrm{int}} \land \neg\varphi$. Fans of the Algebraic Attitude may then argue that that no sense can be made of the claim that $\mathcal{A}$ is 'preferable' to $\mathcal{B}$, or vice versa. However, given the Metaresources Transcendental Argument of Chapter 9, this sort of argument walks a dangerously thin path between falsity and incoherence. To recap that Argument very briefly: we must describe $\mathcal{A}$ and $\mathcal{B}$ within some model theory; that model theory will interpret a sizeable chunk of arithmetic; so, within that model theory, we can easily define the notion of a standard model of arithmetic and state that these models are 'preferable'. Of course, insofar as we are denied a firm grip on the model theory, we are denied a firm grip on 'preferability'; but, insofar as we are denied a firm grip on the model theory, it is doubtful that we can understand the claim that $\mathrm{PA}_{\mathrm{int}}$ has multiple different models, and hence the supposed *motivation* for the Algebraic Attitude. In sum: insofar as we can make sense of this attempt to motivate the Algebraic Attitude, we can see it is wrong.

---

[21] Koellner (2009: 91) uses this phrase to describe the view that all arithmetical sentences have a determinate truth value. One of our aims in this chapter, though, is to defer invocations of the tricksy notion of *truth* for as long as possible.

New life for deductive approaches

To summarise this section, the Intolerance Theorem does two things. First, it furnishes internalists with an explication of the difference between group theory (or similar algebraic theories) and arithmetic. Second, together with the considerations of Chapter 9, it places pressure on the Algebraic Attitude towards arithmetic.

This is extremely significant. Considered as deductive theories, both group theory and arithmetic are incomplete. It is only a short step from that observation, to the thought that we *must* invoke semantics if we want to place these two theories on opposite sides of an algebraic / univocal divide. But the Intolerance Theorem 10.3 shows that this further thought is *not* forced upon us. As such, the Intolerance Theorem 10.3 breathes fresh life into a deductive-centric approach to arithmetic which otherwise might have seemed dead, post-Gödel.

10.7 Situating internalism in the landscape

Given its focus on deduction, internalism might well seem like a version of *if-thenism*, which just happens to focus on conditionals of the form 'if PA_{int}, then ...'.

That impression is simply *wrong*. Unlike if-thenists, internalists *affirm* PA_{int} unconditionally. As such, internalists happily affirm the antecedents of those conditionals, and so also affirm the consequents. They do not just say: *if* PA_{int}, then there are infinitely many prime numbers. They unconditionally say: there *are* infinitely many prime numbers. The arithmetised Intolerance Theorem 10.3, proved within PA_{int} itself, is something internalists *unconditionally* affirm.

Indeed, with this point clarified, it might seem that internalism is in fact a species of *objects-platonism*. After all, our internalist explicitly and unconditionally claims that *there are* infinitely many prime numbers.

However, as we defined the position in §2.3, objects-platonism *also* involves the claim that numbers are not of our creation, and not spatio-temporal. We would be a little surprised if an internalist claimed that we *do* create the numbers, or that they *are* spatio-temporal, but it simply is not clear that internalists must say that they are *not*. In §10.5, we introduced PA_{int} as a *canonical* theory of arithmetic, on the grounds that intolerance shows that merely orthographical differences between arithmetical vocabularies are irrelevant to arithmetic itself. But we also emphasised that PA_{int} takes no stance on whether 0 is the man Julius Caesar, or the abstract object Goodness, or something else. Now, Caesar is spatio-temporal, whereas Goodness is abstract. Moreover, *if* there are any properties with infinitely many instances and Caesar and Goodness both exist, *then* provably there are arithmetical internal-structures such that:[22]

[22] By considerations in §10.1, DI_2 entails the existence of an arithmetical internal-structure. Now, by

$$\mathrm{PA}_{\mathrm{int}[\mathrm{A}]} \wedge 0_{\mathrm{A}} = \mathrm{Caesar} \wedge \mathrm{PA}_{\mathrm{int}[\mathrm{B}]} \wedge 0_{\mathrm{B}} = \mathrm{Goodness}$$

As such, a claim like '$(\forall x : Num)x$ is abstract' need *not* be agreed on all sides, modulo subscripts. A very natural form of internalism can therefore insist on taking no stance on whether the numbers are abstract or concrete, and can therefore dismiss (rather than negate) a central doctrine of objects-platonism. More generally, this version of internalism can claim to be liberated from *any* metaphysical questions concerning the nature of arithmetic.[23] Perhaps this is a point in internalism's favour.

Finally, because internalists both insist that arithmetic is *true* and emphasise the use of deductive second-order logic, one might think that internalism is a variety of *logicism*. Whilst it certainly has affinities with certain versions of logicism, one particular difference is crucial. In brief: logicists have epistemological ambitions, and so must explain how we can *know* that there is an ω-sequence; internalists have doxological ambitions, and are aiming to articulate what it *means* to say that there is an ω-sequence. In more detail: in this chapter, we showed how to obtain an internally categorical theory of arithmetic, $\mathrm{PA}_{\mathrm{int}}$, from the claim that there is a property with infinitely many instances, DI_2. The internalist should not, though, be regarded as making any particular claim about the epistemological status of DI_2. Her point is just that, using DI_2, we can articulate a theory of arithmetic which is demonstrably intolerant and so, she insists, univocal.

All told, then, internalism about arithmetic is neither formalism, nor object-platonism, nor logicism. Though we revisit this in §12.3, *internalism is its own thing*.

10.8 Moderate internalists

In this chapter, we have outlined the very idea of internalism, as a philosophical reaction to the internal categoricity and intolerance of arithmetic. In the introduction to this chapter, though, we billed internalism as an alternative to modelism, motivated by the incompatibility of modelism with *moderation*. So, we need to show that internalism is compatible with *moderation*. In particular, we need to explain why a moderate can lay claim to Theorems 10.2 and 10.3.

Recall that moderation involves a particular conception of 'creatures like us', which eschews appeal to mathematical intuition, quasi-perception, or anything similar. Now, creatures like us certainly can become competent in using natural deduction systems. So, moderates surely face no special problem in proving Theorem 10.2. Our proof, given in §10.B, is pretty short. Admittedly, that proof is not completely formalised, but an entirely formal proof would not have been that much

---

repeated applications of the Comprehension Schema, we can find an object-language surrogate for the Push-Through Construction of §2.1. We leave this to the reader.

[23] But it is doubtful that they can so easily discharge the question of whether DI_2 holds, i.e. whether there *is* a property with infinitely many instances.

longer. So moderates—like anyone else—can go through it, line by line, and check that everything is in order with it.

The more interesting issue concerns the Intolerance Theorem 10.3. This says that any sentence of a certain shape is a deductive theorem. It is, then, *metalinguistic* in character. Its proof (in §10.B) is similarly metalinguistic, since it proceeds by induction on the complexity of formulas.

It is worth emphasising that internalists need not fear the use of metalanguages altogether. Internalists only oppose *semantic* ascent, and the statement and proof of Theorem 10.2 require no model-theoretic notions, only proof-theoretic ones. Indeed, for just this reason, Theorem 10.2 can be encoded without any real loss as a single deductive theorem of PA_{int} *itself*. In more detail: in §5.A and §7.11 we noted that questions about sentences and proofs can be interpreted *arithmetically*. As such, PA_{int} proves an arithmetised version of Theorem 10.3, i.e. a formalisation of:

for any code of a formula $\varphi(NzS)$ *whose quantifiers are N-restricted and whose free variables are displayed, there is a code of a deduction of the following:*
$\forall NzS(\text{PA}(NzS) \rightarrow \varphi(NzS)) \vee \forall NzS(\text{PA}(NzS) \rightarrow \neg\varphi(NzS))$.

So any internalist who is prepared to affirm PA_{int} can assert a single-sentence ersatz of Theorem 10.3, without even engaging in *metalinguistic* ascent.

This shows that internalists can invoke Theorem 10.3. However, there is a specific complexity which arises when we consider *moderate* internalists. That complexity arises as follows (and it should be directly compared with the issue which arose for moderate formalists in §7.11). Theorem 10.3 involves reasoning *about* deductive systems. To reason effectively *about* them, one needs (intuitively) to have a clear grasp on what they *are*. But deductive systems concern manipulations of sentence-types; these can be of arbitrary finite length, containing vastly more lines than there are atoms in the universe. Indeed, no reasonable theory of deductive systems can be complete (when considered deductively itself). So: what guarantees that a moderate internalist has managed to get hold of the right notions of an 'arbitrary deduction', when she offers Theorem 10.3?

Fortunately, the moderate internalist can give a short response to this problem. The proof of Theorem 10.3 will still go through *even if* (*per impossibile*) she 'somehow' got hold of the 'wrong' notion of an 'arbitrary deduction'. As we just noted: PA_{int} itself proves an arithmetised version of the Intolerance Theorem 10.2. Moreover, the proof of that arithmetised result is finitely long. So, moving back from the arithmetical to the syntactic: Theorem 10.3 goes through even if we are 'somehow' discussing the 'wrong' notion of an 'arbitrary deduction'. Consequently, the moderate internalist can legitimately invoke Theorem 10.3 without 'first' having to supply some guarantee that she is discussing the 'right' notion of an 'arbitrary deduction'.

The situation is, therefore, that if anyone can embrace internalism, then moderates can. And if internalism is ultimately viable, then it allow moderates to explain how creatures like us can come to possess arithmetical concepts which are sharp enough to be univocal. That, we think, is the potential philosophical promise of the internal categoricity of arithmetic.

10.A Connection to Parsons

In this philosophical appendix, we want to explain how Parsons' work relates to the contents of this chapter.[24] The appendix can safely be omitted, since it does not affect any of the philosophical claims that we made earlier. However, discussing Parsons both helps to contextualise our discussion, and helps to illustrate what internal categoricity results can and cannot achieve.

Parsons' approach

Where we have used second-order logic throughout this chapter, Parsons restricts himself to first-order logic.

Parsons asks us to imagine two characters, Kurt and Michael, who are both committed to PA. For simplicity, we imagine that their arithmetical vocabularies are subscripted, 'K' for Kurt, and 'M' for Michael. Parsons imagines that Kurt and Michael start to communicate with one another, to the point that both become fluent in the other's language, and incorporate each other's vocabulary into their own language. Having expanded their vocabularies appropriately, Parsons holds that Kurt should understand his (first-order) Induction Schema in a sufficiently 'open-ended' way that it has instances which contain both Kurt and Michael's vocabulary; and similarly for Michael. As such, Kurt can recursively define the following functor, f, from his numbers to Michael's:

$$f(0_{\mathrm{K}}) := 0_{\mathrm{M}} \qquad\qquad f(S_{\mathrm{K}}(x)) := S_{\mathrm{M}}(f(x))$$

Kurt can then prove, using his (open-ended) arithmetical resources, that f is bijective and preserves structure. And whatever Kurt can do, Michael can do too.[25]

The philosophical upshot is supposed to be something like this. Whenever we encounter someone else who is using PA, we can use this method to assure ourselves that we are (in some sense) engaged in the same enterprise.

[24] Many thanks to Charles Parsons for discussion of all this.

[25] Parsons (1990a: 34–5, 2008: 281–2).

First-order logic versus second-order logic

As mentioned earlier, though, Parsons' approach is first-order. And this leads to an *expressive* problem. Lacking second-order resources, it is not immediately clear what general result we are supposed to be pointed to, by considering the specific interaction between Kurt and Michael.

Parsons himself glosses the relevant general result as follows.

> Suppose our language contains a singular term '0', a one-place functor '*S*' and a predicate '*N*', and also another such triplet '0′', '*S*′', '*N*′'. Suppose that the elementary Peano axioms hold for each. In keeping with Skolem's recursive arithmetic, we can introduce by primitive recursion a functor, *f*, with [certain properties].[26]

However, given the infinity of 'the elementary Peano axioms', to say that these axioms '*hold* for each' triplet essentially requires semantic ascent. For internalists who want to avoid semantic ascent, that is unfortunate.

So let us consider an *alternative* formal result which might be relevant to the case of Kurt and Michael. As mentioned, Parsons wants Kurt and Michael to approach Induction 'open-endedly'. We could make this concrete as follows. Suppose Kurt claims that he is committed to the Induction Schema:

$$[\varphi(0) \wedge \forall y\,(\varphi(y) \rightarrow \varphi(S(y)))] \rightarrow \forall y \varphi(y)$$

But suppose he follows this up by insisting that φ, here, ranges over *any property* that might be picked out by *any possible language*. Kurt's line of thought has an appropriate level of generality to it; but it again invokes semantic ascent. Indeed, it essentially amounts to the claim that φ should be able to take, as a value, any set of natural numbers. This is to fall back on (a syntactic fragment of) full second-order logic, and an *external* categoricity result, in just the way discussed under option (a) in §7.10. (NB: this is *not* the result that Parsons himself wants to invoke.)

The main advantage of the move to deductive second-order logic, then, is that it allows us to formulate a relevant result, Theorem 10.2, in a crisp manner and without any threat of semantic ascent. Of course, some will worry that the use of *second-order* logic is just too high a price. However, the most vocal objections against second-order logic involve qualms about its *semantics*, and any such objections are irrelevant to its *deductive* employment. The one concern that we *cannot* dismiss so lightly concerns the essential use of *Impredicative* Comprehension (see §§10.2 and 10.c). But, insofar as Parsons (or others) have concerns about Impredicative Comprehension, they should have exactly similar concerns about the idea that Kurt and Michael understand Induction in an 'open-ended' fashion, since their ability to invoke each others' vocabulary plays a similar technical role in Parsons' reasoning

[26] Parsons (2008: 281).

as Impredicative Comprehension plays in Theorem 10.2. As such, we commend second-order logic and Theorem 10.2 to Parsons, as a clean route to achieving his philosophical ends.

Indeed, Parsons' imagined interaction between Kurt and Michael is easy to handle using second-order resources, and at that point their interaction simply becomes an extremely vivid illustration of arithmetic's *univocity* (see §10.6). In more detail: imagine that Kurt and Michael are working with *second-order* arithmetic. They can simply plug $Num_K 0_K Succ_K$ and $Num_M 0_M Succ_M$ into the quantifiers of Theorem 10.2, and obtain:

$$[\text{PA}(Num_K 0_K Succ_K) \land \text{PA}(Num_M 0_M Succ_M)] \rightarrow \exists R\, \text{IsoN}_{K \triangleright M}(R)$$

Now, presumably Michael can easily affirm $\text{PA}(Num_M 0_M Succ_M)$; and, after sufficient time with Kurt, he can also affirm $\text{PA}(Num_K 0_K Succ_K)$; whereupon he obtains the required internal-isomorphism. Kurt can do the same. Equally, using Theorem 10.3, Kurt and Michael can conclude that their languages differ *only* in the imposed subscripts, so that they can ditch their subscripts and use PA_{int}.

Limits on what theorems can show

We now wish to consider an objection which Field specifically raises against Parsons' discussion of Kurt and Michael. That said, we will be slightly unfair to Field, and act as if Field's objection were raised against our 'second-orderisation' of Kurt and Michael.

In this context, Field's objection amounts to doubting whether *Michael* can justifiably affirm $\text{PA}(Num_K 0_K Succ_K)$.[27] To be sure, Michael may have heard Kurt repeatedly claim 'I accept unrestricted induction', and he may have observed that Kurt uses his language just like Michael himself does. But if Kurt's second-order quantifier were somehow *restricted*, then Michael would not be allowed to assert $\text{PA}(Num_K 0_K Succ_K)$. And then Michael could not prove the existence of a second-order isomorphism between Kurt's numbers and his.[28]

One might think that the worry can be resolved just by noting that Michael and Kurt can produce *literally* the same proof of Theorem 10.2, that they can discuss it, confirm its correctness together, and so forth. According to Field, however, to say this would be to beg the question: it would *assume* that Michael and Kurt already share a language, which is precisely what they were attempting to *establish* by appealing to Theorem 10.2.[29]

---

[27] See Field (2001: 358–60).

[28] In particular, Michael would be unable to invoke 'induction on Num_K' to prove the left-conjunct of (*in*:3) in the K- and M-subscripted version of Theorem 10.2.

[29] Parsons' own response to Field focusses on the fact that Michael should not be regarded as a 'radical interpreter' of Kurt. This is somewhat different than the response we will make. However, it is at

We think that Field's complaint here is *partially* right. Suppose that a model-theoretical sceptic has suggested that Michael and Kurt are discussing non-isomorphic *Henkin* models of second-order arithmetic. We cannot answer that sceptic by pointing out that Kurt and Michael have produced literally the same proof, line by line. For, if the sceptical scenario obtained, then that same proof would mean different things in their respective mouths, for it would concern non-isomorphic models.

This point is worth emphasising. But it is also worth emphasising that it is irrelevant to our internalist's imagined use of internal categoricity. We stressed in §10.3 that internal categoricity results cannot be used to pin down an $\mathscr{L}$-structure, or to rule out Henkin models, or whatever.

Here is another sense in which Field's complaint is correct: no amount of theorem-proving can guarantee that Kurt might not one day do something which makes Michael do a double-take, and exclaim 'but then your induction axiom was restricted after all, for you have rejected this instance of induction!' However, it is worth noting that this point has nothing much to do with the *induction* axiom. Similarly, no amount of theorem-proving can guarantee that Kurt might not one day do something which makes Michael exclaim 'but then you did not mean conjunction by "*and*" after all, for you accepted φ and accepted ψ but now you are refusing to accept $\varphi \land \psi$!'

Indeed, the inability of a *theorem* to provide us with certain kinds of guarantee has nothing specifically to do with *mathematics* or *logic*. To take an entirely humdrum case: Kurt and Michael might have both used the word 'green' to apply to similar things for a very long time, until one day one of them starts using the word 'green' where the other uses 'red', causing Michael to exclaim 'but then you did not mean greenness by "*green*" after all!' No *theorem* can block these sorts of concerns, which are particular instances of a much more general worry: scepticism about meaning, or rule-following scepticism. This is not to say that these concerns are philosophically uninteresting. It is just to say that we have nothing special to say about them here, and nor do we feel under any *duty* to say something special. To put arithmetical vocabulary on as firm a doxological footing as any other vocabulary is all anyone could really want.

10.B Proofs of internal categoricity and intolerance

In this technical appendix, we prove the results used in this chapter. Throughout, we will use $X \sqsubseteq Y$ to abbreviate $\forall \bar{v}(X(\bar{v}) \rightarrow Y(\bar{v}))$.

least somewhat consonant with internalism, as we have outlined it; in particular, internalism sits well with Parsons' claim that 'language as used is prior to semantic reflection on it' (2008: 285).

We start with the result of §10.1, that DI_2 proves the existence of an arithmetical internal-structure. Our proof here is just that of Väänänen and Wang:[30]

Proposition (10.1): $DI_2 \vdash \exists NzS\, PA(NzS)$

Proof. In this proof, we reuse the notation $\Gamma(X)$, N, z and S, from §10.1.

We first show that $\Gamma(N)$. The definition of Γ trivially entails $N(z)$, so it suffices to show that $(\forall x : N)\forall y(R(x,y) \rightarrow N(y))$. So suppose $N(a)$ and $R(a,b)$. Since $N(a)$ we have $\forall X[\Gamma(X) \rightarrow X(a)]$, whereupon by definition of Γ we have $\forall X[\Gamma(X) \rightarrow X(b)]$, and so $N(b)$. So $\Gamma(N)$ holds.

Now we check that each of the conjuncts of $PA(NzS)$ holds, given our definitions of N, z and S.

(*pa:res*). From $\Gamma(N)$, we immediately have $N(z)$. Then the existence and uniqueness of successors in N follows from the inductive clause of $\Gamma(N)$ and the way in which S is defined by restricting R to N.

(*pa:q1*). For reductio, suppose $N(a)$ and $S(a) = z$. Since $S \sqsubseteq R$, it follows that z is in R's range, contradicting the fact that z and R witness DI_2.

(*pa:q2*). R is injective since it is a witness to DI_2, and so S is injective since $S \sqsubseteq R$.

(*pa:ind*). Fix G such that $G(z) \wedge (\forall x : N)[G(x) \rightarrow G(S(x))]$. Using Comprehension to take the intersection of G and N, we may assume that $G \sqsubseteq N$. Since S is the restriction of R to N, it follows that $\Gamma(G)$. Now fix a such that $N(a)$; so $\forall X(\Gamma(X) \rightarrow N(a))$, and so $G(a)$. ☐

Our proof of Theorem 10.2, again following Väänänen and Wang, is similar:[31]

Theorem (Parsons, 10.2):

$$\vdash \forall N_1 z_1 S_1 N_2 z_2 S_2 \left([PA(N_1 z_1 S_1) \wedge PA(N_2 z_2 S_2)] \rightarrow \exists R\, IsoN_{1 \triangleright 2}(R)\right)$$

Proof. We define a formula $H(X)$ in one free two-place relation-variable, X, where 'H' is a good mnemonic for 'hereditary':

$$H(X) := X(z_1, z_2) \wedge (\forall v : N_1)(\forall y : N_2)[X(v,y) \rightarrow X(S_1(v), S_2(y))] \quad (10.1)$$

By Comprehension in the deductive system, there is a relation R such that:

$$\forall v \forall y(R(v,y) \leftrightarrow \forall X[H(X) \rightarrow X(v,y)]) \quad (10.2)$$

Now we verify conditions (*in:1*)–(*in:4*), described in §10.2.

[30] Väänänen and Wang (2015: Theorem 2). But there is an obvious affinity to Dedekind's own proof; see footnote 3.

[31] Väänänen and Wang (2015: Theorem 1). Alternatively, one could simply transcribe Shapiro's (1991: 82–3) proof of Dedekind's Theorem 7.3 into a deductive system for second-order logic. This contrasts with our (much shorter) proof of Dedekind's Theorem in §7.4, which invoked the natural numbers in the metatheory.

(*in:1*). By Comprehension, there is R^* such that

$$\forall x \forall y\,(R^*(x,y) \leftrightarrow [R(x,y) \wedge N_1(x) \wedge N_2(y)])$$

Intuitively, R^* is like what we might write as $R \cap (N_1 \times N_2)$ in a set-theoretic metatheory. Clearly $R^* \sqsubseteq R$. We will show that $\mathrm{H}(R^*)$, i.e.:

$$R^*(z_1, z_2) \wedge (\forall v : N_1)(\forall y : N_2)[R^*(v,y) \rightarrow R^*(S_1(v), S_2(y))]$$

To see that $R^*(z_1, z_2)$, we must show that $N_1(z_1)$ and $N_2(z_2)$ and $R(z_1, z_2)$. The first two hold by assumption; and by (10.1) we have $\forall X(\mathrm{H}(X) \rightarrow X(z_1, z_2))$, so that $R(z_1, z_2)$ by (10.2).

Next, suppose that $N_1(v)$ and $N_2(y)$ and $R^*(v,y)$; we must show that $R^*(S_1(v), S_2(y))$. First, suppose that $\mathrm{H}(X)$; since $R^*(v,y)$, we also have $R(v,y)$ and so $X(v,y)$ by (10.2), so that $X(S_1(v), S_2(y))$ by (10.1). Generalising, $\forall X(\mathrm{H}(X) \rightarrow X(S_1(v), S_2(y)))$, so that $R(S_1(v), S_2(y)))$ by (10.2). Since $N_1(v)$ and $N_2(y)$, obviously $N_1(S_1(v))$ and $N_2(S_2(v))$. So now $R^*(S_1(v), S_2(y))$.

This shows that $\mathrm{H}(R^*)$. By (10.2), it follows that $R \sqsubseteq R^*$. Since $R^* \sqsubseteq R$, we have that R and R^* are coextensive. This establishes (*in:1*), and also that $\mathrm{H}(R)$.

(*in:2*). By Comprehension there is an F such that $\forall v(F(v) \leftrightarrow \exists! y R(v,y))$. Invoking induction in $\mathrm{PA}(N_1 z_1 S_1)$, we have:

$$(F(z_1) \wedge (\forall v : N_1)\,[F(v) \rightarrow F(S_1(v))]) \rightarrow (\forall v : N_1)F(v)$$

To show $F(z_1)$, first note that $R(z_1, z_2)$ because $\mathrm{H}(R)$; so it suffices to show uniqueness. So suppose for reductio that $R(z_1, a_2)$ and $a_2 \neq z_2$. By Comprehension there is Q such that $\forall v \forall y(Q(v,y) \leftrightarrow [R(v,y) \wedge \neg(v = z_1 \wedge y = a_2)])$. We will show that $\mathrm{H}(Q)$. Clearly $Q(z_1, z_2)$ since $R(z_1, z_2)$ and since $z_2 \neq a_2$. Suppose now that $N_1(v)$ and $N_2(y)$. Then since both $Q \sqsubseteq R$ and $\mathrm{H}(R)$, we have $R(S_1(v), S_2(y))$. The only way in which $Q(S_1(v), S_2(y))$ could fail is if both $S_1(v) = z_1$ and $S_2(y) = a_2$. But by (*pa:q1*) in $\mathrm{PA}(N_1 z_1 S_1)$ we know $S_1(v) \neq z_1$, so indeed $Q(S_1(v), S_2(y))$ and so $\mathrm{H}(Q)$. Hence $R \sqsubseteq Q$ by (10.2). But $R(z_1, a_2)$ and so $Q(z_1, a_2)$, which is a contradiction. This finishes the argument that $F(z_1)$.

Suppose now that $N_1(v)$ and $F(v)$. Let y be the unique element with $R(v,y)$, so that $N_1(v)$ and $N_2(y)$ by (*in:1*). Then since $\mathrm{H}(R)$ we have $R(S_1(v), S_2(y))$. So to show $F(S_1(v))$ it again suffices to show uniqueness. So suppose for reductio that $R(S_1(v), b_2)$ and $b_2 \neq S_2(y)$. By Comprehension there is P such that $\forall u \forall x(P(u,x) \leftrightarrow [R(u,x) \wedge \neg(u = S_1(v) \wedge x = b_2)])$. We will show that $\mathrm{H}(P)$. Clearly $P(z_1, z_2)$ since $R(z_1, z_2)$ and since $z_1 \neq S_1(v)$ by (*pa:q1*) in $\mathrm{PA}(N_1 z_1 S_1)$. Suppose now that $N_1(u)$ and $N_2(x)$ and $P(u,x)$. Then since both $P \sqsubseteq R$ and $\mathrm{H}(R)$, we have $R(S_1(u), S_2(x))$. The only way in which $P(S_1(u), S_2(x))$ could fail is if $S_1(u) = S_1(v)$ and $S_2(x) = b_2$. By (*pa:q2*) in $\mathrm{PA}(N_1 z_1 S_1)$ this would require that

$u = v$. But then $P(u, x)$ implies $R(u, x)$ which implies $R(v, x)$, and our hypothesis on y implies $y = x$, which in turn implies $S_2(y) = S_2(x) = b_2$, a contradiction. Hence in fact $P(S_1(u), S_2(x))$. This completes the argument that $\mathrm{H}(P)$. Hence $R \sqsubseteq P$ by (10.2). But $R(S_1(v), b_2)$ and so $P(S_1(v), b_2)$, which is a contradiction. This finishes the argument that $F(S_1(v))$.

Now $(\forall v : N_1)F(v)$ by induction in $\mathrm{PA}(N_1 z_1 S_1)$.

(*in:3*). By an exactly similar induction on N_2.

(*in:4*). Simply appeal to (*in:1*) and the fact that $\mathrm{H}(R)$. □

We next prove our Intolerance Theorem. Essentially, this simply amounts to 'internalising' the proof of Theorem 2.3:

Theorem (Intolerance of PA, 10.3): *For any formula $\varphi(NzS)$ whose quantifiers are N-restricted and whose free variables are all displayed:*

$$\vdash \forall NzS(\mathrm{PA}(NzS) \rightarrow \varphi(NzS)) \vee \forall NzS(\mathrm{PA}(NzS) \rightarrow \neg\varphi(NzS))$$

Proof. We start by introducing some notation. When $\mathrm{IsoN}_{1\triangleright 2}(R)$, we let r be a functor induced from R. So $r(x)$ is to be read as 'the unique y such that $R(x, y)$', and $r(\bar{u})$ is to be read as 'the unique $\bar{v}$ such that $R(u_1, v_1)$ for each u_i in $\bar{u}$'. This notation is licensed by the fact that R will be a bijection. We will work deductively, and establish the following:

$$\forall N_1 z_1 S_1 N_2 z_2 S_2 \big(\mathrm{IsoN}_{1\triangleright 2}(R) \rightarrow$$
$$(\forall \bar{u} : N_1)(\forall \overline{X} : N_1)[\varphi(N_1 z_1 S_1, \bar{u}, \overline{X}) \leftrightarrow \varphi(N_2 z_2 S_2, r(\bar{u}), r(\overline{X}))]\big)$$

This will suffice to establish the main theorem, by ignoring $\bar{u}$ and $\overline{X}$, using an internal-isomorphism R from Theorem 10.2, and invoking some simple deductive manipulation.

Atomic cases. Recall that, officially, S is a relation symbol rather than a function symbol. So, within the metatheory, we have three kinds of atomic formula:

- Case 1: φ is $S(x, y)$, perhaps with z replacing one or both of x and y. Now, if $S_1(x, y)$, then via $\mathrm{IsoN}_{1\triangleright 2}(R)$ we immediately get $S_2(r(x), r(y))$, so the deductive system proves $\varphi_1(x, y) \rightarrow \varphi_2(r(x), r(y))$. The converse conditional is similar, with r^{-1} in place of r.
- Case 2: φ is $x = y$, perhaps with z replacing one or both of x and y. This case is exactly analogous.
- Case 3: φ is $X(v)$, perhaps with N replacing X, or z replacing v. In this case, since $r(X) = \{r(v) : X(v)\}$ we can infer from $X(v)$ to $(r(X))(r(v))$, so the deductive system proves $\varphi_1(u, X) \rightarrow \varphi_2(r(u), r(X))$. The converse conditional is similar, using r^{-1} in place of r.

Propositional connectives. These inductive steps are trivial.

First-order quantifiers. Suppose, for induction, that we have:

$$\vdash \forall N_1 z_1 S_1 N_2 z_2 S_2 \big(\text{IsoN}_{1\triangleright 2}(R) \rightarrow (\forall v, \overline{u} : N_1)(\forall \overline{X} : N_1) \\ [\varphi(N_1 z_1 S_1, v, \overline{u}, \overline{X}) \leftrightarrow \varphi(N_2 z_2 S_2, r(v), r(\overline{u}), r(\overline{X}))]\big)$$

Working deductively, suppose $\text{PA}(N_1 z_1 S_1)$ and $\text{PA}(N_2 z_2 S_2)$ and $\text{IsoN}_{1\triangleright 2}(R)$, that $\overline{a}$ are all from N_1, and that there is some b from N_1 with $\varphi(N_1 z_1 S_1, b, \overline{a}, \overline{X})$. By the induction hypothesis we have $\varphi(N_2 z_2 S_2, r(b), r(\overline{a}), r(\overline{X}))$. Moreover, since r maps from N_1 to N_2, we have $N_2(r(b))$. As φ's quantifiers are N_2-restricted, we have $\exists y\, \varphi(N_2 z_2 S_2, y, r(\overline{a}), r(\overline{X}))$; so:

$$\vdash \forall N_1 z_1 S_1 N_2 z_2 S_2 \big(\text{IsoN}_{1\triangleright 2}(R) \rightarrow (\forall \overline{u} : N_1)(\forall \overline{X} : N_1) \\ [\exists y\, \varphi(N_1 z_1 S_1, y, \overline{u}, \overline{X}) \leftrightarrow \exists y\, \varphi(N_2 z_2 S_2, y, r(\overline{u}), r(\overline{X}))]\big)$$

Second-order quantifiers. These are handled similarly. □

10.C Predicative Comprehension

In §10.2 we mentioned that *Im*predicative Comprehension is required for internal categoricity. In this appendix, we explain that claim.

As explained in §1.11, the *Predicative* Comprehension Schema is just the restriction of the (Impredicative) Comprehension Schema to instances where the formula in question contains only *first-order* quantifiers.[32] The canonical models of the Predicative Comprehension Schema are given by taking a first-order structure and letting the second-order quantifiers range over the first-order definable subsets of the structure.[33] The following result roughly says that simply placing two structures side-by-side does not really affect definability, and so does not interestingly change which second-order entities there are in a predicative setting:

Proposition 10.4: *Let $\mathscr{L}_1$ and $\mathscr{L}_2$ be disjoint relational signatures, and let $\mathcal{N}_1$ and $\mathcal{N}_2$ be respectively an $\mathscr{L}_1$-structure and an $\mathscr{L}_2$-structure with disjoint domains. Let $\mathscr{L} = \mathscr{L}_1 \cup \mathscr{L}_2$ and let $\mathcal{A} = \mathcal{N}_1 \cup \mathcal{N}_2$ be the $\mathscr{L}$-structure where for $i \in \{1,2\}$ the interpretation of the n-place $\mathscr{L}_i$ predicate R is given by $R^{\mathcal{A}} = R^{\mathcal{N}_i}$. Then for any $i \in \{1,2\}$, any $n \geq 1$ and any $X \subseteq \mathcal{N}_i^n$: if X is $\mathcal{A}$-definable, then X is $\mathcal{N}_i$-definable.*

We will prove this proposition below. However, it yields a quick proof that internal categoricity requires Impredicativity. More precisely:

32 And, in the formalism introduced in §5.7, the theory $\text{PA}_{1.5}$ is the same thing as PA_2 when the Comprehension Schema is taken to be Predicative.

33 Cf. Hájek and Pudlák (1998: 153) and Simpson (2009: 361–362).

Theorem 10.5: *Theorem 10.2 is not provable using only Predicative Comprehension*

Proof. Let $\mathscr{L}_1$ and $\mathscr{L}_2$ be disjoint relational versions of the signature of PA_2 together with a number-predicate, so $\mathscr{L}_1 = \{Num_1, 0_1, Succ_1\}$ and $\mathscr{L}_2 = \{Num_2, 0_2, Succ_2\}$, with all symbols relational. Let $\mathcal{N}_1$ and $\mathcal{N}_2$ be two disjoint models of PA which are not isomorphic, with $N_1 = Num_1^{\mathcal{N}_1}$ and $N_2 = Num_2^{\mathcal{N}_2}$. As in Proposition 10.4, let $\mathcal{A} = \mathcal{N}_1 \cup \mathcal{N}_2$, and consider the second-order structure $\mathcal{B}$ whose first-order quantifiers range over $\mathcal{A}$ and whose second-order quantifiers range over the first-order definable subsets of $\mathcal{A}$.

We will show that $\mathcal{B} \vDash \text{PA}(Num_1 0_1 Succ_1)$ and $\mathcal{B} \vDash \text{PA}(Num_2 0_2 Succ_2)$. It suffices to verify the Induction Axiom. Without loss of generality, let $X \subseteq \mathcal{N}_1$ be a second-order object in $\mathcal{B}$, so that X is $\mathcal{A}$-definable by definition of $\mathcal{B}$ and X is $\mathcal{N}_1$-definable by Proposition 10.4. Since $\mathcal{N}_1 \vDash \text{PA}$, it satisfies the instance of Induction associated to X, as required. So $\mathcal{B} \vDash \text{PA}(Num_1 0_1 Succ_1)$ and $\mathcal{B} \vDash \text{PA}(Num_2 0_2 Succ_2)$.

Finally, suppose for reductio that Theorem 10.2 is provable with only Predicative Comprehension. Then it would be true on $\mathcal{B}$, and so $\mathcal{N}_1$ and $\mathcal{N}_2$ would be isomorphic, contrary to hypothesis. □

Before proceeding, we should make two brief remarks on this theorem. First, as mentioned in §1.A, the signature of PA_2 is just $\{0, S\}$, whereas the signature of the first-order theory PA adds $<$, $+$ and $\times$. Given merely *Predicative* Comprehension, it might be natural to allow PA_2 to have the same expansive signature as PA. But since our proof of Theorem 10.5 is based on very general considerations about first-order structures, it would go through even with this enriched signature.

Second, Theorem 10.2 can be proven with weaker comprehension principles, *if* one additionally assumes that one of the two domains is a subset of the other.[34] Likewise, as mentioned in §10.A, Parsons proves a version of Theorem 10.2 by allowing for a kind of recursion which can 'reach between' the two domains.[35]

It only remains to prove Proposition 10.4. Though the result is perhaps obvious, its proof requires a long induction on complexity of formulas. For the remainder of this appendix, we fix a structure $\mathcal{A}$ as in the statements of the proposition, we introduce some definitions for certain kinds of definable subsets, and we slowly come to an understanding of how the definable subsets of $\mathcal{A}$ interact with the definable subsets of $\mathcal{N}_1$ and $\mathcal{N}_2$.

Definition 10.6: *A set* $X \subseteq A^n$ *is* partition-definable *iff there is some permutation* $\pi : \{1, ..., n\} \longrightarrow \{1, ..., n\}$ *and some* $m \leq n$ *and some* $\mathcal{N}_1$*-definable* $X_1 \subseteq N_1^m$ *and*

[34] For detail and its reverse mathematical status, see Simpson and Yokoyama (2012).

[35] Parsons (1990a: 34–5, 2008: 281–2); see also Lavine (1999: §5.1).

$\mathcal{N}_2$-definable $X_2 \subseteq N_2^{n-m}$ such that $X = \{(x_1, \ldots, x_n) \in A^n : (x_{\pi(1)}, \ldots, x_{\pi(m)}) \in X_1 \wedge (x_{\pi(m+1)}, \ldots, x_{\pi(n)}) \in X_2\}$.

In the case $m = 0$, this means $X = \{(x_1, \ldots, x_n) \in A^n : (x_{\pi(1)}, \ldots, x_{\pi(n)}) \in X_2\}$ and in the case $m = n$, this means $X = \{(x_1, \ldots, x_n) \in A^n : (x_{\pi(1)}, \ldots, x_{\pi(n)}) \in X_1\}$. Hence $\mathcal{N}_1$-definable sets $X_1 \subseteq N_1^n$ and $\mathcal{N}_2$-definable sets $X_2 \subseteq N_2^n$ are trivially examples of partition-definable subsets of A^n. We start by showing a simple closure property of these sets:

Proposition 10.7: *The partition-definable subsets are closed under finite intersection.*

Proof. The empty set is vacuously partition-definable. Now, suppose we have

$$X = \{(x_1, \ldots, x_n) \in A^n : (x_{\pi(1)}, \ldots, x_{\pi(m)}) \in X_1 \wedge (x_{\pi(m+1)}, \ldots, x_{\pi(n)}) \in X_2\}$$
$$Y = \{(x_1, \ldots, x_n) \in A^n : (x_{\rho(1)}, \ldots, x_{\rho(\ell)}) \in Y_1 \wedge (x_{\rho(\ell+1)}, \ldots, x_{\rho(n)}) \in Y_2\}$$

If $\{\pi(1), \ldots, \pi(m)\} \neq \{\rho(1), \ldots, \rho(\ell)\}$, then without loss of generality suppose that $i = \pi(j) = \rho(k)$ for some $i \leq n$ and $j \leq m$ and $k > \ell$; then if $(x_1, \ldots, x_n)$ is in $X \cap Y$, we would have $(x_{\pi(1)}, \ldots, x_{\pi(m)}) \in X_1 \subseteq N_1^m$ and $(x_{\rho(\ell+1)}, \ldots, x_{\rho(n)}) \in Y_2 \subseteq N_2^{n-m}$ and so $x_i = x_{\pi(j)} = x_{\rho(k)}$ would be in $N_1 \cap N_2$. Hence if $\{\pi(1), \ldots, \pi(m)\} \neq \{\rho(1), \ldots, \rho(\ell)\}$ then $X \cap Y$ is empty and hence partition-definable. Likewise, if $\{\pi(m+1), \ldots, \pi(n)\} \neq \{\rho(\ell+1), \ldots, \rho(n)\}$ then $X \cap Y$ is empty and hence partition-definable. Hence suppose that $\{\pi(1), \ldots, \pi(m)\} = \{\rho(1), \ldots, \rho(\ell)\}$ and $\{\pi(m+1), \ldots, \pi(n)\} = \{\rho(\ell+1), \ldots, \rho(n)\}$, and thus $m = \ell$. Then this is clearly $\mathcal{N}_i$-definable:

$$Z_i = \{(z_1, \ldots, z_m) \in X_i : (z_{(\rho\circ\pi^{-1})(1)}, \ldots, z_{(\rho\circ\pi^{-1})(m)}) \in Y_i\}$$

So that:

$$X \cap Y = \{(x_1, \ldots, x_n) : (x_{\pi(1)}, \ldots, x_{\pi(m)}) \in Z_1 \wedge (x_{\pi(m+1)}, \ldots, x_{\pi(n)}) \in Z_2\}$$

And so $X \cap Y$ is partition-definable. □

We next characterise the $\mathcal{A}$-definable sets in terms of partition-definable sets:

Proposition 10.8: *Every $\mathcal{A}$-definable subset $X \subseteq A^n$ is a finite union of partition-definable sets.*

Proof. This by induction on the complexity of the formula $\varphi(x_1, \ldots, x_n)$ defining X. For any $\mathcal{L}_i$-predicate R, one has that $\{(x_1, \ldots, x_n) \in A^n : \mathcal{A} \models R(a_1, \ldots, a_n)\} = \{(x_1, \ldots, x_n) \in N_i^n : \mathcal{N}_i \models R(a_1, \ldots, a_n)\}$ and so the result trivially follows. For the atomic formula associated to identity, we have the finite union $\{(x_1, x_2) \in A^2 : \mathcal{A} \models x_1 = x_2\} = \{(x_1, x_2) \in N_1^2 : \mathcal{N}_1 \models x_1 = x_2\} \cup \{(x_1, x_2) \in N_2^2 : \mathcal{N}_2 \models x_1 = x_2\}$.

Suppose the result holds for X, Y, so that both X, Y are finite unions of partition-definable sets. Then trivially the result holds for $X \cup Y$.

Suppose that the result holds for X. We show it holds for $A^n \setminus X$. Since the result holds for X, we may write it as the finite union of partition-definable sets $Y, Z, \ldots$. Then $A^n \setminus X$ is the finite intersection of the sets $A^n \setminus Y, A^n \setminus Z, \ldots$. But note that:

$$\begin{aligned} A^n \setminus Y = \{&(x_1, \ldots, x_n) \in A^n : (x_{\pi(1)}, \ldots, x_{\pi(m)}) \notin N_1^m\} \cup \\ \{&(x_1, \ldots, x_n) \in N_1^n : (x_{\pi(1)}, \ldots, x_{\pi(m)}) \notin Y_1\} \cup \\ \{&(x_1, \ldots, x_n) \in A^n : (x_{\pi(m+1)}, \ldots, x_{\pi(n)}) \notin N_2^{n-m}\} \cup \\ \{&(x_1, \ldots, x_n) \in N_2^n : (x_{\pi(m+1)}, \ldots, x_{\pi(n)}) \notin Y_2\} \end{aligned}$$

Obviously the second of these is $\mathcal{N}_1$-definable and the fourth is $\mathcal{N}_2$-definable, so both are partition-definable. Further, the first and third of these are finite unions of partition-definable sets: for, if e.g. $n = 2$ and $\pi(1) = 1$ and $\pi(2) = 2$ then the first of these is equal to $(N_1 \times N_2) \cup (N_2 \times N_1) \cup (N_2 \times N_2)$, and each of these is clearly partition-definable. Hence $A^n \setminus X$ is the intersection of the sets $A^n \setminus Y, A^n \setminus Z, \ldots$, each of which is a finite union of partition-definable sets $\bigcup_i Y_i, \bigcup_j Z_j, \ldots$. Then $A^n \setminus X$ is a finite union of finite intersections of the form $Y_i \cap Z_j \cap \cdots$, which are partition-definable by Proposition 10.7.

Finally, suppose that the result holds for $X \subseteq A^{n+1}$. We must show that the result holds for $\pi(X) = \{(x_1, \ldots, x_n) \in A^n : (\exists x_{n+1} \in A)(x_1, \ldots, x_{n+1}) \in X\}$. Since $\pi(X \cup X') = \pi(X) \cup \pi(X')$, we may without loss of generality assume that X itself is partition-definable. Then we have:

$$\begin{aligned} \pi(X) = \{&(x_1, \ldots, x_n) \in A^n : (\exists x_{n+1} \in N_1)[(x_{\pi(1)}, \ldots, x_{\pi(m)}) \in X_1 \wedge \\ &\qquad (x_{\pi(m+1)}, \ldots, x_{\pi(n+1)}) \in X_2]\} \cup \\ \{&(x_1, \ldots, x_n) \in A^n : (\exists x_{n+1} \in N_2)[(x_{\pi(1)}, \ldots, x_{\pi(m)}) \in X_1 \wedge \\ &\qquad (x_{\pi(m+1)}, \ldots, x_{\pi(n+1)}) \in X_2]\} \end{aligned}$$

If $n + 1 \in \{\pi(1), \ldots, \pi(m)\}$, then the second union is empty and the first is clearly partition-definable. If $n+1 \in \{\pi(m+1), \ldots, \pi(n+1)\}$, then the first union is empty and the second is partition-definable. Either way, $\pi(X)$ is partition-definable. □

We can now prove Proposition 10.4 itself:

Proof of Proposition 10.4. Suppose that $X \subseteq N_1^n$ is $\mathcal{A}$-definable. If X is empty then it is trivially $\mathcal{N}_1$-definable, and so suppose without loss of generality that X is non-empty. By Proposition 10.8, X is the finite union of partition-definable subsets $Y, Z, \ldots$, each of which may likewise be assumed non-empty. So suppose that

$$Y = \{(x_1, \ldots, x_n) \in A^n : (x_{\pi(1)}, \ldots, x_{\pi(m)}) \in Y_1 \wedge (x_{\pi(m+1)}, \ldots, x_{\pi(n)}) \in Y_2\}$$

for some $\mathcal{N}_1$-definable $Y_1 \subseteq N_1^m$ and $\mathcal{N}_2$-definable $Y_2 \subseteq N_2^{n-m}$. Since $\varnothing \neq Y \subseteq X \subseteq N_1^n$, we must have that $m = n$ and hence that

$$Y = \{(x_1, \ldots, x_n) \in A^n : (x_{\pi(1)}, \ldots, x_{\pi(m)}) \in Y_1\}$$

which is obviously $\mathcal{N}_1$-definable. □

11

Internal categoricity and the sets

As outlined in Chapter 10, internalists invoke metamathematics without semantics. They rely upon internal categoricity, rather than external categoricity. As a result, they give up both on the attempt to 'pin down' referents for our theories, and on the attempt to find complete theories. However, they seek solace from the *intolerance* that is associated with internal categoricity.

In Chapter 10, though, we only considered internalism about arithmetic. We now turn our attention to set theory, and to its internal (quasi-)categoricity.[1] So, this chapter stands to Chapter 10 as Chapter 8 stands to Chapter 7. However, this chapter also presents some new results. In particular, we present a theory which is too weak to decide whether there is more than one (pure) set, which is nevertheless internally categorical and intolerant in exactly the same way as PA_{int}.

11.1 Internalising Scott–Potter set theory

We begin by focussing on an internalisation of the Scott–Potter theory of levels, as introduced in §8.5. In that section, we explained that SP_2 is strictly weaker than ZFC_2, since the full models of SP_2 are exactly the stages of the cumulative iterative hierarchy (up to isomorphism). That was an *external* quasi-categoricity result. In this section, we present an *internal* quasi-categoricity result.

We must start by laying down a second-order formula, $\text{SP}(PE)$, which stands to SP_2 as $\text{PA}(NzS)$ stands to PA_2. So, the idea is to relativise SP_2 to a one-place relation-variable P, and a two-place relation-variable E. Since the latter will go proxy for membership, we use E since it reminds us of 'epsilon', and we write $x\ E\ y$ rather than $E(x, y)$. Likewise we define the subset relation $v \subseteq_E u := (\forall w\ E\ v)w\ E\ u$ relative to the membership relation E. When the E is clear from context, we often drop the subscript from $\subseteq_E$ and simply use the ordinary subset symbol $\subseteq$. Sometimes in what follows we will have two candidate membership relations E_1 and E_2, and we will use e.g. $\subseteq_1$ as an abbreviation for $\subseteq_{E_1}$ to enhance readability.

Then, guided by Definition 8.9 of SP_2, we define:[2]

---

[1] See Lavine (1999: 55–66, 89–95) and Väänänen and Wang (2015: 126–8).

[2] As in footnote 32 of Chapter 8, we only use '$\cap$' in defining H to assist readability. More explictly, we would write $H(x) := (\forall z\ E\ x)(\forall v)(v\ E\ z \leftrightarrow (\exists u\ E\ x)(u\ E\ z \wedge (v\ E\ u \vee v \subseteq u)))$.

$$A(x,y) := \forall v(v\,E\,x \leftrightarrow (\exists u\,E\,y)(v\,E\,u \lor v \subseteq u))$$
$$H(x) := (\forall z\,E\,x)A(z, x \cap z)$$
$$L(x) := (\exists y : H)A(x,y)$$

We use these to define a formula with P and E free (we reuse the conventions of the last chapter, of suppressing repeated commas and quantifiers):

$$\text{SP}(PE) := \forall x(\forall y\,E\,x)[P(x) \land P(y)] \land \qquad (sp{:}pure)$$
$$(\forall x,y : P)\,[\forall z\,(z\,E\,x \leftrightarrow z\,E\,y) \rightarrow x = y] \land \qquad (sp{:}ext)$$
$$\forall X[(\exists x : P)\forall z\,(z\,E\,x \leftrightarrow X(z)) \leftrightarrow (\exists y : L)(\forall z : X)z\,E\,y] \qquad (sp{:}levels)$$

Clearly, (*sp:ext*) relativises Extensionality, and (*sp:levels*) relativises Levelling. But (*sp:pure*), and our approach more generally, merits comment.

We do not want to assume that *everything* is a set. This is why we relativise Extensionality to P in (*sp:ext*). Then (*sp:pure*) restricts the implementation of membership, E, to P. But if we informally gloss $P(x)$ as *x is a set*, then our axiomatization will have ruled out the possibility of sets which contain *urelements*, i.e. elements which are not themselves sets. This would be unfortunate: it would prevent us from speaking of 'the set of cows in the field', or even 'the set of natural numbers', unless we *happen* to think that natural numbers (or cows) are sets.[3]

Fortunately, this unhappy situation is *not* forced upon us. We should not gloss $P(x)$ as *x is a set*. Rather, we should gloss $P(x)$ as *x is a pure set*, i.e., intuitively, a set which involves no urelements anywhere in its construction. Read like this, SP(PE) says *nothing at all* about whether there are any impure sets, or what they are like if there are any.[4]

Moreover, the aim of SP(PE) is not to tell us *everything* about pure sets, but just to describe the iterative process of set-formation itself. We can illustrate the point simply. With Knuth, we write $x{\uparrow\uparrow}n$ for *tetration*, i.e. for $x^{x^{\cdot^{\cdot^{\cdot^{x}}}}}$ where 'x' occurs n-times.[5] Then, for any $n \geq 0$, the following theory is consistent:

$$\exists PE[\text{SP}(PE) \land \text{`there are exactly } (2{\uparrow\uparrow}n)\text{-many } Ps\text{'}]$$

After all, by Theorem 8.10, $V_\alpha \vDash \exists PE\,\text{SP}(PE)$ for any ordinal $\alpha \geq 1$. As such, we cannot obtain an internal *categoricity* result for SP(PE), but must aim for internal

[3] See McGee (1997: 49) and Potter (2004: vi, 24, 50–1) for discussions of why we might want to allow urelements into our theorising about sets.

[4] Scott (1974) and Potter (2004) instead offer a theory which has impure sets, whereupon we can define the pure sets as those whose transitive closure contains no urelements. That approach is certainly well-motivated. However, for our purposes, it involves an unnecessary detour, since all our results concern pure sets anyway. Moreover, our approach is strictly weaker, which is in advantageous in that it has a potentially wider field of application (see §11.A).

[5] Formally, $x{\uparrow\uparrow}0 = 1$ and $x{\uparrow\uparrow}(n+1) = x^{x\uparrow\uparrow n}$.

quasi-categoricity. And such a result is, indeed, available. To state the result formally, we must define a formula, $\text{QuasiP}_{1\triangleright 2}(R)$, which intuitively states that R is a (second-order) quasi-isomorphism, i.e. an isomorphism between an 'initial segment of one set-like internal-structure' and the 'whole of the other'. The definition follows the template established in our definition of $\text{IsoN}_{1\triangleright 2}(R)$ from §10.2, but it is a little more cumbersome. We reserve the precise definition for §11.C, where we also prove the following:

Theorem 11.1 (Internal Quasi-Categoricity of SP):

$$\vdash \forall P_1 E_1 P_2 E_2\,([\text{SP}(P_1E_1) \land \text{SP}(P_2E_2)] \rightarrow \exists R\, \text{QuasiP}_{1\triangleright 2}(R))$$

In some sense, this *surely* shows internalists that $\text{SP}(PE)$ is a stellar axiomatization of the very idea of the iterative process of set formation, which deliberately avoids commenting on 'how far' that process runs.

11.2 Quasi-intolerance for pure set theory

We say '*surely*'. In fact, internalists must take some care in spelling out this claim. A naïve thought is that Theorem 11.1 allows us to pin down all initial segments of the hierarchy (or, depending on your point of view, all the pure set hierarchies). This should be compared with the naïve reaction to Theorem 10.2, discussed in §10.3, that PA_{int} pins down the natural numbers. Both reactions are equally naïve, equally tempting, and equally mistaken.

We briefly recap the main points of §10.3. Talk of 'pinning down' is rather loose. If the internalist explicates such talk in terms of semantic ascent, then she has given up on her aim of treating Theorem 11.1 as an *internal* quasi-categoricity result. (In particular, we should repeat that it is *literally a type-confusion* to think that an internal result could directly pin down certain $\mathcal{L}$-structures.) Equally, there are hard Gödelian limits on how completely a concept can be articulated by any deductive theory. As in Chapter 10, then, we think that the best way for internalists to extract promise from internal (quasi-)*categoricity*, is by looking to (quasi-)*intolerance*.

Very roughly, Theorem 11.1 yields *tolerance* regarding the height of internal-structures, but *intolerance* about their width (given any height). Slightly more precisely, it yields a result which we can gloss as follows: *For any second-order formula φ concerning sets of some particular level, all internal-structures which go far enough to take a non-vacuous stance on φ must agree on it.* But the precise statement of this result will take some time.

Let $\pi(PE\ell)$ be any property definable in terms of P and E. We treat 'ℓ' as a first-order variable which intuitively will range over levels, in the sense of L we defined from P and E as in §11.1. Since levels are well-ordered by Proposition 8.24, internal-structures with any π-levels must have a *least* π-level. So, for any formula $\pi(\ell)$, we define a formula $\mu_\pi(PE\ell)$ which intuitively states that ℓ is the E-least level with the property π:

$$\mu_\pi(PE\ell) := \pi(PE\ell) \land L(\ell) \land (\forall \ell' : L)(\ell' \, E \, \ell \to \neg\pi(PE\ell'))$$

Now, by Theorem 11.1, the initial segments of any internal-structures beneath their least π-levels are totally internally isomorphic. So we obtain:

Theorem 11.2 (Quasi-Intolerance of SP): *For any formulas $\pi(PE\ell)$ and $\varphi(PE\ell)$ with all first-order quantifiers bound to elements of ℓ and all second-order quantifiers bound to subsets of ℓ, we have*

$$\vdash \forall PE \forall \ell \left(\left[\mathrm{SP}(PE) \land \mu_\pi(PE\ell) \right] \to \varphi(PE\ell) \right) \lor \\ \forall PE \forall \ell \left(\left[\mathrm{SP}(PE) \land \mu_\pi(PE\ell) \right] \to \neg\varphi(PE\ell) \right)$$

This is the result which internalists can legitimately gloss as follows: *For any second-order formula φ concerning sets of some particular level, all internal-structures which go far enough to take a non-vacuous stance on φ must agree on it.*[6]

Here is an interesting corollary of our quasi-intolerance theorem: *all set-like internal-structures with an $\omega+2^{th}$ level must agree concerning subsets of that level.* In particular: *they must all agree concerning the continuum hypothesis* (as explained in §8.5). Consequently, it is *deductively inconsistent* to assert that there are two internal-structures which both go far enough to take a non-vacuous stance on the continuum hypothesis, but which differ on the status of the continuum hypothesis.

In the next section, we will discuss what (if anything) this shows about the status of the continuum hypothesis. For now, though, we can simply treat the continuum hypothesis as a particularly striking illustration of a very general point.

The immediate import of the Quasi-Intolerance Theorem 11.2 is that allows internalists to locate SP(PE) on the cusp of the algebraic / univocal distinction. Theories in the Scott–Potter-style are multiply applicable by design, and indeed it is deductively consistent to affirm the existence of two internal-structures of different heights (as expressed in the object language). To that extent, SP(PE) resembles *group theory* (see §10.6). However, its quasi-intolerance dramatically curtails the extent of its multiple-applicability. For, provably, the only way for two internal-structures to disagree (modulo subscripts) is for one to stop short of the other. In that respect, SP(PE) resembles *arithmetic*. And this, we suggest, is the sense in

[6] To see why we must bind the quantifiers in these formulas, let $\pi(PE\ell)$ be 'there is no level greater than ℓ'. The last levels of models of SP_2 (when they exist) can be very different, e.g. one can be of size 4 while another is of size 16.

which internalists can say that $\mathrm{SP}(PE)$ is a stellar axiomatization of the very idea of the iterative process of set formation, which deliberately withholds from commenting on 'how far' that process runs. In a slogan: $\mathrm{SP}(PE)$ *is quasi-univocal.*

11.3 The status of the continuum hypothesis

Theorems 11.1 and 11.2 concern a Scott–Potter-style set theory. This is strictly weaker than Zermelo–Fraenkel-style theories. To explain this point, let $\mathrm{ZFC}(PE)$ be an appropriately internalised version of ZFC_2, obtained by relativising the axioms of Definition 1.12 and adding (*sp:pure*) to them. Then since Proposition 8.32(1) states that $\mathrm{ZFC}_2 \vdash \mathrm{SP}_2$, we have that $\vdash \forall PE\,(\mathrm{ZFC}(PE) \rightarrow \mathrm{SP}(PE))$. So we immediately obtain the internal quasi-categoricity of ZFC:[7]

Corollary 11.3:

$$\vdash \forall P_1E_1P_2E_2\,([\mathrm{ZFC}(P_1E_1) \wedge \mathrm{ZFC}(P_2E_2)] \rightarrow \exists R\, \mathrm{QuasiP}_{1\triangleright 2}(R))$$

Equally obviously, we can deductively prove from $\mathrm{ZFC}(PE)$ the existence of an $\omega{+}2^{\text{th}}$ level. So, by the Quasi-Intolerance Theorem 11.2, it is *deductively inconsistent* to assert that there are two internal-structures such that $\mathrm{ZFC}(PE)$ holds of both, but such that the continuum hypothesis holds in one and fails in the other.[8]

Now, in §10.6, we explained how the Arithmetical Intolerance Theorem 10.3 applies pressure to those who regard arithmetic as algebraic. The present observation applies similar pressure to those who regard the continuum hypothesis as *indeterminate.* Indeed, it is not obvious how best to sustain that attitude, in the face of this deductive inconsistency.

As in §10.6, one might try to argue for the indeterminacy of the continuum hypothesis by invoking faithful Henkin semantics. Let $\mathrm{CH}(PE)$ be a suitable formalization of the continuum hypothesis using P and E. Then, just because the continuum hypothesis is deductively independent of ZFC_2,[9] there are faithful Henkin-structures $\mathcal{A}$ and $\mathcal{B}$ such that $\mathcal{A} \vDash \exists PE[\mathrm{ZFC}(PE) \wedge \mathrm{CH}(PE)]$ and $\mathcal{B} \vDash \exists PE[\mathrm{ZFC}(PE) \wedge \neg\mathrm{CH}(PE)]$. Those who think the continuum hypothesis is indeterminate may then argue that no sense can be made of the claim that $\mathcal{A}$ is 'preferable' to $\mathcal{B}$, or vice versa. However, something like the Transcendental Arguments of Chapter 9 suggests that this kind of attempt to defend the indeterminacy of

---

[7] This slightly extends Väänänen and Wang (2015: Theorem 2).

[8] Väänänen and Wang (2015: 128) make the same observation.

[9] This is due to the connection between strongly inaccessible cardinals and ZFC_2 discussed in §8.1, as well as the fact that the continuum hypothesis is independent of ZFC plus the usual large cardinal axioms. See Martin (1976: §3) and Jech (2003: Theorem 21.2 p.390).

the continuum hypothesis walks a dangerously thin path between falsity and incoherence. In brief: within the model theory within which $\mathcal{A}$ and $\mathcal{B}$ are introduced, we can define the idea of a *full* semantics, and so—if we want—state that full structures are 'preferable'. Of course, insofar as we are denied a firm grip on the model theory, we are denied a firm grip on this notion of 'preferability'; but, insofar as we are denied a firm grip on the model theory, it is doubtful that we can understand the claim that there are multiple 'equally preferable' models, some of which satisfy the continuum hypothesis and others which do not, and hence the supposed *motivation* for thinking that the continuum hypothesis is indeterminate. In sum: insofar as we can make sense of this attempt to motivate the indeterminacy of the continuum hypothesis, we have a rebuttal to it.

Before moving on, we should briefly note two things. First, as in the case of arithmetic, Theorem 11.1 essentially uses the *impredicative* Comprehension Schema, so that the ambient second-order deductive framework is crucial to these results. Second, even if Theorem 11.1 is taken to suggest that the continuum hypothesis is determinate, it gives us no reason whatsoever to think that we can ever even obtain a justified belief (let alone knowledge) concerning whether it is right or wrong.

Still, the short moral is that the quasi-intolerance of Scott–Potter-style theories—and hence of all extensions of Zermelo–Fraenkel-style theories—does two things for an *internalist* about pure set theory. First, as in Chapter 10, it allows them to regard pure set theory as quasi-univocal. Second, in tandem with the considerations of Chapter 9, it places serious pressure on certain views concerning the indeterminacy of the continuum hypothesis, and other set-theoretical claims.

11.4 Total internal categoricity for pure set theory

We have been discussing internal *quasi*-categoricity and *quasi*-intolerance. In fact, we can obtain *full-fledged* internal categoricity quite easily. We just need to add one new conjunct to SP(PE) to obtain CSP(PE):

$$\mathrm{CSP}(PE) := \mathrm{SP}(PE) \land \exists f\,[\forall x(\exists! y : P)f(x) = y \land (\forall y : P)\exists! x\, f(x) = y)] \quad (\textit{csp:many})$$

The initials 'CSP' stand for Categorical Scott–Potter set theory, since this theory will be internally categorical (and not just internally *quasi*-categorical). But the new conjunct merits discussion.

Intuitively, (*csp:many*) states that there is a second-order bijection whose domain is all the objects, and whose range is (just) the pure sets. Otherwise put: there are as many pure sets as objects simpliciter. Of course, (*csp:many*) can be trivially satisfied by assuming that *everything* is a pure set. However, if we reject that assumption—perhaps doubting whether cows are sets—then (*csp:many*) will require that there are infinitely many pure sets. Even then, (*csp:many*) will hold provided that the process of set-formation runs far enough that either (*a*) the pure sets 'eventually' outnumber everything else, or (*b*) there are as many pure sets as other things, and infinitely many of both. Almost anyone working in set theory, or its philosophy, will surely accept that one of these disjuncts obtains (though we revisit this in §11.A).

To obtain total internal categoricity, we now define a formula, $\text{IsoP}_{1\triangleright 2}(R)$, which states that R is an isomorphism of set-like internal-structures. We defer the exact formula until §11.D, where we also prove our main result:

Theorem 11.4 (Internal Categoricity of CSP):

$$\vdash \forall P_1 E_1 P_2 E_2 \left([\text{CSP}(P_1 E_1) \land \text{CSP}(P_2 E_2)] \rightarrow \exists R \, \text{IsoP}_{1\triangleright 2}(R) \right)$$

11.5 Total intolerance for pure set theory

As usual, a naïve and tempting thought is that an internal categoricity result shows that 'the structure of the pure sets *can* be categorically characterized'.[10] Indeed, according to Incurvati, this is becoming the 'consensus' view on the significance of such results. But Incurvati is rightly critical of this consensus.

Incurvati's main concern stems from the use of *unrestricted* first-order quantification in proving internal categoricity. For example, in our Theorem 11.4, the principle that there are as many pure sets as objects simpliciter, (*csp:many*), requires unrestricted first-order quantification.[11] Incurvati puts the point as follows. In order for the Theorem to show that the hierarchy is categorically determined, we would need some reason to think that 'universal' models—i.e. ones with utterly unrestricted domains—are *preferable* to 'restricted' models. And, by itself, the Theorem gives no reason to think that.

We agree, and would push this further. The notion of 'characterizing the hierarchy' is obviously informal, but it is naturally understood in terms of semantic ascent. According to this understanding, to 'characterize the hierarchy' would be to produce a formal theory which describes some object(s). (Note that this is what a *modelist* will think about the project of 'characterizing the hierarchy'.) But then, for all the reasons pointed out in §10.3 and §11.2, it is a *type-confusion* to think that an internal categoricity result might 'characterize the hierarchy'. Equally, it is a mistake to think that $\text{CSP}(PE)$ exhausts everything there is to say about pure sets. At the risk of repetition: it fails to decide whether there is *more than one* pure set.

[10] Incurvati (2016: 368). Thanks to Luca Incurvati for discussion on all this.

[11] Incurvati is actually discussing McGee's Theorem 11.6 (see §11.A), and his focus is the unrestricted quantification in McGee's claim that there is a set of all the urelements, (*zfcu:ur*).

In keeping with our general line of thought, then, the best way for internalists to extract juice from internal *categoricity* is via *intolerance*. And, predictably, we have:[12]

Theorem 11.5 (Intolerance of CSP): *For any formula $\varphi(PE)$ whose quantifiers are P-restricted and whose free variables are all displayed:*

$$\vdash \forall PE(\text{CSP}(PE) \rightarrow \varphi(PE)) \vee \forall PE(\text{CSP}(PE) \rightarrow \neg\varphi(PE))$$

Exactly as in §10.4, this (total) intolerance result licenses us in defining a canonical theory of pure sets. So: '*Pure*' will be our canonical pure-set-predicate, and '$\in$' will be our canonical membership-predicate. We then define CSP_{int} as the theory $\text{CSP}(\textit{Pure}, \in)$. The justification for using CSP_{int} is simple: modulo subscripts, all the same pure set-theoretic claims hold of every internal-structure (in the sense of $\text{CSP}(PE)$), so we might as well ditch the subscripts.

It is might seem genuinely startling that CSP_{int} is totally intolerant. For CSP_{int} itself is far too weak to interpret Robinson's Q. And yet, all CSP_{int}-like internal-structures must agree upon *anything* you care to formulate in the language of pure set theory: whether there are infinitely many sets: whether the continuum hypothesis holds; whether projective determinacy holds; whether the generalised continuum hypothesis holds; whether there is a proper class of Mahlo cardinals; anything you like.

Moreover, as in §10.6, the total intolerance of CSP_{int} allows internalists to maintain that pure set theory, like arithmetic, is *entirely* univocal. This is surprising, because—as in the case of arithmetic—it gives internalists a way to explain how their theory can count as univocal, without having to engage in semantic-ascent. But the case of CSP_{int} is *doubly* surprising. After all, semantic-ascent concerning set theory typically allows that a theory like ZFC_2 is only *quasi*-univocal, in the sense that ZFC_2 tolerates differences concerning the height of the hierarchy. But here we are, claiming that CSP_{int} is *totally* univocal.

11.6 Internalism and indefinite extensibility

Now, CSP_{int}'s intolerance generates a deep challenge for anyone who does want to view set-theory as anything less than 'totally' univocal. These difficulties are akin to those discussed surrounding the Algebraic Attitude towards arithmetic (see §10.6), and the indeterminacy of the continuum hypothesis (see §11.3). However, to bring them out, we will consider the supposed *indefinite extensibility* of the concept *set*.

From a model-theoretic perspective, the domain of any $\mathscr{L}$-structure satisfying ZFC_2—even on the full semantics—cannot be 'universal', since the domain's

[12] This follows from Theorem 11.4, exactly as Theorem 10.3 follows from Theorem 10.2.

power set, for example, is not a member of the domain itself. (We essentially made the same point in §8.6, when arguing that any externally categorical set theory would fail to handle the purportedly *all-encompassing* nature of set theory.) As such, if we intend to quantify *unrestrictedly*, then no model of ZFC_2 can be 'intended'; at best, we have increasingly better (but always inadequate) *approximations* of our intent. This is one way to be led to the conclusion that the concept *set* is *indefinite extensibility*. (Indeed, it may well have been Zermelo's, in his 1930.)

Sadly, it is a short step from this line of thought to outright incomprehensibility.[13] Modelists insist on understanding the use of quantifiers model-theoretically; yet the above line of thought suggests that we can never use our quantifiers to quantify unrestrictedly; and so our modelist is led to say: *No sentence containing quantifiers can quantify over everything.* Unfortunately, *that* claim undermines itself: it 'wants' to quantify over *everything*, in order to say that it *cannot*. So, insofar as we understand the claim, we can see that it is false.

As in Chapters 6–9, then, modelists are drawn towards a position which is ineffable. Perhaps modelists will be able to come up with their own way to avoid this mess. But, once again, internalism suggests itself as a likely-looking alternative. Internalists specifically reject the modelist thought, that our use of quantifiers *must* be understood model-theoretically. As such, internalists are under no pressure whatsoever to say: *No sentence containing quantifiers can quantify over everything.*

Indeed, internalists have a rather short line with the issue of 'quantifying over all the sets'. Since they have eschewed semantic ascent, the only sense that internalists can give to (un)restricted quantification concerns the syntactic thought that '$\forall x$' is unrestricted, whereas '$(\forall x : \Xi)$' is restricted. Now, it is indeed important to the proof of Theorem 11.4 that (*csp:many*) involves unrestricted quantification in *that* sense. Equally, though, no one can seriously worry about the possibility of *that* sort of unrestricted quantification.[14]

In sum, CSP_{int}'s intolerance allows internalists to regard set theory as wholly univocal and to bypass entirely the vexed notion of indefinite extensibility. Perhaps this is another point in its favour.

[13] The contents of the next few paragraphs join up with Button (2010); see also the references there.

[14] Anticipating the discussion of §11.A, we suggest that this is all that is going on in McGee's own discussion of unrestricted quantification. McGee states his result as follows: 'Any two models of second-order ZFCU + [(*zfcu:ur*)] with the same universe of discourse have isomorphic pure sets. In particular, any two models of second-order ZFCU + [(*zfcu:ur*)] in which the first-order variables range over everything have isomorphic pure sets' (1997: 55; see also p.53). One might worry that mentioning a 'universe of discourse' flags semantic ascent, undercutting the idea that he is considering *internal* categoricity. However, as mentioned in footnote 8 of Chapter 10, McGee works with class-models but describes them as 'merely figurative'. And, in the setting of class-models, talk of a 'universe of discourse' is a merely figurative way to describe quantifier-restrictions.

11.A Connection to McGee

The material of Chapter 10 was heavily influenced by Parsons. As we explain in this philosophical appendix, the material in this chapter is heavily influenced by McGee.[15] Now, like Parsons in §10.A, McGee hesitates to use second-order logic directly, instead invoking first-order logic with 'open-ended schemas'. As in §10.A, this creates certain difficulties for no obvious gains. So, although it is slightly unfaithful to McGee's paper, for the rest of this appendix we will simply act as if McGee had worked directly within deductive second-order logic.

McGee proved a result which is often described as a proof of *categoricity* for ZFCU. His result concerns an internalised version of second-order Zermelo–Fraenkel set theory with Choice and urelements. Unlike SP(PE) or CSP(PE), this is supposed to be a theory of both pure and impure sets. The internalisation begins as follows:

$$\begin{aligned} \text{ZFCU}(\mathit{SetE}) := {} & \forall x\,[\exists y\, y\, E\, x \rightarrow \mathit{Set}(x)] \wedge && (\mathit{zfcu{:}sets}) \\ & (\exists x : \mathit{Set})\forall z\,[\neg \mathit{Set}(z) \rightarrow z\, E\, x] \wedge && (\mathit{zfcu{:}ur}) \\ & (\forall x, y : \mathit{Set})\,[\forall z(z\, E\, x \leftrightarrow z\, E\, y) \rightarrow x = y] \wedge && (\mathit{zfcu{:}ext}) \\ & \forall x \forall y \exists z \forall w\,[w\, E\, z \leftrightarrow (w = x \vee w = y)] && (\mathit{zfcu{:}pairing}) \\ & \forall x \exists y \forall z [z\, E\, y \leftrightarrow (\exists v\, E\, x) z\, E\, v] \wedge && (\mathit{zfcu{:}union}) \\ & \ldots \end{aligned}$$

Conjuncts (*zfcu:ext*) onwards simply relativise the axioms of ZFC_2 to *Set* and *E*; so we have written out relativised versions of Extensionality, Pairing, and Union, but we leave it to the reader to complete the relativisation (using Definition 1.12 as their guide). Additionally, (*zfcu:sets*) intuitively states that only sets have members. Finally, (*zfcu:ur*) intuitively states that *there is a set of all the urelements*.

Working within ZFCU($SetE$), we can define a property, P, which intuitively holds of the pure sets, i.e. sets which contains no urelements anywhere in their construction.[16] McGee now proves that all ZFCU($SetE$)-like internal-structures have isomorphic pure sets. So, in our terminology, he proves:

Theorem 11.6 (McGee): *Where 'P_1' and 'P_2' are defined within* ZFCU(Set_1E_1) *and* ZFCU(Set_2E_2) *in the manner sketched above:*

$$\vdash \forall Set_1 E_1 Set_2 E_2\,([\text{ZFCU}(Set_1E_1) \wedge \text{ZFCU}(Set_2E_2)] \rightarrow \exists R\, \text{IsoP}_{1\triangleright 2}(R))$$

McGee's Theorem is obviously extremely similar to our Theorem 11.4. Indeed, it follows from ours via a simple observation:

[15] Many thanks to Vann McGee for discussion of all this.

[16] Roughly, the definition is: x is a set and and there are no non-sets in x's transitive closure.

Lemma 11.7: *Where 'P' is defined from 'Set' and 'E' in the manner sketched above:*

$$\vdash \forall SetE\,(\mathrm{ZFCU}(SetE) \rightarrow \mathrm{CSP}(PE))$$

Proof. Assume ZFCU($SetE$). Proposition 8.32(1) shows that SP(PE). It remains to check (*csp:many*), i.e. that there are as many pure sets as objects simpliciter.[17] By (*zfcu:ur*), the urelements form a set x; so every object is an element of some $U_\alpha(x)$, where this is the cumulative hierarchy relativised to x:[18]

$$\begin{aligned} U_0(x) &= x \\ U_{\alpha+1}(x) &= U_\alpha(x) \cup \wp(U_\alpha(x)) \\ U_\alpha(x) &= \bigcup_{\beta<\alpha} U_\beta(x) \text{ if } \alpha \text{ is a limit ordinal} \end{aligned}$$

Since x is a set, work within ZFCU($SetE$) and choose an ordinal β and a set b which is a subset of $V_{\beta+1} \setminus V_\beta$, such that b and x have the same cardinality. By transfinite recursion, we can define a bijection from $U_\alpha(x)$ to $U_\alpha(b)$, and the latter are pure sets. Hence, there is an injection from the objects to the pure sets. Further, the identity map is clearly an injection from the pure sets to the objects. Hence, the Schröder–Bernstein Theorem yields our second-order bijection from the objects to the pure sets.[19] □

McGee's Theorem 11.6 is an immediate corollary of this Lemma plus Theorem 11.4. And of course this is no accident. We arrived at our Theorem 11.4, by considering this question: *What is the weakest theory, which recognisably deals with the iterative notion of set, and for which a result like McGee's can be obtained?* Our answer to that was: CSP(PE). But we now want to give two reasons for thinking that this question was worth asking and answering in the first place.

First: using ZFCU($SetE$) is not entirely uncontroversial. In particular, one can contest the assumption that the non-sets form a set, i.e. (*zfcu:ur*). Consider the view that the ordinals are *sui generis* objects, which are not identical to sets but which can be modelled as certain sets. Since on this view the non-sets will include all the ordinals, presumably the non-sets will not form a set.[20] Then (*zfcu:ur*) will fail. But this viewpoint is wholly compatible with (*csp:many*), for there can be just as many *sui generis* ordinals as pure sets.

Second, and more important: ZFCU($SetE$) is a very rich theory. So, although McGee's total internal categoricity result for ZFCU($SetE$) might at first be surprising, one can quickly become desensitised to it. (One can find oneself thinking:

17 Cf. McGee (1997: 63–4).

18 See Jech (1973: 45, 2003: 250).

19 Schröder–Bernstein is provable in the deductive second-order deductive system (see Shapiro 1991: 102).

20 Thanks to Neil Barton for suggesting this possible objection to (*zfcu:ur*). See also Menzel (2014).

'well, ZFC_2 was *nearly* externally categorical anyway, so the result is not so suspris-ing'.) We hope that using $CSP(PE)$ will keep you *sensitised* to the phenomenon of total internal categoricity. So, at the risk of repetition: $CSP(PE)$ *is astonishingly weak*. After all, for any $n \geq 0$, the following theory is consistent:

$$\exists PE[CSP(PE) \land \text{'there are exactly } (2\uparrow\uparrow n)\text{-many } Ps\text{'}]$$

11.B Connection to Martin

In this second philosophical appendix, we relate our Theorem 11.4 to an *informal* argument, suggested by Martin, that the iterative concept of (pure) sets is *totally* categorical (rather than quasi-categorical). Likewise, we shall use this opportunity to round out our discussion of Martin, started in §8.5, by mentioning his views on second-order logic, as well as his position on the instantiation of the set concept.

Martin's informal argument begins by considering 'structures M_1 and M_2 both of which meet the strong concept of pure set'.[21] Following Zermelo's proof of his quasi-categoricity Theorem 8.8, Martin informally argues that M_1 and M_2 must at least be quasi-isomorphic. He then also assumes that the stages of M_1 and M_2 'are well ordered and that there an absolute infinity of them'; this allows him to argue, in effect, that any two absolutely infinite hierarchies must run as far as each other. And so he concludes 'that M_1 and M_2 themselves are uniquely isomorphic'.[22]

However, it is not clear what M_1 and M_2 *are*. Martin does not say much positive here: he deliberately keeps 'the notion of *structure* loose and informal', though he is explicit that he does not 'want to rule out proper class domains', or even assume 'that structures need themselves be objects in a strict sense'.[23] Additionally, Martin says very little about how to make sense of the assumption that there are absolutely infinitely many stages, although he cites Cantor's discussion of the absolute infinite as a precedent.[24]

One can, however, get around these difficulties by offering an 'internalist take' on Martin's argument. Internalists can read Martin's talk of 'structures' as short-hand for 'internal-structures'. Then, when Martin tells us that M_1 and M_2 are ab-solutely infinite structures satisfying the strong concept of pure set, internalists can take Martin to be assuming (at least) that $CSP(P_1E_1)$ and $CSP(P_2E_2)$. For, what-ever exactly 'absolute infinity' amounts to, it should at least entail that there are as many pure sets as objects simpliciter, and so it should entail (*csp:many*). Internal-ists can then recast Martin's informal argument as a deductive proof of total internal categoricity, i.e. of Theorem 11.4.

[21] Martin (2001: 10).
[22] Martin (2001: 11).
[23] See Martin (2001: 9).
[24] See the end of Martin (2015: §3).

At this point, we should mention a distinctive element of Martin's overall approach to categoricity: he is agnostic on whether the concept of set is instantiated at all. So the most Martin extracts from his categoricity argument is a conditional: *if* the concept of set is instantiated, *then* (low-level) statements are determinate. Martin leaves it open whether we should endorse the antecedent or deny the consequent.[25] But this is easily accommodated by the 'internalist take' on Martin's argument: we can ask whether or not $\exists PE\, \text{CSP}(PE)$; if not, then every statement of the form $\forall PE(\text{SP}(PE) \rightarrow \varphi)$ will be vacuous.

Of course, someone might resist the 'internalist take' on Martin's argument, either because they want the argument to remain *informal*, or because they want the argument to have the flavour of an *external* categoricity result. Fair enough; our only point is that the 'internalist take' deals neatly with all of the earlier worries concerning the argument's informality.

Additionally, someone might worry that the 'internalist take' on Martin's argument involves (deductive) second-order logic. Indeed Martin—like Parsons and McGee—attempts to obtain the force of categoricity considerations without a commitment to second-order logic *per se*.[26] On this issue, though, the balance of considerations seems to us similar to that discussed in §10.A.

11.C Internal quasi-categoricity for SP

In this technical appendix, we prove the internal quasi-categoricity of the sets, in the sense of $\text{SP}(PE)$. We start by defining a formula which says that R is a quasi-isomorphism between P_1 with its membership relation E_1, and P_2 with its membership relation E_2:

$$\begin{aligned}
\text{QuasiP}_{1\triangleright 2}(R) := {} & \forall v \forall y\,(R(v,y) \rightarrow [P_1(v) \wedge P_2(y)]) \wedge && (qp{:}1)\\
& \forall v \forall y \forall z\,([R(v,y) \wedge R(v,z)] \rightarrow y = z) \wedge && (qp{:}2)\\
& \forall y \forall v \forall x\,([R(v,y) \wedge R(x,y)] \rightarrow v = x) \wedge && (qp{:}3)\\
& \forall v \forall y \forall x \forall z\,([R(v,y) \wedge R(x,z)] \rightarrow [v\, E_1\, x \leftrightarrow y\, E_2\, z]) \wedge && \\
& && (qp{:}4)\\
& \forall v\,(\exists y R(v,y) \rightarrow (\forall x \subseteq_1 \text{ord}_1(v)) \exists z R(x,z)) \wedge && (qp{:}5)\\
& \forall y\,(\exists v R(v,y) \rightarrow (\forall z \subseteq_2 \text{ord}_2(y)) \exists x R(x,z)) \wedge && (qp{:}6)\\
& [(\forall v : P_1) \exists y R(v,y) \vee (\forall y : P_2) \exists v R(v,y)] && (qp{:}7)
\end{aligned}$$

Roughly, $(qp{:}1)$ says that R is a relation with domain P_1 and P_2; $(qp{:}2)$ says that R is functional; $(qp{:}3)$ says that R is injective; $(qp{:}4)$ says that R preserves structure;

[25] See Martin (2001: 11, 15, 2015: §3).

[26] See the remark 'it does not depend upon any obscure notion of arbitrary set of numbers' in Martin (2001: 11) and the remarks on full second-order logic in Martin (2015: §2).

and $(qp{:}7)$ says that R exhausts one of P_1 and P_2. We retain the definition of $\mathrm{ord}(x)$ from §8.B (merely tacitly relativising it to E). As such, $x \subseteq_1 \mathrm{ord}_1(v)$ indicates that x 'enters' the P_1-hierarchy no later than v, so that $(qp{:}5)$ says that R's domain is an initial segment of the P_1-hierarchy, and $(qp{:}6)$ indicates that R's range is an initial segment of the P_2-hierarchy.

We just mentioned that we retain the definition of $\mathrm{ord}(x)$ from §8.B. More generally, throughout our proof of quasi-categoricity, we freely invoke the results proved in §8.B concerning SP_2, since these deductions obviously carry over to the internalised context. Here is our proof.

Theorem (Internal Quasi-Categoricity of SP, 11.1):

$$\vdash \forall P_1 \forall E_1 \forall P_2 \forall E_2 \, ([\mathrm{SP}(P_1 E_1) \wedge \mathrm{SP}(P_2 E_2)] \rightarrow \exists R \, \mathrm{QuasiP}_{1 \triangleright 2}(R))$$

Proof. We assume $\mathrm{SP}(P_1 E_1)$ and $\mathrm{SP}(P_2 E_2)$, and start by defining two formulas:

$$\begin{aligned} \Lambda(X, v, y) &:= P_1(v) \wedge P_2(y) \wedge \\ &\qquad (\forall x \, E_1 \, v)(\exists z \, E_2 \, y) X(x, z) \wedge \\ &\qquad (\forall z \, E_2 \, y)(\exists x \, E_1 \, v) X(x, z) \\ \Gamma(X) &:= \forall v \forall y (\Lambda(X, v, y) \rightarrow X(v, y)) \end{aligned}$$

Note that by definition, Λ is closed upwards under inclusion, i.e. reusing the terminology introduced in §10.B: $(\Lambda(X, v, y) \wedge X \sqsubseteq Y) \rightarrow \Lambda(Y, v, y)$. Now, by Comprehension, there is an R such that:

$$\forall v_1 \forall v_2 \, (R(v_1, v_2) \leftrightarrow \forall X [\Gamma(X) \rightarrow X(v_1, v_2)])$$

Intuitively, R is the intersection of all the Γ's, so that $R \sqsubseteq X$ for any X such that $\Gamma(X)$. This will be our required R.

First let us show that $\Gamma(R)$. So suppose that v, y are such that $\Lambda(R, v, y)$; we must show that $R(v, y)$. Let X be such that $\Gamma(X)$; we must show that $X(v, y)$. But since $R \sqsubseteq X$ and $\Lambda(R, v, y)$ we have that $\Lambda(X, v, y)$. Then from $\Gamma(X)$ we conclude that $X(v, y)$. Hence indeed $\Gamma(R)$ holds.

This in turn directly implies the *left-to-right* direction of the following:

$$\forall v_1 \forall v_2 \, (\Lambda(R, v_1, v_2) \leftrightarrow R(v_1, v_2)) \tag{11.1}$$

For the *right-to-left* direction, suppose that $\neg \Lambda(R, a_1, a_2)$. By Comprehension, there is Q such that $\forall x \forall y (Q(x, y) \leftrightarrow (R(x, y) \wedge \neg [x = a_1 \wedge y = a_2])$, so that $\neg Q(a_1, a_2)$. Since R is the intersection of all the Γ's, it suffices to argue that $\Gamma(Q)$. For, suppose that $\Lambda(Q, v, y)$. Since $Q \sqsubseteq R$ we have $\Lambda(R, v, y)$ and since $\Gamma(R)$ we have $R(v, y)$. If $Q(v, y)$ failed, then we would have to have that $v = a_1$ and

$y = a_2$, contradicting that $\Lambda(R, v, y)$ while $\neg\Lambda(R, a_1, a_2)$. This finishes the argument for (11.1).

We can now set to work proving that each conjunct of (*qp:1*)–(*qp:7*) holds.

(*qp:1*). This follows from (11.1) and the fact that $\Lambda(X, v, y)$ implies by definition that $P_1(v) \wedge P_2(y)$.

(*qp:2*). For reductio, suppose a_1 is such that $\exists y \exists z(y \neq z \wedge R(a_1, y) \wedge R(a_1, z))$. So $P_1(a_1)$ by (*qp:1*) and so we can assume that a_1 is ord_1-minimal by Proposition 8.25(1). So for some a_2, b_2 we have $a_2 \neq b_2 \wedge R(a_1, a_2) \wedge R(a_1, b_2)$. Since $a_2 \neq b_2$, by (*qp:1*) and (*sp:ext*) there is some $P_2(c_2)$ such that $c_2 \mathrel{E_2} a_2 \leftrightarrow c_2 \not\mathrel{E_2} b_2$; without loss, assume $c_2 \mathrel{E_2} a_2$ and $c_2 \not\mathrel{E_2} b_2$. Since $R(a_1, a_2)$, by (11.1) we have $\Lambda(R, a_1, a_2)$ and hence $(\forall u_2 \mathrel{E_2} a_2)(\exists u_1 \mathrel{E_2} a_1)R(u_1, u_2)$. So there is some $c_1 \mathrel{E_1} a_1$ such that $R(c_1, c_2)$. Similarly, $\Lambda(R, a_1, b_2)$, so there is some $d_2 \mathrel{E_2} b_2$ such that $R(c_1, d_2)$. Now $c_2 \neq d_2$ because $c_2 \not\mathrel{E_2} b_2$. So $c_2 \neq d_2 \wedge R(c_1, c_2) \wedge R(c_1, d_2)$ with $c_1 \mathrel{E_1} a_1$, contradicting the ord_1-minimality of a_1.

(*qp:3*). Exactly similar.

(*qp:4*). Suppose $R(a_1, a_2)$ and $R(b_1, b_2)$. If $b_1 \mathrel{E_1} a_1$, then by (11.1) there is some $c_2 \mathrel{E_2} a_2$ with $R(b_1, c_2)$, and $b_2 = c_2$ by (*qp:2*), so that $b_2 \mathrel{E_2} a_2$ as required. If $b_2 \mathrel{E_2} a_2$, then (*qp:3*) yields that $b_1 \mathrel{E_1} a_1$.

(*qp:5*). We work by induction on levels$_1$. Let ℓ be a level$_1$, and suppose for induction that:

$$(\forall \ell' \mathrel{E_1} \ell)\forall v \forall y([\text{ord}_1(v) = \ell' \wedge R(v, y)] \rightarrow (\forall x \subseteq_1 \ell')(\exists z \subseteq_2 \text{ord}_2(y))R(x, z))$$

We will show:

$$\forall v \forall y([\text{ord}_1(v) = \ell \wedge R(v, y)] \rightarrow (\forall x \subseteq_1 \ell)(\exists z \subseteq_2 \text{ord}_2(y))R(x, z))$$

Fix v and y such that $\text{ord}_1(v) = \ell$ and $R(v, y)$, and fix $x \subseteq_1 \ell$. Fix $x' \mathrel{E_1} x$; then there is $v' \mathrel{E_1} v$ with $x' \subseteq_1 \text{ord}_1(v') \mathrel{E_1} \ell$, by Proposition 8.25(4). Since $R(v, y)$, we have $\Lambda(R, v, y)$ by (11.1), so there is some $y' \mathrel{E_2} y \subseteq_2 \text{ord}_2(y)$ such that $R(v', y')$. By the induction hypothesis, instantiated with $\text{ord}_1(v')$, v' and y', there is $z' \subseteq_2 \text{ord}_2(y')$ such that $R(x', z')$. Moreover, the link between x' and z' is bijective, given (*qp:2*)–(*qp:3*). Further, we may argue that $z' \mathrel{E_2} \text{ord}_2(y)$: for, $y' \mathrel{E_2} y$ implies $\text{ord}_2(y') \mathrel{E_2} \text{ord}_2(y)$ by Proposition 8.25(3), and together with $z' \subseteq_2 \text{ord}_2(y')$ and the supertransitivity of $\text{ord}_2(y)$, we obtain $z' \mathrel{E_2} \text{ord}_2(y)$. So, for any z', if $(\exists x' \mathrel{E_1} x)R(x', z')$ then $z' \mathrel{E_2} \text{ord}_2(y)$. By Levelling, there is therefore a z in P_2 whose members are exactly such z'. By construction $\Lambda(R, x, z)$, so that $R(x, z)$ by (11.1).

This completes the proof by induction. Now (*qp:5*) follows.

(*qp:6*). Exactly similar.

(*qp*:7). Suppose (*qp*:7) is false for reductio. So there is some element of P_1 not in *R*'s domain. By (*qp*:5) its level$_1$ is also not in *R*'s domain. So there is least level$_1$ ℓ_1 not in *R*'s domain. By minimality, all lower levels$_1$ are in *R*'s domain and by (*qp*:5) any sets$_1$ of lower order$_1$ than ℓ_1 are in *R*'s domain. By a parallel argument using (*qp*:5), there is some least ℓ_2 not in *R*'s range, and all members$_2$ of ℓ_2 are in *R*'s range. So $\Lambda(R, \ell_1, \ell_2)$ and hence $R(\ell_1, \ell_2)$ by (11.1), contradicting our choice of ℓ_1 and ℓ_2. □

Now we use Theorem 11.1 to prove the Quasi-Intolerance of SP:

Theorem (Quasi-Intolerance of SP 11.2): *For any formulas* $\pi(PE\ell)$ *and* $\varphi(PE\ell)$ *with all first-order quantifiers bound to elements of* ℓ *and all second-order quantifiers bound to subsets of* ℓ, *we have*

$$\vdash \forall PE \forall \ell \left(\left[\mathrm{SP}(PE) \wedge \mu_\pi(PE\ell) \right] \rightarrow \varphi(PE\ell) \right) \vee$$
$$\forall PE \forall \ell \left(\left[\mathrm{SP}(PE) \wedge \mu_\pi(PE\ell) \right] \rightarrow \neg\varphi(PE\ell) \right)$$

Proof. We work deductively in our second-order logic. Suppose $\mathrm{SP}(P_1E_1)$ and $\mu_\pi(P_1E_1\ell_1)$ and $\mathrm{SP}(P_2E_2)$ and $\mu_\pi(P_2E_2\ell_2)$. Since $\mathrm{SP}(P_1E_1)$ and $\mathrm{SP}(P_2E_2)$, by the previous theorem, there is *R* such that $\mathrm{QuasiP}_{1\triangleright 2}(R)$. By (*qp*:7), we may assume without loss of generality that *R*'s domain is all of P_1. Let P_2^* be the restriction of P_2 to everything in *R*'s range, and let E_2^* be the restriction of E_2 to P_2^*; then *R* is an isomorphism between P_1E_1 and $P_2^*E_2^*$. Choose ℓ_2' such that $R(\ell_1, \ell_2')$. Emulating the proof of Theorem 10.3, since $\mu_\pi(P_1E_1\ell_1)$ we have $\mu_\pi(P_2^*E_2^*\ell_2')$ and so $\pi(P_2^*E_2^*\ell_2')$. But since the quantifiers in $\pi(P_2^*E_2^*\ell_2')$ are all bounded to ℓ_2' in the manner stipulated in the hypothesis, restriction to P_2^* and E_2^* makes no difference, so that $\pi(P_2E_2\ell_2')$. (Roughly, this is because the quantifiers are bound to ℓ_2' and P^* is supertransitive by (*qp*:6); if so desired, this step of the argument can be formalised along the lines of Proposition 8.28.) Because $\pi(P_2E_2\ell_2')$ and by $\mu_\pi(P_2E_2\ell_2)$, we have $\ell_2 \subseteq_2 \ell_2'$ and hence $P_2^*(\ell_2)$ by supertransitivity. Again, since the quantifiers are bounded, we have that $\pi(P_2E_2\ell_2)$ implies $\pi(P_2^*E_2^*\ell_2)$, so that by $\mu_\pi(P_2^*E_2^*\ell_2')$ we have $\ell_2 = \ell_2'$. Since *R* is an isomorphism between P_1E_1 and $P_2^*E_2^*$ and since $R(\ell_1, \ell_2)$, we can again emulate the proof of Theorem 10.3 to obtain that $\varphi(P_1E_2\ell_1)$ iff $\varphi(P_2^*E_2^*\ell_2)$. Again, since the quantifiers are bounded, this happens iff $\varphi(P_2E_2\ell_2)$. □

11.D Total internal categoricity for CSP

Our next project is to lift the preceding result into full-fledged internal categoricity result. We first define a formula stating that *R* is an isomorphism of sets:

$$\begin{aligned}\text{IsoP}_{1\rhd 2}(R) :=& \forall v \forall y\,(R(v,y) \rightarrow [P_1(v) \land P_2(y)]) \land & (11.2)\\ & (\forall v : P_1)\exists! y R(v,y) \land (\forall y : P_2)\exists! v R(v,y) \land & (11.3)\\ & \forall v \forall y \forall x \forall z\,([R(v,y) \land R(x,z)] \rightarrow [v\, E_1\, x \leftrightarrow y\, E_2\, z]) & (11.4)\end{aligned}$$

We next need a specific version of Cantor's Theorem:

Proposition 11.8: CSP_{int} *deductively proves the following. Where ℓ is any level, temporarily define:*

$$A_\ell(x) := (\exists \ell' : L)(\ell' \in \ell \land x \subseteq \ell')$$
$$B_\ell(x) := x \subseteq \ell$$

Then there is no second-order injection from B_ℓ to A_ℓ.

Proof. For reductio, suppose g is such an injection. By Comprehension, use g to build a second-order surjection f whose domain is A_ℓ and whose range is B_ℓ. By Comprehension and (*sp:levels*), we have a set $d = \{x \in \ell : x \notin f(x)\}$. Since $d \subseteq \ell$, we have $B_\ell(d)$. Since f is a surjection, $d = f(a)$ for some a such that $A_\ell(a)$. Since $A_\ell(a)$, we have some $\ell' \in \ell$ such that $a \subseteq \ell'$, so that $a \in \ell$ by supertransitivity of ℓ. So, absurdly: $a \in d$ iff $a \notin f(a)$ iff $a \notin d$. □

The desired result is now straightforward:

Theorem (Internal categoricity of CSP, 11.4):

$$\vdash \forall P_1 E_1 P_2 E_2\,([\text{CSP}(P_1 E_1) \land \text{CSP}(P_2 E_2)] \rightarrow \exists R\, \text{IsoP}_{1\rhd 2}(R))$$

Proof. Assume that $\text{CSP}(P_1 E_1)$ and $\text{CSP}(P_2 E_2)$. By Theorem 11.1, we have some R such that $\text{QuasiP}_{1\rhd 2}(R)$. For reductio, suppose that R exhausts P_1 but not P_2. Let ℓ be the least level_2 not in R's range. Then R is the graph of an injection h from P_1 to A_ℓ by (*qp:6*). Further, the identity map f is an injection from B_ℓ to the objects. Further, by (*csp:many*), there is an injection g from the objects to P_1. Hence the composition $h \circ g \circ f$ is an injection from B_ℓ to A_ℓ, contrary to the previous proposition. □

In §11.4, we commented that the total categoricity and intolerance of CSP_{int} might seem rather startling. On a technical level, though, we should perhaps emphasise that this kind of result is not really specific to set theory. To see this, consider two well-known external categoricity results:[27]

[27] See Marker (2002: Theorem 2.4.1 p.48), Jech (2003: Theorem 4.3 p.38), and Dasgupta (2014: 160, 165). The canonical instance of (i) is the rational numbers, and the canonical instance of (ii) is the real numbers. Here 'complete' signals the existence of suprema, as in the usual axiomatizations of the real numbers, and 'separable' means that there is a countable dense linear order which intersects any interval in the original linear order.

(i) all countable dense linear orders without endpoints are isomorphic;
(ii) all complete separable dense linear orders without endpoints are isomorphic.

Now consider a theory CLDO which consists of:

(a) the axioms of dense linear orders without endpoints;
(b) an axiom much like (*csp:many*) stating that domain of the linear order is bijective with the objects; and
(c) a 'disjunctive' axiom stating that the linear order is either countable or both complete and separable.

The standard proofs of external categoricity carry over to show that CLO_{int} is internally categorical and hence intolerant. Now, clearly CLO_{int} itself fails to decide whether the order is countable, or both complete and separable, in that either option is deductively consistent with CLO_{int}. Still, CLO_{int} is intolerant, because we have outsourced to the universe the decision as to which option holds: if there are only countably many things then (b) rules out that the linear order is complete and separable, whereas if there are uncountably many things then (b) rules out that the linear order is countable. So CLO_{int} is intolerant, because it explicitly connects the size of the order with the size of the universe. Likewise, CSP_{int} is intolerant because axiom (*csp:many*) says that there are just as many pure sets as there are objects.

11.E Internal quasi-categoricity of ordinals

We have established the main results discussed in this chapter. Having come this far, though, we can prove the internal (quasi-)categoricity of the *ordinals* with almost no extra effort. We start by internalising a theory of ordinals, relative to a one-place relation-variable O and a two-place relation-variable $<$:

$$\begin{aligned} \mathrm{O}(O,<) := {} & \forall x \forall y (x < y \rightarrow [O(x) \wedge O(y)]) \wedge && (\textit{o:res}) \\ & (\forall x, y : O)(x < y \vee x = y \vee y < x) \wedge && (\textit{o:tri}) \\ & \forall X[(\forall x : O)[(\forall y < x)X(y) \rightarrow X(x)] \rightarrow (\forall x : O)X(x)] && (\textit{o:ind}) \end{aligned}$$

And now we can prove internal quasi-categoricity, i.e. intuitively that there is a second-order isomorphism between initial segments of two internal-structures of ordinals, $\mathrm{O}(O_1, <_1)$ and $\mathrm{O}(O_2, <_2)$:

Theorem 11.9 (Internal quasi-categoricity for ordinals):

$$\vdash \forall O_1 \forall <_1 \forall O_2 \forall <_2 ([\mathrm{O}(O_1, <_1) \wedge \mathrm{O}(O_2, <_2)] \rightarrow \exists R\, \mathrm{QuasiO}_{1 \triangleright 2}(R))$$

where we define:

$$\begin{aligned}
\text{QuasiO}_{1 \triangleright 2}(R) := {} & \forall v \forall y\, (R(v,y) \rightarrow [O_1(v) \wedge O_2(y)]) \wedge && (qo{:}1)\\
& \forall v \forall y \forall z\, ([R(v,y) \wedge R(v,z)] \rightarrow y = z) \wedge && (qo{:}2)\\
& \forall y \forall v \forall x\, ([R(v,y) \wedge R(x,y)] \rightarrow v = x) \wedge && (qo{:}3)\\
& \forall v \forall x \forall y \forall z\, ([R(v,y) \wedge R(x,z)] \rightarrow [v <_1 x \leftrightarrow y <_2 z]) \wedge && \\
& && (qo{:}4)\\
& \forall v\, (\exists y R(v,y) \rightarrow (\forall x <_1 v) \exists y R(x,y)) \wedge && (qo{:}5)\\
& \forall y\, (\exists v R(v,y) \rightarrow (\forall z <_2 y) \exists v R(v,z)) \wedge && (qo{:}6)\\
& [(\forall x : O_1) \exists y R(x,y) \vee (\forall y : O_2) \exists x R(x,y)] && (qo{:}7)
\end{aligned}$$

Proof. Exactly as in Theorem 11.1, we define two formulas:

$$\begin{aligned}
\Lambda(X, v, y) := {} & O_1(v) \wedge O_2(y) \wedge \\
& (\forall x <_1 v)(\exists z <_2 y) X(x,z) \wedge \\
& (\forall z <_2 y)(\exists x <_1 v) X(x,z) \\
\Gamma(X) := {} & \forall v \forall y (\Lambda(X, v, y) \rightarrow X(v,y))
\end{aligned}$$

Using Comprehension, and reasoning as in Theorem 11.1, we have an R such that:

$$\begin{aligned}
& \forall v_1 \forall v_2 (R(v_1, v_2) \leftrightarrow \forall X[\Gamma(X) \rightarrow X(v_1, v_2)] \\
& \forall v_1 \forall v_2 (R(v_1, v_2) \leftrightarrow \Lambda(R, v_1, v_2)) && (11.5)
\end{aligned}$$

We now prove each conjunct of (*qo:1*)–(*qo:7*), following the proofs of (*qp:1*)–(*qp:7*). Only a couple of the clauses merit distinct comment.

(*qo:2*). For reductio, suppose a_1 is such that $\exists y \exists z (y \neq z \wedge R(a_1, y) \wedge R(a_1, z))$. So $O_1(a_1)$ by (*qo:1*) and so we can assume that a_1 is $<_1$-minimal by (*o:ind*). So for some a_2, b_2 we have $a_2 \neq b_2 \wedge R(a_1, a_2) \wedge R(a_1, b_2)$. Since $a_2 \neq b_2$, by (*qo:1*) and (*o:tri*) either $a_2 <_2 b_2$ or $b_2 <_2 a_2$; without loss of generality, assume $b_2 <_2 a_2$. Since $R(a_1, a_2)$, by (11.5) we have $\Lambda(R, a_1, a_2)$ and so $(\forall z <_2 a_2)(\exists x <_1 a_1) R(x, z)$. So, there is some $b_1 <_1 a_1$ such that $R(b_1, b_2)$. Similarly, since $\Lambda(R, a_1, b_2)$ there is some $c_2 <_2 b_2$ such that $R(b_1, c_2)$. So $b_2 \neq c_2 \wedge R(b_1, b_2) \wedge R(b_1, c_2)$ with $b_1 <_1 a_1$, contradicting a_1's $<_1$-minimality.

(*qo:5*) *and* (*qo:6*). If $R(a_1, a_2)$, then $\Lambda(R, a_1, a_2)$ by (11.5). □

12
Internal categoricity and truth

So far, we have considered internalism about both arithmetic and set theory. The very idea of internalism is speculative. But in this third and final chapter on internalism, we will push into even more speculative terrain. The framing question is: *Can internalists say anything about mathematical truth?*

As we will quickly see, the question is independently interesting. However, it is also important to understanding the *literature* on internal categoricity. In particular, McGee explicitly claims that his Theorem 11.6 shows that every sentence concerning pure sets has a determinate truth value. If he is right, then presumably our Theorem 11.4 does the same. Equally, presumably Theorem 10.2 shows that every sentence concerning numbers has a determinate truth value. All of this would be extremely exciting, for reasons we shall explore in 12.1.

Unfortunately, though, McGee never explains *how* internal categoricity might yield the determinacy of truth. Moreover, considerations from §10.3 raise serious obstacles to the very idea that an internal categoricity result could ever tell us anything about truth. So the aim of this chapter is to ask whether internalists can eke out *something* here. Along the way, we will introduce internalism about *model theory*, and thereby attain a new understanding of Putnam's internal realism.

12.1 The promise of truth-internalism

We begin by both motivating an internalist's attempt to retain truth-talk, and showing why, if it succeeded, it would be an incredible thing. We will initially focus on the case of arithmetic, though the same points will apply to pure set theory.

There is an intuitive route from intolerance towards truth-talk. Using the notation introduced in §10.6, suppose that one person advances $\text{PA}_{\text{int}[\text{A}]} \wedge \varphi_{\text{int}[\text{A}]}$ and another person advances $\text{PA}_{\text{int}[\text{B}]} \wedge \neg\varphi_{\text{int}[\text{B}]}$. Even though they are using different arithmetical vocabularies, the intolerance of PA_{int} shows that, on pain of deductive inconsistency, both parties must hold that one of them is *right* and that the other is *wrong*. Or—making the obvious connection—only one of them said something *true* and the other said something *false*. In short, intolerance leads quite naturally to truth-talk.[1]

[1] This connects loosely with Price's suggestions that 'Truth is the grit that makes our individual opinions engage with one another', and that 'The distinguishing mark of genuine assertion is [...] that by de-

Suppose—*just for now*—that the internalist can develop this point further, and convincingly defend the following thought:

The truth-internalist manifesto. *The intolerance of arithmetic shows that every sentence in the language of arithmetic has a determinate truth value. Specifically, Theorem 10.3 shows that every arithmetical sentence has a determinate truth value.*

Delivering truth-internalism would be a remarkable achievement, in every sense.

To spell out why: anyone who thinks that every arithmetical sentence has a determinate truth value faces an obvious doxological question: *How are creatures like us capable of speaking an arithmetical language with that property?*[2] Modelists will offer an answer along these lines: *We use our language to pick out a particular isomorphism type; and the determinacy of truth follows from the elementary equivalence of isomorphic models.* But this is just what the *moderate* cannot say. She must, after all, reject modelism, for all the reasons given in Chapters 7–9. She does not think that our language picks out an isomorphism type. What gives the doxological question its bite, then, is the moderate's acceptance of the idea that truth values are not determined by some external thing, like an isomorphism type, but by something which we 'do, say, or think.'[3]

Once upon a time, the following answer to the question seemed available: *Arithmetical truth* JUST IS *deductive provability from a nicely specifiable theory of arithmetic; all we need to do is set down the right theory and master the deductive system.* Unfortunately, Gödel's incompleteness theorems show that this response is fatally flawed. Since any reasonable deductive system of arithmetic is incomplete, identifying arithmetical truth with anything computably enumerable means giving up on the idea that every arithmetical sentence has a determinate truth value.

At this point, though, the doxological question which we have just raised might start to become *desperate*, for any moderate who wants to claim that every arithmetical sentence has a determinate truth value. Indeed, it might well seem that it is simply *impossible* to answer the question without giving up on moderation.[4]

---

fault, difference is taken as a sign of *fault*' (2003: 169, 183). However, it is worth emphasising the differences. Price suggests that the concept of truth is exhausted by combining some kind of friction-generating norm with the disquotational concept of truth. But there is a general question one can put to Price: *When should there be disagreement, and why?* Price cannot simply dismiss this question, for it is clearly part of our concept of disagreement that it has a normative dimension: we can all think of cases of furious friction where the participants are merely talking past each other and so *should stop* disagreeing; or, conversely, of cases where people think they are smoothly agreeing when in fact they *should* be disagreeing. And, until he tells us how to answer that question, Price cannot claim to have exhausted the concept of truth. (Cf. Tiercelin 2013: 663–4.) Now, we do not want to say anything here about the concept of truth in *general*. However, our internalist can invoke Theorem 10.3 to explain *why* disagreement is required in the arithmetical case.

[2] This is the explicit focus of McGee (1997), and implicit in Dummett (1963).

[3] This locution, in this context, is due to McGee (1997: 60, 2000: 59–60).

[4] Cf. Button (2013: 56–7).

The promise of truth-internalism, then, is that it will answer the doxological question, using only doxological resources, without falling afoul of Gödelian incompleteness. For, the truth-internalist about arithmetic attempts to answer the doxological question as follows: *Our use of a particular deductive system—namely deductive second-order logic together with* DI_2*—settles (somehow) that every arithmetical sentence has a determinate truth value.*

Note that the truth-internalist's answer to the doxological question immediately raises a further question: *But what settles* WHICH *truth value each sentence has?* Given the deductive incompleteness of $\mathrm{PA}_{\mathrm{int}}$, the truth-internalist cannot think that the matter is always settled by considering what is deducible within $\mathrm{PA}_{\mathrm{int}}$. Equally, since she eschews semantic ascent, she cannot say that the matter is settled by the relationship between our words and the behaviour of things 'out there' (such as an intended model). So it seems like truth-internalists must accept the possibility that, *our use of a deductive system settles* THAT *every sentence is either true or false, but sometimes nothing settles* WHICH *it is.* This is rather disquieting.

The point is perhaps even more disquieting, if we consider truth-internalism about pure set theory rather than arithmetic. Truth-internalists about pure set theory will want to say: *our use of a deductive system—namely deductive second-order logic together with* $\mathrm{CSP}_{\mathrm{int}}$*—settles (somehow) that every (pure) set-theoretical sentence has a determinate truth value.* Quite remarkably, then, they think that using a system which is too weak to *decide* whether there is more than one pure set nonetheless settles *that* every pure-set-theoretical sentence is either true or false. (And in passing: if this really *is* a problem for truth-internalists who appeal to $\mathrm{CSP}_{\mathrm{int}}$, then it is equally a problem for McGee, who appeals to the much stronger theory $\mathrm{ZFCU}_{\mathrm{int}}$. We imagine this sort of worry: *McGee's approach must be faulty, since if it worked for* $\mathrm{ZFCU}_{\mathrm{int}}$*, it would also work for* $\mathrm{CSP}_{\mathrm{int}}$*, and that's ridiculous!*)

All this, however, just amounts to trading incredulity at the very idea of truth-internalism against grandiose promisory notes from the truth-internalist). We cannot make any real progress until we start to consider *how* one might progress from intolerance to the determinacy of truth.

In the next three sections, then, we unpack truth-internalism in three different ways. To repeat: all of this is extremely speculative. So we should also emphasise: even if *truth*-internalism fails, the internalism outlined in Chapters 10 and 11 may yet succeed.

12.2 Truth operators

We start with an attempt to defend truth-internalism via the introduction of truth *operators*. We will focus on arithmetic and the theory $\mathrm{PA}_{\mathrm{int}}$, but all of the points of this section apply equally to $\mathrm{CSP}_{\mathrm{int}}$.

Defining new operators

Let us say that an *arithmetical*^{Num} *sentence* is a sentence in PA_{int}'s signature whose quantifiers are *Num*-restricted. Where φ is any arithmetical^{Num} sentence, let $\varphi^*(NzS)$ be the formula which results by replacing each instance of '*Num*' with '*N*', each instance of '0' with '*z*', and every instance of '*Succ*' with '*S*'. Now, for each arithmetical^{Num} sentence φ, we define some new object language operators:

(a) $\mathsf{t}\varphi$ iff $\forall NzS(\text{PA}(NzS) \rightarrow \varphi^*(NzS))$
(b) $\mathsf{f}\varphi$ iff $\forall NzS(\text{PA}(NzS) \rightarrow \neg\varphi^*(NzS))$
(c) $\mathsf{i}\varphi$ otherwise

These explicit definitions allow us to restate our Intolerance Theorem 10.3 as follows: *for any arithmetical*^{Num} *sentence* φ, $\vdash (\mathsf{t}\varphi \vee \mathsf{f}\varphi)$.[5]

So far, this is just playing with definitions. The interesting philosophical question is whether the truth-internalist can legitimately gloss (a)–(c) as follows:

(a*) *it is true that* φ iff φ holds in all internal-structures
(b*) *it is false that* φ iff $\neg\varphi$ holds in all internal-structures
(c*) *it is indeterminate whether* φ otherwise

The right-hand-side of each of these claims simply amounts to offering the standard internalist gloss of 'φ holds in all internal-structures', as set out in §10.1. The left-hand-side, however, requires that these newly-defined operators have something to do with the intuitive notions of *truth* and *falsity*. (It is worth adding that these glosses (a*)–(c*) are obviously just an internalist's version of *supervaluationism*. We have encountered supervaluationism many times before—in §2.5, §7.2, and §8.4—precisely because it is is an extremely natural approach to truth and falsity in mathematical settings where one wants to leave room for indeterminacy, at least at the outset.)

We want, then, to consider two questions. First: are internalists allowed to gloss these defined operators in terms of *truth*? Second: would this vindicate truth-internalism? We consider these questions in that order.

Truth with language–object relations

There is an obvious in-principle barrier to the very *idea* of thinking that the operators t, f and i have anything to do with *truth*.

It is common to hold that the very *idea* of truth requires a *bona fide* language–object relation. Certainly this seems right outside of mathematics: the sentence

---

[5] Note: since the operators are explicitly defined, they add no expressive power to the language. Note, also, that they cannot be iterated. The operators are defined only for arithmetical^{Num} sentences. When φ is an arithmetical^{Num} sentence, the quantifiers in the sentence $\forall NzS(\text{PA}(NzS) \rightarrow \neg\varphi^*(NzS))$ are *not* *Num*-restricted. So $\mathsf{t}\varphi$ is *not* an arithmetical^{Num} sentence, and hence $\mathsf{tt}\varphi$ is ill-defined.

'Venus rotates' is true iff the planet Venus rotates, so the *sentence*'s truth value depends upon the behaviour of the *planet* Venus. Objects-modelists think that mathematical truth similarly requires a *bona fide* language–object relation: they think that an arithmetical sentence is true iff it is satisfied in certain $\mathcal{L}$-structures. And, quite generally, if truth and falsity *do* require a *bona fide* language–object relation, then glossing the defined object-language operators t and f in terms of *truth* is hopelessly wrong-headed.

This point is important, and we do not want to downplay its significance. Equally, though, we do not want to suggest that it is a knockdown objection. An analogy may help. Consider the thought: *if God is dead, then everything is meaningless.* That is too hasty: there may be a kind of meaning—maybe even fresh kinds of meaning—in the absence of God. Similarly: there may be a kind of truth—maybe even fresh kinds of truth—in the absence of language–object relations.

Unfortunately, then, the general point is as easy to state as it is unfathomable. The viability of this approach turns on an answer to the question: *What is truth, either in mathematics or elsewhere?* This will be a recurrent theme of the chapter. We know of no way around it, and we know of no way of answering it.

Expressive limitations

Still, at least for the sake of argument, let us allow the internalist to gloss her new operators t, f and i in terms of *truth*. This will allow her to rephrase Theorem 10.3 as follows:

(1) for any arithmetical^{Num} sentence φ, $\vdash$ (it is true that $\varphi \vee$ it is false that φ)

However, what the internalist wanted was *truth-internalism*, i.e. the conclusion:

(2) for any arithmetical^{Num} sentence φ, either φ is true or φ is false.

But, unfortunately for her, there is no direct path from (1) to (2).[6] Obviously (1) is different from (2), since (2) involves the sign '$\vdash$'. Now, the internalist might think that she can simply *delete* that sign, since she presumably accepts everything that she can prove. But simply deleting that sign would yield:

($1_{\#}$) for any arithmetical^{Num} sentence φ, either it is true that φ or it is false that φ

And this is not the same as (2); indeed, it is not even *grammatical*. The essential problem is that 'it is true that...' is an *operator* which applies to sentences, whereas '... is true' is a *predicate* which applies to objects. So one can say, for example, 'it is true that there are infinitely many twin prime pairs', but it is ungrammatical to say 'it is true that the Twin Primes Conjecture'. But an instance of ($1_{\#}$) would indeed be the ungrammatical sentence:

---

[6] Many thanks to Rob Trueman for convincing us of this, via this argument.

($1^*_\#$) either it is true that the Twin Primes Conjecture or it is false that the Twin Primes Conjecture.

The problem is simple. Even if internalists can legitimately gloss their operators in terms of truth and falsity, this does not even give them the resources to *formulate* truth-internalism, let alone endorse it.

12.3 Internalism about model theory and internal realism

The immediate upshot of the last section is that truth-internalism requires the use of truth-predicates, and hence it requires semantic ascent. But this might seem to threaten the very idea of truth-*internalism*. After all, as outlined in Chapters 10–11, internalism eschews semantic ascent altogether. In fact, we think that internalists *can* embark on a project of semantic ascent, without compromising their internalism. But this requires some real delicacy.

Putnam's internal realism

We take our guide here from Putnam:

> Nor does the [internalist] have to foreswear *forever* the notion of a model. He has to foreswear reference to models in his account of *understanding*; but, once he has succeeded in understanding a rich enough language to serve as a metalanguage for some theory T (which may itself be simply a sublanguage of the metalanguage, in the familiar way), he can define 'true in T' à la Tarski, he can talk about "models" for T, etc.[7]

We suggest that internalist should extract, from this passage, the following thought. Suppose we embrace internalism about *model theory itself*. Then nothing prevents us from talking about 'models of T', or truth, provided that this all takes *within* the (deductively understood) model theory.

That is the guiding idea, and we will elaborate upon it in this section. More generally, we think that internalism about model theory will deliver something like Putnam's *internal realism* in the philosophy of mathematics.[8] So it will help to lay down a few more of Putnam's remarks.

Putnam sketched only an extremely brief outline of his internal realism, as it might apply in the case of philosophy of mathematics.[9] This sketch occupies the

[7] Putnam (1980: 479). Putnam has 'intuitionist (or, more generally, the "nonrealist" semanticist)' where we have '[internalist]'.

[8] Cf. Parsons' (1990b: 39, 2008: 288) claim that there is some link between internal categoricity and Putnam's internal realism.

[9] Moreover in the general case, internal realism meant different things at different times to Putnam. For a general discussion, see Button (2013: chs.8–11).

last side or so of his 'Models and Reality', and it is worth quoting two chunks. First, Putnam outlined the problem:

> To adopt a theory of meaning according to which a language whose whole use is specified still lacks something—viz. its 'interpretation'—is to accept a problem which can only have crazy solutions. To speak as if *this* were my problem, 'I know how to use my language, but, now, how shall I single out an interpretation?' is to speak nonsense. Either the use *already* fixes the 'interpretation' or nothing can.[10]

Then, Putnam sketched his solution:

> [...] the metalanguage is completely understood, and so is the object language [...]. Even though the model referred to satisfies the theory, etc., it is 'unintended'; we recognize that it is unintended *from the description through which it is given* [...]. Models are not lost noumenal waifs looking for someone to name them; they are constructions within our theory itself, and they have names from birth.[11]

We do not suggest that Putnam had our internalism in mind when he wrote these passages: he cannot have done, for he never discussed internal categoricity. However, we think that an internalist can make very good sense of Putnam's claims.

From set theory to model theory

In Chapter 11, we outlined how to develop internalism about pure set theory. This straightforwardly allows us to develop internalism about model theory.

In common with almost every branch of mathematics, model theory is largely carried out 'informally': the proofs are discursive; they omit tedious steps; etc. However, we can easily make sense of the idea that, 'officially', model theory is implemented within set theory. After all, model-theorists freely use set-theoretic vocabulary and set-theoretic axioms to describe and construct models. Indeed, all of the definitions of Chapter 1 could in principle be rewritten formally, using the vocabulary of some (second-order) set theory.

So, in what follows, $\mathrm{MT}_{\mathrm{int}}$ will be some suitably internalised set theory, to be used for model theoretic purposes. We need not pin down its specifics, beyond insisting on the following:

(1) $\mathrm{MT}_{\mathrm{int}}$ deals with a pure set property, *Pure*, and a membership relation, $\in$.
(2) $\mathrm{MT}_{\mathrm{int}} \vdash \mathrm{CSP}_{\mathrm{int}}$, so that $\mathrm{MT}_{\mathrm{int}}$ is internally categorical with respect to its pure sets.
(3) $\mathrm{MT}_{\mathrm{int}}$ is a single sentence, so that we can continue to use it during internal categoricity results, in the form of conditionals like $\forall PE(\mathrm{MT}(PE) \rightarrow \ldots)$

[10] Putnam (1980: 481–2).

[11] Putnam (1980: 482).

(4) MT_{int} proves that there are infinitely many sets, so that it has the resources to carry out basic reasoning concerning arithmetic and syntax. (Note that CSP_{int} alone lacks this feature.)

As such, MT_{int} has the vocabulary and conceptual resources for developing model theory as a branch of pure mathematics. And our internalist about model theory states that model theory is 'officially' to be regarded as being implemented deductively within MT_{int}.

An internalist about model theory can now follow Putnam in saying that models, or $\mathcal{L}$-structures, are 'constructions within our theory itself'. To be clear, though, this claim is not a bit of constructivist *metaphysics*.[12] Rather, this claim is almost tautologous. The point is that the very definition of an $\mathcal{L}$-structure is offered within MT_{int}, and any talk of 'construction' of $\mathcal{L}$-structures is an heuristic shorthand for deductive work carried out within MT_{int}.

Dealing with 'interpretations'

At the risk of repetition, internalists understand MT_{int} in a purely *deductive* fashion. The focus on deduction may lead some to mistake internalism for *if-thenism*. But this would be a mistake, for all the reasons mentioned in §10.7. Internalism is not if-thenism, or formalism, or anything in the ballpark, since internalists about model theory affirm MT_{int}, and all its consequences, *unconditionally*.

Still, our internalist's insistence on deductively understanding MT_{int} raises an obvious question: *Every deductive theory has many interpretations if it has any; so how do you pin down* MT_{int}*'s* PARTICULAR *interpretation?* Ultimately, our internalist should reply with a question of her own: *Why should I* WANT *to pin down a particular interpretation of* MT_{int}*?* But that blunt rejoinder needs some unpacking.

Internalists have at least one good way to understand the claim that MT_{int} has many interpretations if it has any. Working within MT_{int}, they can prove a version of the Löwenheim–Skolem Theorem. As such, they can show that any second-order theory has many Henkin interpretations if it has any (as those notions are understood within MT_{int}). Moreover, this applies to the theory MT_{int} itself, considered as an object theory (within a metatheory of MT_{int} itself).[13] All of this is perfectly good pure mathematics. But, understood just *as* pure mathematics, internalists have no obvious reason to think that it has much philosophical import.

The obvious thought—made painfully familiar in Chapters 7 and 8—is that the existence of these Henkin interpretations poses some threat. But the threat must not be overstated. It is a threat to *modelism*, and only to modelism. The existence of multiple interpretations threatens modelists, since they think that our understand-

[12] Contrast this with our discussion of Putnam's internal realism in §2.4.

[13] Note: for Gödelian reasons, MT_{int} will not prove the existence of an $\mathcal{L}$-structure satisfying MT_{int}.

ing of a theory is somehow *inadequate* until we have pinned down certain interpretations, in the sense of (equivalence classes of) $\mathscr{L}$-structures. None of this, though, poses any threat to *internalism*. Having rejected modelism, internalists flat-out deny that our understanding of a theory is *inadequate* until we have pinned down certain $\mathscr{L}$-structures. They insist that *nothing is missing* from a deductively-specified understanding of MT_{int}.

Moreover, their insistence is not just idle posturing. Since $\text{MT}_{\text{int}} \vdash \text{CSP}_{\text{int}}$, we know that MT_{int} is totally internally categorical and hence totally intolerant for its 'pure' part. So, as in §11.5, internalists can and will argue that a 'merely' *deductive* understanding of MT_{int} is sufficient for the articulation of univocal model-theoretical concepts. With Putnam, then, internalists about model theory will say: *'I know how to use my language', and there is no further relevant issue concerning how to 'single out an interpretation'.*

Internalists about model theory might continue to bolster their position by turning from Putnam to Dummett. As internalists see matters, anyone who thinks that the existence of multiple Henkin structures as a *threat* is operating

> [...] with the notion of a model as if it were something that could be given to us independently of any description: as a kind of intuitive conception which we can survey in its entirety in our mind's eye, even though we can find no description which determines it uniquely. This has nothing to do with the concept of a model as that concept is legitimately used in mathematics. There is no way in which we can be 'given' a model save by being given a description of that model.[14]

In particular, according to internalists about model theory, the description of an $\mathscr{L}$-structure should always be given within MT_{int}. So to be 'given' a model is to be given a description in a theory we *already* understand. As such, she concludes, Henkin models pose no threat to our understanding of MT_{int}.

What to say about non-standard models

The preceding remarks concerned $\mathscr{L}$-structures satisfying MT_{int}. Let us now consider what the internalist about model theory should say about models of arithmetic, particularly non-standard models.

Of course, there are no non-standard arithmetical internal-structures. That follows immediately from Theorem 10.2, as applied to the second-order theory PA_{int}.

However, working within MT_{int}, internalists can prove versions of the Compactness Theorem and Dedekind's external categoricity theorem. As such, within MT_{int}, they can conclude: all full structures satisfying PA_2 are isomorphic, but there are non-standard Henkin structures which satisfy PA_2. Indeed, from PA_{int},

[14] Dummett (1963: 191).

they can immediately infer that *Num*, 0 and *Succ* determine an $\mathscr{L}$-structure $\mathcal{M}$ which satisfies PA_2: its domain is the first-order *set* m such that $\forall x(Num(x) \leftrightarrow x \in m)$; then $0^{\mathcal{M}} = 0$, and $S^{\mathcal{M}}$ is the first-order function given by restricting *Succ* to m. So, *within* MT_{int}, we deductively prove: all full models of PA_2 are isomorphic.

Suppose, now, that internalists are 'given' some non-standard Henkin model, $\mathcal{N}$, of PA_2. They will repeat the sentiment from Dummett: there is no way to have been 'given' $\mathcal{N}$, except by description within MT_{int}. And now they will echo Putnam, and say that 'we recognize that it is unintended *from the description through which it is given*.'[15] For example, *within* MT_{int}, we can see from $\mathcal{N}$'s description that it contains a non-standard number, or that $N_1^{rel} \neq \wp(N)$, or some-such.

This is why internalists about model theory can join Putnam in saying that 'Models are not lost noumenal waifs looking for someone to name them; they are constructions within our theory itself, and they have names from birth.'[16] This is not a metaphysical expression of constructivism. Rather, models have 'names from birth', in the simple sense that models are 'given' by *description* in the model theory. As such, internalism about model theory provides an elegant explication of some of Putnam's most beautiful but cryptic remarks about the philosophy of mathematics.

Arithmetical truth with language-object relations

With all of this in place, there is nothing to stop *internalists about model theory* from becoming *truth-internalists about arithmetic.*

They should begin by offering supervaluational clauses, within MT_{int}, for each φ in the language of PA_2:

(a) φ is true iff $PA_2 \vDash \varphi$
(b) φ is false iff $PA_2 \vDash \neg\varphi$
(c) φ is indeterminate otherwise

Within MT_{int}, they can then affirm that every sentence φ in the language of PA_2 is either true or false, just by running through (within MT_{int}) the supervaluational reasoning that we ran through in the early parts of Chapter 7. Since internalists about model theory affirm MT_{int} unconditionally, they can can then conclude: *every arithmetical sentence has a determinate truth value.*

Against this, modelists will point out that MT_{int} is arithmetically incomplete (if it is consistent; see §§5.A and 10.3). Consequently, there are Henkin models of MT_{int} which satisfy $Con(MT_{int})$, within which it is right to say '$Con(MT_{int})$ is true', and arithmetically unsound Henkin models which satisfy $\neg Con(MT_{int})$, and so within which it is right to say '$Con(MT_{int})$ is false'. On these grounds, modelists will complain that a supervaluational definition of arithmetical truth *within* MT_{int} cannot

[15] Putnam (1980: 482).
[16] Putnam (1980: 482).

really achieve what it is supposed to.[17]

Internalists about model theory will reply—as above—that they can only understand claims about the existence of Henkin models *within* $\mathrm{MT}_{\mathrm{int}}$, so that the existence of such models is irrelevant to the understanding of $\mathrm{MT}_{\mathrm{int}}$ *itself*, and hence irrelevant to the supervaluational definition of arithmetical truth. So it goes.

Modelists would do better to shift their line of attack. They should instead complain that internalists have no right to regard the supervaluationally-defined 'truth-predicate' as having anything to do with *truth*. Granted, *within* $\mathrm{MT}_{\mathrm{int}}$, it is correct to describe this predicates as 'connecting words with objects'. But, according to modelists, this does not show that we are dealing with a *bona fide* language–object relation. The problem, according to them, is that the claim 'this predicate connects words with objects' is *merely* being presented as another deductively understood part of $\mathrm{MT}_{\mathrm{int}}$. As such, modelists will complain that internalists never *really* allow that mathematical language is beholden to non-linguistic entities.

We are not sure how to arbitrate this debate. Indeed, as in §12.2, the viability of *truth*-internalism about arithmetic—here, via plain vanilla internalism about model theory—turns on an answer to the question: *What is truth, either in mathematics or elsewhere?* As before, we do not have an answer to that question.

Expressive limitations

So far we have established the following. Internalism about set theory can be turned into internalism about model theory. This provides us with a new way to understand Putnam's internal realism in the philosophy of mathematics. Furthermore—though this is even more controversial—it *may* license truth-internalism about arithmetic.

Even then, though, plain vanilla internalism about model theory will not license *truth*-internalism about *model theory* or *set theory*. Whilst plain vanilla internalism about model theory might license truth-internalism about many theories which can be embedded into the cumulative hierearchy, it offers no progress at all towards the claim that every sentence of (pure) model theory has a determinate truth value. Indeed, this point is essentially entailed by the internalist's own responses to a modelist who asks about the 'intended' interpretation of $\mathrm{MT}_{\mathrm{int}}$. Nothing which counts (within $\mathrm{MT}_{\mathrm{int}}$) *as* a model of $\mathrm{MT}_{\mathrm{int}}$ could be 'intended', for we could prove (within $\mathrm{MT}_{\mathrm{int}}$) that there are objects outside of that model's domain. (This is essentially to repeat the point, that we should not hope for our set theory to be *externally* categorical; see §8.6.) But then, the internalist about model theory cannot gloss the truth of $\mathrm{MT}_{\mathrm{int}}$-sentences in terms of the notion of satisfaction she defines within $\mathrm{MT}_{\mathrm{int}}$.

[17] Cf. Meadows (2013: 539–40).

12.4 Truth in higher-order logic

In fact, truth-internalism about model theory *can* be developed using higher-order logic, and can *only* be developed there.

To begin, we must explain why truth-internalism about model theory requires higher-order logic. Suppose we stick within second-order logic, and aim to formulate a theory, $\mathrm{MT}^{+}_{\mathrm{int}}$, which intuitively consists in adding a truth-predicate, *Tr*, *for* $\mathrm{MT}_{\mathrm{int}}$ to $\mathrm{MT}_{\mathrm{int}}$. We did something like this in Lemma 9.2 of §9.A, where we showed that any consistent (first-order) theory can be augmented with a truth-predicate (for the old language). There, however, we governed the behaviour of the truth-predicate by adding infinitely many new axioms. In the context of internal categoricity results, however, this is unacceptable. We will need to say things like $\forall PE(\mathrm{MT}^{+}(PE) \rightarrow \dots)$, which requires that $\mathrm{MT}^{+}_{\mathrm{int}}$, like $\mathrm{MT}_{\mathrm{int}}$, should be a *single* sentence.[18]

Unfortunately, Tarski's Indefinability Theorem 9.1 entails that $\mathrm{MT}^{+}_{\mathrm{int}}$ cannot merely be a *second-order* theory. Suppose otherwise, for reductio. So let $\varphi(Tr, \overline{R})$ be some first- or second-order sentence, possibly containing some new vocabulary $\overline{R}$, such that, for each first- or second-order sentence σ in $\mathrm{MT}_{\mathrm{int}}$'s signature:

$$\begin{aligned} \mathrm{MT}_{\mathrm{int}} + \varphi(Tr, \overline{R}) &\vdash Tr(\ulcorner \sigma \urcorner) \leftrightarrow \sigma \\ \text{i.e., } \mathrm{MT}_{\mathrm{int}} &\vdash \varphi(Tr, \overline{R}) \rightarrow [Tr(\ulcorner \sigma \urcorner) \leftrightarrow \sigma] \\ \text{i.e., } \mathrm{MT}_{\mathrm{int}} &\vdash \forall X \forall \overline{Y}(\varphi(X, \overline{Y}) \rightarrow [X(\ulcorner \sigma \urcorner) \leftrightarrow \sigma]) \end{aligned}$$

It follows that, for each first- or second-order sentence σ in $\mathrm{MT}_{\mathrm{int}}$'s signature:

$$\mathrm{MT}_{\mathrm{int}} + \exists X \exists \overline{Y} \varphi(X, \overline{Y}) \vdash \forall X \forall \overline{Y}(\varphi(X, \overline{Y}) \rightarrow X(\ulcorner \sigma \urcorner)) \leftrightarrow \sigma$$

So $\mathrm{MT}_{\mathrm{int}} + \exists X \exists \overline{Y} \varphi(X, \overline{Y})$ has a truth predicate, and is therefore inconsistent by Tarski's Theorem; so $\mathrm{MT}^{+}_{\mathrm{int}}$ itself is inconsistent. As such, truth-internalists who want to discuss truth for $\mathrm{MT}_{\mathrm{int}}$ must look beyond second-order logic.

Given their internalism, there are not very many places they can look. They must continue to work deductively, in a system richer than $\mathrm{MT}_{\mathrm{int}}$. But we have just seen that the system cannot simply be enriched by adding some new second-order axioms. So the system must be richer because it adds new *logical* resources. Would-be truth-internalists must, then, ascend to higher-order logic.

Having explained that this is the *only* path available to them, we should add that it is not a path they should *fear*. Crucially, the (truth-)internalist always ultimately insists on working *deductively*. Consequently, any concerns one might have about the *semantics* for higher-order logics need not trouble her. And it would be an odd view, which permitted *second*-order deductive systems, but objected to n^{th}-order deductive systems, for any greater n.

[18] The same observation rules out Rayo and Uzquiano's (1999) approach.

In fact, it turns out that we can prove the existence of a satisfaction relation, for first-order and second-order sentences, using (just) *fifth*-order logic and a theory of arithmetic like PA_{int}.[19] We explain the details of this in §12.A. Having set up satisfaction, we can then prove the existence of 'supervaluational' truth and falsity properties. The details are not pretty. However, we can lay down clauses which roughly say that, for any sentence φ in SP_{int}'s signature whose quantifiers are *Pure*-restricted:

(a) $True(\ulcorner\varphi\urcorner)$ iff $\forall PE(\text{CSP}(PE) \rightarrow \varphi$ is satisfied when '*Pure*' is interpreted as P and '$\in$' is interpreted as $E)$

(b) $False(\ulcorner\varphi\urcorner)$ iff $\forall PE(\text{CSP}(PE) \rightarrow \neg\varphi$ is satisfied when '*Pure*' is interpreted as P and '$\in$' is interpreted as $E)$

(c) $Indeterminate(\ulcorner\varphi\urcorner)$ otherwise

Then, where $\vdash^5$ indicates that we are working deductively in fifth-order logic, for any sentence φ in SP_{int}'s signature whose quantifiers are *Pure*-restricted:

$$\text{PA}_{\text{int}} \vdash^5 True(\ulcorner\varphi\urcorner) \lor False(\ulcorner\varphi\urcorner)$$

All of this, though, is just mathematical logic. The question is whether it is of any *use* to a would-be truth-internalist about model theory or set theory. The question is, whether she can use this to claim (for example): *every first- or second-order sentence in pure set theory has a determinate truth value.* As usual, she will face two kinds of objection.

First: she will face the objection that she has no right to gloss her defined properties as having anything to do with *truth.* The dialectic here will be the same as before, and we have nothing much to add to it..

Second: she will face an expressive limitation. *Even if* we grant that she has genuinely got a notion of *truth* for first-order and second-order sentences, in so doing she has started to use third-, fourth- and fifth-order sentences. And she has no way, yet, to talk about truth *for those.*

At this point, the preceding considerations generalise.[20] As we show in §12.A, $(n+3)^{\text{th}}$-order deductive logic proves the existence of (supervaluational) truth and falsity properties, which enable us to state that any pure sentence of order $\leq n$ has a determinate truth value. But, on the inevitable pain of Tarski's Indefinability Theorem 9.1, we cannot find a truth-property which handles sentences of *every* order

[19] Recall that PA_{int} gives us numbers but, unlike PA_2, it does not require that *everything* is a number. We mention this, as it is important for us that we can build 'models' from arbitrary objects. We should also add that we are not claiming to have a proof that it is impossible to prove the existence of satisfaction relations for second-order sentences in either *third*-order logic or *fourth*-order logic; but the only ways that we know of which avoid ascending through so many types involve either expanding the theoretical resources or slightly restricting the result (see footnote 25, below).

[20] McGee (1997: 47–52) himself seems to reach for internal categoricity and *second-order* theories, *rather* than look to a *third-order* definition of satisfaction, on pain of the regress engendered by Tarski's Theorem. It is perhaps ironic that the regress should come back around.

simultaneously. As such, the truth-internalist cannot hope to say anything like: *every sentence of any order has a determinate truth value.*

There are, then, inevitable expressive limitations on what a truth-internalist can hope to achieve. But the way that they arise in the framework of higher-order logic gives more depth to the anti-internalist complaint, that the internalist's 'truth-properties' have nothing to do with *truth.* After all, she has a second-order-truth property, a third-order-truth property, an nth-order-truth property, and so on. But she does not have, and cannot have, a *single* property.[21]

This problem has been raised before, by Davidson, against the idea that Tarski somehow 'defined the concept of truth':

> Tarski did not define the concept of truth, even as applied to sentences. Tarski showed how to define a truth predicate for each of a number of well-behaved languages, but his definitions do not, of course, tell us what these predicates have in common.[22]

Against the truth-internalist's use of higher-order logic, this complaint can be pushed even further. It is not just that the truth-internalist has not told us what the various truth-properties have in common. On pain of Tarski's Indefinability Theorem, the truth-internalist *cannot even pick out* all and only the truth properties using a single formula. This is easily seen: if Φ held of all and only the truth properties, then $(\exists X : \Phi)X(v)$ would be a universal truth-predicate, again violating Tarski's Theorem. As such, the truth-internalist cannot make any claim of the form: *all and only the truth-properties have such-and-such a feature in common.* At least, she cannot do so without leaving higher-order deductive logic behind and so, it seems, without abandoning the distinctively *deductive* approach characteristic of internalism.

12.5 Two general issues for truth-internalism

In this chapter, we considered three ways for an internalist to attempt to recover certain amounts of truth-talk. We saw that she inevitably encounters two issues. First: there is the question of whether, focussing on deduction, she has any right to regard what she is doing as connecting with *truth.* Second: there are unavoidable expressive limitations concerning her ability to recover truth-talk.

We doubt that these observations are sufficient to *sink* truth-internalism. However, these two observations must be acknowledged and accommodated by anyone

[21] Linnebo and Rayo (2012) explore the idea of continuing through *transfinite* types. But even if we did countenance α^{th}-order sentences for arbitrary ordinals α, the same problem will persist: there will be no *single* truth property to deal with sentences of all (transfinite) orders. We also have some serious reservations about the very idea of transfinite type theory. Types are naturally regarded as essentially *grammatical,* and they arise from considering the ways in which different kinds of expressions combine to create entire sentences. But no sentence is infinitely long. So there is neither any way to 'get' an 'infinite type', nor any need for such a thing.

[22] Davidson (1990: 285).

who wants to use internal categoricity results to demonstrate something of the form 'every sentence of such-and-such a sort a determinate truth value'.

12.A Satisfaction in higher-order logic

In this appendix, we show how define satisfaction for an nth-order language within an $n+3^{\text{th}}$-order language. The result is essentially Tarski's, who also came up with all of the key ideas. Indeed, whilst we outlined a 'Tarskian' approach to *model-theoretic* semantics in §1.3, Tarski's own approach in 1933 was *higher-order*. We will follow his approach closely.

Syntax and deduction for higher-order languages

First-order languages have first-order variables: v, x, and so forth. Second-order languages retain first-order variables, but add second-order variables: V, X, and so forth. Ascending to n^{th}-order languages requires the addition of kth-order variables, for all $1 \leq k \leq n$.

A second-order variable has a certain number of 'gaps', to be filled by first-order variables (or other terms). A third-order variable has a certain number of 'gaps', but these must be filled by first-order or second-order expressions. So, where V is a particular third-order variable, we might have that $\mathrm{V}(X_1, X_2, y)$ is well-formed. The 'gaps', however, are regarded as suitable only for certain expressions; so in this example, $\mathrm{V}(X_1, y, y)$ will be ill-formed.[23] Generally, an $n+1^{\text{th}}$-order variable has 'gaps' which must be filled by lower-order expressions, at least one of which is n^{th}-order. An nth-order formula is one whose highest-order variable is n^{th}-order. An nth-order language is one which allows for k^{th}-order formulas for all $1 \leq k \leq n$.

Since our focus is on deduction, we will not provide a model-theoretic semantics for n^{th}-order languages. This can be done straightforwardly, though, by generalising the semantics for second-order sentences (see §§1.10–1.11).

The deductive system for n^{th}-order languages expands on the system laid down in §1.C. The rules for quantifiers are exactly as for lower-order languages: we simply need to keep track of the orders of the expressions. We also expand the Comprehension Schema, for each order $n \geq 2$:

The nth-order Comprehension Schema. *$\exists V \forall \bar{x}(\varphi(\bar{x}) \leftrightarrow V(\bar{x}))$, where V is an nth-order variable, for every formula $\varphi(\bar{x})$ which does not contain V and whose highest-order free variable is $n-1^{th}$-order.*

[23] Indeed, here we need to take care: if X_1 and X_2 are second-order variables with *different* number of places, then $\mathrm{V}(X_1, X_1, y)$ will also be ill-formed.

In these terms, our original Comprehension Schema from §1.C is the second-order Comprehension Schema (as one would hope). The deductive system for n^{th}-order logic then has k^{th}-order Comprehension and rules governing quantification for k^{th}-order variables, for all $1 \leq k \leq n$. We write $T \vdash^n \varphi$ to indicate that φ is deducible from T in this n^{th}-order deductive system.

Initial approach, and Tarski's trick

Our general aim is to show that an $n{+}3^{\text{th}}$-order deductive system allows us to define satisfaction for n^{th}-order languages. However, we will only prove a single instance of this: we will define satisfaction for second-order languages within a fifth-order deductive system. This illustrates the general point, but—as we will see—it is quite painful enough.

We assume that the non-logical signature, $\mathscr{L}$, of our target second-order language is *relational*. This is no real loss, since constants and function symbols can be simulated using predicates and identity. Moreover, the case that we most care about is the relational signature $\{Pure, \in\}$. For simplicity, we will also assume that the $\mathscr{L}$-formulas only contain the logical connectives $\neg$, $\wedge$, and $\forall$.

To define satisfaction, we must both be able to code the $\mathscr{L}$-formulas and perform induction on their syntax via their coding. To guarantee this, our definitions will take place within PA_{int}.[24] We write $\ulcorner\varphi\urcorner$ for the Gödel-number of φ, on some fixed Gödel-numbering (for a refresher of what this means, see §5.A).

Our approach to defining satisfaction is 'Tarskian', in the sense of §1.3, that we will handle quantifiers and variables using functions which assign 'interpretations' to variables. However, in the setting of higher-order logic (rather than model theory), we encounter a technical difficulty.

In defining satisfaction, we will need to deal with atomic formulas like $V(x_1, ..., x_m)$. An obvious approach would be to assign $\ulcorner V\urcorner$ to some second-order m-place entity $\sigma(\ulcorner V\urcorner)$, assign each of $\ulcorner x_1\urcorner, ..., \ulcorner x_m\urcorner$ to first-order entities $\sigma(\ulcorner x_1\urcorner), ..., \sigma(\ulcorner x_m\urcorner)$, and then say, roughly: *σ satisfies* $\ulcorner V(x_1, ..., x_m)\urcorner$ *iff* $\sigma(\ulcorner V\urcorner)(\sigma(\ulcorner x_1\urcorner), ..., \sigma(\ulcorner x_m\urcorner))$. However, since relation-variables can be of arbitrary numbers of places, we would have to offer distinct clauses for *any* natural number m. But then our definition of satisfaction would require infinitely many clauses, and we would have failed in our goal of defining satisfaction for second-order languages using a *single* formula (see §12.4).

Tarski both identified this obstacle and showed how to avoid it. His trick was to use second-order entities as proxies for sequences of first-order objects.[25] Intu-

[24] In this regard, we follow certain aspects of Tarski's work very closely. Tarski essentially uses a theory of types with an axiom of infinity. As we explained in §10.1, anyone prepared to add an axiom of infinity, i.e. DI_2, to higher-order logic must be happy to employ PA_{int}.

[25] See Tarski (1933: §4), particularly his discussion of the differences between languages of the 2nd and

itively, instead of thinking of an m-place relation as a set of m-tuples:

$$\{(a_{1,1}, \ldots, a_{1,m}), (a_{2,1}, \ldots, a_{2,m}), \ldots\}$$

Tarski suggested that we ascend an order, and consider the relation as a set of *functions* from the numbers 1 through m to the first-order entities, i.e.:

$$\{\{(1, a_{1,1}), \ldots, (m, a_{1,m})\}, \{(1, a_{2,1}), \ldots, (m, a_{2,m})\}, \ldots\}$$

But this idea will take a little spelling out.

Let σ be a function which maps each code of a first-order variable to some first-order object. (So, σ itself is a second-order entity.) Let a be the Gödel-number of some finite sequence of first-order variables, i.e. $a = \ulcorner x_{i_1}, \ldots, x_{i_m} \urcorner$. Then, working in PA_{int} and invoking Comprehension, we can show the existence of some unique relation R_a such that, for all v and every natural number j, we have $R_a(j, v)$ iff $v = \sigma(\ulcorner x_{i_j} \urcorner)$; so, intuitively, R_a is:

$$\{(1, \sigma(\ulcorner x_{i_1} \urcorner)), \ldots, (m, \sigma(\ulcorner x_{i_m} \urcorner))\}$$

More generally, if σ maps each code of a first-order variable to some first-order object, then there is (provably) a unique function, $\sigma^{\uparrow}$, with the following property: for any a which codes a finite sequence of first-order variables, $\sigma^{\uparrow}(a) = R_a$. That is: $\sigma^{\uparrow}$ maps each code of a finite sequence of first-order variables to a second-order object, which acts as a proxy for the sequence of assignments that σ itself makes for each of the variables in the sequence. (Note that $\sigma^{\uparrow}$ itself is a third-order entity.)

Now let Σ be a function which takes as inputs Gödel-numbers of second-order n-place variables, and outputs a third-order entity which a single 'gap' for a two-place relation on first-order objects. (So, Σ itself is a fourth-order entity.) We can now spell out satisfaction of along these lines: *Σ and σ together satisfy* $\ulcorner V(x_1, \ldots, x_m) \urcorner$ *iff* $\Sigma(\ulcorner V \urcorner)(\sigma^{\uparrow}(\ulcorner x_1, \ldots, x_m \urcorner))$. Since $\Sigma(\ulcorner V \urcorner)$ is of the same *logical* type for any $\ulcorner V \urcorner$, this does not require separate clauses for each value of m.

Defining the entities for satisfaction

We will now implement this idea in detail. In what follows, the following rough heuristic will be helpful: treat D as a *domain*; Λ as an *interpretation* of the $\mathscr{L}$-predicates; σ as an assignment to first-order variables; and Σ as an assignment to second-order variables.

3rd kinds. This trick is what leads us to do semantics for n^{th}-order logic in $n+3^{\text{th}}$-order logic. There are alternatives. For example, if we both (a) refuse to admit variables with more than (e.g.) three places, and also (b) employ *set*-theoretic resources on top of type-theoretic, then we need only climb to $n+1^{\text{th}}$-order logic. This is Linnebo and Rayo's approach (2012: 299–308), although they *also* allow for transfinite types (see footnote 21, above).

We say $Ass_1(D\sigma)$ iff σ is a function which maps every v which codes a first-order variable to a particular entity among D.

We say $Ass_2(D\Sigma)$ iff Σ is a function which maps every code of a relation-variable to a third-order entity such that, if v codes an m-place relation-variable and $\Sigma(v)(X)$, then X is a map from the numbers 1 through m to some subclass of D. This implements Tarski's trick, ensuring that (codes for) m-place relation-variables are ultimately handed m arguments.

Similarly, we say $Mod(D\Lambda)$ iff Λ is a function which maps every code of an $\mathscr{L}$-predicate to a third-order entity such that, if v codes an m-place predicate and $\Lambda(v)(X)$, then X is a map from the numbers 1 through m to some subclass of D

Clearly, all three of these formulas can be explicitly defined using PA_{int} in our fifth-order system. We now introduce a new abbreviation:

$$Tars(D\Lambda\Sigma\sigma) := Ass_1(D\sigma) \wedge Ass_2(D\Sigma) \wedge Mod(D\Lambda)$$

In effect, this tells us that D, Λ, σ, and Σ have the right shape to allow us to start considering Tarskian satisfaction for $\mathscr{L}$-formulas.

The satisfaction clauses

Our ultimate aim is to define a formula which says something roughly like this: *v codes an $\mathscr{L}$-sentence which is true on domain D with interpretation Λ*. To get there, we will define a fifth-order formula which roughly says this: *v codes an $\mathscr{L}$-formula which is satisfied with domain D, interpretation Λ, and assignments Σ and σ*. For readability, we will write this as $\langle D\Lambda\Sigma\sigma \mathrel{\sim\!\!\!\!\vdash} v\rangle$ rather than e.g. $V(D, \Lambda, \Sigma, \sigma, v)$; so, in what follows, we will treat '$\mathrel{\sim\!\!\!\!\vdash}$' as a fifth-order variable.

Our definition of $\langle D\Lambda\Sigma\sigma \mathrel{\sim\!\!\!\!\vdash} v\rangle$ simply involves running through each of Tarski's recursion clauses in a higher-order framework. Let $For(v)$ be a predicate indicating that v codes an $\mathscr{L}$-formula, and let us use '$(\forall \ulcorner x = y \urcorner : For)\ldots$', for example, to abbreviate 'for any v among For, if v codes some atomic identity claim relating variables x and y, then….' (Obviously all this can be spelled out precisely using PA_{int}.) Then we start by offering recursion clauses for atomic formulas:

$$(\forall \ulcorner x = y \urcorner : For)[\langle D\Lambda\Sigma\sigma \mathrel{\sim\!\!\!\!\vdash} \ulcorner x = y \urcorner\rangle \leftrightarrow \sigma(\ulcorner x \urcorner) = \sigma(\ulcorner y \urcorner)] \qquad (ho{:}{=})$$

$$(\forall \ulcorner V(\bar{x}) \urcorner : For)[\langle D\Lambda\Sigma\sigma \mathrel{\sim\!\!\!\!\vdash} \ulcorner V(\bar{x}) \urcorner\rangle \leftrightarrow \Sigma(\ulcorner V \urcorner)\sigma^{\dagger}(\ulcorner \bar{x} \urcorner)] \qquad (ho{:}V)$$

$$(\forall \ulcorner R(\bar{x}) \urcorner : For)[\langle D\Lambda\Sigma\sigma \mathrel{\sim\!\!\!\!\vdash} \ulcorner R(\bar{x}) \urcorner\rangle \leftrightarrow \Lambda(\ulcorner V \urcorner)\sigma^{\dagger}(\ulcorner \bar{x} \urcorner)] \qquad (ho{:}R)$$

We next supply clauses for sentential connectives:

$$(\forall \ulcorner \neg\varphi \urcorner : For)[\langle D\Lambda\Sigma\sigma \mathrel{\sim\!\!\!\!\vdash} \ulcorner \neg\varphi \urcorner\rangle \leftrightarrow \neg\langle D\Lambda\Sigma\sigma \mathrel{\sim\!\!\!\!\vdash} \ulcorner \varphi \urcorner\rangle] \qquad (ho{:}\neg)$$

$$(\forall \ulcorner \varphi \wedge \psi \urcorner : For)[\langle D\Lambda\Sigma\sigma \mathrel{\sim\!\!\!\!\vdash} \ulcorner \varphi \wedge \psi \urcorner\rangle \leftrightarrow$$
$$(\langle D\Lambda\Sigma\sigma \mathrel{\sim\!\!\!\!\vdash} \ulcorner \varphi \urcorner\rangle \wedge \langle D\Lambda\Sigma\sigma \mathrel{\sim\!\!\!\!\vdash} \ulcorner \psi \urcorner\rangle)] \qquad (ho{:}\wedge)$$

To handle the quantifiers, we follow Tarski, in considering assignments which differ on exactly one element. We write $\mathit{Diff}_1(\sigma, \tau, x)$ when both $\mathit{Ass}_1(D\sigma)$ and $\mathit{Ass}_1(D\tau)$ and also σ and τ agree everywhere except perhaps on the assignment to the *single* entity x. Similarly, we write $\mathit{Diff}_2(\Sigma, \mathrm{T}, x)$ when both $\mathit{Ass}_2(D\Sigma)$ and $\mathit{Ass}_2(D\mathrm{T})$ and also Σ and T agree everywhere except perhaps on the assignment to the *single* entity x. Obviously, all of this can be explicitly defined using $\mathrm{PA}_{\mathrm{int}}$ in our fifth-order logic. We now offer:

$$(\forall \ulcorner \varphi(x) \urcorner : \mathit{For})[\langle D\Lambda\Sigma\sigma \Vdash \ulcorner \forall x \varphi(x) \urcorner \rangle \leftrightarrow \forall \tau(\mathit{Diff}_1(\sigma, \tau, \ulcorner x \urcorner) \rightarrow \langle D\Lambda\Sigma\tau \Vdash \ulcorner \varphi(x) \urcorner \rangle)] \qquad (ho{:}\forall_1)$$

$$(\forall \ulcorner \varphi(V) \urcorner : \mathit{For})[\langle D\Lambda\Sigma\sigma \Vdash \ulcorner \forall V \varphi(V) \urcorner \rangle \leftrightarrow \forall \mathrm{T}(\mathit{Diff}_2(\Sigma, \mathrm{T}, \ulcorner V \urcorner) \rightarrow \langle D\Lambda\mathrm{T}\sigma \Vdash \ulcorner \varphi(V) \urcorner \rangle)] \qquad (ho{:}\forall_2)$$

If we now conjoin $(ho{:}{=})$, $(ho{:}V)$, $(ho{:}R)$, $(ho{:}\wedge)$, $(ho{:}\neg)$, $(ho{:}\forall_1)$, and $(ho{:}\forall_2)$, we obtain a formula which intuitively tells us that something fifth-order, $\Vdash$, acts as a satisfaction relation.

Existence and uniqueness of this relation

The next stage is to prove the *existence* and *uniqueness* of such a fifth-order entity. We will do this by induction on complexity, and our approach draws inspiration from Takeuti's construction of the semantics for PA within PA_2.[26]

The first task is to define a formula, $\mathit{Sat}(D\Lambda\Sigma\sigma, m, \Vdash)$, which says that $\Vdash$ acts as a satisfaction relation for formulas of complexity $\leq m$. So, let $co(v)$ be a $\mathrm{PA}_{\mathrm{int}}$-formula such that, when v codes an $\mathscr{L}$-formula, $co(v)$ is the number of instances of logical constants (i.e. $\forall$, $\neg$, and $\wedge$) in that formula. Then let $\mathit{Sat}(D\Lambda\Sigma\sigma, m, \Vdash)$ be the formula which says that each of $(ho{:}{=})$, $(ho{:}V)$, $(ho{:}R)$, $(ho{:}\wedge)$, $(ho{:}\neg)$, $(ho{:}\forall_1)$, and $(ho{:}\forall_2)$ hold *for all formulas v with $co(v) \leq m$*. We now prove that, for any complexity, a suitable satisfaction relation, $\Vdash$, exists:[27]

Lemma 12.1:

$$\mathrm{PA}_{\mathrm{int}} \vdash^5 (\forall m : \mathit{Num})\exists \Vdash \forall D\Lambda\Sigma\sigma[\mathit{Tars}(D\Lambda\Sigma\sigma) \rightarrow \mathit{Sat}(D\Lambda\Sigma\sigma, m, \Vdash)]$$

Proof. The proof is by induction on m within $\mathrm{PA}_{\mathrm{int}}$. For the case $m = 0$, by fifth-order Comprehension there is some $\Vdash$ with the following property: whenever $\mathit{Tars}(D\Lambda\Sigma\sigma)$ and for any v, we have $\langle D\Lambda\Sigma\sigma \Vdash v \rangle$ iff

---

[26] Takeuti (1987: 183–7).

[27] As elsewhere in these chapters, we abbreviate e.g. $\forall D \forall \Lambda \forall \Sigma \forall \sigma$ with $\forall D\Lambda\Sigma\sigma$.

$$\begin{aligned}&For(v) \wedge Tars(D\Lambda\Sigma\sigma) \wedge co(v) = 0 \wedge \\ &\quad\big[(\exists \ulcorner x = y \urcorner : For)[v = \ulcorner x = y \urcorner \wedge \sigma(\ulcorner x \urcorner) = \sigma(\ulcorner y \urcorner)] \vee \\ &\quad(\exists \ulcorner V(\bar{x}) \urcorner : For)[v = \ulcorner V(\bar{x}) \urcorner \wedge \Sigma(\ulcorner V \urcorner)\sigma^{\uparrow}(\ulcorner \bar{x} \urcorner)] \vee \\ &\quad(\exists \ulcorner R(\bar{x}) \urcorner : For)[v = \ulcorner R(\bar{x}) \urcorner \wedge \Lambda(\ulcorner R \urcorner)\sigma^{\uparrow}(\ulcorner \bar{x} \urcorner)]\big]\end{aligned}$$

so that $Sat(D\Lambda\Sigma\sigma, 0, \Vdash)$. Now fix m, and suppose for induction that we have some $\Vdash$ such that for any $Tars(D\Lambda\Sigma\sigma)$ we have $Sat(D\Lambda\Sigma\sigma, m, \Vdash)$. By fifth-order Comprehension, there is some $\Vdash_*$ with the following property: for all $D\Lambda\Sigma\sigma$ and all v we have $\langle D\Lambda\Sigma\sigma \Vdash_* v\rangle$ iff

$$\begin{aligned}&For(v) \wedge Tars(D\Lambda\Sigma\sigma) \wedge \\ &\Big[\big(co(v) \leq m \wedge \langle D\Lambda\Sigma\sigma \Vdash v\rangle\big) \vee \\ &\quad\big(co(v) = m + 1 \wedge \\ &\quad[(\exists \ulcorner \varphi \wedge \psi \urcorner : For)[v = \ulcorner \varphi \wedge \psi \urcorner \wedge \langle D\Lambda\Sigma\sigma \Vdash \ulcorner \varphi \urcorner\rangle \wedge \langle D\Lambda\Sigma\sigma \Vdash \ulcorner \psi \urcorner\rangle] \vee \\ &\quad(\exists \ulcorner \neg\varphi \urcorner : For)[v = \ulcorner \neg\varphi \urcorner \wedge \neg\langle D\Lambda\Sigma\sigma \Vdash \ulcorner \varphi \urcorner\rangle] \vee \\ &\quad(\exists \ulcorner \varphi(x) \urcorner : For)[v = \ulcorner \forall x \varphi(x) \urcorner \wedge \\ &\qquad\qquad \forall \tau(Diff_1(\sigma, \tau, \ulcorner x \urcorner) \rightarrow \langle D\Lambda\Sigma\tau \Vdash \ulcorner \varphi(x) \urcorner\rangle)] \vee \\ &\quad(\exists \ulcorner \varphi(V) \urcorner : For)[v = \ulcorner \forall V \varphi(V) \urcorner \wedge \\ &\qquad\qquad \forall \mathrm{T}(Diff_2(\Sigma, \mathrm{T}, \ulcorner V \urcorner) \rightarrow \langle D\Lambda\mathrm{T}\sigma \Vdash \ulcorner \varphi(V) \urcorner\rangle)]]\big)\Big]\end{aligned}$$

so that $Sat(D\Lambda\Sigma\sigma, m + 1, \Vdash_*)$. The result follows by induction in PA_{int}. □

We next show that these $\Vdash$-relations are extensionally unique, where defined:

Lemma 12.2:

$$\begin{aligned}&PA_{int} \vdash^5 (\forall v : For)\forall \Vdash \forall \Vdash_* \\ &\quad\big[\forall D\Lambda\Sigma\sigma\big(Tars(D\Lambda\Sigma\sigma) \rightarrow \\ &\qquad\qquad [Sat(D\Lambda\Sigma\sigma, co(v), \Vdash) \wedge Sat(D\Lambda\Sigma\sigma, co(v), \Vdash_*)]\big) \rightarrow \\ &\quad\forall D\Lambda\Sigma\sigma\big(Tars(D\Lambda\Sigma\sigma) \rightarrow [\langle D\Lambda\Sigma\sigma \Vdash v\rangle \leftrightarrow \langle D\Lambda\Sigma\sigma \Vdash_* v\rangle]\big)\big]\end{aligned}$$

Proof. Fix $\Vdash$ and $\Vdash_*$ and suppose, for induction on $co(v)$, that the claim holds whenever $co(v) < m$; now suppose $co(v) = m + 1$ and reason by cases. We will illustrate the case when v codes $\forall x \psi(x)$, leaving the rest to the reader. Suppose that whenever $Tars(D\Lambda\Sigma\sigma)$ we have both $Sat(D\Lambda\Sigma\sigma, m + 1, \Vdash)$ and $Sat(D\Lambda\Sigma\sigma, m + 1, \Vdash_*)$; then by ($ho{:}\forall_1$) we have both of these:

$$\begin{aligned}\langle D\Lambda\Sigma\sigma \Vdash \ulcorner \varphi \urcorner\rangle &\leftrightarrow \forall \tau[Diff_1(\sigma, \tau, \ulcorner x \urcorner) \rightarrow \langle D\Lambda\Sigma\tau \Vdash \ulcorner \psi(x) \urcorner\rangle) \\ \langle D\Lambda\Sigma\sigma \Vdash_* \ulcorner \varphi \urcorner\rangle &\leftrightarrow \forall \tau[Diff_1(\sigma, \tau, \ulcorner x \urcorner) \rightarrow \langle D\Lambda\Sigma\tau \Vdash_* \ulcorner \psi(x) \urcorner\rangle)\end{aligned}$$

Moreover, we have both $Sat(D\Lambda\Sigma\tau, m, \mathrel{\approx})$ and $Sat(D\Lambda\Sigma\tau, m, \mathrel{\approx}_*)$ whenever $Tars(D\Lambda\Sigma\tau)$. So, since $co(\ulcorner\psi(x)\urcorner) = m$, by the induction hypothesis we have:

$$\langle D\Lambda\Sigma\tau \approx \ulcorner\psi(x)\urcorner\rangle \leftrightarrow \langle D\Lambda\Sigma\tau \approx_* \ulcorner\psi(x)\urcorner\rangle$$

And hence $\langle D\Lambda\Sigma\sigma \approx \ulcorner\forall x\psi(x)\urcorner\rangle \leftrightarrow \langle D\Lambda\Sigma\sigma \approx_* \ulcorner\forall x\psi(x)\urcorner\rangle$, as required. □

Taking Lemmas 12.1 and 12.2 together, by fifth-order Comprehension, there is a *unique* satisfaction relation which covers $\mathscr{L}$-formulas of any complexity. In what follows, we use $\approx$ for this; so

$$(\forall m : Num)\forall D\Lambda\Sigma\sigma[Tars(D\Lambda\Sigma\sigma) \rightarrow Sat(D\Lambda\Sigma\sigma, m, \approx)]$$

Now, recall that our *ultimate* aim is to define an expression which roughly says: *v codes an $\mathscr{L}$-sentence which is true on domain D with interpretation* Λ. With that aim in mind, where $Sen(v)$ indicates that v codes an $\mathscr{L}$-sentence, we define:

$$TrueIn(D\Lambda : v) := Sen(v) \wedge Mod(D\Lambda) \wedge \forall\Sigma\sigma\big[Tars(D\Lambda\Sigma\sigma) \rightarrow \langle D\Lambda\Sigma\sigma \approx v\rangle\big]$$

It just remains to show that this definition does what we would want it to.

Checking this works

For each $\mathscr{L}$-formula φ, let $\varphi^{D\Lambda}$ be the result of first restricting all of φ's quantifiers (of any order) to D, and then, for each $\mathscr{L}$-predicate R, replacing every instance of any formula $R(x_{i_1}, \ldots, x_{i_m})$ with a statement that $\Lambda(\ulcorner R\urcorner)$ applies to the unique relation given by $\{(1, x_{i_1}), \ldots, (m, x_{i_m})\}$. (This can be defined explicitly using Comprehension, and simply mirrors Tarski's trick.) We aim to prove this:

Theorem 12.3: *For each $\mathscr{L}$-sentence φ:*

$$\mathrm{PA_{int}} \vdash^5 \forall D\Lambda(Mod(D\Lambda) \rightarrow [TrueIn(D\Lambda : \ulcorner\varphi\urcorner) \leftrightarrow \varphi^{D\Lambda}])$$

For, if we can prove this result, then we really do have a formula which 'says' that φ is true given 'domain' D and 'interpretation' Λ.

In order to prove Theorem 12.3, we will first prove a lemma which says (roughly) that agreement on the assignments to all the variables in a formula φ entails agreement on satisfaction of φ:

Lemma 12.4: *For any $\mathscr{L}$-formula $\varphi(\overline{V}, \overline{x})$ with all free variables displayed:*

$$\mathrm{PA}_{\mathrm{int}} \vdash^5 \forall D\Lambda\Sigma\sigma\mathrm{T}\tau\Big(\big[\mathit{Tars}(D\Lambda\Sigma\sigma) \wedge \mathit{Tars}(D\Lambda\mathrm{T}\tau) \wedge \bigwedge_{i=1}^{j} \Sigma(\ulcorner V_i \urcorner) = \mathrm{T}(\ulcorner V_i \urcorner) \wedge \bigwedge_{i=1}^{k} \sigma(\ulcorner x_i \urcorner) = \tau(\ulcorner x_i \urcorner)\big] \to \big[\langle D\Lambda\Sigma\sigma \approx \ulcorner \varphi(\overline{V}, \overline{x}) \urcorner \rangle \leftrightarrow \langle D\Lambda\mathrm{T}\tau \approx \ulcorner \varphi(\overline{V}, \overline{x}) \urcorner \rangle\big]\Big)$$

Proof. This is by induction on complexity in our syntactic metatheory.

The case of atomic formulas is immediate from (*ho*:=), (*ho*:V), and (*ho*:R). The cases of conjunction and negation are trivial. We will explain the case of first-order quantifiers; second-order quantifiers are similar.

Let $\varphi(\overline{V}, \overline{x})$ be $\forall y \psi(\overline{V}, \overline{x}, y)$. Fix Σ and T which agree on each $\ulcorner V_i \urcorner$. Fix σ and τ which agree on each $\ulcorner x_i \urcorner$ but need not agree on $\ulcorner y \urcorner$. Without loss of generality, suppose $\langle D\Lambda\Sigma\sigma \approx \ulcorner \forall y \psi(\overline{V}, \overline{x}, y) \urcorner \rangle$ and let τ' be such that $\mathit{Diff}_1(\tau, \tau', \ulcorner y \urcorner)$. Let σ' be such that $\sigma'(\ulcorner y \urcorner) = \tau'(\ulcorner y \urcorner)$ and $\sigma'(v) = \sigma(v)$ for all $v \neq \ulcorner y \urcorner$; then by (*ho*:$\forall_1$) we have that $\langle D\Lambda\Sigma\sigma' \approx \ulcorner \psi(\overline{V}, \overline{x}, y) \urcorner \rangle$. Since σ' and τ' agree on each $\ulcorner x_i \urcorner$ and on $\ulcorner y \urcorner$, by our induction hypothesis $\langle D\Lambda\Sigma\tau' \approx \ulcorner \psi(\overline{V}, \overline{x}, y) \urcorner \rangle$. So $\langle D\Lambda\Sigma\tau \approx \ulcorner \forall y \psi(\overline{V}, \overline{x}, y) \urcorner \rangle$ by (*ho*:$\forall_1$) again. □

Using this, we will prove a result which is strictly more general than Theorem 12.3 (indeed, it is essentially just a version of Theorem 12.3 which accepts parameters). To state the result, though, we need some notation. Let † be a functor which, applied to a second-order entity, V, yields the third-order entity $V^{\dagger}$ corresponding to Tarski's trick. That is, $V^{\dagger}(R)$ iff: there are $a_1, \ldots, a_n$ such that $V(a_1, \ldots, a_n)$ and R is the relation $\{(1, a_1), \ldots, (n, a_n)\}$. Armed with this, here is our desired result:

Lemma 12.5: *For each $\mathscr{L}$-formula $\varphi(\overline{V}, \overline{x})$ with all free variables displayed:*

$$\mathrm{PA}_{\mathrm{int}} \vdash^5 \forall D\Lambda\Big(\mathit{Mod}(D\Lambda) \to (\forall \overline{V} : D)(\forall \overline{x} : D)\Big[\varphi(\overline{V}, \overline{x})^{D\Lambda} \leftrightarrow \forall \Sigma\sigma\Big(\big[\mathit{Tars}(D\Lambda\Sigma\sigma) \wedge \bigwedge_{i=1}^{j} \Sigma(\ulcorner V_i \urcorner) = V_i^{\dagger} \wedge \bigwedge_{i=1}^{k} \sigma(\ulcorner x_i \urcorner) = x_i\big] \to \langle D\Lambda\Sigma\sigma \approx \ulcorner \varphi(\overline{V}, \overline{x}) \urcorner \rangle\Big)\Big]\Big)$$

Proof. This is by induction on complexity in our syntactic metatheory.

We start with atomic formulas of the form $R(\overline{x})$, leaving other atomic cases to the reader. Suppose that $\mathit{Mod}(D\Lambda)$, and fix suitable $\overline{b}$ from D. First suppose $R(\overline{b})^{D\Lambda}$, i.e. that $\Lambda(\ulcorner R \urcorner)$ applies to the relation given by $\{(1, b_1), \ldots, (n, b_n)\}$. Where σ is any assignment mapping each $\ulcorner x_i \urcorner$ to b_i, we have $\langle D\Lambda\Sigma\sigma \approx \ulcorner R(\overline{x}) \urcorner \rangle$ by (*ho*:R). And the converse is similar.

We now consider the induction cases. Conjunction is trivial. To handle negation, suppose $\varphi(\overline{V}, \overline{x})$ is $\neg\psi(\overline{V}, \overline{x})$. Suppose $Mod(D\Lambda)$ and fix $\overline{A}$ and $\overline{b}$ from D. By elementary manipulations on our induction hypothesis and $(ho{:}\neg)$, we have:

$$\neg\psi(\overline{A}, \overline{b})^{D\Lambda} \leftrightarrow \exists\Sigma\exists\sigma\big(Tars(D\Lambda\Sigma\sigma) \wedge \bigwedge_{i=1}^{j} \Sigma(\ulcorner V_i \urcorner) = A_i^{\circ} \wedge \bigwedge_{i=1}^{k} \sigma(\ulcorner x_i \urcorner) = b_i \wedge \\ \langle D\Lambda\Sigma\sigma \approx \ulcorner \neg\psi(\overline{V}, \overline{y}) \urcorner \rangle\big)$$

The desired result now follows since, by Lemma 12.4, the right-hand-side of this biconditional is equivalent to:

$$\forall\Sigma\forall\sigma\big(\big[Tars(D\Lambda\Sigma\sigma) \wedge \bigwedge_{i=1}^{j} \Sigma(\ulcorner V_i \urcorner = A_i^{\circ} \wedge \bigwedge_{i=1}^{k} \sigma(\ulcorner x_i \urcorner) = b_i\big] \rightarrow \\ \langle D\Lambda\Sigma\sigma \approx \ulcorner \neg\psi(\overline{V}, \overline{y}) \urcorner \rangle\big)$$

We now consider first-order quantifiers, leaving second-order quantifiers to the reader. As before, fix suitable $D\Lambda$, $\overline{A}$, and $\overline{b}$. Our induction hypothesis gives:

$$(\forall y : D)\Big[\psi(\overline{A}, \overline{b}, y)^{D\Lambda} \leftrightarrow \forall\Sigma\forall\sigma\big(\big[Tars(D\Lambda\Sigma\sigma) \wedge \bigwedge_{i=1}^{j} \Sigma(\ulcorner V_i \urcorner) = A_i^{\circ} \wedge \\ \bigwedge_{i=1}^{k} \sigma(\ulcorner x_i \urcorner) = b_i \wedge \sigma(\ulcorner y \urcorner) = y\big] \rightarrow \\ \langle D\Lambda\Sigma\sigma \approx \ulcorner \psi(\overline{V}, \overline{x}, y) \urcorner \rangle\big)\Big]$$

First suppose that $(\forall y \psi(\overline{A}, \overline{b}, y))^{D\Lambda}$, i.e. $(\forall y : D)\psi(\overline{A}, \overline{b}, y)^{D\Lambda}$. Where Σ maps each $\ulcorner V_i \urcorner$ to A_i°, and σ maps each $\ulcorner x_i \urcorner$ to b_i, and τ is any assignment such that $Diff_1(\sigma, \tau, \ulcorner y \urcorner)$, the induction hypothesis yields $\langle D\Lambda\Sigma\tau \approx \ulcorner \psi(\overline{V}, \overline{x}, y) \urcorner \rangle$. Hence $\langle D\Lambda\Sigma\sigma \approx \ulcorner \forall y \psi(\overline{V}, \overline{x}, y) \urcorner \rangle$ by $(ho{:}\forall_1)$. Conversely, suppose $\neg(\forall y \psi(\overline{A}, \overline{b}, y))^{D\Lambda}$, i.e. $(\exists y : D)\neg\psi(\overline{A}, \overline{b}, y)^{D\Lambda}$. Fix c from D such that $\neg\psi(\overline{A}, \overline{b}, c)^{D\Lambda}$. By the induction hypothesis there is some Σ mapping each $\ulcorner V_i \urcorner$ to A_i°, and some σ mapping each $\ulcorner x_i \urcorner$ to b_i and $\ulcorner y \urcorner$ to c, with $\neg \langle D\Lambda\Sigma\sigma \approx \ulcorner \psi(\overline{V}, \overline{x}, y) \urcorner \rangle$. Now $\neg \langle D\Lambda\Sigma\sigma \approx \ulcorner \forall y \psi(\overline{V}, \overline{x}, y) \urcorner \rangle$ as $Diff_1(\sigma, \sigma, \ulcorner y \urcorner)$. □

Generalising to n^{th}-order logics

We have worked in fifth-order logic and considered second-order formulas. But it is easy to see that these restrictions is inessential. For each n, we could have worked in $n{+}3^{\text{th}}$-order logic, to formulate similar results for n^{th}-order languages.

That said, spelling this out in any *detail* would be genuinely nightmarish. In the second-order case, Tarski's trick made us treat sequences of first-order variables as relations between first-order objects (numbers) and first-order objects (assignments to individual first-order variables). Lifting the same trick to arbitrary n^{th}-order languages will involve treating sequences of variables of order $< n$ as relations between first-order objects (numbers) and $n-1^{\text{th}}$-order entities. Such raising-of-orders is always *possible*: intuitively, instead of considering a, we can consider $\{\{\{\ldots\{a\}\ldots\}\}\}$, thereby raising the order without losing any information. But we shudder to consider, for example, the explicit details of satisfaction for a 14th-order language.

Done correctly, though, for arbitrary n we can define a predicate *TrueIn-n* which has the same properties as above, i.e.:

Theorem 12.6: *For any natural number n and each $\mathcal{L}$-sentence φ of order $\leq n$:*

$$\text{PA}_{\text{int}} \vdash^{n+3} \forall D\Lambda(Mod(D\Lambda) \rightarrow [TrueIn(D\Lambda : \ulcorner\varphi\urcorner) \leftrightarrow \varphi^{D\Lambda}])$$

Combining this with internal categoricity

Recall that the technicalities in this appendix were motivated by considering the total internal categoricity of CSP_{int}. So let us now return to that motivation.

In the past three chapters, we have explained that internal categoricity entails intolerance. In fact, as we formulated it, 'intolerance' was restricted to *second*-order sentences. But, we can easily tweak the proof of Theorem 10.3 to obtain:

Theorem 12.7: *For any natural number n and each pure n^{th}-order formula $\varphi(PE)$ with all free variables displayed and whose quantifiers (of all orders) are P-restricted:*

$$\vdash^{n} \forall PE(\text{CSP}(PE) \rightarrow \varphi(PE)) \vee \forall PE(\text{CSP}(PE) \rightarrow \neg\varphi(PE))$$

Now, using the techniques in this appendix, we can define a truth-predicate for $\{Pure, \in\}$-sentences φ of order $\leq n$. First, given second-order P and E, let Λ^{PE} be the interpretation which sends $\ulcorner Pure\urcorner$ to $P^{\dagger}$ and $\ulcorner\in\urcorner$ to $E^{\dagger}$. Then say:

(a) *True-n*$(\ulcorner\varphi\urcorner)$ iff $\forall PE(\text{CSP}(PE) \rightarrow TrueIn(P\Lambda^{PE} : \ulcorner\varphi\urcorner))$
(b) *False-n*$(\ulcorner\varphi\urcorner)$ iff $\forall PE(\text{CSP}(PE) \rightarrow TrueIn(P\Lambda^{PE} : \ulcorner\neg\varphi\urcorner))$
(c) *Indeterminate-n*$(\ulcorner\varphi\urcorner)$ iff otherwise

Then, from Theorems 12.6 and 12.7, we straightforwardly obtain a result which vindicates the technical claims we made in §12.4:

Theorem 12.8: *For any n and any $\{Pure, \in\}$-sentence φ of order $\leq n$ whose quantifiers (of all orders) are Pure-restricted:*

$$\text{PA}_{\text{int}} \vdash^{n+3} True\text{-}n(\ulcorner\varphi\urcorner) \vee False\text{-}n(\ulcorner\varphi\urcorner)$$

13
Boolean-valued structures

Part B of this book has explored the interplay between moderation, modelism, scepticism, various external and internal categoricity results, and our ability to pin down mathematical entities and concepts. In this chapter, we want to explore similar issues which arise at a 'deeper' level. These issues concern our ability to pin down *semantics* itself. We scratched the surface of this issue, when we discussed whether a moderate can invoke full (rather than Henkin) semantics for second-order logic (see Chapters 7–8). But in this chapter, we ratchet things up a gear.

From the outset of this book, we have spoken in bivalent terms, describing a model as making a sentence true or false. We have made room for indeterminacy only insofar as we have considered supervaluation. The immediate message of this chapter is that we could have pursued model theory in *radically* non-bivalent terms, using Boolean-valued models.

After a brief discussion of *two-valued* semantics, we provide an introduction to the theory of Boolean algebras, and show how Boolean algebras supply us with *many-valued* semantics. With a multiplicity of different semantics on offer, this immediately raises a doxological question: *How can we pin down the semantics for our language?* The question has a familiar ring to it, and it sustains deep comparison with the Modelist's Doxological Challenge, as discussed in Chapters 6–8. We explore the parallels in detail in this chapter. We close the chapter with a discussion of Suszko's Thesis, which can be used to raise sceptical concerns about the intended interpretation of the semantic values, from the 'opposite' direction.

13.1 Semantic-underdetermination via Push-Through

Suppose we want to pin down a *consequence relation* between theories and individual sentences. Informally, we might think of these as premises and a conclusion; formally, we are looking to pin down a relation on $\wp(\text{Sent}) \times \text{Sent}$, for some set Sent of sentences.

For example, we might want to pin down the consequence relation for classical *sentential* logic. Let us temporarily symbolise this with $\vDash$. As in §1.c, we can lay down rules of inference rules for the sentential connectives. Since these rules are (provably) sound and complete for classical sentential logic, $T \vDash \varphi$ iff $T \vdash \varphi$. So our inference rules pin down the relation $\vDash$ in extension.

However, *the inference rules cannot pin down the usual semantics* which are ordinarily used to define $\vDash$. This follows immediately from a simple Push-Through Construction (see §2.1). The usual semantics for classical sentential logic is given by treating the sentential connectives as truth-functions with two possible values, True and False. A simple Push-Through Construction on the truth-functions will show us that the two values could instead have been: False and True (i.e. the same values, but taken the other way around); or Julius Caesar and Love; or, indeed, *any* two distinct objects. Indeed, the very first permutation argument, due to Frege, was put in precisely these terms.[1]

This elementary observation yields an argument that the inference rules of the logical vocabulary fail to determine the the semantic structure for our language:

The semantic-underdetermination argument. *Inference rules can only manage to define the relation $\vdash$. But the relation $\vdash$ is sound and complete for many different semantic structures. So: inference rules do not determine the meanings of logical expressions.*

Still, in this form, the semantic-underdetermination argument is not very scary. We might well grant that the inference rules cannot, on their own, determine whether truth is the True, or Goodness, or Love, but hope that they *can* determine that there are *exactly* two truth values which interact with each other in *exactly* the ways described by the usual truth-tables. If they could do that, then the inference rules of classical sentential logic would determine the semantic structure *categorically* (i.e. up to isomorphism). And one might well think that this is all the determinacy of meaning that the logical vocabulary *ought* to yield. (Compare this with the now familiar point, first raised in §2.2, that for many mathematical purposes we only care about identity up to isomorphism.)

Alas, this is not to be. We will show this developing a *Boolean-valued* framework, and presenting some interestingly non-standard semantic structures.

13.2 The theory of Boolean algebras

We start by setting out the theory of Boolean algebras. The signature of Boolean algebras has a one-place function, $-$, two two-place functions, $\cdot$ and $+$, and two constants, 1 and 0.[2] We now say:

---

[1] Frege (1893: §10).

[2] This makes it the language of rings. Many authors use $\neg$ for $-$, $\vee$ for $+$, and $\wedge$ for $\cdot$ (see e.g. Hodges 1993: 38). We have avoided this, since we want to avoid the potential confusion between object-language connectives and a signature's function-symbols. Some authors also do not stipulate that $0 \neq 1$, and so admit one-element Boolean algebras.

Definition 13.1: *The theory of Boolean algebras is given by the universal closures of the following axioms:*

$$0 \neq 1$$

$$\begin{aligned} x \cdot x &= x & \qquad x + x &= x \\ x \cdot y &= y \cdot x & x + y &= y + x \\ (x \cdot y) \cdot z &= x \cdot (y \cdot z) & (x + y) + z &= x + (y + z) \\ (x \cdot y) + y &= y & (x + y) \cdot y &= y \\ x \cdot (y + z) &= (x \cdot y) + (x \cdot z) & x + (y \cdot z) &= (x + y) \cdot (x + z) \\ x \cdot -x &= 0 & x + -x &= 1 \end{aligned}$$

A Boolean algebra *is any structure which satisfies the theory of Boolean algebras.*

As an example, $\wp(A)$ supplies a Boolean algebra whenever $A \neq \emptyset$, by interpreting $\cdot$ as intersection, $+$ as union, $-$ as complementation, 0 as $\emptyset$, and 1 as A.

Indeed, the theory of Boolean algebras should look very familiar. Try the following. Systematically replace the variables x, y, z with sentential variables, φ, ψ, θ. Then systematically replace each instance of $-$ with $\neg$, each $\cdot$ with $\wedge$, each $+$ with $\vee$, 1 with your favourite tautology of sentential logic, 0 with your favourite contradiction, and $=$ with $\leftrightarrow$. The resulting statements are now familiar tautologies of sentential logic.

Here is a slightly different perspective on the same point. Let *Two* be a structure whose domain is the usual set of truth values, $\{\text{True}, \text{False}\}$, where $1^{\mathit{Two}} = \text{True}$ and $0^{\mathit{Two}} = \text{False}$ and where $-$, $+$ and $\cdot$ are respectively interpreted as the usual truth-functions implemented by $\neg$, $\vee$ and $\neg$. It is easy to see that *Two* is a Boolean algebra. And this is our first glimpse of the central idea of this chapter, that Boolean algebras can be used to set up semantic frameworks.

Boolean algebras can also be thought of as partial-orders of a certain sort, with the order $x \leq y$ given by $x \cdot y = x$. This corresponds to the elementary fact of classical sentential logic, that $\varphi \rightarrow \psi$ iff $(\varphi \wedge \psi) \leftrightarrow \varphi$. Using this partial order, we can define a particularly important species of Boolean algebras:

Definition 13.2: *A* complete Boolean algebra, $\mathcal{B}$, *is one such that any nonempty set of elements $X \subseteq B$ has a* supremum, *written $\sum X$, satisfying the following two properties:*

Upper bound. *$x \leq \sum X$ for all $x \in X$.*
Least upper bound. *If $x \leq y$ for all $x \in X$, then $\sum X \leq y$.*

As these names suggest, $\sum X$ is an upper bound of the set X and it is the least such one.[3] We can equivalently characterise completeness by insisting that every nonempty set X has an *infimum*, written $\prod X$, which satisfies these two properties:

Lower bound. *$\prod X \leq x$ for all $x \in X$.*
Greatest lower bound. *If $y \leq x$ for all $x \in X$, then $y \leq \prod X$.*

A simple and important example of complete Boolean algebra is $\wp(A)$, for any A, where the partial order corresponds to the subset relation, so that the supremum is the union and the infimum is the intersection.

Using the partial order, we also define the notion of a filter, which will be hugely important in this chapter and the next:

Definition 13.3: *Let $\mathcal{B}$ be a Boolean algebra. A* filter *on $\mathcal{B}$ is any set $F \subseteq B$ such that:*

(1) $0 \notin F$
(2) *if $a \in F$ and $b \in F$, then $a \cdot b \in F$*
(3) *if $a \in F$ and $a \leq b$, then $b \in F$*

An ultrafilter *is any filter which also obeys:*

(4) *for any $a \in B$: either $a \in F$ or $-a \in F$*

A simple example of a filter is the upward closure $\{b \in B : a \leq b\}$ of any non-zero element a of the Boolean algebra $\mathcal{B}$. Another example of a filter, this time on $\wp(X)$, is the set of cofinite subsets of X, which is sometimes called the *Fréchet filter* (we defined cofiniteness in §5.B). But these two examples are not ultrafilters. For a simple example of an ultrafilter on $\wp(X)$, consider $\{A \subseteq X : b \in A\}$ for any $b \in X$. In general, there are fewer natural examples of ultrafilters than filters.[4]

13.3 Boolean-valued models

In this chapter, our interest in Boolean algebras comes via their use in defining Boolean-valued models. The first step in this direction involves sprucing up the notion of a structure, as laid down in Definition 1.2, so that it projects its values onto a complete Boolean algebra:[5]

---

[3] This name invites comparison with real numbers, and a small amount of experience suffices to make one aware of the handful of basic inference patterns involving suprema which are common to the real numbers and complete Boolean algebras.

[4] However, by the Ultrafilter Theorem 14.4, any filter can be extended to an ultrafilter. Indeed, ultrafilters can be equivalently characterised as filters which are maximal in that they are not properly contained in any other filter (see Givant and Halmos 2009: 171, 175; Jech 2003: 74).

[5] In the context of classical logic, the primary use of Boolean algebras is in set theory (see e.g. Bell 2005), where the definition is often restricted to the signature whose only primitive is $\in$ (an exception is Grishin 2011). However, it is easy to generalise the definition; cf. the use of *Heyting* (pre-)algebras in Troelstra and Dalen (1988: Definitions 6.2–6.3 pp.710–11) and Rin and Walsh (2016: Definition 6.1).

Definition 13.4: *Let $\mathcal{B}$ be a complete Boolean algebra and let $\mathscr{L}$ be a signature. Then a $\mathcal{B}$-valued $\mathscr{L}$-structure, $\mathcal{M}$, consists of*

- *a non-empty set, M, which is the underlying domain of $\mathcal{M}$,*
- *an object $c^{\mathcal{M}} \in M$ for each constant symbol $c \in \mathscr{L}$,*
- *a map $[\![R(\cdot, \ldots, \cdot)]\!]^{\mathcal{M}} : M^n \longrightarrow B$ for each n-place relation symbol $R \in \mathscr{L}$,*
- *a map $f^{\mathcal{M}} : M^n \longrightarrow M$ for each n-place function symbol $f \in \mathscr{L}$, and*
- *a map $[\![\cdot = \cdot]\!]^{\mathcal{M}} : M^2 \longrightarrow \mathcal{B}$*

Where no confusion can arise, we omit the superscript on these maps. These maps must satisfy the following, for all n-place relation symbols $R \in \mathscr{L}$, all n-place function symbols $f \in \mathscr{L}$, and all $a, b, c, a_1, \ldots, a_n, b_1, \ldots, b_n \in M$:

$$[\![a = a]\!] = 1$$
$$[\![a = b]\!] = [\![b = a]\!]$$
$$[\![a = b]\!] \cdot [\![b = c]\!] \leq [\![a = c]\!]$$
$$[\![a_1 = b_1]\!] \cdot \ldots \cdot [\![a_n = b_n]\!] \cdot [\![R(a_1, \ldots, a_n)]\!] \leq [\![R(b_1, \ldots, b_n)]\!]$$
$$[\![a_1 = b_1]\!] \cdot \ldots \cdot [\![a_n = b_n]\!] \leq [\![f(a_1, \ldots, a_n) = f(b_1, \ldots, b_n)]\!]$$

Now, where *Two* is the two-element algebra from before, a *Two*-valued $\mathscr{L}$-structure is essentially just an $\mathscr{L}$-structure, in the standard sense. There is just a slightly baroque twist that, instead of considering whether $(a_1, \ldots, a_n) \in R^{\mathcal{M}}$, as in the standard case, we consider whether $[\![R(a_1, \ldots, a_n)]\!]^{\mathcal{M}} = \text{True}$, in the *Two*-valued case.

Having defined Boolean-valued structures, though, we can now set up a Boolean-valued semantics. Our aim is to define the semantic value, $[\![\varphi]\!]^{\mathcal{M}}$, of a formula φ in a Boolean-valued structure $\mathcal{M}$.

Much as in Definition 1.5 of §1.5, where $\mathcal{M}$ is a $\mathcal{B}$-valued structure, we let $\mathcal{M}^\circ$ be the $\mathcal{B}$-valued structure obtained by adding new constant symbols c_a with $c_a^{\mathcal{M}^\circ} = a$ for every $a \in M$. We interpret each $\mathscr{L}(M)$-term, t, as a function $t^{\mathcal{M}^\circ} : M^n \longrightarrow M$, exactly as in §1.5. We then say that $\mathcal{M}^\circ$ assigns values from $\mathcal{B}$ to the $\mathscr{L}(M)$-sentences according to the following rules:

$$[\![t_1 = t_2]\!]^{\mathcal{M}^\circ} = \left[\!\!\left[t_1^{\mathcal{M}^\circ} = t_2^{\mathcal{M}^\circ}\right]\!\!\right]^{\mathcal{M}^\circ} \qquad [\![R(t_1, \ldots, t_n)]\!]^{\mathcal{M}^\circ} = \left[\!\!\left[R(t_1^{\mathcal{M}^\circ}, \ldots, t_n^{\mathcal{M}^\circ})\right]\!\!\right]^{\mathcal{M}^\circ}$$
$$[\![\neg\varphi]\!]^{\mathcal{M}^\circ} = -[\![\varphi]\!]^{\mathcal{M}^\circ}$$
$$[\![\varphi \wedge \psi]\!]^{\mathcal{M}^\circ} = [\![\varphi]\!]^{\mathcal{M}^\circ} \cdot [\![\psi]\!]^{\mathcal{M}^\circ} \qquad [\![\varphi \vee \psi]\!]^{\mathcal{M}^\circ} = [\![\varphi]\!]^{\mathcal{M}^\circ} + [\![\psi]\!]^{\mathcal{M}^\circ}$$
$$[\![\exists x\varphi(x)]\!]^{\mathcal{M}^\circ} = \sum_{a \in M} [\![\varphi(c_a)]\!]^{\mathcal{M}^\circ} \qquad [\![\forall x\varphi(x)]\!]^{\mathcal{M}^\circ} = \prod_{a \in M} [\![\varphi(c_a)]\!]^{\mathcal{M}^\circ}$$

The clauses for the quantifiers explain why Definition 13.4 requires that $\mathcal{B}$ is a *complete* Boolean algebra: we need to be sure that $\sum X$ and $\prod X$ exist for any nonempty set of elements X. We then stipulate that $[\![\varphi]\!]^{\mathcal{M}} = [\![\varphi]\!]^{\mathcal{M}^{\circ}}$ for every (mere) $\mathscr{L}$-sentence φ. And this allows us to define a relation which holds between theories and sentences. Where $D \subsetneq B$, and $[\![T]\!]^{\mathcal{M}} = \{[\![\psi]\!]^{\mathcal{M}} : \psi \in T\}$, and $T \cup \{\varphi\}$ is any first-order theory:

$T \overset{\mathcal{B}}{\underset{D}{\models}} \varphi$ iff: for every $\mathcal{B}$-valued $\mathscr{L}$-structure $\mathcal{M}$, if $[\![T]\!]^{\mathcal{M}} \subseteq D$ then $[\![\varphi]\!]^{\mathcal{M}} \in D$

This naturally generalises a very common way to think about logical consequence. If we had a two-valued semantics given by the Boolean algebra *Two*, and took $D = \{\text{True}\}$, then logical consequence would amount to *preservation of truth*, in this sense: if every sentence in T is assigned True, then φ is also assigned True. In this more general setting, we allow that many different semantic values (i.e. the members of D) can be regarded as worthy of preservation. We call D the set of *designated* values, and this allows us to say that logical consequence amounts to *preservation of designation*.

It is easy to extend the Boolean approach into second-order logic, by presenting a Boolean-valued generalisation of second-order Henkin semantics:

Definition 13.5: *A* Henkin $\mathcal{B}$-valued $\mathscr{L}$-structure *is a $\mathcal{B}$-valued $\mathscr{L}$-structure which is additionally equipped with, for each natural number n:*

- *A set M_n^{rel}, all of whose members are n-place functions $g : M^n \longrightarrow B$, obeying* $[\![a_1 = b_1]\!] \cdot \ldots \cdot [\![a_n = b_n]\!] \cdot g(a_1, \ldots, a_n) \leq g(b_1, \ldots, b_n)$
- *A set M_n^{fun}, all of whose members are n-place functions $g : M^n \longrightarrow M$, obeying* $[\![a_1 = b_1]\!] \cdot \ldots \cdot [\![a_n = b_n]\!] \leq [\![g(a_1, \ldots, a_n) = g(b_1, \ldots, b_n)]\!]$

To extend satisfaction into second-order logic, we simply add some more recursion clauses. (As with the Robinsonian approach to Henkin semantics in the ordinary case, $R_g^{\mathcal{M}^{\circ}} = g$ for each $g \in M_n^{\text{rel}}$, and $f_g^{\mathcal{M}^{\circ}} = g$ for each $g \in M_n^{\text{fun}}$.)

$$[\![\exists X\varphi(X)]\!]^{\mathcal{M}^{\circ}} = \sum_{g \in M_n^{\text{rel}}} [\![\varphi(R_g)]\!]^{\mathcal{M}^{\circ}} \qquad [\![\forall X\varphi(X)]\!]^{\mathcal{M}^{\circ}} = \prod_{g \in M_n^{\text{rel}}} [\![\varphi(R_g)]\!]^{\mathcal{M}^{\circ}}$$

$$[\![\exists p\varphi(p)]\!]^{\mathcal{M}^{\circ}} = \sum_{g \in M_n^{\text{fun}}} [\![\varphi(f_g)]\!]^{\mathcal{M}^{\circ}} \qquad [\![\forall p\varphi(p)]\!]^{\mathcal{M}^{\circ}} = \prod_{g \in M_n^{\text{fun}}} [\![\varphi(f_g)]\!]^{\mathcal{M}^{\circ}}$$

Finally, we call a Henkin $\mathcal{B}$-valued $\mathscr{L}$-structure *faithful* iff $[\![\varphi]\!] \in D$ where φ is any instance of the Comprehension Schema or Choice Schema, as defined in §1.11. Then, where $T \cup \{\varphi\}$ is any second-order theory, we write:

$T \overset{\mathcal{B}}{\underset{D}{\models}} \varphi$ iff: for every faithful $\mathcal{B}$-valued $\mathscr{L}$-structure $\mathcal{M}$, if $[\![T]\!]^{\mathcal{M}} \subseteq D$ then $[\![\varphi]\!]^{\mathcal{M}} \in D$

In what follows, we often speak indifferently about first-order and second-order logics. However, when considering second-order logic in this chapter, we always have *faithful* Henkin $\mathcal{B}$-valued structures in mind (though we discuss 'full' second-order

$\mathcal{B}$-valued structures in §13.B). So a phrase like 'let $\mathcal{M}$ be a (faithful Henkin) $\mathcal{B}$-valued structure' can be read, either as talking about a (first-order) $\mathcal{B}$-valued structure, or a (second-order) faithful Henkin $\mathcal{B}$-valued structure.

13.4 Semantic-underdetermination via filters

We saw in §13.1 that the natural deduction rules for first-order logic (or faithful Henkin second-order logic) fail to determine (*a*) *what* the semantic values are. Using the Boolean framework of §13.3, we will now prove results which show that the natural deduction rules also fail to determine (*b*) *how many* semantic values there are and (*c*) exactly how those semantic values *interact* with the logical vocabulary of connectives and quantifiers.

These limitative results concerning Boolean algebras illustrate the 'weakness' of both first-order logic and faithful Henkin second-order logic. In this regard, the results are comparable to the Compactness and Löwenheim–Skolem Theorems. These latter two results are very well known; and we think that the results concerning Boolean algebras deserve to be much better known.[6] They are of comparable technical interest: where the Compactness and Löwenheim–Skolem theorems form part of the backbone of model theory, the results concerning Boolean algebras connect with set-theoretic forcing.[7] Additionally, they are of comparable philosophical interest: where the Compactness and Löwenheim–Skolem theorems supplied the impetus for Chapters 7–8, the results concerning Boolean algebras give teeth and claws to the semantic-underdetermination argument.

Any filter will do

Our first result concerning Boolean algebras says roughly this: everything works out fine if your designated values form a filter on a complete Boolean algebra. Here is the precise statement:[8]

Theorem 13.6: *If $\mathcal{B}$ is a complete Boolean algebra with $D \subseteq B$, then these are equivalent:*

(1) *D is a filter on $\mathcal{B}$*
(2) *$T \vdash \varphi$ iff $T \overset{\mathcal{B}}{\underset{D}{\vDash}} \varphi$, for any theory $T \cup \{\varphi\}$*

The proof of this result involves grinding through the soundness of each of the inference rules with respect to $\overset{\mathcal{B}}{\underset{D}{\vDash}}$; we leave this grind to §13.A. But Theorem 13.6 shows

[6] Cf. Smiley (1996: 8).

[7] See e.g. Bell (2005).

[8] This generalises e.g. Church (1944: 494, 1953: 41–2), Bell (2005: Theorem 1.17 pp.24–6), Kaye (2007: Theorem 7.10 pp.84–5), and Button (2016b: Theorem 1), and many other discussions.

that $\vdash$ is sound and complete for many different, non-isomorphic, semantic structures. In a slogan: *any filter on any complete Boolean algebra will do.*

This is a bit abstract, so here is a very simple illustration of the point. Consider the four-element Boolean algebra, $\mathcal{Four}$, whose Hasse diagram is as follows:[9]

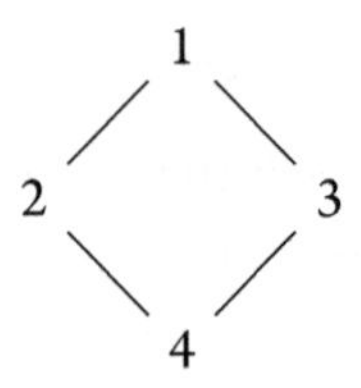

Trivially, $\{1\}$ is a filter on $\mathcal{Four}$. So $\overset{\mathcal{Four}}{\underset{\{1\}}{\vDash}}$ is coextensive with $\vdash$, by Theorem 13.6. So our inference rules fail to determine that our semantics has only two values, rather than four. Indeed, since there are arbitrarily large complete Boolean algebras, our inference rules fail to pin down much at all about the number of semantic values. And this yields a rather more interesting version of the semantic-underdetermination argument, than that first raised by the Push-Through Construction in 13.1.[10]

Designation Principles and Truth Principles

Even if the inference rules cannot pin down the semantic structure up to isomorphism, we might hope that they can pin down the relationship between designated and undesignated values. In particular, we might hope to secure the following *Designation Principles:*[11]

Not-Designation. *$\neg\varphi$ is designated iff φ is undesignated*
i.e. $[\![\neg\varphi]\!] \in D$ iff $[\![\varphi]\!] \notin D$
And-Designation. *$\varphi \wedge \psi$ is designated iff both φ and ψ are designated*
i.e. $[\![\varphi \wedge \psi]\!] \in D$ iff both $[\![\varphi]\!] \in D$ and $[\![\psi]\!] \in D$
Or-Designation. *$\varphi \vee \psi$ is designated iff either φ or ψ is designated*
i.e. $[\![\varphi \vee \psi]\!] \in D$ iff either $[\![\varphi]\!] \in D$ or $[\![\psi]\!] \in D$
All-Designation. *$\forall x\varphi(x)$ is designated iff $\varphi(c_a)$ is designated for all $a \in M$*
i.e. $[\![\forall x\varphi(x)]\!] \in D$ iff $[\![\varphi(c_a)]\!] \in D$ for all $a \in M$
Exists-Designation. *$\exists x\varphi(x)$ is designated iff $\varphi(c_a)$ is designated for some $a \in M$*
i.e. $[\![\exists x\varphi(x)]\!] \in D$ iff $[\![\varphi(c_a)]\!] \in D$ for some $a \in M$

---

[9] This example occurs in Church (1944: 494). The diagram depicts the order relation: $x \leq y$ iff there is a continuous *upward* path from x to y. So: $-2 = 3$, $2 \cdot 3 = 0$, and $2 + 3 = 1$.

[10] Carnap (1943: 73ff) and Church (1944: 494, 1953: 50) were the first philosophers to discuss Boolean algebras in this connection, and to consider this version of the semantic-underdetermination argument.

[11] We can formulate similar principles for the second-order quantifiers. The proof of Theorem 13.8 straightforwardly shows that they are satisfied when D is a principal ultrafilter.

Indeed, if we could secure these principles, then we could recover a version of bivalence. For, if we simply read 'designated' as 'true', then the Designation Principles become the *Truth Principles*:

Not-Truth. *$\neg\varphi$ is true iff φ is false*
And-Truth. *$\varphi \wedge \psi$ is true iff both φ and ψ are true*
Or-Truth. *$\varphi \vee \psi$ is true iff either φ or ψ is true*
All-Truth. *$\forall x\varphi(x)$ is true iff $\varphi(c_a)$ is true for all $a \in M$*
Exists-Truth. *$\exists x\varphi(x)$ is true iff $\varphi(c_a)$ is true for some $a \in M$*

These are precisely the principles that fans of bivalence will want to secure. Indeed, if we *can* secure these Designation/Truth Principles, then whatever point there might be in distinguishing among the 'different ways of being true' (i.e. the different members of D) and among the 'different ways of being false' (i.e. the different members of $B \setminus D$), it would not come through in our logic.[12] At that point, one might be able to argue that this is all the determinacy of meaning which inference rules *should* secure.

Unfortunately, the case of $\mathcal{Four}$ shows that the usual inference rules do *not* secure the Designation Principles. If $[\![\varphi]\!]^{\mathcal{Four}} = 2$, then $[\![\neg\varphi]\!]^{\mathcal{Four}} = -[\![\varphi]\!]^{\mathcal{Four}} = -2 = 3$, and neither 2 nor 3 is a member of our set of designated values, $\{1\}$. So our $\mathcal{Four}$-valued semantic structure violates Not-Designation. It is easy to see that it similarly violates Or-Designation and Exists-Designation. So: the classical inference rules cannot secure the Designation Principles.

In fact, within the Boolean framework, there is a simple necessary and sufficient condition on the Designation Principles. To explain it, we need a definition:

Definition 13.7: *A filter D on a Boolean algebra $\mathcal{B}$ is* principal *iff there is some $a \in B$ such that $D = \{b \in B : a \leq b\}$.*

Then we have the following (again, the proof is in §13.A):

Theorem 13.8: *If $\mathcal{B}$ is a complete Boolean algebra and $D \subseteq B$ is a filter on $\mathcal{B}$, then:*

(1) *D is an ultrafilter on $\mathcal{B}$ iff Not-Designation, And-Designation and Or-Designation hold in every (faithful Henkin) $\mathcal{B}$-valued structure*
(2) *D is a principal ultrafilter on $\mathcal{B}$ iff all the Designation Principles hold in every (faithful Henkin) $\mathcal{B}$-valued structure*

Combining Theorem 13.8 with 13.6, we immediately obtain:

[12] See Dummett (1959: 154–5).

Corollary 13.9: *If $\mathcal{B}$ is a complete Boolean algebra and $D \subseteq B$ is a filter but not a principal ultrafilter on $\mathcal{B}$, then $T \vdash \varphi$ iff $T \models^{\mathcal{B}}_{D} \varphi$, for any theory $T \cup \{\varphi\}$, but some Designation Principle fails in some (faithful Henkin) $\mathcal{B}$-valued structure.*

In summary: our inference rules fail to pin down the semantic values True and False, by the Push-Through Construction. They fail to pin down a semantic structure up to isomorphism, by Theorem 13.6. Indeed, they fail even to secure the Designation Principles, by Corollary 13.9. These results give real bite to our initial semantic-underdetermination argument.

In the rest of this chapter, we consider several responses to the semantic-underdetermination argument, and show how it gives rise to similar issues as those in Chapters 6–11.

13.5 Semanticism

First and foremost, the semantic-underdetermination argument raises a problem for *inferentialism*: the view that the meaning of the logical expressions *is* fully determined by their inference rules. But we will start by raising some problems for inferentialism's main rival, *semanticism*. According to semanticists, logical expressions are given their meanings by their semantic conditions. Our aim is to show that semanticism faces the same sorts of issues as did modelism.

Concerning the Push-Through Construction

Our semanticist might start by resisting even the level of underdetermination threatened by the Push-Through Construction in §13.1. So: she might insist that she can unproblematically refer to the usual two semantic values, True and False. She might then claim that the meanings of the logical expressions can be given just by *stipulating* their semantic conditions.

This response to the semantic-underdetermination argument is extremely similar to certain responses to the permutation argument (see Chapter 2). When the Push-Through Construction is used to show that the truth values of the sentences in a language fail to determine that language's interpretation, a natural reply is that something *else* fixes the interpretation: that there are *preferable* interpretations. Now that the Push-Through Construction is being used to show that the inference rules fail to determine the semantics, our semanticist has replied that something *else* fixes the semantics.

This analogy with the permutation argument is worth pursuing. The objects-platonist owed us an account of how we refer to *abstract* mathematical entities. Similarly, whatever semantic values are, exactly, they are presumably *abstract*:

metaphors aside, one does not trip over True, or bump up against False. So we must ask our semanticist a doxological question: *How can we determinately refer to True and False?* And the Push-Through Construction gives the question bite.

Of course, this is not a knockdown argument against semanticists who insist that they *can* unproblematically refer to True and False. By comparison: the permutation argument does not show that *all* objects-platonists must accept that mathematical language is radically referentially indeterminacy. Rather, it shows that *moderate* objects-platonists must accept that indeterminacy. The same point applies here. Faced with the Push-Through Construction, *moderate* semanticists must accept that we cannot refer to the *particular* entities, True and False.

Concerning Theorem 13.6

Moderate semanticists may yet insist that they can provide the logical expressions with exactly as precise a meaning as they *should* have. In particular, they may insist that they can lay down the meaning of the logical vocabulary *up to isomorphism.*

By comparison: a moderate objects-platonist faced with the Push-Through Construction may well decide to become a moderate objects-modelist, who aims only to pin down the natural numbers *up to isomorphism.* We first mentioned this option in §2.5, we outlined a problem facing this moderate objects-modelist in Chapter 7: no theory with a finitary proof system can pin down the natural numbers up to isomorphism (see §7.9). The moderate semanticist faces a similar problem: by Theorem 13.6, our inference rules cannot pin down the semantic structure up to isomorphism.

But a response *seems* to be available to moderate *semanticists* which was not available to moderate *objects-platonists.* For there *is* a categorical first-order theory of two-element Boolean algebras: just let CBA be the theory of Boolean algebras (from Definition 13.1) together with the axiom $\forall x(x = 0 \vee x = 1)$. Moderate semanticists may, then, deploy CBA in order to try to pin down the semantic structure up to isomorphism.

Sadly, though, this appeal to CBA amounts to *just more theory,* in exactly the way in which we used that phrase in Chapters 2 and 7–9. Certainly CBA is categorical on the *usual* semantics for first-order logic: within that semantics, all of CBA's models are isomorphic to the structure $\mathcal{T}wo$. But CBA is *not* categorical for *Boolean-valued* semantics. Where $\mathcal{B}$ is *any* complete Boolean algebra, we can easily treat $\mathcal{B}$ itself as a $\mathcal{B}$-valued model of CBA. More precisely (saving the proof for §13.D):

Proposition 13.10: *For any complete Boolean algebra $\mathcal{B}$, there is a $\mathcal{B}$-valued model of* CBA *whose domain is B.*

So, let $\mathcal{B}$ be any complete Boolean algebra with more than two elements, and let D be any filter on $\mathcal{B}$. By Theorem 13.6, $\vdash$ is coextensive with $\overset{\mathcal{B}}{\underset{D}{\vDash}}$. But there is a two-element $\mathcal{B}$-valued model of CBA, provided by *Two*, and a $|B|$-element $\mathcal{B}$-valued model of CBA, provided by Proposition 13.10. So CBA is *not* categorical on the semantics provided by $\mathcal{B}$ and D.

Consequently, the semanticist can insist on the categoricity of CBA, only if she can somehow insist that the two-valued semantics is *preferable* to any other semantics. She now faces a familiar challenge. Whatever she says in attempting to explain the preferability of the two-valued semantics—indeed, even as she tries to say that it has *two* values—Proposition 13.10 shows that we can interpret her as describing an algebra with *four* values (for example). Indeed, her attempt to insist on the preferability of two-valued semantics is *just more theory*, in precisely Putnam's pejorative sense. In this regard, the moderate semanticist shares the same fate as a moderate modelist who wants to insist that full second-order semantics is *preferable* to Henkin semantics (see §§7.7–7.10).

As such, moderate semanticists must cede the point concerning Theorem 13.6, and accept that they cannot pin down the semantic structure up to isomorphism.

Concerning Corollary 13.9

A similar argument shows that moderate semanticists cannot succeed in delivering the Designation Principles of §13.4. By Theorem 13.8, to insist upon the Designation Principles within the Boolean framework is to insist that the designated values form a principal ultrafilter. It is hard to see how semanticists could claim to have a handle on the semantics to *this* extent, but no further.

To reinforce this point, we can invoke the just-more-theory manoeuvre again. If the moderate semanticist says that the semantic values form a principal ultrafilter, we should understand her as trying to say that certain semantic structures are *preferable*. But in saying this, she must produce a theory describing principal ultrafilters on Boolean algebras. The most obvious way to do this would be to augment the theory of Boolean algebras with some new axioms concerning a new one-place predicate, D, which are intended to tell us that (the interpretation of) D is a principal ultrafilter. Unfortunately for her, if unsurprisingly, we can supply a $\mathcal{B}$-valued model of *this* theory, which does not interpret D as anything like a principal ultrafilter (see §13.D).

As such, moderate semanticism fails. Although *im*moderate semanticism may, for all this, succeed, in what follows, we set aside semanticism altogether, and consider three *inferentialist* responses to the semantic-underdetermination argument. These responses all try to retain the idea that the meaning of the logical connectives *is*—somehow—determined by their inference rules after all.

13.6 Bilateralism

Our first inferentialist response is *bilateralism*, as advanced by Smiley.[13] Bilateralists attempt to resist the semantic-underdetermination argument by broadening the idea of what counts as an inference rule. They do this by adding two *force indicators* to their metalanguage: an indicator for *assertion*, symbolised with $\uparrow$, and an indicator for *rejection*, symbolised with $\downarrow$. Bilateralists then lay down inference rules which employ these force indicators. The hope is that these *richer* inference rules, decorated with force indicators, can deal with semantic-underdetermination.

Bilateralism fails in this regard. To succeed, bilateralism requires a very particular relationship between the force indicators and the semantics for the language. And, as we will explain, the attempt to pin down that relationship is just more theory.

Decorated proofs

We start by outlining the bilateralist inference rules. First, we present the inference rules for the connectives of first-order logic. These rules may usefully be compared with those of §1.c. (The same kind of caveat applies as in §1.c, that in $\forall\mathrm{I}\updownarrow$ and $\exists\mathrm{I}\updownarrow$ the symbol c must occur in no undischarged assumption.)[14]

$$\frac{\downarrow\varphi}{\uparrow\neg\varphi}\,\neg\mathrm{I}\updownarrow \qquad \frac{\uparrow\neg\varphi}{\downarrow\varphi}\,\neg\mathrm{E}\updownarrow$$

$$\frac{\uparrow\varphi \quad \uparrow\psi}{\uparrow(\varphi\wedge\psi)}\,\wedge\mathrm{I}\updownarrow \qquad \frac{\uparrow(\varphi\wedge\psi)}{\uparrow\varphi}\,\wedge\mathrm{E}\updownarrow \qquad \frac{\uparrow(\varphi\wedge\psi)}{\uparrow\psi}\,\wedge\mathrm{E}\updownarrow$$

$$\frac{\downarrow\varphi \quad \downarrow\psi}{\downarrow(\varphi\vee\psi)}\,\vee\mathrm{I}\updownarrow \qquad \frac{\downarrow(\varphi\vee\psi)}{\downarrow\varphi}\,\vee\mathrm{E}\updownarrow \qquad \frac{\downarrow(\varphi\vee\psi)}{\downarrow\psi}\,\vee\mathrm{E}\updownarrow$$

$$\frac{\uparrow\varphi(c)}{\uparrow\forall x\varphi(x)}\,\forall\mathrm{I}\updownarrow \qquad \frac{\uparrow\forall x\varphi(x)}{\uparrow\varphi(c)}\,\forall\mathrm{E}\updownarrow$$

$$\frac{\downarrow\varphi(c)}{\downarrow\exists x\varphi(x)}\,\exists\mathrm{I}\updownarrow \qquad \frac{\downarrow\exists x\varphi(x)}{\downarrow\varphi(c)}\,\exists\mathrm{E}\updownarrow$$

$$\frac{}{\uparrow c=c}\,{=}\mathrm{I}\updownarrow \qquad \frac{\uparrow s=t \quad \uparrow\varphi(s)}{\uparrow\varphi(t)}\,{=}\mathrm{E}\updownarrow \qquad \frac{\uparrow t=s \quad \uparrow\varphi(s)}{\uparrow\varphi(t)}\,{=}\mathrm{E}\updownarrow$$

As an heuristic, the rule $\neg\mathrm{I}\updownarrow$ might be read roughly as follows: if it is correct to reject φ, then it is correct to assert $\neg\varphi$.[15] Next, we have four versions of reductio:

[13] See Smiley (1996: 6–9). Note that our discussion of bilateralism is entirely restricted to its supposed role in answering the semantic-underdetermination argument; there may be other motivations for bilateralism which are not threatened by our discussion.

[14] Incurvati and P. Smith (2010: 5) and Hjortland (2014: 452) lay down the sentential rules; we have added the obvious rules for quantifiers and identity.

[15] But this is *just* a rough gloss; we are not suggesting that bilateralists must take any particular stance on

$$\dfrac{\dfrac{}{\uparrow\varphi}\, n \quad \vdots \quad \uparrow\psi \quad \downarrow\psi}{\downarrow\varphi}\ \text{Raa}\updownarrow, n \qquad \dfrac{\dfrac{}{\uparrow\varphi}\, n \quad \vdots \quad \downarrow\psi \quad \uparrow\psi}{\downarrow\varphi}\ \text{Raa}\updownarrow, n \qquad \dfrac{\dfrac{}{\downarrow\varphi}\, n \quad \vdots \quad \uparrow\psi \quad \downarrow\psi}{\uparrow\varphi}\ \text{Raa}\updownarrow, n \qquad \dfrac{\dfrac{}{\downarrow\varphi}\, n \quad \vdots \quad \downarrow\psi \quad \uparrow\psi}{\uparrow\varphi}\ \text{Raa}\updownarrow, n$$

These are all the decorated rules of our decorated proof system for first-order logic. We can easily expand the system to second-order logic, by presenting analogous inference rules for the second-order quantifiers, and treating any asserted Comprehension Schema or Choice instance as an axiom (as with $=\text{I}\updownarrow$).

We write $\updownarrow\varphi$ to indicate that the sentence φ has been decorated with a single force indicator, but we are indifferent as to which. Using this terminology, we say:

> $T \Vdash \updownarrow\varphi$ iff there is a decorated proof whose only undischarged assumptions are among T and which ends with $\updownarrow\varphi$

In a very straightforward sense, decorated proofs provide an elegant alternative to undecorated proofs. To show this, we define a metalinguistic operation, $\flat$, which flattens decorations as follows:

$$\flat(\uparrow\varphi) := \varphi \qquad\qquad \flat(\downarrow\varphi) := \neg\varphi$$

As usual, $\flat(T) = \{\flat(\updownarrow\varphi) : \updownarrow\varphi \in T\}$. We now have:

Theorem 13.11: *$T \Vdash \updownarrow\varphi$ iff $\flat(T) \vdash \flat(\updownarrow\varphi)$, for any decorated theory $T \cup \{\updownarrow\varphi\}$*

The proof simply consists in mechanically turning decorated proofs into undecorated ones, and vice versa; we relegate it to §13.E.

Semantic underdetermination again

However, Theorems 13.6 and 13.11 jointly entail that decorated proofs are sound and complete for any semantics given by any filter on any complete Boolean algebra:

Corollary 13.12: *If $\mathcal{B}$ is a complete Boolean algebra and $D \subseteq B$ is a filter on $\mathcal{B}$, then $T \Vdash \updownarrow\varphi$ iff $\flat(T) \overset{\mathcal{B}}{\underset{D}{\vDash}} \flat(\updownarrow\varphi)$, for any decorated theory $T \cup \{\updownarrow\varphi\}$*

And now we must ask how bilateralism could possibly *hope* to offer any resistance against the semantic-underdetermination argument. After all, Corollary 13.12 seems

the normativity of logic. However, the gloss also helps to show that bilateral proof systems are equivalent to *multiple-conclusion logics*. Multiple-conclusion logics allow that the conclusion of an argument need not be a single sentence, but a set of sentences. So we can write e.g. '$T \vdash S$', where both T and S are sets of sentences. To obtain the equivalence between multiple-conclusion logics and bilateralism, we need simply gloss '$T \vdash S$' as 'it is incorrect to assert everything in T whilst rejecting everything in S' (see Smiley 1996; Hjortland 2014). This equivalence means that all of the criticisms we applied to bilateralism apply (almost unchanged) against an inferentialist who wants to invoke multiple-conclusion logic.

to threaten bilateralism in exactly the same way that Theorem 13.6 threatens 'undecorated' inferentialism.

Bilateralists will reply that Corollary 13.12 employs an operator, $\flat$, which steamrollers overs the distinction between rejecting a sentence and asserting its negation, in the sense that $\flat(\downarrow\varphi) = \flat(\uparrow\neg\varphi)$. They will then insist that the *preferable* semantics should acknowledge the important distinction between rejecting a sentence and asserting its negation. And so, they will continue, their decorated proof-system provides novel resources for resisting the semantic-underdetermination argument.

In saying this, though, bilateralists must tread very carefully. Suppose a bilateralist simply insists that asserting φ amounts to assigning True to φ, and that rejecting φ amounts to assigning False to φ. Then she has indeed aimed to pin down the standard, two-valued semantics. Equally, though, she has simply become a peculiarly decorated *semanticist*, and we direct her back to §13.5.

Instead, bilateralists should insist that there is a particular relationship between force indicators and designation. In particular, they should suggest that it is correct to assert φ iff φ is designated, and correct to reject φ iff φ is undesignated. To explore this suggestion, we will formalise it following Murzi and Hjortland.[16] So, where $\mathcal{M}$ is a $\mathcal{B}$-valued structure and $D \subseteq B$, bilateralists will advance two Correctness Principles:

Asserted-correctness. *$\uparrow\varphi$ is D-correct in $\mathcal{M}$ iff $[\![\varphi]\!]^{\mathcal{M}} \in D$*
Rejection-correctness. *$\downarrow\varphi$ is D-correct in $\mathcal{M}$ iff $[\![\varphi]\!]^{\mathcal{M}} \notin D$*

And they will then define logical consequence as *preservation of D-correctness*:

$T \overset{\mathcal{B}}{\underset{D}{\vDash}} \updownarrow\varphi$ iff: for every $\mathcal{B}$-valued model $\mathcal{M}$, if every $\updownarrow\psi \in T$ is D-correct in $\mathcal{M}$, then $\updownarrow\varphi$ is D-correct in $\mathcal{M}$

Moreover, this notion of logical consequence connects immediately with the Designation Principles (see §13.E):

Corollary 13.13: *If $\mathcal{B}$ is a complete Boolean algebra and $D \subseteq B$ is a filter on $\mathcal{B}$, then these are equivalent:*

(1) *$T \overset{\mathcal{B}}{\underset{D}{\vDash}} \updownarrow\varphi$ iff $T \vdash \updownarrow\varphi$*
(2) *Not-Designation, And-Designation and Or-Designation hold for any (faithful Henkin) $\mathcal{B}$-valued structure*

So: *if* bilateralists can insist that the semantic structure must respect the Correctness Principles, then they can secure *some* of the Designation Principles, by Corollary 13.13.

[16] Murzi and Hjortland (2009: 485–6) and Hjortland (2014: 253).

Note, at once, that this ambition falls well short of pinning down the semantic structure up to isomorphism.[17] Indeed, it fails even to secure *all* the Designation Principles. Let $\mathcal{B}$ be a complete Boolean algebra and let D be a *non*-principal ultrafilter on $\mathcal{B}$. By Theorem 13.8(1), the semantic structure guarantees all of Not-, And- and Or-Designation. So, by Corollary 13.13, using $\mathcal{B}$ and D will respect the bilateralist's Correctness Principles. But because D is not principal, by Theorem 13.8(2), there is a (faithful Henkin) $\mathcal{B}$-valued structure $\mathcal{M}$ where All-Designation and Exists-Designation fail. If we now gloss 'designated' as 'true', as in §13.4, this is to say that we can find a (faithful Henkin) $\mathcal{B}$-valued structure, $\mathcal{M}$, where $\varphi(c_a)$ is true for every $a \in M$, but where $\forall x\varphi(x)$ is *false*. This is surely an unacceptable form of semantic-underdetermination if any is. But bilateralism is utterly powerless to rule it out.

We must also ask whether bilateralists *can* legitimately insist on their Correctness Principles. To see the problem, we should ask why the bilateralist's rejection-correctness principle is preferable to the following alternative:

Rejection-correctness (alternative). *$\downarrow\varphi$ is D-correct in $\mathcal{M}$ iff $[\![\neg\varphi]\!]^{\mathcal{M}} \in D$*

This alternative principle would again collapse the difference between rejecting φ and asserting $\neg\varphi$. Consequently, defining $\overset{\mathcal{B}}{\underset{D}{\models}}$ using this alternative principle would return us to the setting where *any* filter on any complete Boolean algebra would yield an adequate semantic structure. Bilateralists need, then, to explain why we should favour their *original* rejection-correctness principle over this alternative.

To achieve this, it is not enough for bilateralists to show that there *is* some natural-language difference between rejecting a sentence and asserting its negation.[18] They must also show that this natural-language difference should have very *specific* consequences for our formal semantics. And showing that is a much taller order.

Indeed, the problem now facing bilateralism is entirely predictable. Bilateralists want to insist that a somewhat constrained formal semantics is *preferable* to a marginally more general alternative. (Compare the situation facing a moderate modelist who wanted to insist that full second-order logic is *preferable* to Henkin semantics, as discussed in Chapter 7.) But anything they say to emphasise the difference between rejecting a sentence and asserting its negation, can be underpinned by a semantic structure which flattens that difference. In short: as it stands, any attempt to insist on the Correctness Principles is *just more theory*, in precisely Putnam's sense.

[17] Slightly confusingly, several authors in this literature say that bilateralism promises to deliver 'categoricity', in effect to mean merely that bilateralism promises to deliver the (sentential) Designation Principles (see e.g. Smiley 1996: 8–9; Hjortland 2014: 450–1), rather than that it pins down a semantic structure up to isomorphism.

[18] Much of the debate around bilateralism has focussed simply upon this question, e.g. Smiley (1996: 1–4), Incurvati and P. Smith (2010: 10), and Hjortland (2014).

13.7 Open-ended-inferentialism

We set aside bilateralism, then, and turn to a second kind of inferentialism, which we call *open-ended inferentialism.* Open-ended inferentialists insist that inference rules are importantly open-ended, and that this open-endedness is inadequately reflected in the semantic-underdetermination argument.[19] We will explain how invoking open-endedness promises to narrow down the semantic structures; however, without further supplement, invoking open-endedness is just more theory.

New connectives and inference rules

Famously, the ordinary sentential connectives are expressively adequate for the ordinary two-valued semantics: that is, every truth-functional connective is definable using the usual connectives.[20] By contrast, when we move to $\mathcal{B}$-valued semantics, we have the option to introduce some genuinely new connectives. And adding such connectives can cause some of our existing inference rules to become unsound.

Intuitively, in calling a rule *unsound*, we mean that applying the rule can take you from designation to undesignation. Specifically, in this section we will focus on the rules $\rightarrow$I, $\vee$E, $\neg$I, $\exists$E and $\forall$I, as laid down in §1.c, and we define their soundness with respect to a consequence relation $\Vdash$ as follows:[21]

$\rightarrow$I is sound iff: *if* $T \cup \{\varphi\} \Vdash \psi$, *then* $T \Vdash \varphi \rightarrow \psi$
$\vee$E is sound iff: *if* both $T \cup \{\varphi\} \Vdash \rho$ and $T \cup \{\psi\} \Vdash \rho$, *then* $T \cup \{\varphi \vee \psi\} \Vdash \rho$
$\neg$I is sound iff: *if* $T \cup \{\varphi\} \Vdash \bot$, *then* $T \Vdash \neg\varphi$
$\exists$E is sound iff: *if* $T \cup \{\varphi(c)\} \Vdash \psi$ and c does not occur anywhere in T, $\varphi(x)$ or ψ, *then* $T \cup \{\exists x\varphi(x)\} \Vdash \psi$.
$\forall$I is sound iff: *if* $T \Vdash \varphi(c)$ and c does not occur in T, *then* $T \Vdash \forall x\varphi(x)$

Clearly all of these rules are sound for the ordinary, two-valued semantics. Indeed, one usually proves the soundness of the entire deductive system with respect to that semantics by proving the soundness of each individual rule and then performing an induction on the length of a deduction. But, when we move to arbitrary Boolean semantics, we can add connectives which would make these five rules unsound.

Here is a simple example. Let $\mathcal{B}$ be a complete Boolean algebra, let D be a filter on $\mathcal{B}$ which is not principal, and let $h : B \longrightarrow B$ be given by:

$$h(a) = \begin{cases} a & \text{if } a \in D \\ -a & \text{if } -a \in D \\ 1 & \text{otherwise} \end{cases}$$

---

[19] The contents of this subsection expand upon Button (2016b).

[20] See e.g. Gamut (1991: 56) and Humberstone (2011: 403ff).

[21] We treat $\bot$ as an abbreviation for anything of the form $\varphi \wedge \neg\varphi$; so $[\![\bot]\!] = [\![\varphi \wedge \neg\varphi]\!] = [\![\varphi]\!] \cdot - [\![\varphi]\!] = 0$.

We now introduce a new one-place sentential connective, $\heartsuit$, which implements h, i.e. we stipulate that $[\![\heartsuit\varphi]\!] = h\,[\![\varphi]\!]$. We let $\vDash_{\heartsuit}$ be the consequence relation defined using the algebra $\mathcal{B}$ augmented with the function h, where D is the set of designated values, for the sentences of a logic which expands ordinary first-order logic by adding $\heartsuit$ to the logical vocabulary. Now the presence of $\heartsuit$ renders $\forall$I unsound:

Proposition 13.14: *The inference rule $\forall I$ is unsound for $\vDash_{\heartsuit}$.*

Proof. It suffices to show that $\vDash_{\heartsuit} \heartsuit F(a)$ but $\nvDash_{\heartsuit} \forall x \heartsuit F(x)$. By construction, $[\![\heartsuit F(c_a)]\!] = h\,[\![F(a)]\!] \in D$ for any a in any $\mathcal{B}$-valued structure. But now let $\mathcal{M}$ be a $\mathcal{B}$-structure obtained by augmenting $\mathcal{B}$ itself with a predicate F such that $[\![F(a)]\!] = a$ for all $a \in B$. Then $[\![\forall x \heartsuit F(x)]\!] = \prod_{a \in M} h\,[\![F(c_a)]\!] = \prod D \notin D$, since D is not principal. □

To obtain the unsoundness of the other four rules we mentioned, we will switch example. Let D be a filter on $\mathcal{B}$ which is not an *ultra*filter, and define $k : B \longrightarrow B$ as follows:

$$k(a) = \begin{cases} a & \text{if either } a \in D \text{ or } -a \in D \\ -a & \text{otherwise} \end{cases}$$

We now introduce a new one-place sentential connective, ᛟ, which implements k, i.e. we stipulate that $[\![\text{ᛟ}\varphi]\!] = k\,[\![\varphi]\!]$. We define $\vDash_{\text{ᛟ}}$ similarly to $\vDash_{\heartsuit}$, and obtain the unsoundness of several rules:

Proposition 13.15: *The inference rules $\rightarrow$I, $\vee$E, $\neg$I and $\exists$E are unsound for $\vDash_{\text{ᛟ}}$.*

Proof. $\rightarrow$I, $\vee$E and $\neg$E are covered elsewhere.[22] For $\exists$E, it suffices to show that $F(c) \wedge \neg\text{ᛟ}F(c) \vDash_{\text{ᛟ}} \bot$ but $\exists x(F(x) \wedge \neg\text{ᛟ}F(x)) \nvDash_{\text{ᛟ}} \bot$.

For the first part, just note that for any a in any $\mathcal{B}$-valued structure we have: $[\![F(c_a) \wedge \neg\text{ᛟ}F(c_a)]\!] = [\![F(a)]\!] \cdot -k\,[\![F(a)]\!] \notin D$. To see this, let $[\![F(a)]\!] = d$. Then if either $d \in D$ or $-d \in D$, we have $d \cdot -k(d) = d \cdot -d = 0$; and otherwise, $d \cdot -k(d) = d \cdot --d = d \notin D$.

For the second part, we define a structure where $[\![\exists x(F(x) \wedge \neg\text{ᛟ}F(x))]\!] = 1$. Invoking the fact that D is not an ultrafilter, choose $d \in B$ such that $d \notin D$ and $-d \notin D$. Let $\mathcal{M}$ be a $\mathcal{B}$-valued structure containing elements a and b such that $[\![F(a)]\!] = d$ and $[\![F(b)]\!] = -d$. Now $[\![F(c_a) \wedge \neg\text{ᛟ}F(c_a)]\!] = d \cdot -k(d) = d$ and $[\![F(c_b) \wedge \neg\text{ᛟ}F(c_b)]\!] = -d \cdot -k(-d) = -d$, so that $[\![\exists x(F(x) \wedge \neg\text{ᛟ}F(x))]\!] = 1$. □

The general point is simple. By allowing arbitrary filters on arbitrary complete Boolean algebras, we can render certain inference rules unsound.

[22] See Button (2016b: Theorem 2).

Inference rules as open-ended

Open-ended-inferentialist will seize upon these rule-violations, in an attempt to explain what is wrong with the original semantic-underdetermination argument.

As good inferentialists, open-ended-inferentialists maintain that inference rules determine the meaning of logical vocabulary. But, they continue, an important part of the inference rules $\rightarrow$I, $\vee$E, $\neg$I, $\exists$E, and $\forall$I is that they are to hold *always and without exception*. That is: *it is impossible to add connectives to our language which would ever make it illegitimate to use the rules* $\rightarrow$*I*, $\vee$*E*, $\neg$*I*, $\exists$*E or* $\forall$*I*. Now suppose, they continue, that the designated values constitute a filter, but—for reductio—not a principal ultrafilter. Then either h or k, as defined above, would be *bona fide* functions on the semantic values. So, presumably either $\heartsuit$ or 🎗 would be a *bona fide* sentential connective, and nothing would prevent them from being added to our language. But, open-ended inferentialist will say, we *are* prevented from adding these connectives to our language. For, adding these connectives would make certain inference rules unsound, even though they have been laid down so as to hold *always* and without exception. As such, they conclude, the designated values must form a *principal ultrafilter*. And so, by Theorem 13.8, within the Boolean setting this secures all of the Designation Principles.[23]

This is quite some achievement; but open-ended-inferentialists need not stop there. It turns out that $\vDash_{\text{🎗}}$ does not *merely* flout certain rules of inference. It also violates the principle *Substitutivity of Equivalents*; i.e. the principle that substituting a subsentence for a logically equivalent subsentence never affects entailment. For it is clear that φ and $\text{🎗}\varphi$ are logically equivalent, in the sense that $\varphi \vDash_{\text{🎗}} \text{🎗}\varphi$ and $\text{🎗}\varphi \vDash_{\text{🎗}} \varphi$; however $\vDash_{\text{🎗}} \varphi \rightarrow \varphi$ but $\nvDash_{\text{🎗}} \varphi \rightarrow \text{🎗}\varphi$, so that Substitutivity of Equivalents is unsound in $\vDash_{\text{🎗}}$. So, suppose the open-ended inferentialist also insists: *It is impossible to add connectives to our language which would make it illegitimate to use the Substitutivity of Equivalents.* This stipulation again seems in the spirit of inferentialism: it mentions only inferential concerns, and inferentialists can insist upon Substitutivity of Equivalents as a constraint on inference (perhaps as a structural rule). However, assessed at the semantic level, this rule uniquely determines that, if our semantic structure is a complete Boolean algebra, then it must be a *two-valued* Boolean algebra.[24] That is, within the Boolean setting, we obtain the *categoricity* of the semantic structure: we pin it down *up to isomorphism*.[25]

---

[23] Cf. Button (2016b: Theorems 3–4).

[24] See Button (2016b: Theorem 6).

[25] But there is an easy way to introduce a non-standard semantic structure by *abandoning* the Boolean setting. Let $\mathcal{Hyp}$ be an object which gets its own very special (non-Boolean) semantics: it assigns *every* sentence True. Say that $\mathcal{M}$ is a *hyperstructure* iff either $\mathcal{M}$ is a *Two*-valued structure or $\mathcal{M} = \mathcal{Hyp}$. We now lay down our deviant semantics:

$$T \vDash_{T} \varphi \text{ iff: for every hyperstructure } \mathcal{M}, \text{ if } [\![T]\!]^{\mathcal{M}} = \{\text{True}\}, \text{ then } [\![\varphi]\!]^{\mathcal{M}} = \text{True}$$

Open-endedness and impossibility

We can see, then, why open-ended-inferentialism is promising. Unfortunately, we think it falls short of delivering on that promise.

The open-ended-inferentialist attempted to rule out certain interpretations via a modal stipulation: *It is impossible to add connectives to our language....* In order to have the desired effect, this must be a very specific sort of *impossibility*. For example: suppose this *impossibility* does not indicate that there *is* no connective which implements the function k, but that we are somehow blocked from *formulating* a connective which implements that function. Then a semantics where the designated values fail to form an ultrafilter may yet be appropriate after all.

For comparison, consider how we might react to non-standard models of PA or PA_2. Within the metalanguage, we can say that these unintended models violate an instance of induction which is not 'comprehended' in the unintended model. (That is just one way to explain the relationship between the existence of non-standard Henkin models and Dedekind's Categoricity Theorem 7.3.) Having said this, we might go on to say that the non-standard models are *unintended*, because induction is supposed to be *totally open-ended*. The logical option (a) of §7.10 was an attempt to implement this idea of open-endedness, and the motivating idea behind it was that induction should hold in 'all possible extensions of our language'. But, as we explained at the time, this idea runs headlong into the just-more-theory manoeuvre: to explain what 'all possible extensions of our language' amounts to, we must already have grasped the semantic notions which are employed in full second-order logic. This is not to say that it is *wrong* to appeal to logical option (a), any more than it is *wrong* to use full second-order logic. It is just to say that the response cannot save the *moderate* modelist, since from her point of view it will count as *just more theory*.

The issue for the open-ended inferentialist is essentially the same. It is not that open-ended inferentialism is *wrong*. It is just that, unless more can be said, there is a question as to why the open-ended inferentialist's mention of 'possibility', when considering the definability of new connectives, should be sufficiently encompassing that it *genuinely* rules out non-standard semantic structures.

13.8 Internal-inferentialism

With two unsuccessful attempts to save the moderate inferentialist behind us, we turn to a position which we call *internal-inferentialism*. The position is interesting in

A moment's reflection will convince you that the presence of $\mathcal{H}yp$ makes no difference to logical consequence, i.e. that $T \vDash_{\mathcal{T}} \varphi$ iff $T \vdash \varphi$. However, this semantics specifically violates the principles of Not-Truth / Not-Designation. And nothing that the open-ended-inferentialist has said about her inference rules rules out this (mildly deviant) approach to semantics. For more , see Smiley (1996: 7–8), Murzi and Hjortland (2009), and Incurvati and P. Smith (2010: 6–7).

its own right, but it also connects with the internalism about model theory outlined in §12.3.

We want to begin by agreeing with Raatikainen's remark, that there is not much mileage in any version of inferentialism which simply *gives up* on the use of semantic predicates *altogether*.[26] Perhaps we could give up on truth-talk in certain *specific* regions. But, as a matter of fact, we *do* employ semantic terminology in wide swathes of our daily lives and in our philosophy. And when we do so, we typically make use of the Truth Principles which we set down in §13.4.[27] Now, inferentialists cannot think that their inference rules secure the Truth Principles; this follows from our discussion of Corollary 13.9 in §13.4. So, if the inferentialist says nothing more, then she will have to *abandon* the Truth Principles altogether. But then it seems like she will have to give up altogether on truth-talk as we know it. And that would be a disaster. In short: internalists must hope to win the right to use semantic predicates.

Internal-inferentialists approach this issue, by saying that they understand semantic predicates themselves in an *inferentialist* fashion. In particular, they insist that the meaning of the truth-predicate is determined by its inferential rules, and that these rules directly license the Truth Principles. The simplest way in which this would happen is if the Truth Principles *themselves* were among the rules governing the truth-predicate. Alternatively, the Principles might be entailed by certain *other* inferential rules. We will say more on this below. However, the general idea is to treat truth-predicates inferentially. Not only is this *prima facie* reasonable, but it also sits well with the spirit of inferentialism about logico-cum-semantic notions.

This response immediately gives rise, though, to a serious question. *Every consistent theory not only has many models, in the usual sense, but also has 'models' underpinned by arbitrary filters on $\mathcal{B}$-valued $\mathcal{L}$-structures. And this includes your truth-talk. So how can producing it pin down any* PARTICULAR *semantic structure?*

To explain how internal-inferentialists reply to this question, we revisit internalism about model theory, from §12.3. We saw that internalists about model theory will be asked how they pin down a particular 'interpretation'—in the sense of an isomorphism type—of their model theory, MT_{int}. In response, they simply deny the *need* to pin down any such thing, and insist that nothing is missing from a deductive understanding of MT_{int}.

Internal-inferentialists should take a leaf from the same book. They have been asked how to pin down a particular semantic structure, and they will answer by denying the *need* to do so. They claim to understand semantic vocabulary in *purely* inferentialist terms,[28] and—as inferentialists–boggle at the (semanticist) sugges-

---

[26] See Raatikainen (2008: 285–7).

[27] Or, rather: we do so for a very broad class of sentences, though we might allow exceptions to handle vagueness, or the semantic paradoxes, or whatever. We will not repeat this caveat in what follows.

[28] We should emphasise that the internal-inferentialist need not hold that the meaning of *every* word can be exhausted via its inference rules. On the contrary: the meaning of 'pink cupcake' must ultimately

tion that something is missing from this understanding.

This 'resolutely internal' version of inferentialism surely represents the best line of defence for moderate inferentialists against the semantic-underdetermination argument. But we should stress that it is only the *start* of a defence. Everything will turn on how, exactly, internal-inferentialists specify the rules which are to license the Truth Principles. We already mentioned that the Principles could be introduced directly. Equally, they could be introduced for large fragments of the inferentialist's home language via either of the approaches discussed in §12.3–12.4. Indeed, in principle there are at least as many ways to be an internalist-inferentialist as there are formal theories of truth, and we cannot hope to survey that space of possibilities.[29] So—for now—our investigation of the interplay between inferentialism and Boolean-valued semantics must come to an end.

13.9 Suszko's Thesis

In this final section, we will consider something like the *inverse* of the problem that occupied us in this chapter. In brief: almost any relation of logical consequence can be given a *two-valued* semantics, and this gives rise to what is known as *Suszko's Thesis*. Our earlier work puts us in an excellent position both to explain this Thesis, and to explain why we reject it.

Two-valued, designatedly-valued, and tarskian relations

Let Sent be some set of objects, thought of as *sentences*. Let $\Vdash$ be some subset of $\wp(\text{Sent}) \times \text{Sent}$; so it is the kind of relation where one can say $T \Vdash \varphi$, for some 'theory' T and some 'consequence' φ. For most of this chapter, we have considered consequence relations defined via designated values on complete Boolean algebras. Relaxing this idea of designated values as far as possible, we obtain this:

Definition 13.16: *We say that* $\Vdash$ *is* designatedly-valued *iff:*

(1) *there are sets* B *and* D *with* $D \subseteq B$*; and*
(2) *there is a class,* V*, of functions* $\text{Sent} \longrightarrow B$*; and*
(3) $T \Vdash \varphi$ *iff: if* $v(T) \subseteq D$ *then* $v(\varphi) \in D$*, for all* $v \in V$*.*

If $\Vdash$ *is designatedly-valued with* $B = \{1, 0\}$ *and* $D = \{1\}$*, we say that* $\Vdash$ *is* two-valued*.*

For example, classical first-order consequence is two-valued, as V is supplied to us by considering all the $\mathscr{L}$-structures.

connect with various *non*-inferential facts, such as that we cook, seek out, and eat such things. The *inferentialism* can plausibly be restricted to certain areas of vocabulary, e.g. logic / mathematics / semantics.

[29] See e.g. Halbach (2011).

The preceding Definition approached logical consequence in terms of semantic values. A different approach would simply be to stipulate some extremely basic properties that our consequence relation ought to have. With that in mind:[30]

Definition 13.17: *We say that* $\Vdash$ *is* tarskian *iff it obeys all of:*

> *reflexivity:* $\{\varphi\} \Vdash \varphi$
> *weakening: if* $T \Vdash \varphi$*, then* $T \cup S \Vdash \varphi$
> *idempotency: if* $\{\psi : T \Vdash \psi\} \Vdash \varphi$*, then* $T \Vdash \varphi$

This definition simply specifies some extremely innocuous rules for a relation of logical consequence. The only rule meriting comment in *idempotency*, which formalises the idea that closing a theory under consequence yields no *new* consequences.

Unsurprisingly, many things called 'logics' are tarskian. Strikingly, though, our three notions are all equivalent:[31]

Theorem 13.18: *The following are equivalent:*

> *(1)* $\Vdash$ *is two-valued*
> *(2)* $\Vdash$ *is designatedly-valued*
> *(3)* $\Vdash$ *is tarskian*

Proof. *(1)* ⇒ *(2)*. Trivial.

(2) ⇒ *(3)*. Suppose $\Vdash$ is designatedly-valued. Evidently *reflexivity* holds. To establish *weakening*: suppose $T \Vdash \varphi$; so any $v \in V$ with $v(T) \subseteq D$ has $v(\varphi) \in D$; hence any $v \in V$ with $v(T \cup S) \subseteq D$ has $v(\varphi) \in D$, so that $T \cup S \Vdash \varphi$. To establish *idempotency*, suppose that $\{\psi : T \Vdash \psi\} \Vdash \varphi$; now fix $v \in V$ such that $v(T) \subseteq D$; then $v(\psi) \in D$ for any ψ such that $T \Vdash \psi$; so $v(\{\psi : T \Vdash \psi\}) \subseteq D$, and hence $v(\varphi) \in D$ by supposition.

(3) ⇒ *(1)*. Suppose $\Vdash$ is tarskian. For each $S \subseteq$ Sent, define a function v_S : Sent $\longrightarrow \{1, 0\}$ by $v_S(\varphi) = 1$ if $S \Vdash \varphi$, and $v_S(\varphi) = 0$ if $S \nVdash \varphi$. Now let $V = \{v_S : S \subseteq \text{Sent}\}$, and stipulate:

$$T \VDash \varphi \text{ iff: for all } v \in V, \text{if } v(T) \subseteq \{1\} \text{ then } v(\varphi) = 1$$

Evidently $\VDash$ is two-valued. So it suffices to show that $T \Vdash \varphi$ iff $T \VDash \varphi$.

Left-to-right. Suppose $T \Vdash \varphi$ and $v_S(T) \subseteq \{1\}$. By definition, $S \Vdash \psi$ for each $\psi \in T$. So $T \subseteq \{\psi : S \Vdash \psi\}$ and hence $\{\psi : S \Vdash \psi\} \Vdash \varphi$ by *weakening*. So $S \Vdash \varphi$ by *idempotency*, and hence $v_S(\varphi) = 1$.

---

[30] See Malinowski (1993: 32, (T0)–(T2)) Caleiro, Carnielli, et al. (2005: 177, (CR1)–(CR3)). We follow Caleiro, Carnielli, et al. (2005: 177) in using the term 'tarskian' for their conjunction.

[31] See Suszko (1975b), Malinowski (1993: 72–3), and Caleiro, Carnielli, et al. (2005: 178).

Right-to-left. Suppose $T \Vdash\!\models \varphi$. So if $v_(T) \subseteq \{1\}$ then $v_T(\varphi) = 1$. But $T \Vdash \varphi$ for each $\varphi \in T$, by *reflexivity* and *weakening*; so $v_T(T) \subseteq \{1\}$. So $v_T(\varphi) = 1$, i.e. $T \Vdash \varphi$. □

Suszko's Thesis and compositionality

Theorem 13.18 simplifies a result due to Suszko, who once claimed that it showed that 'every logic is (logically) two-valued', so that 'there are but two logical values, true and false'.[32] Since this is how Suszko's line of thought is typically reported in the secondary literature, we call this claim *Suszko's Thesis*. (However, Suszko's *own* views are more complicated than this name suggests, as we will explain in the next subsection.)

There are good reasons to resist the move from Theorem 13.18 to Susko's Thesis. In particular, Suszko's Thesis is sometimes incompatible with compositionality. To explain why, we will revisit our nasty connective, $\wr$, as defined in §13.7. By construction, $\models_\wr$ is designatedly-valued: the class of valuations is just given by the class of $\mathcal{B}$-valued $\mathscr{L}$-structures. So $\models_\wr$ is two-valued by Theorem 13.18. Spelling this out: there is a class of functions, V, from the set of sentences in the language of first-order logic augmented with $\wr$, to the set $\{1, 0\}$, such that:

$T \models_\wr \varphi$ iff: for every $v \in V$, if $v(T) \subseteq \{1\}$ then $v(\varphi) = 1$

Now, since $\not\models_\wr (\wr\varphi \to \varphi)$, there must be some $v \in V$ with $v(\wr\varphi \to \varphi) = 0$. Suppose, for reductio, that there is also two-place function $\multimap$ on $\{1, 0\}$ such that, for all sentences φ and ψ:

$$v(\varphi \to \psi) = v(\varphi) \multimap v(\psi)$$

Because $\varphi \models_\wr \wr\varphi$ and $\wr\varphi \models_\wr \varphi$, we have $v(\wr\varphi) = v(\varphi)$. Equally, because $\models_\wr (\varphi \to \varphi)$, we must have $v(\varphi \to \varphi) = 1$. But now we quickly reach contradiction:

$$1 = v(\varphi \to \varphi) = v(\varphi) \multimap v(\varphi) = v(\wr\varphi) \multimap v(\varphi) = v(\wr\varphi \to \varphi) = 0$$

So: no class of valuations which witnesses the two-valuedness of $\models_\wr$ treats the sentential connective $\to$ as *truth-functional*. As such, 'reducing' an apparently many-valued logic to a two-valued logic can leave us unable to provide a truth-functional account of the meanings of the sentential connectives of that logic. But then we seem left without a *compositional* account of the meaning of the logical connectives. Fans of many-valued logics will, then, have as good reason to reject Suszko's Thesis as they had to insist on compositionality.

[32] Suszko (1977: 378), Caleiro, Carnielli, et al. (2005: 175).

In response to these kinds of worries, Caleiro, Carnielli, et al. have suggested that Susko's Thesis is compatible with a 'generalized' notion of compositionality.[33] Focussing on the case of truth-functional logic, they define a kind of sentential connective called a *separator*. They then show that, for any logic which contains sufficiently many separators, 'the value of a formula is... (uniquely) determined from the values of separators applied to its immediate subformulas'. But there are two good reasons to doubt that such 'generalized' compositionality is of much value.

First: this 'generalized' notion of compositionality falls a long way short of what we wanted from (genuine) compositionality. In particular, in the truth-functional case, we would want the following: given a finite vocabulary, the values of all of the infinitely many sentences in that vocabulary are determined by the values of finitely many (quite simple) sentences (compare the discussion of compositionality in §1.8). This is precisely what the case of $\vDash_{\text{🎗}}$ shows us we *cannot* have.

Second: a given logic may not contain enough separators to allow for 'generalized' compositionality. To combat this, Caleiro, Marcos, and Volpe note that one can always add new connectives to a logic,[34] and that doing so will never disrupt any existing entailments between the sentences which do not contain the new connectives. However, as we saw in our discussion of ♡ and 🎗 in §13.7, adding new connectives to a logic can render certain general rules of inference unsound (in their full generality). Consequently, there can be philosophically motivated resistance to adding new connectives to a logic.

Suszko's own view

For all these reasons, Suszko's Thesis seems unpromising. In fact, Suszko's own viewpoint was slightly more complicated than the name 'Suszko's Thesis' suggests.

In earlier work, Suszko was developing an intensional logic which he took to be a competitor to standard modal logics. His logic expanded ordinary sentential logic with an operator for propositional identity and was able to interpret certain modal logics. But, expressing sympathy with Quine's concerns about modal logic, Suszko wanted to emphasise the classical features of his logic. Since the most obvious semantics for his intensional logic were many-valued, he deployed Theorem 13.18 to show that his logic also had a two-valued semantics.[35]

However, Theorem 13.18 is entirely general in character, and not specific to the particular intensional logics that Suszko was then considering. And so, drawing

---

[33] Caleiro, Carnielli, et al. (2005: 184), Caleiro and Marcos (2009: 270), and Caleiro, Marcos, and Volpe (2014: 2, 14–15, 17, 42).

[34] Caleiro, Marcos, and Volpe (2014: Proposition 2.12).

[35] Suszko (1971: 38), Bloom and Suszko (1971: 80, 1972: 306), and Suszko (1975a: 203–6) describe how the intensional logic recovers certain modal inferences, while Suszko (1975a: 169, 204) records various Quinean sympathies. See Suszko (1975a: 187–92) for the application of considerations like Theorem 13.18 to his intensional logics.

an analogy with the Church–Turing Thesis, Suszko suggested a more general conclusion.[36] In the early days of the theory of computation, many different models of computation were discovered—Turing machines, the general recursive functions, various lambda-calculi—which were subsequently shown to be extensionally equivalent, in that they generated the same class of functions on the natural numbers. Suszko suggested that the equivalence provided by Theorem 13.18 had a similar role to play: it showed that there is a core, absolute notion of entailment which underlies certain extensionally equivalent formalisms. Suszko then suggested that, just as it would be useless to insist upon the primacy of Turing machines over general recursive functions, so it would be useless to insist upon two-valued logics over many-valued logics (or vice-versa). Or, as Suszko put it, 'discussions of intended interpretations' are 'fruitless.'[37]

Suszko, then, advanced a kind of scepticism about the 'intended' semantical underpinning of certain logics. By this point in the book, such scepticism should be almost painfully familiar. However, another theme which has run more quietly through Part B of this book: as a good rule of thumb, it is impossible to extract philosophical juice from a piece of pure mathematics without invoking some philosophical thesis (see in particular §7.5). In particular, a technical result can only motivate a kind of scepticism in the light of some philosophical thesis. For example, moderate modelism combined with the Löwenheim–Skolem results pushes us towards a kind of scepticism about our ability to pin down certain 'structures', and moderate semanticism combined with the results of this chapter pushes us towards a kind of scepticism about our ability to pin down our semantics.

The only philosophical impetus which Suszko mentions for his scepticism, though, is an analogy with computability. And that analogy misses its mark. The formal equivalence between different notions of computability shows that we can happily employ any of several different formalisms. But this does not indicate that every formalism is *philosophically* on a par. Indeed, when presenting arguments in favour of the Church–Turing Thesis, many people have found *Turing's* formal notion of computability more helpful than *Church's* (on the grounds, for example, that it offers some kind of perspicuous decomposition of what computors could do in principle). Similarly, in Chapter 1, we outlined three different approaches to the semantics for classical languages. Since they are extensionally equivalent, we noted that we can happily use any of the three approaches. But we also suggested that the in-principle availability of the Hybrid approach has certain philosophical benefits: it makes room for compositionality without generating the antinomy of the variable and without straining the notion of a language to breaking point. Similarly here. Since Theorem 13.18 states that any tarskian logic is also designatedly-valued

---

[36] See Suszko (1975a: 189).

[37] Suszko (1975a: 191).

and indeed two-valued, we may have a certain level of flexibility concerning how to characterise our favoured logic(s). Still, one of these characterisations may be particularly philosophically perspicuous. Indeed, *pace* Suszko, a many-valued characterisation which allows for compositionality may be *particularly* salient.

13.A Boolean-valued structures with filters

In this appendix, we prove the main results associated Boolean-valued structures and the semantic-underdetermination argument of §§13.1–13.4.

Theorem (13.6): *If $\mathcal{B}$ is a complete Boolean algebra with $D \subseteq B$, then these are equivalent:*

(1) *D is a filter on $\mathcal{B}$*
(2) *$T \vdash \varphi$ iff $T \underset{D}{\overset{\mathcal{B}}{\models}} \varphi$, for any theory $T \cup \{\varphi\}$*

Proof. *(2) ⇒ (1).* By Definition 13.3, there are three ways D could fail to be a filter, and each leads to a violation of equivalence. For example, if $0^{\mathcal{B}} \in D$ then $\underset{D}{\overset{\mathcal{B}}{\models}} \varphi \wedge \neg\varphi$ for any sentence φ, since $[\![\varphi \wedge \neg\varphi]\!] = [\![\varphi]\!] \cdot -[\![\varphi]\!] = 0^{\mathcal{B}} \in D$. The other two cases are similar.[38]

(1) ⇒ (2). Assuming D is a filter, we must show that $T \vdash \varphi$ iff $T \underset{D}{\overset{\mathcal{B}}{\models}} \varphi$.

Right-to-left. Suppose $T \underset{D}{\overset{\mathcal{B}}{\models}} \varphi$. So in particular, there is no $\mathcal{B}$-valued structure $\mathcal{M}$ such that $[\![\psi]\!]^{\mathcal{M}} \in \{1, 0\}$ for every $\mathscr{L}$-sentence ψ, and that $[\![T]\!]^{\mathcal{M}} = \{1\}$, but $[\![\varphi]\!]^{\mathcal{M}} = 0$. So there is no $\mathscr{L}$-structure $\mathcal{N}$ (in the sense of Definition 1.2) such that $\mathcal{N} \models T$ but $\mathcal{N} \not\models \varphi$. So $T \vdash \varphi$ by the Completeness Theorem 4.24.

Left-to-right. If $T \vdash \varphi$, then $T_0 \vdash \varphi$ for some finite $T_0 \subseteq T$. We aim to show that this proof-system is sound for our $\mathcal{B}$-valued semantics, in the following sense:

> If $\varphi_1, \ldots, \varphi_n \vdash \psi$, then $[\![\varphi_1]\!]^{\mathcal{M}} \cdot \ldots \cdot [\![\varphi_n]\!]^{\mathcal{M}} \leq [\![\psi]\!]^{\mathcal{M}}$ for any $\mathcal{B}$-valued structure $\mathcal{M}$

We will establish this by induction on the length of the derivation. Since $T_0 \vdash \varphi$, it will follow that $\prod [\![T_0]\!]^{\mathcal{M}} \leq [\![\varphi]\!]^{\mathcal{M}}$ for any $\mathcal{M}$. Since D is a filter and T_0 is finite, if $[\![T_0]\!]^{\mathcal{M}} \subseteq D$ then $\prod [\![T_0]\!]^{\mathcal{M}} \in D$ by Definition 13.3(2). Since D is a filter $[\![\varphi]\!]^{\mathcal{M}} \in D$ by Definition 13.3(3). So $T_0 \underset{D}{\overset{\mathcal{B}}{\models}} \varphi$, and hence $T \underset{D}{\overset{\mathcal{B}}{\models}} \varphi$.

It remains to prove our claim, i.e. to prove that applying any rule in our proof-system weakly increases the assigned value. The case of the sentential connectives is easy and is covered elsewhere.[39]

---

[38] See Button (2016b: Theorem 1).

[39] Bell (2005: Theorem 1.17 pp.24–6), Kaye (2007: Theorem 7.10 pp.84–5), and Button (2016b: Theorem 1).

To deal with existential quantifiers, we treat $\exists x$ as abbreviating $\neg\forall x\neg$ and note that $[\![\exists x\varphi(x)]\!]^{\mathcal{M}} = \Sigma_{a\in M}[\![\varphi(c_a)]\!]^{\mathcal{M}^{\circ}} = -\prod_{a\in M} -[\![\varphi(c_a)]\!]^{\mathcal{M}^{\circ}} = [\![\neg\forall x\neg\varphi(x)]\!]^{\mathcal{M}}$.

We next show that applying $\forall$I weakly increases the assigned value. So, suppose we have a derivation showing $U \vdash \varphi(c)$ with c not occurring in U, and for induction suppose that $\prod [\![U]\!]^{\mathcal{M}} \leq [\![\varphi(c)]\!]^{\mathcal{M}}$ for any $\mathcal{M}$ interpreting c anyhow. Then, invoking the greatest-lower bound property, $\prod [\![U]\!]^{\mathcal{M}} \leq \prod_{a\in M}[\![\varphi(c_a)]\!]^{\mathcal{M}^{\circ}} = [\![\forall x\varphi(x)]\!]^{\mathcal{M}}$ for any complete Boolean algebra, as required.

To show that $\forall$E weakly increases the assigned value, simply note that $[\![\forall x\varphi(x)]\!]^{\mathcal{M}} = \prod_{a\in M}[\![\varphi(a)]\!]^{\mathcal{M}} \leq [\![\varphi(c_a)]\!]^{\mathcal{M}}$ for each $a \in M$.

We now turn to the rules for identity. To show that =I is innocuous, observe that $[\![a = a]\!]^{\mathcal{M}} = 1$ for any $a \in M$, so that $[\![t = t]\!]^{\mathcal{M}} = 1$ for any term t.

To show $[\![t_1 = t_2]\!]^{\mathcal{M}} \cdot [\![\varphi(t_1)]\!]^{\mathcal{M}} \leq [\![\varphi(t_2)]\!]^{\mathcal{M}}$, we do a subinduction on the complexity of formulas. A simple induction, using the stipulations concerning the interpretation of function-symbols, shows that $[\![t_1 = t_2]\!]^{\mathcal{M}} \leq [\![s(t_1) = s(t_2)]\!]^{\mathcal{M}}$ for any term $s(x)$. So now let $\varphi(x)$ be a formula $R(v_1, \ldots, s(x), \ldots, v_n)$; then for any $a_1, \ldots, a_n$ from M:

$$\begin{aligned}
&[\![t_1 = t_2]\!] \cdot [\![R(a_1, \ldots, s(t_1), \ldots, a_n)]\!] \\
&\leq [\![s(t_1) = s(t_2)]\!] \cdot [\![R(a_1, \ldots, s(t_1), \ldots, a_n)]\!] \\
&= [\![a_1 = a_1]\!] \cdot \ldots \cdot [\![s(t_1) = s(t_2)]\!] \cdot \ldots \cdot [\![a_n = a_n]\!] \cdot [\![R(a_1, \ldots, s(t_1), \ldots, a_n)]\!] \\
&\leq [\![R(a_1, \ldots, s(t_2), \ldots, a_n)]\!]
\end{aligned}$$

Next, let $\varphi(x)$ be a formula $s(x) = r$ for terms r and $s(x)$; then:

$$\begin{aligned}
[\![t_1 = t_2]\!] \cdot [\![s(t_1) = r]\!] &\leq [\![s(t_1) = s(t_2)]\!] \cdot [\![s(t_1) = r]\!] \\
&= [\![s(t_2) = s(t_1)]\!] \cdot [\![s(t_1) = r]\!] \\
&\leq [\![s(t_2) = r]\!]
\end{aligned}$$

This handles the base cases of the induction. Now suppose for induction that $[\![t_1 = t_2]\!]^{\mathcal{M}} \cdot [\![\psi(t_1, c)]\!]^{\mathcal{M}} \leq [\![\psi(t_2, c)]\!]^{\mathcal{M}}$ for any $\mathcal{M}$ interpreting c in any way. Then, for any complete Boolean algebra $\mathcal{M}$:

$$\begin{aligned}
[\![t_1 = t_2]\!]^{\mathcal{M}} \cdot [\![\forall x\psi(t_1, x)]\!]^{\mathcal{M}} &= [\![t_1 = t_2]\!]^{\mathcal{M}} \cdot \prod_{a\in M} [\![\psi(t_1, c_a)]\!]^{\mathcal{M}^{\circ}} \\
&\leq \prod_{a\in M} [\![\psi(t_2, c_a)]\!]^{\mathcal{M}^{\circ}} \\
&= [\![\forall x\psi(t_2, x)]\!]^{\mathcal{M}}
\end{aligned}$$

The remaining cases for the sentential connectives are similar. This completes the induction, establishing that applying =E weakly increases the assigned valued.

Finally, we turn to the second-order rules. The case of second-order quantification is exactly similar to first-order quantification, and the Comprehension and Choice Schemas hold by stipulation in any faithful $\mathcal{B}$-valued structure. □

Theorem (13.8): *If $\mathcal{B}$ is a complete Boolean algebra and $D \subseteq B$ is a filter on $\mathcal{B}$, then:*

(1) *D is an ultrafilter on $\mathcal{B}$ iff Not-Designation, And-Designation and Or-Designation hold in every (faithful Henkin) $\mathcal{B}$-valued structure*

(2) *D is a principal ultrafilter on $\mathcal{B}$ iff all the Designation Principles hold in every (faithful Henkin) $\mathcal{B}$-valued structure*

Proof. (1) is dealt with elsewhere,[40] so we focus on (2). Left-to-right of All-Designation and right-to-left of Exists-Designation hold under no special condition; the interesting cases are right-to-left of All-Designation and left-to-right of Exists-Designation. We prove Exists-Designation; All-Designation is similar.

Necessity. Let D be a nonprincipal ultrafilter, so $\prod D \notin D$ and hence $-\prod D \in D$. Let $\mathcal{M}$ be a $\mathcal{B}$-valued $\{F\}$-structure as follows:

- $\mathcal{M}$'s underlying domain is $M = D$
- $[\![F(a)]\!]^{\mathcal{M}} = -a$ for all $a \in M = D$
- $[\![a = b]\!]^{\mathcal{M}} = 1$ if $a = b$, and $[\![a = b]\!]^{\mathcal{M}} = 0$ otherwise

By construction, $[\![F(c_a)]\!] \notin D$ for any $a \in M$. However

$$[\![\exists x F(x)]\!] = \sum_{a \in M} [\![F(a)]\!] = \sum_{a \in D} -a = -\prod D \in D$$

Sufficiency. Let D be a principal ultrafilter; so $\prod D \in D$. Suppose there is a $\mathcal{B}$-valued structure $\mathcal{M}$ such that $[\![\varphi(c_a)]\!] \notin D$ for all $a \in M$; then $-[\![\varphi(c_a)]\!] \in D$ since D is an ultrafilter; so

$$\prod D \leq \prod_{a \in M} -[\![\varphi(c_a)]\!] = -\sum_{a \in M} [\![\varphi(c_a)]\!] = -[\![\exists x \varphi(x)]\!]$$

Since $\prod D \in D$ by hypothesis and since D is closed upwards, $-[\![\exists x \varphi(x)]\!] \in D$ and so $[\![\exists x \varphi(x)]\!] \notin D$. □

13.B Full second-order Boolean-valued structures

In Definition 13.5, we outlined a *Henkin*-style second-order $\mathcal{B}$-valued semantics. However, there is an obvious Boolean generalisation of *full* semantics (as ordinarily defined; see §§1.10–1.11). We start by defining *full* $\mathcal{B}$-valued structures:

Definition 13.19: *A $\mathcal{B}$-valued structure $\mathcal{M}$ is* full *iff it obeys the following:*

(1) *M_n^{rel} is the set of all functions $g : M^n \longrightarrow B$, for all n*

(2) *M_n^{fun} is the set of all functions $g : M^n \longrightarrow M$, for all n*

(3) *if $a \neq b$ then $[\![a = b]\!] = 0$, for all $a, b \in M$*

[40] See Button (2016b: Corollary 5).

The motivations for clauses (1) and (2) are clear. Clause (3) then ensures both that $[\![a_1 = b_1]\!] \cdot \ldots \cdot [\![a_n = b_n]\!] \cdot g(\bar{a}) \leq g(\bar{b})$ for any $g \in M_n^{\mathrm{rel}}$, and that $[\![a_1 = b_1]\!] \cdot \ldots \cdot [\![a_n = b_n]\!] \leq [\![g(\bar{a}) = g(\bar{b})]\!]$ for any $g \in M_n^{\mathrm{fun}}$.[41]

Given Theorem 13.6, one might be tempted to make this conjecture: *Where $\mathcal{B}$ is any complete Boolean algebra with a filter D, Boolean entailment $\overset{\mathcal{B}}{\underset{D}{\models}}$ restricted to full $\mathcal{B}$-valued structures is equivalent to full second-order logic.* But that conjecture would be unwise. The proof of Theorem 13.6 relied upon the existence of a sound and complete deduction-system for faithful Henkin second-order logic, but *full* second-order logic has no such deduction-system. And in fact Ikegami and Väänänen have proved that this conjecture is outright *false*.[42]

To obtain a Boolean-valued logic which *is* exactly as powerful as (ordinary) full second-order logic, we must restrict our attention again to *principal ultrafilters* on $\mathcal{B}$-valued structures:

Proposition 13.20: *Let $\mathcal{B}$ be a complete Boolean algebra with D a principal ultrafilter on $\mathcal{B}$. Where $T \cup \{\varphi\}$ is any second-order theory, these are equivalent:*

(1) $T \models \varphi$, where throughout the proof $\models$ is understood in terms of (ordinary) full second-order logic, as defined in §1.10.

(2) $T \overset{\mathcal{B}}{\underset{D}{\models}} \varphi$, where throughout the proof $\overset{\mathcal{B}}{\underset{D}{\models}}$ is understood as restricted to full $\mathcal{B}$-valued structures. That is, we are writing $T \overset{\mathcal{B}}{\underset{D}{\models}} \varphi$ to indicate: for every full $\mathcal{B}$-valued $\mathcal{L}$-structure $\mathcal{M}$, if $[\![T]\!]^{\mathcal{M}} \subseteq D$ then $[\![\varphi]\!]^{\mathcal{M}} \in D$.

Proof. (1) $\Rightarrow$ (2). Suppose $T \overset{\mathcal{B}}{\underset{D}{\not\models}} \varphi$, and let $\mathcal{M}$ be a full $\mathcal{B}$-valued structure where $[\![T]\!]^{\mathcal{M}} \subseteq D$ but $[\![\varphi]\!]^{\mathcal{M}} \notin D$. Using Robinsonian notions for full second-order logic, we expand this to a full $\mathcal{B}$-valued structure $\mathcal{M}^{\blacksquare}$. This has a constant c_a for each $a \in M$ with $c_a^{\mathcal{M}^{\blacksquare}} = a$, a predicate R_g for each $g : M^n \longrightarrow B$, and a function symbol f_g for each $g : M^n \longrightarrow M$. We will define (in the ordinary sense) a full second-order $\mathcal{L}^{\blacksquare}$-structure $\mathcal{N}$, with the following property:

$$[\![\varphi]\!]^{\mathcal{M}^{\blacksquare}} \in D \text{ iff } \mathcal{N} \models \varphi, \text{ for any } \mathcal{L}^{\blacksquare}\text{-sentence } \varphi \tag{13.1}$$

It will follow immediately that $\mathcal{N} \models T$ but $\mathcal{N} \not\models \varphi$, so that $T \not\models \varphi$, as required.

To define $\mathcal{N}$, we simply quotient $\mathcal{M}$ by identity-according-to-$\mathcal{M}$. In detail, let $[a] = \{b \in M : [\![a = b]\!]^{\mathcal{M}} \in D\}$, and define $\mathcal{N}$ as follows, for all $\mathcal{L}^{\blacksquare}$-constant

[41] To see how these could fail without clause (3), let $\mathcal{B}$ be a complete Boolean algebra with more than two elements, and define a $\mathcal{B}$-valued structure $\mathcal{A}$ with domain B which interprets identity as $[\![x = y]\!]^{\mathcal{A}} = ((x \cdot y) + (-x \cdot -y))$. (The proof that that $\mathcal{A}$ is a $\mathcal{B}$-valued $\varnothing$-structure can be extracted from §13.D). Let D be a principal ultrafilter on $\mathcal{B}$, and let $g(x) = 1$ if $x \in D$, and $g(x) = 0$ otherwise. Pick some element $a \in D \setminus \{1\}$. Then $[\![a = 0]\!] = -a > 0 = [\![1 = 0]\!] = [\![g(a) = g(0)]\!]$, and also $[\![a = 0]\!] \cdot g(a) = -a \cdot 1 = -a > 0 - g(0)$.

[42] Ikegami and Väänänen (2015: 173).

symbols c, all n-place $\mathscr{L}^{\blacksquare}$-predicates R, all n-place $\mathscr{L}^{\blacksquare}$-function symbols f and all $\bar{a}$ from M:

$$\begin{aligned}\mathcal{N}\text{'s underlying domain is } N &= \{[a] : a \in M\}\\ c^{\mathcal{N}} &= [c^{\mathcal{M}}]\\ R^{\mathcal{N}} &= \{([a_1], \ldots, [a_n]) : [\![R(a_1, \ldots, a_n)]\!]^{\mathcal{M}} \in D\}\\ f^{\mathcal{N}}\left(\overline{[a]}\right) &= [f^{\mathcal{M}}(\bar{a})]\end{aligned}$$

Since $\mathcal{M}$ is a $\mathcal{B}$-valued structure and D is a filter, $\mathcal{N}$ is well-defined.

A simple induction on complexity confirms that $t^{\mathcal{N}}\left(\overline{[a]}\right) = [t^{\mathcal{M}}(\bar{a})]$ for any $\bar{a} \in M^n$ and any $\mathscr{L}^{\blacksquare}$-term t. We now prove claim (13.1) by induction.

Atomic case: identity. The claim holds for identity statements in $\mathscr{L}^{\blacksquare}$ since $[\![t_1^{\mathcal{M}}(\bar{a}) = t_2^{\mathcal{M}}(\bar{b})]\!]^{\mathcal{M}^{\blacksquare}} \in D$ iff $[t_1^{\mathcal{M}}(\bar{a})] = [t_2^{\mathcal{M}}(\bar{b})]$ iff $t_1^{\mathcal{N}}\left(\overline{[a]}\right) = t_2^{\mathcal{N}}\left(\overline{[b]}\right)$.

Atomic case: relations. It suffices to show that $[\![R(a_1, \ldots, a_n)]\!]^{\mathcal{M}^{\blacksquare}} \in D$ iff $([a_1], \ldots, [a_n]) \in R^{\mathcal{N}}$. Left-to-right is immediate from the definition of $R^{\mathcal{N}}$. For the converse, suppose $([a_1], \ldots, [a_n]) \in R^{\mathcal{N}}$; then $[\![R(b_1, \ldots, b_n)]\!]^{\mathcal{M}^{\blacksquare}} \in D$ for some b_is with $[\![b_i = a_i]\!]^{\mathcal{M}} \in D$. Since $\mathcal{M}$ is a $\mathcal{B}$-valued structure:

$$[\![b_1 = a_1]\!]^{\mathcal{M}} \cdot \ldots \cdot [\![b_n = a_n]\!]^{\mathcal{M}} \cdot [\![R(b_1, \ldots, b_n)]\!]^{\mathcal{M}^{\blacksquare}} \leq [\![R(a_1, \ldots, a_n)]\!]^{\mathcal{M}^{\blacksquare}}$$

But all the identity-expressions on the left are designated; and since D is a filter, $[\![R(a_1, \ldots, a_n)]\!]^{\mathcal{M}^{\blacksquare}} \in D$, as required.

Negation. Invoking the induction hypothesis and that D is an ultrafilter:

$$[\![\neg\varphi]\!]^{\mathcal{M}^{\blacksquare}} \in D \text{ iff } [\![\varphi]\!]^{\mathcal{M}^{\blacksquare}} \notin D \text{ iff } \mathcal{N} \nvDash \varphi \text{ iff } \mathcal{N} \vDash \neg\varphi$$

First-order quantification. We have:

$$\begin{aligned}[\![\forall x\varphi(x)]\!]^{\mathcal{M}^{\blacksquare}} \in D &\text{ iff } \prod_{a \in M} [\![\varphi(c_a)]\!]^{\mathcal{M}^{\blacksquare}} \in D\\ &\text{ iff } [\![\varphi(c_a)]\!]^{\mathcal{M}^{\blacksquare}} \in D \text{ for all } a \in M\\ &\text{ iff } \mathcal{N} \vDash \varphi(c_a) \text{ for all } a \in M\\ &\text{ iff } \mathcal{N} \vDash \forall x\varphi(x)\end{aligned}$$

The first biconditional follows from semantics for the quantifier. The second biconditional holds because D is principal. The third holds by the induction hypothesis. The last holds because $N = \{c_a^{\mathcal{N}} : a \in M\}$.

Second-order quantification. These are exactly similar.

(2) ⇒ (1). Suppose $T \nvDash \varphi$, and let $\mathcal{N}$ be an $\mathscr{L}$-structure where $\mathcal{N} \vDash T$ and $\mathcal{N} \nvDash \varphi$. Let $\mathscr{L}^{\blacksquare}$ be the signature of $\mathcal{N}^{\blacksquare}$. We begin by defining a Henkin $\mathcal{B}$-valued

$\mathcal{L}^{\blacksquare}$-structure $\mathcal{H}$. Its domain $H = N$, and identity is given by the insistence that if $a = b$ then $[\![a = b]\!]^H = 1$ and if $a \neq b$ then $[\![a = b]\!]^H = 0$. We then stipulate: $c^{\mathcal{H}} = c^{\mathcal{N}}$ for all $\mathcal{L}^{\blacksquare}$-constants c; $f^{\mathcal{H}} = f^{\mathcal{N}}$ for all $\mathcal{L}^{\blacksquare}$-function-symbols f; and $[\![R(a_1, \ldots, a_n)]\!] = 1$ if $(a_1, \ldots, a_n) \in R^{\mathcal{N}^{\blacksquare}}$ and $[\![R(a_1, \ldots, a_n)]\!] = 0$ otherwise, for all $\mathcal{L}^{\blacksquare}$-predicates R. Finally, H_n^{rel} is the set of functions picked out by any n-place $\mathcal{L}^{\blacksquare}$-predicate, and H_n^{fun} is the set of functions picked out by any n-place $\mathcal{L}^{\blacksquare}$-function-symbol. It is trivial from the definition of $\mathcal{H}$ that:

$$[\![\varphi]\!]^{\mathcal{H}} \in D \text{ iff } \mathcal{M} \vDash \varphi, \text{ for any } \mathcal{L}^{\blacksquare}\text{-sentence } \varphi \tag{13.2}$$

We now define a *full* $\mathcal{B}$-valued $\mathcal{L}^{\blacksquare}$-structure, $\mathcal{E}$. This is the same as $\mathcal{H}$ in every regard, except that E_n^{rel} and E_n^{fun} are now as in Definition 13.19. For each $g \in E_n^{\text{rel}}$ we let $g^{\chi} \in H_n^{\text{rel}}$ be the function given by $g^{\chi}(\bar{a}) = 1$ iff $g(\bar{a}) \in D$ and $g^{\chi}(\bar{a}) = 0$ iff $g(\bar{a}) \notin D$. Observe that $H_n^{\text{fun}} = E_n^{\text{fun}}$ and that $H_n^{\text{rel}} = \{g^{\chi} : g \in E_n^{\text{rel}}\}$. We now claim that, for any $\mathcal{L}^{\blacksquare}$-formula $\varphi(\overline{V})$ with all free (relational-)variables displayed, and all $g \in M_n^{\text{rel}}$:

$$[\![\varphi(\overline{R_g})]\!]^{\mathcal{E}} \in D \text{ iff } [\![\varphi(\overline{R_{g^{\chi}}})]\!]^{\mathcal{H}} \in D \tag{13.3}$$

This is proved by a simple induction on complexity. The crucial clause concerns quantification over relations, and this holds because our filter is principal (cf. the proof for first-order quantification in the previous part of this proof). Now we just combine (13.2) and (13.3) to obtain that $[\![T]\!]^{\mathcal{E}} \subseteq D$ and $[\![\varphi]\!]^{\mathcal{E}} \notin D$. □

13.c Ultrafilters, ultraproducts, Łoś, and compactness

The proof of Proposition 13.20 put us in a position to explain one of the most celebrated uses of ultrafilters within model theory, namely, in *ultraproducts*. Our approach to setting up ultraproducts is unusual, since it uses Boolean-valued structures; given this framework, though, our approach is surprisingly swift.

Let I be an index set and let M_i be a set for each $i \in I$. Then $\text{Prod}_{i \in I} M_i$ is the set of all choice functions on $\{M_i : i \in I\}$.[43] Where each M_i is the underlying domain of some $\mathcal{L}$-structure $\mathcal{M}_i$, we can use this to define a $\wp(I)$-valued $\mathcal{L}$-structure, $\mathcal{P}$, by stipulating, for all $\mathcal{L}$-constant symbols c, all $\mathcal{L}$-predicates R, all $\mathcal{L}$-function symbols f, all $g_1, \ldots, g_n \in \text{Prod}_{i \in I} M_i$, and all $i \in I$:

[43] I.e. the set of all functions $f : I \longrightarrow M_i$ such that $f(i) \in M_i$ for all $i \in I$. Other authors use $\prod_{i \in I} M_i$ for $\text{Prod}_{i \in I} M_i$, but this would interfere with our infimum-notation.

$$\begin{aligned}
\mathcal{P}\text{'s underlying domain is } P &= \text{Prod}_{i\in I}M_i \\
c^{\mathcal{P}}(i) &= c^{\mathcal{M}_i} \\
(f^{\mathcal{P}}(g_1, \dots, g_n))(i) &= f^{\mathcal{M}_i}(g_1(i), \dots, g_n(i)) \\
[\![R(g_1, \dots, g_n)]\!]^{\mathcal{P}} &= \{i \in I : (g_1(i), \dots, g_n(i)) \in R^{\mathcal{M}_i}\} \\
[\![g_1 = g_2]\!]^{\mathcal{P}} &= \{i \in I : g_1(i) = g_2(i)\}
\end{aligned}$$

It is easy to check that $\mathcal{P}$ is a $\wp(I)$-valued $\mathcal{L}$-structure. We now have:

Lemma 13.21: *Let I, each $\mathcal{M}_i$, and $\mathcal{P}$ be as above. For any first-order $\mathcal{L}$-formula $\varphi(\bar{v})$ and any $\bar{g}$ from* $\text{Prod}_{i\in I}M_i$*:*

$$[\![\varphi(\bar{g})]\!]^{\mathcal{P}^\circ} = \left\{i \in I : \mathcal{M}_i \vDash \varphi\left(\overline{g(i)}\right)\right\}$$

Proof. This is a simple induction on complexity. Every case is trivial, except quantification, where we have:

$$\begin{aligned}
[\![\exists v\varphi(\bar{g}, v)]\!]^{\mathcal{P}^\circ} &= \sum_{h\in\text{Prod}_{i\in I}M_i} [\![\varphi(\bar{g}, h)]\!]^{\mathcal{P}^\circ} \\
&= \bigcup\left\{[\![\varphi(\bar{g}, h)]\!]^{\mathcal{P}^\circ} : h \in \text{Prod}_{i\in I}M_i\right\} \\
&= \bigcup\left\{\left\{i \in I : \mathcal{M}_i \vDash \varphi\left(\overline{g(i)}, h(i)\right)\right\} : h \in \text{Prod}_{i\in I}M_i\right\} \\
&= \left\{i \in I : \mathcal{M}_i \vDash \exists v\varphi\left(\overline{g(i)}, v\right)\right\}
\end{aligned}$$

The third identity invokes the induction hypothesis. The fourth identity holds because $\text{Prod}_{i\in I}M_i$ has all choice functions on $\{M_i : i \in I\}$. □

We now turn $\mathcal{P}$ into an ultraproduct, by quotienting through identity-according-to-$\mathcal{P}$. The idea is as in Proposition 13.20. So let I, each $\mathcal{M}_i$, and $\mathcal{P}$ be as above. Let D be an ultrafilter on $\wp(I)$, and let $[g] = \{h \in \text{Prod}_{i\in I}M_i : [\![g = h]\!]^{\mathcal{P}} \in D\}$. As in Proposition 13.20,[44] we define an $\mathcal{L}$-structure, $\mathcal{U}$, as follows:

$$\begin{aligned}
\mathcal{U}\text{'s underlying domain is } U &= \{[g] : g \in \text{Prod}_{i\in I}M_i\} \\
c^{\mathcal{U}} &= [c^{\mathcal{P}}] \\
R^{\mathcal{U}} &= \{([g_1], \dots, [g_n]) : [\![R(g_1, \dots, g_n)]\!]^{\mathcal{P}} \in D\} \\
f^{\mathcal{U}}\left(\overline{[g]}\right) &= [f^{\mathcal{P}}(\bar{g})]
\end{aligned}$$

We say that $\mathcal{U}$ is an *ultraproduct*, and we can obtain a famous result:

[44] There is an important difference: in Proposition 13.20, we considered full second-order Boolean-valued structures; but $\mathcal{P}$ is essentially first-order, and clause (3) of Definition 13.19 typically fails of $\mathcal{P}$.

Theorem 13.22 (Łoś's Ultraproduct Theorem): *Let I, each $\mathcal{M}_i$, D, and $\mathcal{U}$ be as above. For any first-order $\mathcal{L}$-formula $\varphi(\bar{v})$ and any $\bar{g}$ from* $\text{Prod}_{i \in I} M_i$:

$$\mathcal{U} \vDash \varphi\left(\overline{[g]}\right) \textit{ iff } \left\{i \in I : \mathcal{M}_i \vDash \varphi\left(\overline{g(i)}\right)\right\} \in D$$

Proof. Again, this is an induction on complexity. For all cases except the quantifiers, we simply follow the proof of Proposition 13.20, invoking Lemma 13.21. In the case of quantifiers, we must depart from the proof of Proposition 13.20, since we have not assumed that D is principal. But here:

$$\begin{aligned}\mathcal{U} \vDash \exists v \varphi\left(\overline{[g]}, v\right) &\text{ iff } \mathcal{U} \vDash \varphi\left(\overline{[g]}, [h]\right), \text{ for some } h \in \text{Prod}_{i \in I} M_i \\ &\text{ iff } \left\{i \in I : \mathcal{M}_i \vDash \varphi\left(\overline{g(i)}, h\right)\right\} \in D, \text{ for some } h \in \text{Prod}_{i \in I} M_i \\ &\text{ iff } \left\{i \in I : \mathcal{M}_i \vDash \exists v \varphi\left(\overline{g(i)}, v\right)\right\} \in D\end{aligned}$$

The second biconditional invokes the induction hypothesis; the third appeals, as in Lemma 13.21, to the fact that $\text{Prod}_{i \in I} M_i$ has all the choice functions and so, in particular, a function h which selects witnesses wherever possible. □

We can now reprove the Compactness Theorem 4.1 with ultraproducts (and appealing to Theorem 14.4 of §14.2):[45]

Corollary 13.23: *Let T be a set of first-order $\mathcal{L}$-sentences, let I be the set of all finite subsets of T, and for each $i \in I$ let $\mathcal{M}_i \vDash i$. Then there is an ultrafilter D over $\wp(I)$ such that $\mathcal{U} \vDash T$, where $\mathcal{U}$ is defined from I and D as above.*

Proof. We begin with two definitions:

$$\begin{aligned} u(\varphi) &= \{i \in I : \varphi \in i\}, \text{for each } \varphi \in T \\ F &= \{u(\varphi) : \varphi \in T\}\end{aligned}$$

Let $i = \{\varphi_1, \ldots, \varphi_n\} \subseteq T$; then $i \in u(\varphi_1) \cap \ldots \cap u(\varphi_n)$, i.e. F is finitely-meetable, in the terminology of Theorem 14.4. So there is an ultrafilter $D \supseteq F$. For each $\varphi \in T$ we now have $u(\varphi) \subseteq \{i \in I : \mathcal{M}_i \vDash \varphi\} \in D$, since D is closed upwards; so that $\mathcal{U} \vDash \varphi$, by Łoś's Theorem 13.22, where $\mathcal{U}$ is the ultraproduct constructed from D. □

13.D The Boolean-non-categoricity of CBA

We now prove the results connected with semanticism and §13.5. Recall that CBA augments the theory of Boolean algebras with the claim $\forall x(x = 0 \vee x = 1)$.

---

[45] See Chang and Keisler (1990: Corollary 4.1.11).

Proposition (13.10): *For any complete Boolean algebra $\mathcal{B}$, there is a $\mathcal{B}$-valued model of* CBA *whose domain is B.*

Proof. The idea is to regard $\mathcal{B}$ *itself* as a $\mathcal{B}$-valued model of CBA. So: let $\mathcal{A}$ have the domain B, and let $\mathcal{A}$ interpret all of the symbols $1, 0, -, \cdot$ and $+$ exactly as in $\mathcal{B}$. Finally, let $\mathcal{A}$ interpret identity as follows:

$$[\![x = y]\!]^{\mathcal{A}} = ((x \cdot y) + (-x \cdot -y))$$

where the expression on the right is interpreted on $\mathcal{B}$.

To confirm that $\mathcal{A}$ really *is* a $\mathcal{B}$-valued model, we need to check that:

$$\begin{aligned}
[\![x = x]\!] &= 1 \\
[\![x = y]\!] &= [\![y = x]\!] \\
[\![x = y]\!] \cdot [\![y = z]\!] &\leq [\![x = z]\!] \\
[\![x = y]\!] &\leq [\![-x = -y]\!] \\
[\![x_1 = y_1]\!] \cdot [\![x_2 = y_2]\!] &\leq [\![x_1 \cdot x_2 = y_1 \cdot y_2]\!] \\
[\![x_1 = y_1]\!] \cdot [\![x_2 = y_2]\!] &\leq [\![x_1 + x_2 = y_1 + y_2]\!]
\end{aligned}$$

This is easily done: given our interpretation of $=$ in $\mathcal{A}$, we can replace each instance of $\xi = \chi$ in the above with $((\xi \cdot \chi) + (-\xi \cdot -\chi))$. Under this reinterpretation, all six (in)equalities are theorems of the theory of Boolean algebras. For example, the fourth inequality becomes:

$$((x \cdot y) + (-x \cdot -y)) \leq ((-x \cdot -y) + (--x \cdot --y))$$

which is obviously true of any Boolean algebra. So, all six reinterpreted statements hold in $\mathcal{B}$. Since $\mathcal{A}$ interprets $-$, $\cdot$ and $+$ exactly as $\mathcal{B}$ does, all six of the *original* (in)equalities are therefore true.

To check that $[\![\forall x(x = 0 \vee x = 1)]\!]^{\mathcal{A}} = 1$, it suffices to observe the following, for each $x \in B$:

$$\begin{aligned}
[\![c_x = 0 \vee c_x = 1]\!]^{\mathcal{A}^\circ} &= [\![c_x = 0]\!]^{\mathcal{A}^\circ} + [\![c_x = 1]\!]^{\mathcal{A}^\circ} \\
&= ((x \cdot 0) + (-x \cdot -0)) + ((x \cdot 1) + (-x \cdot -1)) \\
&= (0 + -x) + (x + 0) = 1
\end{aligned}$$

Similar checks will show that $[\![\text{CBA}]\!] = \{1\}$. For example, since $\mathcal{B}$ is a Boolean algebra, $x \cdot (y + z) = (x \cdot y) + (x \cdot z)$ for any $x, y, z \in B$. Furthermore, $\mathcal{A}$ interprets $\cdot$ and $+$ exactly as $\mathcal{B}$ does, and interprets identity so that $[\![w = w]\!] = 1$ for any w. Hence $[\![c_x \cdot (c_y + c_z) = (c_x \cdot c_y) + (c_x \cdot c_z)]\!] = 1$, and so $[\![\forall x \forall y \forall z[x \cdot (y + z) = (x \cdot y) + (x \cdot z)]]\!] = 1$. □

It is easy to check that $\mathcal{T}wo$ provides a $\mathcal{B}$-valued model of CBA, whatever $\mathcal{B}$ happens to be. It follows that, whilst CBA is categorical with the standard semantics for first-order logic, it is not categorical on any other (non-isomorphic) Boolean-valued semantics.

In §13.5, we also considered augmenting the theory of Boolean algebras with new axioms concerning a new one-place predicate, D, which are intended to say that (the interpretation of) D is a principal ultrafilter. Here are the axioms:

$$\begin{array}{ll} \neg D(0) & \\ \forall x \forall y\,([D(x) \wedge D(y)] \rightarrow D(x \cdot y)) & \forall x\,(D(x) \vee D(-x)) \\ \forall x \forall y\,([D(x) \wedge x \cdot y = x] \rightarrow D(y)) & \exists x \forall y\,(D(y) \leftrightarrow x \cdot y = x) \end{array}$$

The left column details a filter, the right column details a principal ultrafilter. Let DBA be the result of adding these axioms to the theory of Boolean algebras (or, indeed, to CBA). It is easy to check that, if $\mathcal{M}$ is a model (in the ordinary sense) of DBA, then $D^{\mathcal{M}}$ is a principal ultrafilter on $\mathcal{B}$.

However, we can obtain a $\mathcal{B}$-valued model of DBA, just by augmenting our model $\mathcal{A}$, from the proof of Proposition 13.10, with the function $[\![D(x)]\!]^{\mathcal{A}} = x$ for all x.[46] It is easy to confirm that $[\![x = y]\!] \cdot [\![D(x)]\!] \leq [\![D(y)]\!]$, since this reduces to the claim that $x \cdot y \leq y$. To confirm that $[\![\text{DBA}]\!]^{\mathcal{A}} = \{1\}$, the only interesting case is $[\![\exists x \forall y\,(D(y) \leftrightarrow x \cdot y = x)]\!]$, which holds because $[\![\forall y\,(D(y) \leftrightarrow 1 \cdot y = 1)]\!] = 1$. And, as we claimed in §13.5, there is no sense in claiming that $\mathcal{A}$'s interpretation of 'D', i.e. the function $[\![D(\cdot)]\!]^{\mathcal{A}}$, is (anything like) a principal ultrafilter on $\mathcal{A}$. For $[\![D(\cdot)]\!]^{\mathcal{A}}$ is just the *identity map* on $\mathcal{A}$.

13.E Proofs concerning bilateralism

We conclude with the results connected with bilateralism and §13.6.

Theorem (13.11): *$T \updownarrow\!\!\vdash \updownarrow\varphi$ iff $\natural(T) \vdash \flat(\updownarrow\varphi)$, for any decorated theory $T \cup \{\updownarrow\varphi\}$*

Proof. Left-to-right. We show that each decorated rule corresponds to a (derived) undecorated rule. For example:

$$\dfrac{\dfrac{}{\downarrow\varphi}{}^{n} \\ \vdots \\ \uparrow\psi \qquad \downarrow\psi}{\uparrow\varphi}\ \text{Raa}\updownarrow, n \qquad\qquad \dfrac{\dfrac{\dfrac{}{\neg\varphi}{}^{n} \\ \vdots \\ \psi \qquad \neg\psi}{\neg\neg\varphi}\ \neg\text{I}, n}{\varphi}\ \text{DNE}$$

[46] Compare the definition of the 'canonical name' for the generic ultrafilter in forcing, e.g. Kunen (1980: 190), Bell (2005: 93), and Jech (2003: 214).

Using such derived rules, the decorated proof witnessing $T \mathrel{\dagger\!\!\vdash} \updownarrow\varphi$ can be mechanically translated into an undecorated proof witnessing $\natural(T) \vdash \flat(\updownarrow\varphi)$.

Right-to-left. We first define an operation on decorated sentences, by $a(\uparrow\varphi) = \uparrow\varphi$ and $a(\downarrow\varphi) = \uparrow\neg\varphi$. We then show that if $\natural(T) \vdash \flat(\updownarrow\varphi)$ then $a(T) \mathrel{\dagger\!\!\vdash} a(\updownarrow\varphi)$, just by showing that each undecorated rule corresponds to a derived decorated rule, e.g.:

$$\dfrac{\exists x\varphi(x) \qquad \begin{array}{c}\overline{\varphi(c)}^{\,1}\\ \vdots\\ \psi\end{array}}{\psi}\ \exists\text{E}, 1 \qquad\qquad \dfrac{\uparrow\exists x\varphi(x) \qquad \dfrac{\dfrac{\overline{\downarrow\psi}^{\,2} \qquad \begin{array}{c}\overline{\uparrow\varphi(c)}^{\,1}\\ \vdots\\ \uparrow\psi\end{array}}{\downarrow\varphi(c)}\ \text{Raa}\updownarrow, 1}{\downarrow\exists x\varphi(x)}\ \exists\text{I}\updownarrow}{\uparrow\psi}\ \text{Raa}\updownarrow, 2$$

Using such derived rules, the undecorated proof witnessing $\natural(T) \vdash \flat(\updownarrow\varphi)$ can be mechanically translated into a decorated proof witnessing $a(T) \mathrel{\dagger\!\!\vdash} a(\updownarrow\varphi)$.

It now suffices to show that if $a(T) \mathrel{\dagger\!\!\vdash} a(\updownarrow\varphi)$ then $T \mathrel{\dagger\!\!\vdash} \updownarrow\varphi$. To see that this holds, first note that if $\uparrow\psi \in T$ then $a(\uparrow\psi) = \uparrow\psi$, and if $\downarrow\psi \in T$ then $\downarrow\psi \mathrel{\dagger\!\!\vdash} \uparrow\neg\psi = a(\downarrow\psi)$ by $\neg$E$\updownarrow$; so if $\updownarrow\psi \in T$ then $T \mathrel{\dagger\!\!\vdash} a(\updownarrow\psi)$. Hence T proves all the decorated sentences among $a(T)$, and hence if $a(T) \mathrel{\dagger\!\!\vdash} a(\updownarrow\varphi)$ then $T \mathrel{\dagger\!\!\vdash} a(\updownarrow\varphi)$. So $T \mathrel{\dagger\!\!\vdash} \updownarrow\varphi$, invoking $\neg$E$\updownarrow$ again if necessary. □

Corollary (13.13): *If $\mathcal{B}$ is a complete Boolean algebra and $D \subseteq B$ is a filter on $\mathcal{B}$, then these are equivalent:*

(1) *$T \overset{\mathcal{B}}{\underset{D}{\models}} \updownarrow\varphi$ iff $T \mathrel{\dagger\!\!\vdash} \updownarrow\varphi$, for any decorated theory $T \cup \{\updownarrow\varphi\}$*

(2) *Not-Designation, And-Designation and Or-Designation hold for any (faithful Henkin) $\mathcal{B}$-valued structure*

Proof. It is easy to see that And-Designation holds under no special conditions, and that Not-Designation entails Or-Designation. So it suffices to check Not-Designation.

(1) ⇒ (2). Left-to-right of Not-Designation holds because D is a filter. For the right-to-left direction, let $\mathcal{M}$ be a $\mathcal{B}$-valued structure such that $[\![\neg\varphi]\!] \notin D$. So $\downarrow\neg\varphi$ is D-correct in $\mathcal{M}$. Now:

$$\dfrac{\downarrow\neg\varphi \qquad \dfrac{\overline{\downarrow\varphi}^{\,1}}{\uparrow\neg\varphi}\ \neg\text{I}\updownarrow}{\uparrow\varphi}\ \text{Raa}\updownarrow, 1$$

Since $\mathrel{\dagger\!\!\vdash}$ is sound, $\uparrow\varphi$ is D-correct in $\mathcal{M}$, i.e. $[\![\varphi]\!] \in D$.

(2) ⇒ (1). Let $\mathcal{M}$ be any (faithful Henkin) $\mathcal{B}$-valued structure. By Not-Designation, $\updownarrow\psi$ is D-correct in $\mathcal{M}$ iff $[\![\flat(\updownarrow\psi)]\!] \in D$. So $T \overset{\mathcal{B}}{\underset{D}{\models}} \updownarrow\varphi$ iff $\natural(T) \overset{\mathcal{B}}{\underset{D}{\models}} \flat(\updownarrow\varphi)$, i.e. iff $\natural(T) \vdash \flat(\updownarrow\varphi)$ by Theorem 13.6, i.e. iff $T \mathrel{\dagger\!\!\vdash} \updownarrow\varphi$ by Theorem 13.11. □

C

Indiscernibility and classification

Introduction to Part C

The main topics of this third part of the book are *indiscernibility* and *classification*.

The chief aim of Chapter 15 is to explore Leibniz's principle of the Identity of Indiscernibles. Model theory supplies us with the resources to distinguish between many *different* notions of indiscernibility, and Chapter 15 examines how these distinctions can be used to explicate different versions of Leibniz's famous principle. After outlining the technicalities, we pour some cooling water on the topic. Model theory allows us to make questions about the identity of indiscernibles precise; but its sheer flexibility also makes it quite hard to get too excited about the identity of indiscernibles, at least within the philosophy of mathematics.

Chapter 15 approaches indiscernibility using the model-theoretic notion of a *type*. Briefly, a type is the collection of formulas satisfied by an element of some elementary extension. The contemporary study of types treats them as the points of a certain kind of topological space, called a *Stone space*. We explore this in Chapter 14, and hint at the richness of moving back-and-forth between algebraic and topological perspectives. This is the first intimation that there is a fertile interaction between broadly geometrical notions and model-theoretic ones (a theme of Chapter 17). Moreover, we can map these two perspectives—and the possibility of moving between them—over to a more *metaphysical* setting, using them to illuminate the question of whether propositions should be regarded as sets of possible worlds or vice-versa. We prove that these two rival metaphysical approaches are biinterpretable (in the sense of §5.4), and discuss the philosophical significance of this result.

In Chapter 16, we use the notions of indiscernibility which we introduced in Chapters 14 and 15 to investigate how to *classify quantifiers* as logical / non-logical. We start by introducing *generalised* quantifiers. We then introduce the Tarski–Sher thesis, which states that quantifiers are logical provided they exhibit a certain kind of invariance. We argue that intuitions about 'non-discrimination' are insufficient to establish Tarski–Sher. Then, by introducing various infinitary logics, we raise difficulties for further attempts to establish Tarski–Sher.

In Chapter 17, we consider contemporary model-theoretic programs in classification. We begin by proposing a wholly general philosophical framework for understanding classification programs within mathematics. We then turn to Shelah's famous results on the number of non-isomorphic of models of theories, and discuss how this work can be seen as an instance of classification in our proposed sense. It is worth noting that every theory which is classifiable in Shelah's sense is *stable*, where stability is a notion which we introduced in Chapter 14 in terms of a restriction on the size of the type space. We close the chapter by discussing Zilber's ambitious proposal for the classification of uncountably categorical theories (i.e. theories which have only one model up to isomorphism in a given uncountable cardinality). Whereas categoricity was seen as a potential philosophical desideratum in Part B, Zilber's programme regards uncountable categoricity as a kind of 'extreme classification'.

Readers who only want to dip into particular topics of Part C can consult the following Hasse diagram of dependencies between the sections:

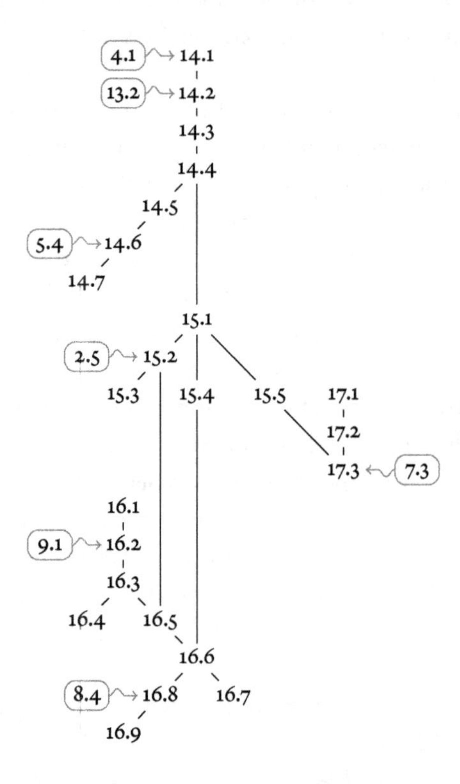

We should note that the mathematical content in Part C is generally much more advanced than in Parts A and B. In earlier chapters, we could usually afford to place proofs in the appendices, on the understanding that the reader who understood the statement of the results could potentially take the proof itself 'on trust'. However, in Part C, we will need to comment on several of the proofs themselves.

14
Types and Stone spaces

We first introduced compactness in Chapter 4. The name 'compactness' came out of the blue, but its roots are in topology. In this chapter, we pursue these topological ideas, invoking Boolean algebras and introducing Stone spaces. We already introduced the notion of Boolean algebras in Chapter 13, and we outline the required topological notions in §14.A. We consider several different ways to think about compactness—topological, algebraic, and model-theoretic—and outline their mathematical and philosophical relationships to one another.

Introducing these topological notions allows us to discuss certain philosophical positions concerning propositions and possible words. On the one hand, Stalnaker has suggested that we should treat propositions as sets of possible worlds; on the other, Adams has suggested that we should treat possible worlds as maximally consistent sets of propositions. Stone's Duality Theorem naturally suggests that these two approaches are *notational* variants, in some sense. Developing this, we show that the apparently rival perspectives are *biinterpretable*, in the sense of Chapter 5.

14.1 Types for theories

To introduce the model-theoretic notion of a *type*, we begin with types for *theories*.

Definition 14.1: *Let T be an $\mathcal{L}$-theory, let $v_1, \dots, v_n$ be variables, and let p be any set of $\mathcal{L}$-formulas whose free variables are among $v_1, \dots, v_n$.*

We say that p is an n-type *over T iff: $p \cup T$ is satisfiable.*

We say that p is a complete n-type *over T iff: p is an n-type and either $\varphi \in p$ or $\neg\varphi \in p$ for all $\mathcal{L}$-formulas φ whose free variables are among $v_1, \dots, v_n$.*

The set of complete n-types over T is $\text{Types}_n(T)$.

Here, to say that $p \cup T$ is *satisfiable* is just to say that we can find a model of T which interprets n new constant symbols $\bar{c}$ to make $\varphi(\bar{c})$ true for all formulas $\varphi(\bar{v})$ from p.[1] As a limiting case, if p is a complete 0-type over T, then p is just a complete, consistent $\mathcal{L}$-theory extending T. So, if we wanted, we could rephrase Lindenbaum's Lemma 4.22 as follows: if T is consistent, then $\text{Types}_0(T) \neq \varnothing$.

[1] In detail: if p is a set of $\mathcal{L}$-formulas whose free variables are among $v_1, \dots, v_n$, then to say that p is satisfiable is to say that, when we expand $\mathcal{L}$ to $\mathcal{L}^*$ by adding exactly n new constant symbols $c_1, \dots, c_n$, and when we let p^* be the result of respectively replacing v_i with c_i, then p^* is satisfiable in the usual sense.

The case that gives types their name, though, is when we consider n-types for $n > 0$. In particular, we can use the notion of a type to characterise elements within a structure. For example, suppose we started with the theory PA, and considered the set of formulas:

$$q = \{v_1 > S^n(0) : n \geq 0\} \qquad \text{(type:nonstandard)}$$

By the work done in §4.1, we know that $\text{PA} \cup q$ is satisfiable, so that q is a 1-type over PA. We also know that there is a structure $\mathcal{N} \vDash \text{PA}$ with an element $a \in N$ such that $\mathcal{N} \vDash \varphi(a)$ for all $\varphi \in q$. In such cases, we say that *a realises the type q in $\mathcal{N}$*.

14.2 An algebraic view on compactness

We now connect types with Boolean algebras, as introduced in §13.2. The connection begins with a method for 'reading off' Boolean algebras from theories. Let $\mathscr{L}$ be any signature, and let T be a satisfiable $\mathscr{L}$-theory. We quotient the $\mathscr{L}$-formulas with free variables among $v_1, \ldots, v_n$ by logical equivalence relative to T, i.e.:

$$[\varphi]_l = \{\psi : T \vDash \varphi \leftrightarrow \psi\}$$

We now form T's n^{th} Lindenbaum algebra, $\text{Lind}_n(T)$.[2] Its domain is the set of these equivalence classes, i.e.:

$$\{[\varphi]_l : \varphi \text{ is an } \mathscr{L}\text{-formula with free variables among } v_1, \ldots, v_n\}$$

and it has algebraic functions defined as follows:

$$-[\varphi]_l = [\neg\varphi]_l$$
$$[\varphi]_l \cdot [\psi]_l = [\varphi \wedge \psi]_l \qquad [\varphi]_l + [\psi]_l = [\varphi \vee \psi]_l$$
$$0 = [\varphi \wedge \neg\varphi]_l \qquad 1 = [\varphi \vee \neg\varphi]_l$$

We leave it to the reader to check this simple fact:

Proposition 14.2: *If T is satisfiable, then* $\text{Lind}_n(T)$ *is a Boolean algebra.*

When we defined Boolean algebras in §13.2, we also defined filters and ultrafilters (see Definition 13.3). These have familiar realisations in Lindenbaum algebras: in effect, $\text{Lind}_n(T)$'s filters are the extensions of T that are closed under logical consequence,[3] and its ultrafilters are the complete, satisfiable extensions of T. The following result puts this precisely:

[2] See e.g. Hodges (1993: 280).

[3] We say S is closed under consequence iff: $\varphi \in S$ iff $S \vDash \varphi$.

Lemma 14.3: *For any satisfiable $\mathscr{L}$-theory T:*

(1) *if F is a filter on* $\text{Lind}_n(T)$, *then $\bigcup F \supseteq T$ is satisfiable and closed under consequence*

(2) *if $S \supseteq T$ is satisfiable and closed under consequence, then $\{[\varphi]_l : \varphi \in S\}$ is a filter on* $\text{Lind}_n(T)$

(3) *p is a complete n-type on T iff $\{[\varphi]_l : \varphi \in p\}$ is an ultrafilter on* $\text{Lind}_n(T)$

Proof. We leave most of this to the reader, proving only that, if F is a filter, then $\bigcup F$ is satisfiable. Consider any finite $\{\varphi_1, ..., \varphi_n\} \subseteq \bigcup F$. Since F is a filter, $[\varphi_1]_l \cdot ... \cdot [\varphi_n]_l \in F$. Equally, since F is a filter, $[\varphi_1]_l \cdot ... \cdot [\varphi_n]_l \neq 0$, i.e. $[\varphi_1 \wedge ... \wedge \varphi_n]_l \neq [\varphi \wedge \neg\varphi]_l$, i.e. $T \nvDash (\varphi_1 \wedge ... \wedge \varphi_n) \leftrightarrow (\varphi \wedge \neg\varphi)$. Hence $T \cup \{\varphi_1, ..., \varphi_n\}$ is consistent. By the Compactness Theorem 4.1, $T \cup \bigcup F$ is satisfiable. □

We should pause. To guarantee that $\bigcup F$ is satisfiable, we appealed to the fact that, since F is a filter, it is closed under finite meet without containing 0. Squinting slightly, this property looks like (logical) compactness. And these two ideas are very closely connected, for Boolean algebras obey this analogue of the Compactness Theorem:

Theorem 14.4 (Ultrafilter Theorem): *Let $\mathcal{B}$ be a Boolean algebra. Say that $E \subseteq B$ is finitely-meetable iff $b_1 \cdot ... \cdot b_n \neq 0$, for all $b_1, ..., b_n \in E$. If E is finitely-meetable, then there is an ultrafilter F on $\mathcal{B}$ with $E \subseteq F$.*

Proof sketch. Use exactly the same strategy as for the proof of Lindenbaum's Lemma 4.22, replacing *consistency* with *finitely-meetability*. □

Recall that Lindenbaum's Lemma can be used as a key step in proving the Compactness Theorem. And it is no accident that there is a common core to these results: the Ultrafilter Theorem is *equivalent* to the Compactness Theorem.[4] So we have an algebraic viewpoint on compactness, to compare with our logical viewpoint.

14.3 Stone's Duality Theorem

We now aim for a *topological* viewpoint on compactness. We start by recalling some basic topological notions (for more background, see §14.A):

Definition 14.5: *Let $\mathcal{X} = (X, \tau)$ be any topological space.*

We say that $U \subseteq X$ is open iff $U \in \tau$, that U is closed iff $(X \setminus U) \in \tau$; that U is clopen *iff U is both closed and open.*

[4] Jech (1973: Theorem 2.2). They are *theorems* in that both are provable in (e.g.) ZFC. They are *equivalent* in that both are independent from ZF, but adding either as a new axiom entails the other.

We say that $\gamma \subseteq \tau$ *is an* open cover *of* X *iff* $\bigcup \gamma = X$.

We say that $\mathcal{X}$ *is* compact *iff every open cover of* X *has a finite subcover, i.e. for any open cover* γ, *there is some finite cover* $\gamma_0 \subseteq \gamma$.

We say that $\mathcal{X}$ *is* totally separated *iff for any* $x \neq y \in X$ *there is some clopen* $U \in \tau$ *such that* $x \in U$ *but* $y \notin U$.

We say that $\mathcal{X}$ *is a* Stone space *iff* $\mathcal{X}$ *is totally separated and compact.*[5]

The name here is in honour of Stone, who proved that Boolean algebras are *dual* to Stone spaces. His duality result amounts, roughly, to the following. Each Boolean algebra can naturally be transformed into a Stone space. Similarly, each Stone space can naturally be transformed into a Boolean algebra. Moreover, if we perform two of these transformations consecutively, then we end up back where we started (up to isomorphism). And finally: these transformations preserve various interesting maps between Boolean algebras and between Stone spaces.

The purpose of this section is to outline this duality in more detail. To begin, we must explain what the 'transformations' are. One of them is simple.

Definition 14.6: *Let* $\mathcal{X}$ *be a topological space with domain* X. *Let* Clopen$(\mathcal{X})$ *be an algebra whose domain is the set of* $\mathcal{X}$*'s clopen sets, with algebraic operations on these clopen sets defined as follows:*

$$-U = X \setminus U$$

$$U \cdot V = U \cap V \qquad\qquad U + V = U \cup V$$

$$0 = \varnothing \qquad\qquad 1 = X$$

It is straightforward to show that, if $\mathcal{S}$ is a topological space, then Clopen$(\mathcal{S})$ is a Boolean algebra.

It takes more skill to move in the opposite direction, from algebras to topologies. The general idea is to construct a space whose points are the *ultrafilters* on the algebra. The topology is then built from basic open sets, which are determined simply by the points of the Boolean algebra. Here is the idea more formally:

Definition 14.7: *Let* $\mathcal{B}$ *be a Boolean algebra. Let* Ultra$(\mathcal{B})$ *be the set of ultrafilters on* $\mathcal{B}$. *For each* $b \in B$, *let* $\mathrm{u}(b) = \{F \in \mathrm{Ultra}(\mathcal{B}) : b \in F\}$. *Call these the basic sets. We form a topological space whose underlying domain is the set of ultrafilters,* Ultra$(\mathcal{B})$, *and whose open sets are exactly the arbitrary unions of sets of basic sets, i.e.* U *is open in the space iff* $U = \bigcup_{i \in I} \mathrm{u}(b_i)$ *for some index set* I. *By standard abuse of notation, we call this topology simply* Ultra$(\mathcal{B})$.

[5] Johnstone (1982: 69–70) offers various alternative, equivalent definitions.

Note that, for each ultrafilter $F \in \text{Ultra}(\mathcal{B})$ and each $b \in B$, we always have

$$b \in F \text{ iff } F \in \text{u}(b) \qquad (\textit{ultra:flip})$$

Moreover, since each $F \in \text{Ultra}(\mathcal{B})$ is 'complete', in the sense that either $b \in F$ or $-b \in F$ for each $b \in B$, we always have

$$F \notin \text{u}(b) \text{ iff } b \notin F \text{ iff } -b \in F \text{ iff } F \in \text{u}(-b) \qquad (\textit{ultra:clopen})$$

so that $\text{u}(b)$ is always clopen. The force of this comes out in the following result:

Lemma 14.8: Ultra$(\mathcal{B})$ *is a Stone space.*

Proof. We must first check that Ultra$(\mathcal{B})$ is a topology. It suffices to check that $\{\text{u}(b) : b \in B\}$ is a basis (see Proposition 14.22). Clearly every ultrafilter is in $\bigcup\{\text{u}(b) : b \in B\}$. Now if $\text{u}(a)$ and $\text{u}(b)$ are basic open sets with $F \in \text{u}(a) \cap \text{u}(b)$, then $a \in F$ and $b \in F$ by (*ultra:flip*). Since F is an ultrafilter, $a \cdot b \in F$, i.e. $F \in \text{u}(a \cdot b)$ by (*ultra:flip*) as required.

Totally separated. Let $F \neq G$ be ultrafilters on $\mathcal{B}$. Since they are distinct, we have some $b \in B$ such that $b \in F$ and $b \notin G$, i.e. $F \in \text{u}(b)$ and $G \notin \text{u}(b)$ by (*ultra:flip*). Since $\text{u}(b)$ is always clopen as in (*ultra:clopen*), Ultra$(\mathcal{B})$ is totally separated.

Compact. It suffices to consider covers from the basis (see Lemma 14.25). Let γ be a cover, i.e. $\bigcup \gamma = \text{Ultra}(\mathcal{B})$. For reductio, suppose the following set is finitely-meetable:

$$G = \{-b : \text{u}(b) \in \gamma\}$$

By the Ultrafilter Theorem 14.4, there is an ultrafilter $F \supseteq G$. But then $F \notin \bigcup \gamma$, contradicting the fact that γ is a cover. So G is not finitely-meetable, i.e. there are $-b_1, \ldots, -b_n \in G$ such that $-b_1 \cdot \ldots \cdot -b_n = 0$. So for any $F \in \text{Ultra}(\mathcal{B})$, there is some $i \leq n$ such that $-b_i \notin F$, i.e. such that $b_i \in F$ by (*ultra:clopen*). Now

$$\gamma_0 = \{\text{u}(b_1), \ldots, \text{u}(b_n)\}$$

is a finite cover, and $\gamma_0 \subseteq \gamma$. □

Note that the *algebraic* analogue of compactness, namely the Ultrafilter Theorem 14.4, here yields the topological compactness of Ultra$(\mathcal{B})$.

Combining the preceding results—and adding in a few more observations which we will *not* prove—we obtain Stone's full-fledged duality result (we define continuity and homeomorphism in §14.A):[6]

Theorem 14.9 (Stone Duality): *The categories of Boolean algebras and Stone spaces are dual. In particular:*

[6] Stone (1936); for a contemporary proof, see Coppelberg (1989: ch.3).

(1) *If* $\mathcal{B}$ *is a Boolean algebra, then* Ultra($\mathcal{B}$) *is a Stone space, and* $\mathcal{B} \cong$ Clopen(Ultra($\mathcal{B}$)), *with* $b \mapsto \mathrm{u}(b)$ *an isomorphism.*
(2) *If* S *is a Stone space, then* Clopen(S) *is a Boolean algebra and* $S \cong$ Ultra(Clopen(S)), *with* $x \mapsto \{p \text{ is clopen} : x \in p\}$ *a homeomorphism.*
(3) *If* $\mathcal{B}, C$ *are Boolean algebras and* $f : B \longrightarrow C$ *is a homomorphism, then* f^{-1} *is a continuous map* Ultra(C) $\longrightarrow$ Ultra($\mathcal{B}$).
(4) *If* $\mathcal{X}, \Upsilon$ *are Stone spaces and* $g : X \longrightarrow Y$ *is continuous, then* g^{-1} *is a homomorphism* Clopen(Υ) $\longrightarrow$ Clopen($\mathcal{X}$). □

Stone's Duality Theorem 14.9 is a beautiful result. It is one of the earliest results of category theory, and highlights deep connections between different areas of mathematics. Johnstone takes Stone's work to suggest that:

> [...] abstract algebra cannot develop to its fullest extent without the infusion of topological ideas, and conversely if we do not recognize the algebraic aspects of the fundamental structures of analysis our view of them will be one-sided.[7]

But to plumb those depths would remove us even further from the core questions of this book, concerning model theory; so we return to them.

14.4 Types, compactness, and stability

Stone's Duality Theorem 14.9 allows us to take the ultrafilters on any Boolean algebra and form a Stone space whose points are those ultrafilters. In §14.1, though, we saw that the ultrafilters on a theory's Lindenbaum algebra are, in effect, that theory's complete types. Composing these thoughts, we can form a Stone space whose points are a theory's complete types.

Here is the idea in detail. If T is any theory, then $\mathrm{Lind}_n(T)$ is a Boolean algebra, and $\mathrm{Ultra}(\mathrm{Lind}_n(T))$ is a Stone space. By Lemma 14.3(3), the ultrafilters on $\mathrm{Lind}_n(T)$ are, in effect, T's complete n-types. So we can generate a Stone space which is homeomorphic to $\mathrm{Ultra}(\mathrm{Lind}_n(T))$, just by replacing each ultrafilter F with the complete type $\bigcup F$. Otherwise put: there is a very natural topology on $\mathrm{Types}_n(T)$, called the *Stone topology*. Abusing notation again, we refer to this topology by its carrier set, so we simply call it $\mathrm{Types}_n(T)$, where its open sets are provided by taking every set $\mathrm{u}(\varphi) = \{p \in \mathrm{Types}_n(T) : \varphi \in p\}$ as basic.[8] It follows immediately from Lemma 14.8 that $\mathrm{Types}_n(T)$ is totally separated and compact. But, given our particular interest in compactness, we will also prove the compactness of $\mathrm{Types}_n(T)$ directly:

---

[7] Johnstone (1982: xxi).

[8] Marker (2002: 119) denotes $\mathrm{u}(\varphi)$ by $[\varphi]$; Hodges (1993: 280) denotes $\mathrm{u}(\varphi)$ by $\|\varphi\|$.

Direct proof of compactness of $\text{Types}_n(T)$. It suffices to consider covers from the basis (see Lemma 14.25). Let γ be a cover such that every member of γ is some $\text{u}(\varphi)$. Suppose, for reductio, that the following is a type:

$$d = \{\neg\varphi : \text{u}(\varphi) \in \gamma\}$$

By Lindenbaum's Lemma 4.22, there is a complete type $p \supseteq d$. But then $p \notin \bigcup\gamma$, contradicting the fact that γ is a cover. So d is not a type, i.e. $T \cup d$ is inconsistent. By the Compactness Theorem 4.1, there is some finite $d_0 \subseteq d$ such that $T \cup d_0$ is inconsistent. So for any $p \in \text{Types}_n(T)$, there is some $\neg\varphi \in d_0$ such that $\neg\varphi \notin p$, i.e. such that $\varphi \in p$ (since p is complete). Now

$$\gamma_0 = \{\text{u}(\varphi) : \neg\varphi \in d_0\}$$

is a finite cover, and $\gamma_0 \subseteq \gamma$. □

In the general case, we used the Ultrafilter Theorem 14.4 to prove the compactness of the associated space. In this specific case we rely upon its logical analogue, the Compactness Theorem 4.1. This is no surprise, and it highlights that the Stone topology on the space of types *inherits* its topological compactness directly from the compactness of first-order logic. Schematically, the point is that 'every inconsistent theory has a finite inconsistent subtheory' entails 'every open cover has a finite subcover'. In §14.5, we obtain a converse to this result, showing that the compactness of the topology entails the compactness of the underlying logic.

In practice, the types that are most interesting are those associated with the complete theory of a model $\mathcal{M}$, which we define as follows:

$$\text{Th}(\mathcal{M}) = \{\varphi \text{ is a sentence in } \mathcal{M}\text{'s signature} : \mathcal{M} \vDash \varphi\}$$

This naturally generalises, to admit parameters. Much as in Definition 1.5 from §1.5, where $\mathcal{M}$ is an $\mathscr{L}$-structure and $A \subseteq M$, let $\mathscr{L}(A)$ be the language obtained by augmenting $\mathscr{L}$ with a new constant symbols c_a for each $a \in A$, and let $\mathcal{M}_A$ be an $\mathscr{L}(A)$-expansion of $\mathcal{M}$ obtained just by setting $c_a^{\mathcal{M}_A} = a$. Then we define:

$$\text{Th}(\mathcal{M}_A) = \{\varphi \text{ is an } \mathscr{L}(A)\text{-sentence} : \mathcal{M}_A \vDash \varphi\}$$

and we can consider the types over $\text{Th}(\mathcal{M}_A)$. So an n-type over $\text{Th}(\mathcal{M}_A)$ is a set p of $\mathscr{L}(A)$-formulas with free variables among $v_1, \ldots, v_n$ such that $\text{Th}(\mathcal{M}_A) \cup p$ is satisfiable. For readability, we denote the set of complete n-types over $\text{Th}(\mathcal{M}_A)$ by $\text{Types}_n(\mathcal{M}_A) = \text{Types}_n(\text{Th}(\mathcal{M}_A))$.[9]

[9] Many authors use variants of the less-descriptive but easier-to-write notation $S_n^{\mathcal{M}}(A)$ for the type-space. See Marker (2002: 115), Hodges (1993: 280), and Pillay (1983: 2).

The simplest elements of the type space are those given by elements from the underlying model. More specifically, given a sequence of elements $\overline{b}$ from $\mathcal{M}$, we say that $\text{tp}^{\mathcal{M}}(\overline{b}/A)$ is the set of $\mathcal{L}(A)$-formulas with free variables among $v_1, \ldots, v_n$ which are satisfied by $\overline{b}$ in $\mathcal{M}_A$. (In the case where $A = \varnothing$, we just write $\text{tp}^{\mathcal{M}}(\overline{b})$ for $\text{tp}^{\mathcal{M}}(\overline{b}/A)$.) Another way to put this is to say that $\text{tp}^{\mathcal{M}}(\overline{b}/A)$ is the complete n-type of $\text{Types}_n(\mathcal{M}_A)$ that is *realised* by $\overline{b}$.

A model need not realise all of its types. For instance, recall the type of a non-standard natural number, (*type:nonstandard*): there are many completions of this type, but clearly none of them are realised in the standard model of the natural numbers. However, for every type, there is an elementary extension of the given model realising that type. (This is proved by a simple argument from Compactness, which is very similar to the proof of Proposition 4.18.)[10]

As we saw earlier, $\text{Types}_n(\mathcal{M}_A)$ naturally yields a Stone topology. That topology has the following basic open sets, as φ ranges over $\mathcal{L}(A)$-formulas whose free variables are among $v_1, \ldots, v_n$:

$$\text{u}(\varphi(\overline{v})) = \{p \in \text{Types}_n(\mathcal{M}_A) : \varphi \in p\}$$

When $\mathcal{L}(A)$ is countable, this space has an extremely familiar topological structure. In this case, there is an enumeration $\varphi_1, \ldots, \varphi_m, \ldots$ of the $\mathcal{L}(A)$-formulas whose free variables are among $v_1, \ldots, v_n$. Then, since each complete n-type either contains φ_m or $\neg\varphi_m$, we can visualise the complete n-types as paths through an infinite binary tree, the first three levels of which are displayed below:

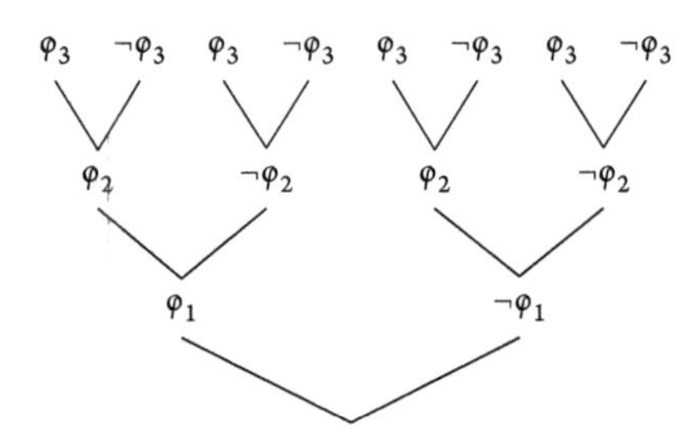

Every complete n-type corresponds to a path through the tree. (However, some paths do not correspond to n-types: for example, if φ_2 is $\neg\varphi_1$, then no infinite path through which goes through φ_1 and φ_2 is a type, since it would be unsatisfiable.) On this way of visualising the complete n-types, the basis for the topology can be given by taking finite unions of the 'cones' above each consistent node on the tree (i.e., all the complete n-types that contain the formula on the node). We can then give this topology a *metric* which, intuitively, says that the higher in the tree two types part ways, the closer together they are (metrics are defined in Definition 14.23).

[10] For details, see Marker (2002: Proposition 4.1.3).

Specifically, we define the distance between two types to be 2^{-m}, where m is the least number such that one type goes through a node containing φ_m while the other goes through a node containing $\neg\varphi_m$.[11]

To illustrate the usefulness of this topological perspective, note that it lets us gauge the size of $\text{Types}_n(\mathcal{M}_A)$ when $\mathscr{L}(A)$ is countable. Cantor proved that every closed subset of the real line is either countable or has the cardinality of the continuum, and it was later observed that this result holds for a large class of metric spaces, including the infinite binary tree.[12] Now, $\text{Types}_n(\mathcal{M}_A)$ is a closed subset of this tree, since closed subsets of the infinite binary tree correspond to subtrees of the infinite binary tree.[13] So:

Theorem 14.10: *If $\mathscr{L}(A)$ is countable, then* $\text{Types}_n(\mathcal{M}_A)$ *is either countable or has cardinality continuum.*

This depends crucially on the assumption that $\mathscr{L}(A)$ is countable, since we defined a metric on $\text{Types}_n(\mathcal{M}_A)$ via an enumeration of the $\mathscr{L}(A)$-formulas. Now, we have already seen one example where $\text{Types}_n(\mathcal{M}_A)$ has cardinality continuum. Let $\mathcal{M}$ be the natural numbers in the signature $\{0, 1, +, \times, <\}$, let $A = \varnothing$, let π_n be the nth prime number, and let '$a \mid b$' mean that a divides b. Then we can consider the following generalisation of (*type:nonstandard*), for any $X \subseteq \mathbb{N}$:

$$q_X = \{v_1 > S^n(0) : n \geq 0\} \cup \{S^{\pi_n}(0) \mid v_1 : n \in X\} \cup \{S^{\pi_n}(0) \nmid v_1 : n \notin X\}$$

By Compactness, q_X is a type for any $X \subseteq \mathbb{N}$; but $q_X = q_Y$ iff $X = Y$. Hence, since each subset of the natural numbers may be uniquely associated to a complete type, the number of complete types has cardinality continuum.

Before describing a situation where $\text{Types}_n(\mathcal{M}_A)$ is countable, we will define a more general notion which intuitively says that a theory has as few types as possible:

Definition 14.11: *Let T be a complete theory with an infinite model and let λ be an infinite cardinal. We say that T is λ-*stable *iff for every model $\mathcal{M}$ of T and every set $A \subseteq M$ of cardinality $\leq \lambda$, the set* $\text{Types}_1(\mathcal{M}_A)$ *has cardinality $\leq \lambda$.*[14] *We say that T is* stable *iff T is λ-stable for some infinite cardinal λ.*

---

[11] For more on this topological perspective on the type space, see Laskowski (2006: §1).

[12] More specifically, it holds for all Polish spaces, which are topological spaces whose topology can be given by a complete metric and which contain a countable dense set. This special case of the continuum hypothesis is provable in ZFC; see Kechris (1995: 32).

[13] See Kechris (1995: §2B).

[14] Note that λ-stability is equivalent to the condition that $\text{Types}_n(\mathcal{M}_A)$ has cardinality $\leq \lambda$ for *all* $n \geq 1$. So it is simply for convenience that the 'official' definition of stability only concerns 1-types rather than n-types in general. See Pillay (1983: 5).

That is, a theory is stable if it has comparatively few types, even as one allows the parameters to grow. Hence, an ω-stable theory in a countable signature will be one for which $\text{Types}_n(\mathcal{M}_A)$ is countable whenever A is countable. For instance, the complex numbers in the signature $\{0, 1, +, \times\}$ yield an interesting example of an ω-stable theory, but we postpone discussion of this until §15.5.

14.5 Bivalence and compactness

We began §14.4 by showing that the compactness of first-order logic entails the compactness of the Stone space generated from a first-order theory. We will now present a 'converse' of this result. To do this, we must develop a framework which does not 'build in' compactness. Our suggested framework is abstract, but *bivalent*. Having developed the framework, we then define something *like* a Stone space (though again, not wanting to assume that it is compact). Finally, we show that the Stone-like space is compact iff the logic is.

We start by defining our abstract, bivalent framework:

Definition 14.12: *A* bivalent-calculus, *C, is a domain of objects, C, together with a privileged object $\bot$, a one-place total function $\sim$, and a two-place relation $\Vdash$ which is a subset of $\wp(C) \times C$. Every bivalent-calculus obeys:*

$$\{a\} \Vdash a \qquad (\textit{reflexivity})$$

$$\textit{if } T \Vdash a, \textit{ then } T \cup U \Vdash a \qquad (\textit{weakening})$$

$$\textit{if } \{a\} \Vdash b \textit{ and } \{b\} \Vdash a, \textit{ then } a = b \qquad (\textit{decluttering})$$

$$T \cup \{\sim a\} \Vdash \bot \textit{ iff } T \Vdash a \qquad (\textit{reductio})$$

$$\textit{if } T \nVdash \bot, \textit{ then there is some story } S \supseteq T \qquad (\textit{lindenbaum})$$

where a story *on a bivalent-calculus, C, is a set $S \subseteq C$ which obeys:*

$$S \nVdash \bot \qquad (\textit{consistency})$$

$$\textit{for all } a \in C: \textit{ either } a \in S \textit{ or } \sim a \in S \qquad (\textit{maximality})$$

We should think of $\bot$ as contradiction, $\Vdash$ as entailment, and $\sim$ as negation. A bivalent-calculus could be augmented with functions behaving like conjunction, disjunction, and anything else you like; but this is the core. These operations are governed by very simple rules: (*reflexivity*), (*weakening*), and (*reductio*) are familiar from classical propositional logic, and need no further comment.[15] The rule

[15] We encountered (*reflexivity*) and (*weakening*) in the context of tarskian relations; see definition 13.17. Indeed, $\Vdash$ is also idempotent, and hence tarskian. *Proof.* Suppose $\{b \in C : T \Vdash b\} \Vdash a$ and let $S \supseteq T$ be a story. If $T \Vdash b$ then $S \Vdash b$ by (*weakening*); so $S \supseteq \{b \in C : T \Vdash b\}$. Hence $S \Vdash a$ by (*weakening*), and now $T \Vdash a$ by Lemma 14.13(3).

(*decluttering*) states that propositions are individuated by their structure within a network of entailments. For present purposes, this merely saves us the bother of forming equivalence classes of elements (as when we construct a theory's Lindenbaum algebra). The only rule meritting real comment is (*lindenbaum*), together with the notion of a *story*. Stories are our analogues of ultrafilters. This is why we have given suggestive names to the conditions on being a story. In these terms, (*lindenbaum*) effectively states that any consistent set is a subset of some story, i.e. some maximally consistent set.[16]

To make bivalent-calculi more familiar, here are some simple observations:

Lemma 14.13: *For any bivalent-calculus C and any $a \in C$*

(1) $\{a, \sim a\} \Vdash \bot$

(2) *If S is a story on C then $\{a, \sim a\} \nsubseteq S$*

(3) *If $a \in S$ for every story $S \supseteq T$, then $T \Vdash a$*

Proof. (1). By (*reflexivity*) and (*reductio*).

(2). By (1), (*weakening*) and (*consistency*).

(3). Suppose $T \nVdash a$. So $T \cup \{\sim a\} \nVdash \bot$ by (*reductio*) and hence there is some story $S \supseteq T \cup \{\sim a\}$ by (*lindenbaum*). So $a \notin S$ by (2). □

All of this should be comforting. But to make the ideas *really* familiar, here is a simple way to obtain bivalent-calculi. Given a fixed language for first-order logic, form its 0th Lindenbaum algebra from the empty theory; so the elements are just $[\varphi]_l = \{\psi : \vDash \varphi \leftrightarrow \psi\}$. Now just interpret:

$$\bot = 0$$
$$\sim[\varphi]_l = -[\varphi]_l = [\neg\varphi]_l$$
$$[T]_l \Vdash [\varphi]_l \text{ iff } T \vDash \varphi$$

It is easy to see that this yields a bivalent-calculus. Crucially, though, the rules for bivalent-calculi take no stance on *compactness*. In more detail,

Definition 14.14: *We say that a bivalent-calculus C is* compact *iff, for all $T \cup \{\varphi\} \subseteq C$, we have: $T \Vdash a$ iff $T_0 \Vdash a$ for some finite $T_0 \subseteq T$.*

[16] The inspiration behind Definition 14.12 is a relaxation of a framework used by Stalnaker, in the philosophical setting we discuss in §14.7. (*decluttering*) is effectively endorsed by Stalnaker (1976: (I) pp.72–3, 2012: (P6) ch.2 §1). (*lindenbaum*) should be compared with Stalnaker's principle (P5): 'Every consistent set of propositions is a subset of a maximal consistent set' (2012: ch.2 §1). Stalnaker has (minor) reservations about (P5), since 'one might think that for every proposition x, no matter how specific, there are always further propositions that are incompatible with each other, but each is compatible with x.' But this threatens (P5) only if we follow Stalnaker in accepting this further principle: for any set of propositions S, there is a single proposition equivalent to S. The notion of a bivalent-calculus takes no particular stance on that further principle (see footnote 21).

The bivalent-calculus obtained from first-order logic will be compact, just because first-order logic is. But suppose we instead obtain a bivalent-calculus, in similar fashion, from some (fixed language for) *full* second-order logic. Since that logic is not compact (see §7.5), the associated bivalent-calculus will not be either. In short: there are compact and non-compact bivalent-calculi.

It is easy to generate a 'space' from a bivalent-calculus. We essentially follow the steps which took us from a Boolean algebra $\mathcal{B}$ to the Stone space Ultra$(\mathcal{B})$:

Definition 14.15: *For any bivalent-calculus, C, let* Stories(C) *be the set of all stories on C, and let* $\mathrm{w}(a) = \{S \in \mathrm{Stories}(C) : a \in S\}$ *for each $a \in C$. We say that the universe* Uni(C) *of C is a structure whose domain is* Stories(C), *and that each* $\mathrm{w}(a)$ *is a* prop *on* Uni(C).

There are only two real differences between this and the definition of a Stone space: first, we considered stories, rather than ultrafilters; second, we did *not* close the props under arbitrary unions.[17] So Stories(C) need not *quite* be a topology. Still, we can easily provide the usual definition of *compactness* for a topological space, as in Definition 14.5, and obtain what we desired:

Theorem 14.16: *If C is bivalent-calculus: C is compact iff* Uni(C) *is compact.*

Proof. Left-to-right. Suppose C is compact. Let γ be a cover and define

$$T = \{\sim a \in C : \mathrm{w}(a) \in \gamma\}$$

Suppose, for reductio, that $T \nVdash \bot$; by (*lindenbaum*), there is some story $S \supseteq T$; and evidently $S \notin \bigcup \gamma$, contradicting our assumption that γ is a covering. So $T \Vdash \bot$. By the compactness of C, there is some finite $T_0 \subseteq T$ with $T_0 \Vdash \bot$. So for any story S there is some $\sim a \in T_0$ such that $\sim a \notin S$ i.e. $a \in S$ by (*maximality*). Now $\gamma_0 = \{\mathrm{w}(a) : \sim a \in T_0\}$ is a finite subcover of γ.

Right-to-left. Suppose Uni(C) is compact and that $T \Vdash a$, i.e. $T \cup \{\sim a\} \Vdash \bot$ by (*reductio*). So, by (*consistency*) and (*weakening*), there is no story $S \supseteq T \cup \{\sim a\}$. So $\gamma = \{\mathrm{w}(\sim b) : b \in T\} \cup \{\mathrm{w}(a)\}$ is a cover, and hence there is a finite cover $\gamma_0 \subseteq \gamma$. So there is a finite $T_0 = \{b_1, \ldots, b_n\} \subseteq T$ such that every story has at least one of $\sim b_1, \ldots, \sim b_n, a$ as a member. So by Lemma 14.13(2), every story $S \supseteq T_0$ has $a \in S$. Hence $T_0 \Vdash a$ by Lemma 14.13(3). □

Hence the 'topological' compactness of the associated space entails its logical compactness, and vice versa. We thus have what we wanted. Given an assumption of

[17] This is because the props on Uni(C) may not form a (topological) basis; in effect, because we did not include a 'conjunction' operator in our definition of a bivalent-calculus.

bivalence,[18] there is a clear *equivalence* between logical and topological viewpoints on compactness.

By the by, Theorem 14.16 can easily be extended to handle a generalisation of compactness. Say that a (topological) space is κ-*compact* iff every open cover has a subcover of cardinality $< \kappa$. Say that a calculus is κ-compact iff for any T and a we have: $T \Vdash a$ iff $T_0 \Vdash a$ for some T_0 of cardinality $< \kappa$. Then we can easily prove: C is κ-compact iff $\text{Uni}(C)$ is κ-compact.

14.6 A biinterpretation

The notion of a bivalent-calculus formalises a generally bivalent approach to logic, which deliberately takes no particular stance on compactness. Now, just as Stone spaces are the dual of Boolean algebras, we might look for the dual of bivalent-calculi. That dual is easy to define:

Definition 14.17: *A* bivalent-universe, *$\mathcal{X}$, is a domain of objects, X, together with a privileged subset of $\wp(X)$, called the* props *on $\mathcal{X}$, such that:*

$\varnothing$ is a prop	(LNC)
if x and y are members of exactly the same props, then $x = y$	(*kolmogorov*)
p is a prop iff $(X \setminus p)$ is a prop	(*bisecting*)

In these terms, Definition 14.15 states how to move from a bivalent-calculus C to a bivalent-universe $\text{Uni}(C)$. Inspired by Stone, we can move in the opposite direction (cf. Definition 14.6):

Definition 14.18: *For any bivalent-universe, $\mathcal{X}$, let* $\text{Calc}(\mathcal{X})$ *be a structure whose domain is $\mathcal{X}$'s props, such that, where $Q \cup \{p\}$ is any set of props on $\mathcal{X}$:*

$$\bot = \varnothing$$
$$\sim p = X \setminus p$$
$$Q \Vdash p \textit{ iff } \bigcap Q \subseteq p$$

Define $\text{t}(x) = \{p \text{ is a prop on } \mathcal{X} : x \in p\}$ for each $x \in X$. Now, Stone-like, we can obtain the following (we relegate the proof to §14.B):

Theorem 14.19: *(1) If C is a bivalent-calculus, then* $\text{Uni}(C)$ *is a bivalent-universe, and the function* $\text{w} : C \longrightarrow \text{Calc}(\text{Uni}(C))$ *is an isomorphism.*

[18] There are, of course, compact but non-bivalent logics (e.g. intuitionistic logic) and there are associated duality theorems; but we will not explore them in this book.

(2) *If $\mathcal{X}$ is a bivalent-universe, then* Calc($\mathcal{X}$) *is a bivalent-calculus, and the function* $\mathrm{t} : \mathcal{X} \longrightarrow \mathrm{Uni}(\mathrm{Calc}(\mathcal{X}))$ *is an isomorphism.*

This result enables us to move back and forth between bivalent-calculi and bivalent-universes. In fact, we can go further, and regard Theorem 14.19 as stating that *the theory of bivalent-calculi and the theory of bivalent-universes are biinterpretable*, in the sense of Chapter 5. But of course, to read Theorem 14.19 this way, we must first present some suitable *theories*.

So: *bivalent-universe-theory* is formally defined by incorporating Definition 14.17 into a set theory with urelements. We discussed set theory with urelements in Chapter 11. Its signature is $\{Set, \in\}$ where, intuitively, *Set* applies only to sets, and hence not to urelements. To define bivalent-universe-theory, we augment the set theory with new one-place predicates, W and P, and supplement the usual set-theoretic axioms with five new axioms:

$$\begin{aligned}
\text{worlds = urelements, i.e.: } & (\forall x)(W(x) \leftrightarrow \neg Set(x)) \\
\text{props are sets of worlds, i.e.: } & (\forall x : P)(Set(x) \wedge (\forall y \in x)W(x)) \\
(\textsc{lnc})\text{, i.e.: } & (\forall x : Set)(\neg \exists y\, y \in x \rightarrow P(x)) \\
(\textit{kolmogorov})\text{, i.e.: } & (\forall x, y : W)((\forall z : P)(x \in z \leftrightarrow y \in z) \rightarrow x = y) \\
(\textit{bisecting})\text{, i.e.: } & (\forall x : P)(\exists y : P)(\forall z : W)(z \in x \leftrightarrow z \notin y)
\end{aligned}$$

Similarly, *bivalent-calculus-theory* is formally defined by presenting Definition 14.12 within a set theory with urelements. Again, the idea is to augment the set theory with axioms stating: propositions are the only urelements; there is a privileged proposition 0 and an operation $\sim$ on propositions; and entailment is a specific relation between sets of propositions and individual propositions.

The set theory used in the proof of Theorem 14.19 can now be carried out within the object-language of either of these two theories. Consequently, the proof of Theorem 14.19 provides a biinterpretation between the two theories, in the sense of Definition 5.4. For instance, starting in bivalent-calculus-theory, which we view as axiomatising a bivalent calculus C, we can sequentially define Uni(C), and Calc(Uni(C)), and the map $\mathrm{w} : C \longrightarrow \mathrm{Calc}(\mathrm{Uni}(C))$, and then transcribe the proof of Theorem 14.19 within this theory. The same holds, in reverse, for the second part of the theorem, where we start in bivalent-universe-theory.

14.7 Propositions and possible worlds

In Chapter 5, we considered the general question of when two theories, or two structures, should count as 'the same'. One possible answer is that biinterpretable theories are 'the same'. And, certainly, biinterpretation preserves plenty of logical properties. So it might be right to 'identify' bivalent-calculus-theory with bivalent-universe-theory in some settings. However, to close this chapter, we will consider a use to which both theories might be put, which sits at the intersection of logic and metaphysics, and where there might be interesting reasons to *resist* the 'identification' of bivalent-calculus-theory with bivalent-universe-theory.

Explicating two approaches

Philosophers frequently talk of *propositions*, in a sense belonging more to philosophical than to mathematical logic. These propositions are something like (reified) *meanings* of sentences. Equally often, philosophers talk of *possible worlds*. These are something like (reified) *ways things might be*. In the literature, there are two canonical approaches to the relationship between propositions and possible worlds. The first is the *Worlds-First Approach*: propositions are sets of possible worlds. The second is the *Propositions-First Approach*: possible worlds are maximally consistent sets of propositions.[19]

On the face of it, these are rival metaphysical approaches. Our aim is to use the biinterpretation of §14.6 to shed light on this apparent disagreement. But we must start by explaining why bivalent-universe-theory and bivalent-calculus-theory might even be relevant to the disagreement.

Fans of the Propositions-First Approach should embrace bivalent-calculus-theory as a theory of their propositions. After all, if they are asked to spell out their claim that possible worlds are sets of propositions, they are likely to say (for example): a set of propositions is *consistent* so long as it does not entail a contradiction; a set of propositions is *maximal* so long as, given any proposition, it contains either that proposition or its negation; and possible worlds are maximally consistent sets of propositions. This informal explication involves three ideas: there is a way to express contradiction; there is a suitable notion of entailment; and every proposition has a negation.[20] That is precisely what bivalent-calculus-theory formalises.

To be sure, bivalent-calculus-theory is not wholly uncontroversial. First, it builds bivalence into the very idea of a proposition, and not everyone subscribes to bivalence. Second, it involves (*decluttering*): this amounts to identifying propositions

[19] The Propositions-First Approach is most famously associated with Stalnaker (1976). The World-First Approach is most famously associated with Adams (1974). Note that *both* approaches effectively assume that propositions are *necessary* existents. For suppose that propositions *could* have existed, which in fact *do not* exist. Then (*a*) a possible world which would intuitively satisfy a 'merely possible but actually non-existent proposition' might not be characterisable by reference only to propositions which *do* exist, so that the Propositions-First Approach fails. And (*b*) if (as seems plausible) possible worlds are necessary existents, and sets whose members are necessary existents are also necessary existents, then every set of possible worlds is a necessary existent, so that the Worlds-First Approach fails. Indeed, Stalnaker has recently distanced himself from his earlier (1976) Worlds-First Approach, precisely because he now holds 'that propositions themselves may exist only contingently' (2012: ch.2 §1).

[20] See Stalnaker (1976: 71–2, 2012: ch.2 §1).

which entail each other, which is likely to lead to thinking of propositions in a very coarse-grained way. Nonetheless, if we are happy think of propositions in a coarse-grained, bivalent way, then bivalent-calculus-theory explicates the Propositions-First-Approach wonderfully.

Similarly, fans of the Worlds-First Approach should embrace bivalent-universe-theory as a theory of their possible worlds. After all, if they are asked to spell out their claim that propositions are sets of possible worlds, they are likely to say a few things. First: the contradictory proposition is not 'true of' any world; so the empty set of possible worlds is a proposition, namely, the contradictory proposition. Second: different things are 'true of' different worlds; so given two distinct worlds, some proposition must distinguish them. Third: every proposition must have a negation, which consists of exactly those worlds not in the original proposition. And these are just what the notions of (LNC), (*kolmogorov*) and (*bisecting*), from Definition 14.17, articulate.

Unsurprisingly, bivalent-universe-theory is exactly as (un)controversial as bivalent-calculus-theory. First, (*bisecting*) amounts to assuming that propositions behave in a bivalent fashion. Second, (*kolmogorov*) effectively amounts to assuming that propositions are extremely coarse-grained. But the general point is simple. If all parties are happy to assume a coarse-grained, bivalent approach to propositions, then bivalent-calculus-theory and bivalent-universe-theory offer excellent explications of the informal notions of propositional entailment and possible worlds.[21]

Rapprochement between the approaches

We now want to bring Theorem 14.19 to bear on the Worlds-First and Propositions-First Approaches.

For vividness, suppose that *Paige* embraces the Propositions-First Approach. So

[21] Note that, since neither bivalent-calculi nor bivalent-universes assume anything about *compactness*, these explications avoid making controversial assumptions about compactness. This is noteworthy, because it differs from two obvious alternative approaches we might have followed.

We might have explicated propositions in terms of first-order logic or its Lindenbaum algebra, and explicated possible worlds in terms of Stone spaces. (This approach is discussed by Bricker 1983.) In the ensuing philosophical discussion, Stone's Duality Theorem 14.9 would have played the role of our Theorem 14.19. But this explication would have involved firmly *embracing* the compactness of propositional-entailment and the compactness of our possible universe.

Alternatively, we might have explicated propositions in terms of complete, atomic Boolean algebras, and explicated possible worlds in terms of discrete topological spaces. (Where $\mathcal{B}$ is a Boolean algebra, we say that an element $a \in B$ is an *atom* such that if $0 \leq x \leq a$ then x is either 0 or a. We then say that $\mathcal{B}$ is atomic iff all its elements are suprema of some set of atoms. This approach is favoured by Stalnaker 1976: (C) pp.72–3, 2012: (P4) ch.2 §1; Bricker 1983.) In the ensuing philosophical discussion, a well-known result of Lindenbaum and Tarski (see e.g. Coppelberg 1989: Corollary 2.7 p.30) would have played the role of our Theorem 14.19. But this explication would have involved firmly *rejecting* the compactness of propositional-entailment and the compactness of our possible universe. After all, any infinite, discrete topological space is trivially non-compact; and to insist on a complete algebra amounts to assuming that, for any (possibly infinite) set of propositions x, there is a single proposition equivalent to the conjunction of all the propositions in X.

she thinks that some *intended* bivalent-calculus, C, correctly explicates the entailments between *the* propositions. Using C, she can then obtain something which behaves just like a possible universe, $\text{Uni}(C)$. Paige uses this to justify her belief that possible worlds *are* certain sets of propositions.

Equally vividly, but conversely, *Wendy* embraces the Worlds-First-Approach. So she thinks that some *intended* bivalent-universe, $\mathcal{X}$, correctly explicates *the* possible universe. Using $\mathcal{X}$, she can obtain something which behaves just like a bunch of propositions with entailments between them, $\text{Uni}(\mathcal{X})$. Wendy uses this to justify her belief that propositions *are* certain sets of possible worlds.

Wendy and Paige now start talking to each other. To interpret each other, they lean upon the biinterpretation between their respective theories, provided by Theorem 14.19. In fact, this allows them to interpret each other in every detail: not just in their attempts to discuss propositions and worlds directly, but in their attempts to interpret each other, and their attempts to interpret each other's interpretations of each other, and so on, *ad infinitum*.

This (bi)interpretation preserves a lot. For example, by Proposition 5.9, biinterpretability implies *faithful interpretability*. So proofs and disproofs transfer seamlessly back and forth between Paige's and Wendy's perspectives: anything one can prove, the other can too. And there are always two ways to arrive at a proof of a claim: stay within one perspective and prove it directly, or temporarily adopt the other perspective and prove the translation of the claim. Now, in §5.8, we worried that leaning on a faithful interpretation may mean that Paige and Wendy's proofs cannot really count as *distinct*. In the present context, though, that looks quite *desirable*. The whole thrust of this biinterpretation is that the Worlds-First and Propositions-First perspectives are very close.

And they are *very* close. For example, suppose Paige entertains the claim 'entailment should be compact'. Then both Paige and Wendy can think through, and discuss, the theoretical consequences that this claim would have, for both their 'propositions' and their 'worlds'. They can consider whether these are *desirable* consequences and, whilst they may not agree on the answer, they can genuinely *communicate* about it. Conversely, they can both discuss the effects of embracing a claim like 'the possible universe should be a discrete space'. In short, the biinterpretation allows Wendy and Paige to explore, *together*, the pros and cons of embracing or rejecting various proposals.

But, for all its many virtues, Paige and Wendy may find one particular aspect of the biinterpretation somewhat *infelicitous*. Paige thinks that propositions are basic and that worlds are derivative. As such, when she formalises her position using bivalent-calculus-theory, she treats propositions as primitive urelements, at the base of some hierarchy of sets, and she treats worlds as (constructed) sets of urelements, at the next stage of the hierarchy. Wendy, by contrast, thinks that worlds are basic and that propositions are derivative. As such, when she formalises her position using bivalent-universe-theory, she treats worlds as primitive urelements and propositions as (constructed) sets of urelements, at the next stage. Locating these entities at different levels of the hierarchy precisely reflects their disagreement over whether worlds or propositions are *metaphysically prior*. But, when Wendy uses the biinterpretation given by Theorem 14.19 to interpret Paige, she treats what Paige would regard as an urelement (i.e. as a primitive entity) as a set (i.e. as something constructed). The same holds when Paige interprets Wendy. As such, Paige and Wendy may simply insist that the biinterpretation fails to preserve their respective views on *metaphysical priority*.

This, then, is precisely the extent of the rapprochement our biinterpretation will give to Paige and Wendy. They can agree to a large extent, they can communicate with each other on several issues of mutual concern, and they can localise their disagreement to a specific issue concerning metaphysical priority.

We should add that the biinterpretation has even more clout for philosophers who are *unconcerned* by questions of metaphysical priority. (This will include any philosophers who are less concerned with the *metaphysics*, than with the *truth or falsity*, of claims made using either propositions-talk or possible-worlds-talk.)[22] Such philosophers will be able to move seamlessly and indifferently between treating propositions as sets of worlds, and treating worlds as sets of propositions.

14.A Topological background

This appendix arms the reader with the bare topological bones required to follow the mathematics in this chapter. We define a topological space as follows:

Definition 14.20: *A* topology *on a set T is any set $\tau \subseteq \wp(T)$, such that:*

(1) $\varnothing, T \in \tau$
(2) *if* $\gamma \subseteq \tau$, *then* $\bigcup \gamma \in \tau$
(3) *if* $\gamma \subseteq \tau$ *is finite and nonempty, then* $\bigcap \gamma \in \tau$

When τ is a topology on T, we say that (T, τ) is a topological space. The members of T are the points of the space.

We will be fairly sloppy in whether we talk about topologies or about topological spaces. This sloppiness is licensed by the fact that, when τ is a topology on T, we have $\bigcup \tau = T$, and $\langle \bigcup \tau, \tau \rangle$ is the associated space.

Given a topology τ on T, the members of τ are called *open* sets of the space. We say that $C \subseteq T$ is *closed* iff $T \setminus C$ is open, i.e. a member of τ. It it is immediate from

[22] Cf. the conceptual cosmopolitan of Button (2013: ch.19).

this definition that both $\varnothing$ and T are both open and closed. Sets like this, which are both open and closed, are called *clopen* (as we mentioned in Definition 14.5).

As an example: $\{\varnothing, T\}$ is always a topology on T, known as the *trivial* topology. Equally, $\wp(T)$ is always a topology on T, known as the *discrete* topology.

For a more interesting example, we turn to real analysis. Here, sets of the form $(a, b) = \{r \in \mathbb{R} : a < r < b\}$ are often called 'basic open intervals'. To explain this, we introduce the idea of a *basis*:

Definition 14.21: *A* basis β *on T is a set $\beta \subseteq \wp(T)$ such that:*

(1) $\bigcup \beta = T$
(2) for all $U, V \in \beta$ and all $x \in U \cap V$, there is some $W \in \beta$ such that $x \in W \subseteq U \cap V$

Bases are of interest because they generate topologies:

Proposition 14.22: *If β is a basis on T, then $\{\bigcup \gamma : \gamma \subseteq \beta\}$ is a topology on T, called* the *topology generated by β.*

Proof. Where $\tau = \{\bigcup \gamma : \gamma \subseteq \beta\}$, we must check that τ meets each clause of Definition 14.20.

(1). Clearly $\varnothing \subseteq \beta$ so $\bigcup \varnothing = \varnothing \in \tau$, and we are given that $\bigcup \beta = T \in \tau$.

(2). Fix $\gamma \subseteq \tau$. For each $U \in \gamma$ there is some $\beta_U \subseteq \beta$ such that $U = \bigcup \beta_U$. So $\delta = \bigcup\{\beta_U : U \in \gamma\} \subseteq \beta$ and $\bigcup \gamma = \bigcup \delta \in \tau$.

(3). We show that if $U, V \in \tau$ then $U \cap V \in \tau$; the general result follows. If $U, V \in \tau$ then there are $\beta_U, \beta_V \subseteq \beta$ such that $U = \bigcup \beta_U$ and $V = \bigcup \beta_V$. So, for any $x \in U \cap V$, there are $B_U \in \beta_U$ and $B_V \in \beta_V$ such that $x \in B_U \cap B_V$. Since β is a basis, there is some $B_x \in \beta$ such that $x \in B_x \subseteq B_U \cap B_V$. Now $U \cap V = \bigcup_{x \in U \cap V} B_x \in \tau$. □

We can apply Proposition 14.22 to real analysis as follows. Let $\mathbb{R}$ be our set of points. Let ρ be the set of basic open intervals, i.e.: $\rho = \{(a, b) : a \leq b \in \mathbb{R}\}$. It is easy to check that this is a basis on $\mathbb{R}$; so ρ generates a topology on $\mathbb{R}$. Indeed, it generates the *standard* topology on $\mathbb{R}$. The same topology can also be given by a *metric*:

Definition 14.23: *A* metric space *is a pair (T, d), where d is a function from $T \times T$ to the non-negative real numbers, such that, for all $x, y, z \in T$: $d(x, y) = 0$ iff $x = y$; and $d(x, y) = d(y, x)$; and $d(x, y) \leq d(x, z) + d(z, y)$.*

A metric space (T, d) induces a basis, by treating every set $\{y \in T : d(x, y) < r\}$ as basic, for any $x \in T$ and non-negative real r; and by Proposition 14.22 this basis generates a topology. The Euclidean metric on $\mathbb{R}$ is given by $d(x, y) = |x - y|$, and it is easy to see that the topology induced by this metric is the standard topology.

Many results from real analysis can now be realised in a more abstract setting. For example, we can generalise analysis's ε–δ definition of a continuous function. The guiding idea here is that the inverse image of an open set is open:

Definition 14.24: *Let τ_1 be a topology on T_1 and τ_2 be a topology on T_2. A function $f : T_1 \longrightarrow T_2$ is a* continuous *function from τ_1 to τ_2 iff it has the following property: if $U \in \tau_2$ then $f^{-1}(U) = \{a \in T_1 : f(a) \in U\} \in \tau_1$.*

We say that f is a homeomorphism *iff f is a bijection where both f and f^{-1} are continuous. (So a homeomorphism is a bijection such that X is open iff $f(X)$ is open.)*

The central topological notion in this chapter was compactness, as set out in Definition 14.5. The trivial topology is trivially compact, whereas the discrete topology on T is only compact when T is finite: consider the open cover $\{\{x\} : x \in T\}$. The standard topology on $\mathbb{R}$ is not compact—consider the open cover $\{(-r, r) : r \text{ is a positive real}\}$—but the Heine–Borel Theorem states that every closed and bounded subset of the reals is compact (and vice versa). And, as we repeatedly mentioned during this chapter, when a topology is generated by a basis, we can determine whether it is compact just by considering open covers by basic elements.

Lemma 14.25: *Let β be a basis on T. If every cover $\gamma \subseteq \beta$ has a finite subcover, then the topology generated by β is compact.*

Proof. Let τ be the topology generated by β, and let $\gamma \subseteq \tau$ be a cover. For each $U \in \gamma$ there is some $\beta_U \subseteq \beta$ such that $U = \bigcup \beta_U$. So $\delta = \bigcup\{\beta_U : U \in \gamma\} \subseteq \beta$ and $\bigcup \gamma = \bigcup \delta = T$. By supposition, there is some finite subcover $\delta_0 \subseteq \delta$. For each $B \in \delta_0$, choose some $U_B \in \gamma$ such that $B \in \beta_{U_B}$ i.e. $B \subseteq U_B$; now $\{U_B : B \in \delta_0\} \subseteq \gamma$ is a finite cover. □

14.B Bivalent-calculi and bivalent-universes

In this appendix, we prove the central result of §14.6 which connects bivalent-calculi with bivalent-universes:

Theorem (14.19): *(1) If C is a bivalent-calculus, then* $\text{Uni}(C)$ *is a bivalent-universe, and* $\text{w} : C \longrightarrow \text{Calc}(\text{Uni}(C))$ *is an isomorphism.*

(2) If $\mathcal{X}$ is a bivalent-universe, then $\text{Calc}(\mathcal{X})$ *is a bivalent-calculus, and* $\text{t} : \mathcal{X} \longrightarrow \text{Uni}(\text{Calc}(\mathcal{X}))$ *is an isomorphism.*

Proof. *(1).* We start by confirming that $\text{Uni}(C)$ is a bivalent-universe. Concerning

(LNC): if S is a story, then $\bot \notin S$, by (*consistency*) and (*weakening*), so $\mathrm{w}(\bot) = \varnothing$. Concerning (*kolmogorov*): suppose $S \in \mathrm{w}(a)$ iff $T \in \mathrm{w}(a)$, for all props $\mathrm{w}(a)$ on $\mathrm{Uni}(C)$; so $a \in S$ iff $a \in T$, for all $a \in C$, and hence $S = T$. Concerning (*bisecting*): by (*maximality*) and Lemma 14.13(2), for any story S we have: $S \in \mathrm{w}({\sim}a)$ iff ${\sim}a \in S$ iff $a \notin S$ iff $S \notin \mathrm{w}(a)$ iff $S \in \mathrm{Stories}(C) \setminus \mathrm{w}(a)$. So $\mathrm{w}({\sim}a) = \mathrm{Stories}(C) \setminus \mathrm{w}(a)$.

We next confirm that w is a bijection. Clearly w is a surjection. To show injectivity, suppose that $\mathrm{w}(a) = \mathrm{w}(b)$. So for any story $S \supseteq \{a\}$ we have $b \in S$, so that $\{a\} \Vdash b$ by Lemma 14.13(3). Similarly, $\{b\} \Vdash a$. So $a = b$, by (*decluttering*).

To end, we confirm that w preserves structure, i.e. falsum, negation, and entailment. In this argument, we use the natural notation $\mathrm{w}(T) = \{\mathrm{w}(b) : b \in T\}$.

- $\mathrm{w}(\bot) = \varnothing = \bot$, as above
- $\mathrm{w}({\sim}a) = \mathrm{Stories}(C) \setminus \mathrm{w}(a) = {\sim}\mathrm{w}(a)$, as above.
- $T \Vdash a$ iff $\bigcap \mathrm{w}(T) \subseteq \mathrm{w}(a)$. To see this, first observe that the right-hand-side is equivalent to the claim that every story $S \supseteq T$ has $a \in S$. So right-to-left holds by Lemma 14.13(3). Conversely, suppose $T \Vdash a$ and let S be a story with $a \notin S$, so that ${\sim}a \in S$ by (*maximality*); then $S = S \cup \{{\sim}a\} \nVdash \bot$ by (*consistency*), so $S \nVdash a$ by (*reductio*); and so $S \not\supseteq T$ by (*weakening*).

(2). We start by confirming that $\mathrm{Calc}(\mathcal{X})$ is a bivalent-calculus. We trivially have (*reflexivity*), (*weakening*), (*decluttering*), and (*reductio*). Concerning (*lindenbaum*), let Q be a set of props on $\mathcal{X}$ and suppose $\bigcap Q \nsubseteq \varnothing$, i.e. we have some $x \in \bigcap Q$; then it follows that $\mathrm{t}(x) \supseteq Q$. It remains to check that $\mathrm{t}(x)$ is a story on $\mathrm{Calc}(\mathcal{X})$. Concerning (*consistency*), $x \in \bigcap \mathrm{t}(x)$, so that $\bigcap \mathrm{t}(x) \nsubseteq \varnothing$, and so $\mathrm{t}(x) \nVdash \bot$ by definition. For (*maximality*), let p be any prop on $\mathcal{X}$, and observe that either $x \in p$ or $x \in (X \setminus p)$, so that either $p \in \mathrm{t}(x)$ or $(X \setminus p) = {\sim}p \in \mathrm{t}(x)$.

We now confirm t is a bijection. To check injectivity, suppose $\mathrm{t}(x) = \mathrm{t}(y)$; so $x \in p$ iff $y \in p$ for all props p on X; hence $x = y$ by (*kolmogorov*). To check surjectivity, let S be a story on $\mathrm{Calc}(\mathcal{X})$; then there is some $x \in \bigcap S$ by (*consistency*). We claim that $\mathrm{t}(x) = S$. Fix some prop p on $\mathcal{X}$. If $p \in S$, then $x \in p$. Conversely, if $x \in p$, then $(X \setminus p) \notin S$ and hence $p \in S$ by (*maximality*). Hence $x \in p$ iff $p \in S$, and so $\mathrm{t}(x) = S$.

It remains to check that the map t preserves the structure of the bivalent-universes, that is, that it preserves the props. If p is any subset of $\mathcal{X}$ (not necessarily a prop), then the map t induces the map $\mathrm{t}(p) = \{\mathrm{t}(x) : x \in X, x \in p\}$. We must show that if p is a subset of $\mathcal{X}$, then:

$$p \text{ is a prop on } \mathcal{X} \text{ iff } \mathrm{t}(p) \text{ is a prop on } \mathrm{Uni}(\mathrm{Calc}(\mathcal{X}))$$

Appealing in turn to Definition 14.15 and Definition 14.18, the props on $\mathrm{Uni}(\mathrm{Calc}(\mathcal{X}))$ are exactly the sets of the form $\mathrm{w}(p) = \{S \in \mathrm{Stories}(\mathrm{Calc}(\mathcal{X})) : p \in S\}$ as p ranges over props on $\mathcal{X}$. Since t is bijective, each story S on $\mathrm{Calc}(\mathcal{X})$ can be written uniquely as $S = \mathrm{t}(x)$ for some $x \in X$. Hence the props on $\mathrm{Uni}(\mathrm{Calc}(\mathcal{X}))$

are exactly the sets of the form $\mathrm{w}(p) = \{\mathrm{t}(x) : x \in X, p \in \mathrm{t}(x)\}$, as p ranges over props on $\mathcal{X}$. But for props p on $\mathcal{X}$, we have $p \in \mathrm{t}(x)$ iff $x \in p$. So the props on $\mathrm{Uni}(\mathrm{Calc}(\mathcal{X}))$ are exactly the sets of the form

$$\mathrm{w}(p) = \{\mathrm{t}(x) : x \in X, p \in \mathrm{t}(x)\} = \{\mathrm{t}(x) : x \in X, x \in p\} = \mathrm{t}(p)$$

as p ranges over props on $\mathcal{X}$. And so the props on $\mathrm{Uni}(\mathrm{Calc}(\mathcal{X}))$ are exactly the sets of the form $\mathrm{t}(p)$, as p ranges over props on $\mathcal{X}$, as required. □

15

Indiscernibility

Two entities are indiscernible, intuitively speaking, when they cannot be told apart, or discriminated from each other in some way. The purpose of this chapter is to use model theory to shed some light on two broad questions concerning indiscernibles: *Are there any indiscernible objects? And if so, how do we talk about them?* We consider these questions in reverse order, starting with the semantics of indiscernibles in §15.2, and considering their existence in §15.3.

We begin the chapter, though, with some philosophical-cum-technical preliminaries. The reason for this is that there are many, many things one might mean by 'indiscernible', and model theory can helpfully illuminate these different meanings. And we close the chapter by introducing some technical apparatus which relates indiscernibles to infinitary logics (§15.4) and to stability (§15.5). This apparatus also underpins the final two chapters of the book.

15.1 Notions of indiscernibility

As mentioned, 'indiscernible' can mean many things. Indeed, there are at least three degrees of freedom in our meaning. We can vary: (a) the primitive ideology; (b) the background logic; and (c) the grade of discernibility.

Primitive ideology

When attempting to discern two objects, an obvious starting point would be appeal to their (intrinsic) 'properties' or 'qualities'. For example, perhaps we can immediately discriminate between cups and saucers. To track this model-theoretically, we might select a signature involving two one-place predicates, C and S, which respectively pick out the cups and the saucers (on the intended interpretation). However, if our signature only includes these crockery-predicates, then it will not allow us to discern red cups from blue ones. The immediate upshot is that any model-theoretic notion of discernibility will be signature-relative.

We should also be careful not to restrict ourselves—at least, initially—to considering properties or qualities which are only symbolised with one-place predicates. For example, we might want to allow that two otherwise indiscernible cups are discerned from one another by the fact that only one of them is on a saucer. This would involve adding a two-place predicate to stand for *x is on y*. Then the one-place formula $\exists x(S(x) \wedge O(x, v))$ will be satisfied by one cup but not the other (on the intended interpretation).

The moral is: *we must take care concerning which primitive features—modelled via the primitive symbols of our signature—count towards discernibility.*

Background logic

We just considered the formula $\exists x(S(x) \wedge O(x, v))$, a formula of first-order logic with one free variable. Unsurprisingly, we can affect what can be discerned from what, by varying the logic within which we provide such formulas.

To illustrate the point, let $\mathcal{B}$ be the disjoint union of a complete countably-infinite graph with a complete uncountable graph:

$$B = \mathbb{R}$$
$$R^{\mathcal{B}} = \{(n, m) \in \mathbb{N}^2 : n \neq m\} \cup \{(p, q) \in (\mathbb{R} \setminus \mathbb{N})^2 : p \neq q\}$$

Let $\varphi(v)$ be any first-order formula in this signature with no parameters. Then either φ applies to every entity in the structure, or it applies to no entity in the structure. (If this is not immediately clear, we explain why in Theorem 15.7.) However, if we allow φ to take *parameters*, then this no longer holds: only the members of $\mathbb{N}$ have an edge to 0 in $\mathcal{B}$. Equally, if we allowed *second*-order formulas, we would not obtain the same result: only the members of $\mathbb{N}$ have edges to only countably many entities.

The immediate moral, as above, is: *we must take care concerning what logical notions count towards discernibility.* In particular, we must take care about whether we are allowing first- or second-order quantification. Equally, we must take care over whether the discerning formulas are allowed to contain parameters. By default, throughout this chapter, formulas will contain *no* parameters (unless explicitly stated otherwise)

Relatedly, we also need to decide whether or not our logic should include a primitive notion of *identity*. And this point merits careful discussion. To approach the point, we need to step back and ask: *Why might one want to consider the idea of discernibility in the first place?*

The ambitious hope is that considering the idea of discernibility will provide us with a genuinely illuminating answer to the question: *When are objects identical?* To take a simple example: set theory states that sets are identical iff they share all their members. To take a more contentious example: reflecting on physical theories might convince us that nature abhors a (non-trivial) symmetry. The general hope, perhaps, is that some notion of indiscernibility will provide us with a *non-trivial criterion of identity*.

This ambition need not be *reductive*. Someone might simply seek an illuminating *constraint* upon the conditions under which objects can be distinct. That said, some philosophers have hoped to find a *reductive* criterion for identity, looking to replace the identity primitive with some *defined* notion of indiscernibility. This reductive ambition is found among philosophers who have defended some 'Principle of Identity of Indiscernibles', and we explore this in §15.3. However, if our aim *is* reductive, then we obviously cannot help ourselves to a *primitive* identity relation, in the way that standard first-order logic does (for example). Similarly, we are unlikely to want to help ourselves to full second-order logic, since the identity relation is *definable* there, via $x = y \leftrightarrow \forall X(X(x) \leftrightarrow X(y))$.

None of this is to say that full second-order logic, or first-order logic with identity, have no place when it comes to discussing notions of indiscernibility. Rather, we must always be clear on whether or not our logic should contain a primitive notion of identity, and we should be aware that certain philosophical motivations may push us towards using an identity-free logic.

Grades of discernibility

Thus far, we have highlighted that any discussion of discernibility must take care concerning the primitive vocabulary and the background logic. We now explore a third degree of freedom: the *grade* of discernibility.

To speak of a 'grade of discernibility' is somewhat jargonistic. But we can introduce the idea fairly naturally via Black's celebrated dialogue between two characters, Ali and Bob.[1] Bob presents a now famous thought experiment:

> Isn't it logically possible that the universe should have contained nothing but two exactly similar spheres? We might suppose that each was made of chemically pure iron, had a diameter of one mile, that they had the same temperature, colour, and so on, and that nothing else existed. Then every quality and relational characteristic of one would also be a property of the other.[2]

Here, Bob allows us various resources for distinguishing entities from each other: we can consider their shape (spherical), their composition (iron), their size (one mile diameter), their temperature, their colour, and so on. All of these features will, presumably, be symbolised with one-place predicates. But Bob also mentions 'relational characteristics'. Specifically, Bob has in mind the point that both spheres satisfy the one-place formula which symbolises '*being at a distance of two miles, say, from the centre of a sphere one mile in diameter*.'[3] So, when a and b are our two spheres,

[1] Black (1952). Actually, Black simply calls his characters 'A' and 'B'; we have given them full names in the interests of readability.

[2] Black (1952: 156).

[3] Black (1952: 157).

Bob is effectively suggesting that we should simply consider whether a and b satisfy all the same formulas in one free variable (in some suitable signature, and with some suitable background logic). And Bob's point is then that $\varphi(a) \leftrightarrow \varphi(b)$, for all formulas $\varphi(v)$ with only one free variable (on the intended interpretation). As such, Bob insists that the spheres are *indiscernible*.

Ali, however, insists that Bob has omitted something important:

> Each of the spheres will surely differ from the other in being at some distance from that other one, but at no distance from itself—that it to say, it will bear at least one relation to itself—*being at no distance from* [...]—that it does not bear to the other.[4]

Ali's point is that there is some *two*-place formula, $\varphi(x, y)$, such that $\varphi(a, a)$ but $\neg\varphi(a, b)$. And this, Ali thinks, shows that the spheres are *discernible* after all.

To make sense of Ali and Bob's discussion, we define various *grades of indiscernibility*. We start by recalling some notation from §14.4. Given an $\mathscr{L}$-structure $\mathcal{M}$ and a sequence of elements $\bar{a}$, recall that $\text{tp}^{\mathcal{M}}(\bar{a})$ is the set of $\mathscr{L}$-formulas with free variables among $v_1, \ldots, v_n$ which are satisfied by $\bar{a}$ in $\mathcal{M}$. That is, $\text{tp}^{\mathcal{M}}(\bar{a})$ is the complete n-type $p \in \text{Types}_n(\mathcal{M})$ realised by $\bar{a}$. Now, all of that takes place in first-order logic with identity, but we can easily extend the ideas to other logics:

Definition 15.1: *Let $\mathcal{M}$ be an $\mathscr{L}$-structure with a sequence of elements $\bar{a}$. We say:*[5]

$$\text{tp}^{+}_{\mathcal{M}}(\bar{a}) = \{\varphi \textit{ is a first-order } \mathscr{L}\textit{-formula, possibly } \text{with '='} : \mathcal{M} \vDash \varphi(\bar{a})\}$$
$$\text{tp}^{-}_{\mathcal{M}}(\bar{a}) = \{\varphi \textit{ is a first-order } \mathscr{L}\textit{-formula, } \text{without '='} : \mathcal{M} \vDash \varphi(\bar{a})\}$$
$$\text{tp}^{\text{s}}_{\mathcal{M}}(\bar{a}) = \{\varphi \textit{ is a second-order } \mathscr{L}\textit{-formula} : \mathcal{M} \vDash \varphi(\bar{a})\}$$

In the second-order case, we assume the full *semantics.*

The mnemonic that we have introduced here is to use '$-$' to indicate first-order logic without identity, '$+$' to indicate first-order logic with identity, and 's' to indicate (full) second-order logic. We now define some grades of discernibility:[6]

Definition 15.2: *For any $\mathscr{L}$-structure $\mathcal{M}$, say that a and b are:*

(1) one-indiscernibles$^-$ *in $\mathcal{M}$ iff* $\text{tp}^{-}_{\mathcal{M}}(a) = \text{tp}^{-}_{\mathcal{M}}(b)$
(2) one-indiscernibles$^+$ *in $\mathcal{M}$ iff* $\text{tp}^{+}_{\mathcal{M}}(a) = \text{tp}^{+}_{\mathcal{M}}(b)$
(3) one-indiscernibles^{s} *in $\mathcal{M}$ iff* $\text{tp}^{\text{s}}_{\mathcal{M}}(a) = \text{tp}^{\text{s}}_{\mathcal{M}}(b)$
(4) two-indiscernibles$^-$ *in $\mathcal{M}$ iff* $\text{tp}^{-}_{\mathcal{M}}(a, b) = \text{tp}^{-}_{\mathcal{M}}(b, a)$

[4] Black (1952: 157).

[5] Hence, what we referred to in §14.4 with the notation $\text{tp}^{\mathcal{M}}(\bar{a})$, we are now referring to with the notation $\text{tp}^{+}_{\mathcal{M}}(\bar{a})$.

[6] This family of definitions has a long heritage, e.g.: Hilbert and Bernays (1934: §5), Quine (1960: 230–2, 1976), Caulton and Butterfield (2012: §2.1, §3.2), Ketland (2006: 306–7, 2011: Definitions 2.3, 2.5), Ladyman et al. (2012: Definition 3.1 §6.4), and Button (2017: Definition 2.1).

(5) two-indiscernibles$^+$ *in* $\mathcal{M}$ *iff* $\text{tp}^+_{\mathcal{M}}(a, b) = \text{tp}^+_{\mathcal{M}}(b, a)$
(6) two-indiscernibles$^\text{s}$ *in* $\mathcal{M}$ *iff* $\text{tp}^\text{s}_{\mathcal{M}}(a, b) = \text{tp}^\text{s}_{\mathcal{M}}(b, a)$
(7) Leibniz-indiscernibles$^-$ *in* $\mathcal{M}$ *iff* $\text{tp}^-_{\mathcal{M}}(a, a) = \text{tp}^-_{\mathcal{M}}(a, b)$

Note that there is no need to define *Leibniz-indiscernibles*$^+$ or *Leibniz-indiscernibles*$^\text{s}$. After all: if $\text{tp}^\text{s}_{\mathcal{M}}(a, a) = \text{tp}^\text{s}_{\mathcal{M}}(a, b)$, then also $\text{tp}^+_{\mathcal{M}}(a, a) = \text{tp}^+_{\mathcal{M}}(a, b)$, and so in particular $a = a$ iff $a = b$, and hence $a = b$. So these grades of indiscernibility would simply be equivalent to genuine identity.

Since Leibniz-indiscernibility$^-$ is perhaps the least familiar of the previous notions, it is worth noting a few quick equivalents of it:[7]

Lemma 15.3: *For any* $\mathscr{L}$*-structure* $\mathcal{M}$*, the following are equivalent:*

(1) *a and b are Leibniz-indiscernibles$^-$ in $\mathcal{M}$*
(2) $\mathcal{M} \vDash \forall \bar{v} \, (\varphi(a, \bar{v}) \leftrightarrow \varphi(b, \bar{v}))$*, for all atomic formulas φ not containing '='*
(3) $\mathcal{M} \vDash \forall \bar{v} \, (\varphi(a, \bar{v}) \leftrightarrow \varphi(b, \bar{v}))$*, for all first-order formulas φ not containing '='*

Let us now apply these notions to Black's two-sphere world. To do this, we must represent that world with a formal model, $\mathcal{S}$. Since the only point at issue between Ali and Bob concerns the relation 'being at some non-zero distance from', the simplest formal model is just a two-element graph with an edge between nodes iff they stand at a distance from one another, which we can depict thus:

$$a \text{ --- } b$$

We now have $\text{tp}^-_{\mathcal{S}}(a) = \text{tp}^-_{\mathcal{S}}(b)$ but $\text{tp}^-_{\mathcal{S}}(a, a) \neq \text{tp}^-_{\mathcal{S}}(a, b)$. So the two spheres will be one-*indiscernibles*$^-$, but Leibniz-*discernibles*$^-$.

To consider the other five grades defined in Definition 15.2, recall from Theorem 2.3 that isomorphisms preserve even full satisfaction second-order formulas. Evidently, from our description of the example, there is a non-trivial isomorphism $h : \mathcal{S} \longrightarrow \mathcal{S}$, obtained just by swapping the two spheres around. Hence our spheres are two-indiscernibles$^\text{s}$. In more detail: with $h(a) = b$ and $h(b) = a$, we have $\text{tp}^\text{s}_{\mathcal{S}}(a, b) = \text{tp}^\text{s}_{\mathcal{S}}(h(a), h(b)) = \text{tp}^\text{s}_{\mathcal{S}}(b, a)$.

An isomorphism $h : \mathcal{M} \longrightarrow \mathcal{M}$ is called an *automorphism*, or sometimes a *symmetry*, on $\mathcal{M}$.[8] As we just saw, invoking symmetries can provide a handy test for

[7] See Casanovas et al. (1996: 508) and Ketland (2011: Theorem 3.17).

[8] Model theorists prefer 'automorphism'; physicists prefer 'symmetry'. We stick with 'symmetry' partly because it is shorter, but mostly because 'one-symmetricals' is snappier than 'one-automorphics' (see Definition 15.4).

satisfaction of some of the grades of discernibility from Definition 15.2. But we can also use symmetries to define three *grades of symmetry*:[9]

Definition 15.4: *For any $\mathscr{L}$-structure $\mathcal{M}$, say that a and b are:*

(1) one-symmetricals *in $\mathcal{M}$ iff there is a symmetry h on $\mathcal{M}$ such that $h(a) = b$,*
(2) two-symmetricals *in $\mathcal{M}$ iff there is a symmetry h on $\mathcal{M}$ such that $h(a) = b$ and $h(b) = a$,*
(3) Leibniz-symmetricals *in $\mathcal{M}$ iff there is a symmetry h on $\mathcal{M}$ such that $h(a) = b$, $h(b) = a$ and $h(x) = x$ for all $x \notin \{a, b\}$.*

In these terms, the spheres a and b are Leibniz-symmetricals in S.

In defining the notion of an isomorphism, the only object-language symbols which are mentioned are those of the signature; there is no need to mention '='. Nevertheless, the notion of an isomorphism—and hence each grade of symmetry—straightforwardly depends upon the notion of identity. After all, an isomorphism is a bijection, which is to say it maps *unique* objects to *unique* objects, and *vice versa*. If we want to avoid treating identity as a primitive—for philosophical or technical reasons—then the notion of an isomorphism is therefore probably too strong. In looking for a weaker notion, a first thought would be to consider non-bijective functions between structures.[10] But this is insufficiently concessive, since the very idea of a *function* also depends upon the notion of identity, in that each argument yields *exactly one* value. Consequently, we should consider structure-preserving *relations* that may hold between structures. The appropriate notion is this (writing aHb to indicate that $(a, b) \in H$):[11]

Definition 15.5: *Let $\mathcal{M}, \mathcal{N}$ be $\mathscr{L}$-structures. A* relativeness correspondence *from $\mathcal{M}$ to $\mathcal{N}$ is any two-place relation H whose domain is exactly M and whose range is exactly N, such that for all $\mathscr{L}$-constant symbols c, all n-place $\mathscr{L}$-predicates R, all n-place $\mathscr{L}$-function symbols f, and all $a_1Hb_1, \ldots, a_nHb_n$:*

$$c^{\mathcal{M}}Hc^{\mathcal{N}}$$

$$(a_1, \ldots, a_n) \in R^{\mathcal{M}} \textit{ iff } (b_1, \ldots, b_n) \in R^{\mathcal{N}}$$

$$f^{\mathcal{M}}(\bar{a})Hf^{\mathcal{N}}(\bar{b})$$

A relativity *on $\mathcal{M}$ is a relativeness correspondence from $\mathcal{M}$ to $\mathcal{M}$.*

Then, by simple analogy with our three grades of symmetry, we can consider three grades of relativity:

[9] This formulation follows Button (2017: Definition 2.5). For some precedents, see Ketland (2006, 2011) and Ladyman et al. (2012).

[10] The notion of a *strict homomorphism* is often employed in the technical literature.

[11] This is due to Casanovas et al. (1996: Definition 2.5).

Definition 15.6: *For any $\mathscr{L}$-structure $\mathcal{M}$, say that a and b are:*

(1) one-relatives *in $\mathcal{M}$ iff there is a relativity H on $\mathcal{M}$ with aHb*

(2) two-relatives *in $\mathcal{M}$ iff there is a relativity H on $\mathcal{M}$ with aHb and bHa*

(3) Leibniz-relatives *in $\mathcal{M}$ iff there is a relativity H on $\mathcal{M}$ with aHb and bHa and xHx for all x such that neither x, a nor x, b are Leibniz-indiscernibles^{-}*

These are the last of our grades of discernibility. The very obvious moral is that, in discussing discernibility: *we must take care concerning the grade of discernibility.*

We can chart the relative strengths of our different grades of discernibility using a Hasse diagram. According to this notation, we have a connected path down the page between two grades R and S iff being R entails being S. For example, the result below tells us that all two-indiscernibles^{s} are one-indiscernibles^{-}, and that all one-symmetricals are one-indiscernibles^{-}, but that there are structures containing two-indiscernibles^{s} which are not one-symmetricals, and structures containing one-symmetricals which are not two-indiscernibles^{s}. The proof is in §15.A.

Theorem 15.7: *The following Hasse diagram represents the relationships between the grades of discernibility:*

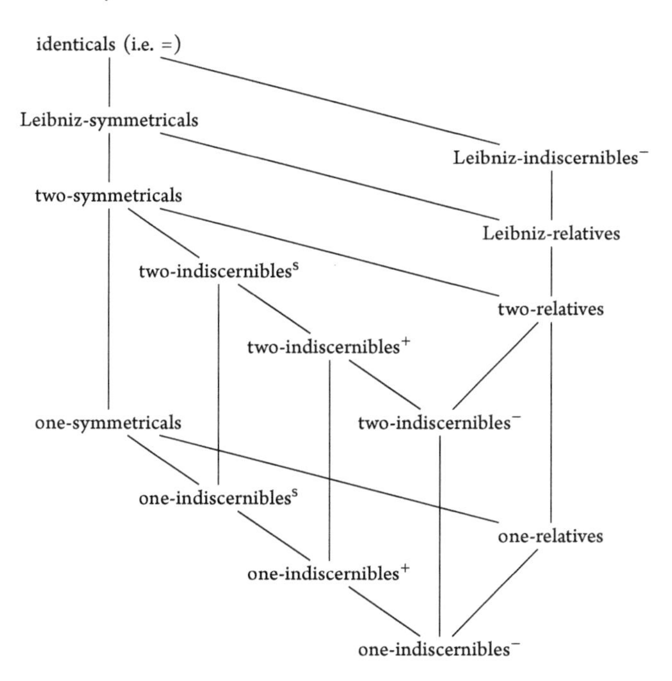

If we restrict our attention to structures with relational *signatures, the diagram is the same, except that* 'Leibniz-indiscernibles^{-}' *sits below* 'identicals' *and above* 'Leibniz-symmetricals'. *(I.e. in relational structures, all Leibniz-indiscernibles are Leibniz-symmetricals, but the converse is false.)*

15.2 Singling out indiscernibles

In the previous section, we set out some preliminaries for a philosophical discussion of indiscernibility. We will now embark on that discussion, beginning with the question: *If there are indiscernibles, then how do we single out any one of them?*

We must start by commenting on the idea of *singling out* an object. This is a rough-and-ready, intuitive, idea. We may want to single out an entity in thought, thinking about one particular object rather than any other. Equally, we may want to single out an entity verbally, perhaps with a singular term which (we hope) refers determinately to exactly one object, or perhaps using a definite description.[12] But this rough-and-ready idea is enough to motivate a difficulty. Where two (or more) objects are indiscernibles, there is an obvious (doxological) question: *How could we possibly single out one of these, rather than the other?*

We cannot single out two-symmetricals

Having posed the question, we must immediately recall the morals of §15.1. We are asking a question which concerns indiscernibility. So, we must determine the appropriate signature, the appropriate logic, and the appropriate grade of discernibility, for our discussion.

We start with the signature. If the idea is to single out an entity in thought, then when we model what is going on, we should allow ourselves symbols standing for anything that can legitimately be entertained in thought. That said, we must take care to avoid trivialising the question. For example: suppose we insisted on adopting a primitive one-place predicate, whose intended interpretation was *the entity uniquely singled out by me right now*. If we are, indeed, capable of uniquely singling out one of two *otherwise* indiscernible objects, then it will indeed be the case that they are also discerned by the fact that only one of them is (uniquely) singled out. But it would obviously put the cart before the horse, to begin with such a predicate and *assume* it has a unique interpretation. For similar reasons, in this context we should probably restrict our attention to *relational* signature. After all, if we allow our signature to contain distinct names for two *otherwise* indiscernible objects, then we again put the cart before the horse in assuming that we can (determinately) name one rather than the other.

We now turn to the logic. We mentioned in §15.1 that there are philosophical contexts in which we should not allow ourselves access to identity (either by treating it as a primitive, or by using full second-order logic). However, this does not seem to be one of those contexts. After all, if a stands in some relation to exactly one object, whilst b stands in some relation to exactly two objects, but a and b are *otherwise* in-

[12] We are taking no stance on whether definite descriptions should sometimes be regarded *as* singular terms, or rendered along more Russellian lines.

discernible, it is plausible that we can single out a rather than b by formalising these relational-cum-numerical facts using an identity predicate.

The more substantial question is whether we should allow ourselves only first-order logic with identity, or full second-order logic, or maybe some stronger logic still. In asking this, we run the risk of reopening some of the wounds from Part B of this book. To bypass that, we will simply argue for the following: *Given some two-symmetricals, it is impossible to single out any one of them.*

Here is why we think this, and why it bypasses the debate about 'which logic is correct'. Let a and b be two-symmetricals in some appropriate relational signature. Then there cannot be *any* formula $\varphi(x, y)$ which enables us to tell them apart, pairwise. That is, for any such formula, we will have $\varphi(a, b)$ iff $\varphi(b, a)$. We can see this if we consider formulas of (full) second-order logic, just by looking at Theorem 15.7. And in fact, if we skip ahead to Corollary 15.10 of §15.4, we see that the same point holds even if we allow for *arbitrarily infinitely long* formulas. This suggests to us that there is *literally nothing* we can say, do, or think, that will allow us to single out a over b.

(It is not obvious, though, that any *weaker* grade of discernibility will do. For, if a and b are *not* two-symmetricals, then Corollary 15.10 of §15.4 entails that there is some infinitary formula $\varphi(x, y)$ such that $\varphi(a, b)$ but $\neg\varphi(b, a)$. So, *if* we have access to infinitary logic, and can narrow our attention down to considering *either a or b*, then we can use φ to single out a rather than b.)

On mental faculties

We have just argued that, once we have selected an appropriate relational signature, we cannot single out any two-symmetricals. We now want to consider some applications of this point, and to defend it against some possible counter-arguments.

In §15.1, we discussed Black's two-sphere world, and treated it as a graph, S. We also noted that there is a symmetry on S, obtained just by swapping the two spheres. So, the spheres are relevantly two-symmetrical. By the preceding argument, then, it is impossible to single out either of the spheres.[13]

Here is a more mathematical illustration of phenomenon. Objects-platonists believe that the complex numbers are abstract objects. But there is a symmetry on the complex plane which sends every $a + bi$ to $a - bi$, and vice versa. So, in particular, the two square-roots of -1 are two-symmetricals in the signature of the complex plane, $\{0, 1, +, \times\}$. And this suggests that they are two-symmetricals in *any* signature which is appropriate for considering our ability to think about complex numbers. Consequently, objects-platonists should accept that is impossible to single out either of the square-roots of -1.

---

[13] Here, then, we are in some agreement with Black's character Bob (1952: 156–7).

Now, this conclusion may well come as no surprise. After all, in Chapter 2, we suggested that the *moderate* objects-platonist must accept that she cannot single out *any* natural number over any other. (Indeed, in a sense, the moral of Chapter 2 is that, when the *moderate* objects-platonist wants to single out abstract objects, the appropriate signature is just the *empty* signature.) However, we would note that not all objects-platonists are *moderate*. So the above point is of particular interest to *immoderate* objects-platonists.

Our verdicts here are in line with the general philosophical consensus.[14] But Priest dissents. He thinks that, since we *can and do* determinately single out i over $-i$, some 'powerful' mental faculty of ours must allow us to do this:[15]

> An act of pure intention can intend an object when there are other indiscriminable objects. How is this possible? That I think, is the nature of the beast. It must be possible, however, because it has actually been done [...] we can intend $+i$ rather than $-i$ [...]. At some stage, some mathematician or committee of mathematicians, must have chosen one of these objects arbitrarily and called it '$+i$'. Acts of pure intention, it would seem, can be very powerful.[16]

We respond to Priest by insisting that postulating this 'powerful' mental faculty achieves nothing. Priest has told us that 'some mathematician or committee of mathematicians must have chosen one of [the square-roots of -1] arbitrarily and called it "$+i$"'. Let us suppose—consistently with Priest's metaphysics-cum-metasemantics—that two *different* mathematicians, working independently from one another, made *different* arbitrary choices at some stage, so that their two uses of 'i' referred to the two different square-roots. Working in isolation from one another, both mathematicians would (of course) have managed just fine. Moreover, when the mathematicians met and discussed complex analysis, they neither would nor could be aware that they had made *different* choices. But in that case, supposing that they made *different* choices plays no role in accounting for mathematical practice; and there cannot really be any reason *beyond* mathematical practice to make the supposition. There is, then, just no reason to postulate that the mathematicians ever *made* these choices, and so no reason here to countenance the existence of a faculty, whose sole purpose is to make such choices possible. (And, in passing, we would offer a similar response to anyone who thinks that one of the square-roots of -1 is more *natural* as the referent of 'i' than of '$-i$': there is just no *point* in postulating this mysterious property of 'naturalness'.)

[14] See Brandom (1996: 298ff), Field (2001: 328), and Shapiro (2012: 381ff).

[15] We should note that Priest's postulates this mental faculty, not *specifically* in order to deal with the complex numbers, but on more general grounds concerning (for example) our ability to think about fictional characters. However, *even if* one has postulated this faculty on other grounds, we would still argue (exactly as in the text) that nothing is achieved by insisting that the faculty is 'activated' in the case of the complex numbers.

[16] Priest (2005: 142).

That is our main response to Priest. But there is a slight wrinkle. Mathematicians seem to treat '*i*' as a name. And when asked to elaborate on the name '*i*', they might say that it means 'the positive square-root of -1'. Both the use of a name, and the use of the definite article, naïvely suggest that we *have* singled out some unique object. So we must deal with this naïve concern.

Cutting a long story short, our response is that it is just *too* naïve. We already made this point in §2.4, but it is worth repeating: we can and do employ grammatically singular expressions, without requiring that they single out any particular entity. Here is a lovely example, which Shapiro uses to make exactly this point:

> Remember the chess set that came with nine white pawns? I could have used an extra king, but I never needed the extra pawn.[17]

The crucial fact is that we can sensibly employ the grammatically singular expression '*the* extra pawn', even though no pawn was in any way 'special', and so no particular pawn was singled out. So it may be with expressions like '*i*' or 'the positive square-root of -1'.

Next, we need a *formal* semantics which can accommodate this fact. Shapiro suggests the supervaluational semantics sketched in §2.5, and we agree. The general idea, recall, is that what is *true* (simpliciter) is what is satisfied in all appropriate models. In the particular case of the complex numbers, the appropriate models would be the *two* distinct signature-expansions of the complex plane where '*i*' is allowed to refer to either of the two different square-roots of -1. Then a sentence φ in this richer signature is true (simpliciter) iff φ is satisfied in *both* signature-expansions. Job done, with minimal fuss.

This is just a particular instance of the supervaluational strategy which we sketched in §2.5, and one point raised there is worth repeating. The use of supervaluational semantics does not make the issue of singling out disappear. Rather, the issue recurs at every higher level. Our distinct structures introduce labels to the complex numbers: one structure picks out one of the square-roots of -1 with '*i*', and the other structure picks out the very same object with '$-i$'. Precisely because we cannot possibly single out one square-root over the other, we cannot say *which* structure labels which square-root with '*i*'. The two (augmented) structures are exactly as indiscernible as the two square-roots of -1 themselves.[18] If we like, we can say 'the structure which labelled the *positive* root with "*i*" is distinct from the structure which labelled the negative root with "*i*"'; but we must understand *this* claim supervaluationally too.

[17] Shapiro (2012: 393, 396; see also 401–5).

[18] Cf. Shapiro (2012: 406–7), particularly his discussion of choice functions and 'circularity'.

15.3 The identity of indiscernibles

The preceding section assumed that certain kinds of indiscernibles exist. It also disarmed one potential objection against their existence. For, whilst we argued we cannot *single out* two-symmetricals (in some appropriate signature), we also showed that this does not raise difficulties for our ability to *discuss* two-symmetricals. (Roughly: we can say everything about them that we *ought* to be able to say.)

There is still, though, room for us to deny that there *are* any indiscernibles. And this leads us towards the debate around *the Principle of the Identity of Indiscernibles*. Our basic argument in this section is that it is quite hard to get too stressed about the Principle of the Identity of Indiscernibles in the philosophy of mathematics.[19]

Many principles of identity of indiscernibles

We start by noting that there is not just *one* principle of identity of indiscernibles. Given the morals of §15.1, there are *many* principles, which differ depending upon the signature, the background logic, and the particular grade of discernibility.

To illustrate this, consider Black's two-sphere world again. This was presented as a potential counterexample to the identity of indiscernibles. Now, given a suitable (relational) signature, the two spheres are Leibniz-symmetricals, for there is a symmetry sending one sphere to the other. So, if Black's two-sphere world is (metaphysically) possible, then the following principle is not (metaphysically) necessary:

Identity of Leibniz-Symmetricals. *Leibniz-symmetricals are identical*

However, each sphere is at no distance from itself, but at some distance from the other sphere. Consequently, the spheres are Leibniz-discernible^{-}, and Black's two-sphere world is compatible with (the necessity of) this principle:

Identity of Leibniz-Indiscernibles^{-}. *Leibniz-indiscernibles^{-} are identical*

Indeed, by Theorem 15.7, the Identity of Leibniz-Indiscernibles^{-} is strictly *weaker* than the Identity of Leibniz-Symmetricals, when we consider relational signatures. And it does seem appropriate to consider relational signatures, when discussing versions of the identity of indiscernibles, since allowing our signature to contain names or functions in this context would prejudge key questions of identity (see the discussion of §15.1).

There are, then, many principles of identity of indiscernibles. The question is, whether there are any good arguments for or against *any* such principle, in mathematics or elsewhere. Unsurprisingly, we will focus on the mathematical case, but we first want to make a brief comment about *physical* indiscernibles.

---

[19] *Tim adds:* so Button (2006) no longer reflects my views. Instead, this section develops Button (2013: 211fn.8, 2017: §3n.8).

Physical indiscernibles and verificationism

In the physical case, Black suggested that *verificationism* might license some principle of identity of indiscernibles. Specifically, his character Ali insists that the existence of indiscernible entities 'would be unverifiable *in principle*.'[20] She is not corrected within Black's dialogue, but Ali is mistaken. Following Hawley, we present a toy example to show that nothing (in principle) prevents us from obtaining evidence that there are distinct indiscernibles.[21]

Let a and b be distinct particles of a certain kind, K, and suppose that a and b are Leibniz-indiscernibles^{-} for a suitable relational signature. (Note that, other than identity, Leibniz-indiscernibility^{-} is the most refined grade of indiscernibility for relational signatures, by Theorem 15.7.) Since a and b are Leibniz-indiscernible^{-}, if we have chosen the signature suitably, then they will have exactly the same locations at all times. Nonetheless, we may be able to obtain evidence *that* there are two (always perfectly overlapping) entities, rather than one. Suppose, for example, that we also have excellent evidence that all particles of kind K have a certain fixed mass, m. Then if we take a mass-reading at a spacetime location containing both a and b, it will be $2m$ rather than m. This reading will provide some (defeasible) evidence that there are two particles of kind K. But then it is wrong to claim that the existence of indiscernible entities 'would be unverifiable *in principle*', at least in this case.

To repeat: that was a toy example. There is much more to say about whether *contemporary* physical theory is best understood as vindicating some form of the identity of indiscernibles. We hope, though, that the logical framework established in §15.1—selecting an appropriate signature, an appropriate logic, and an appropriate grade of discrimination—may be fruitful for philosophers of physics.

Cheap routes to discernibility

We now return from the physical to the mathematical, and consider the question: *Does mathematical practice require that we must believe in indiscernibles?*

Many philosophical programmes are pitched against objects-platonism. If any of them succeed, then of course we should not infer from mathematical practice that there are indiscernible mathematical entities, since we should not infer that there are *any* mathematical entities. On the other hand, though, it might just seem *obvious* that objects-platonists must believe in indiscernibles: if they believe in the complex numbers then, following the discussion of §15.2, it seems that they must believe that the two square-roots of -1 are two-symmetricals. Leitgeb and Ladyman have run a more general argument to this effect, roughly as follows:[22]

---

[20] Black (1952: 155). Ali immediately adds: 'Hence it would be meaningless.'

[21] Hawley (2009: 116).

[22] This is how we understand Leitgeb and Ladyman (2008: 394–5).

(a) Mathematical practice tells us that there are many structures which contain Leibniz-indiscernibles^{-}.

(b) Considerations from mathematical practice trump any epistemological or metaphysical qualms one might have about the existence of Leibniz-indiscernibles^{-}.

Whilst discussing (b) might lead to interesting questions in philosophical methodology, we will focus on criticising (a).

Leitgeb and Ladyman mostly focus on the case of graph theory. They note that practising graph theorists will say, for example, 'there is exactly one graph with two nodes and no edges'. According to Leitgeb and Ladyman, since the graph has no edges at all, its nodes are Leibniz-indiscernibles^{-}.[23] But their conclusion is surely too quick. In §2.4 and §15.2 we emphasised that it is too naïve to assume that grammatically singular phrases must single out *particular* entities. The point applies to a grammatically singular phrase like 'the graph with two nodes and no edges': we can use the phrase without thinking that there is *exactly one* such graph.

This is not merely a 'philosophical' point; it is a point made by practising graph-theorists. Here, for example, is a quote from Tutte's *Graph Theory*:

Pure graph theory is concerned with those properties of graphs that are invariant under isomorphism, for example the number of vertices, the number of loops, the number of links, and the number of vertices of a given valency. It is therefore natural for a graph theorist to identify two graphs that are isomorphic. For example, all link-graphs are isomorphic, and therefore he speaks of the 'link-graph' as though there were only one. Similarly one hears of 'the null graph', 'the vertex graph', and 'the graph of the cube'. When this language is used, it is really an isomorphism class (also called an abstract graph) that is under discussion.[24]

Tutte is not alone here; many graph theorists say similar things.[25] But if we are really considering the isomorphism class of all two-element graphs with no edges—i.e. simply the class of all two-element domains—then it makes no sense to say that *the* nodes of *the* graph are Leibniz-indiscernibles^{-}. After all, as we noted back in §2.2, *every* object is treated as a node by *some* graph.

To really emphasise this point, say that a *haecceitistic property* is a property which applies to exactly one object, and necessarily only to that object. Now, if every entity has a haecceititistic property, then *every* entity is one-discernible^{-} from every other in the haecceitistic signature. It is entirely compatible with Tutte's description of graph theoretic practice that there *is* such a haecceitistic signature; it would just be a signature which graph-theorists *ignore*, in favour of more spartan signatures.

[23] Leitgeb and Ladyman (2008: 392).

[24] Tutte (1984: 6). Leitgeb and Ladyman (2008: 390fn.3) cite the passage, but they do not really discuss the threat it poses to their viewpoint.

[25] See e.g. the list in De Clercq (2012: 668).

To be clear, *even if* such haecceitistic properties exist, they will not help us to *single out* any particular objects. After all, the 'haecceitistic signature' will not be something which we could hope to master, in thought or speech. As such, the dialectic of §15.2 will not be affected by countenancing the existence, in the abstract, of this haecceitistic signature. The point is simply this: if, metaphysically speaking, there are haecceitistic properties, then the principle of the *Identity of One-Indiscernibles*$^{-}$, as understood for the haecceitistic signature, will be true to the metaphysics.

Leitgeb and Ladyman, or others, might complain that appealing to haecceitistic properties would be *ad hoc*. For example, they might suggest that haecceities have only been invoked in order to save some version of the identity of indiscernibles, on spurious metaphysical grounds, and not because they are *needed* by our best account of mathematical practice. Here, they would be leaning back on (b).

In response, note that we introduced haecceitistic properties only to make our discussion of Tutte more vivid. More important, though: it is genuinely unclear that haecceities *are* an *ad hoc* addition. The ambient set theory within which we do our model theory trivially licenses the following: for every entity, a, there is a set whose unique member is a. Such singleton sets serve precisely the role of haecceitistic properties:[26] they serve as potential extensions of one-place predicates in our structures, and these extensions are precisely our formal surrogates for properties. So, far from being at odds with mathematical practice, mathematical practice can be regarded as *requiring* haecceitistic properties. Maybe, indeed, mathematical practice yields a version of the identity of indiscernibles *on the cheap*.

Here is a second way to think that mathematical practice 'cheaply' yields some version of the identity of indiscernibles. It is commonplace to insist that (classical) mathematical practice can all be embedded within the set theoretic hierarchy. So, a set-theoretic reductionist—who insists that each branch of (classical) mathematics ultimately reduces to (pure) set theory—can invoke the set-theoretic Axiom of Extensionality to obtain the Identity of Leibniz-Indiscernibles^{-} in the signature whose only primitive is $\in$.

Doubtless, some philosophers will complain that these 'cheap' routes to a version of the identity of indiscernibles get matters 'back to front'. They will complain that the distinctness of a and b should *ground* the fact that a and b are discernible (in some way), and not vice versa.[27] However, this complaint is orthogonal to our question. We have only been considering *whether* indiscernible objects are identical. We are not here especially interested in whether *identity* or *indiscernibility* is 'metaphysically prior' (or whether that question even makes sense).

To be clear: we have not tried to argue that some version of the identity of indiscernibles *is* correct. We have simply noted that mathematical practice does not

[26] Alternatively, we might say that they entail the existence of haecceitistic properties, since a uniquely possesses the property of being a member of $\{a\}$.

[27] Versions of this argument are discussed by Hawley (2009: 108–11).

really seem to steer us in either direction. Indeed, for everything said so far, it is not at all clear that objects-platonists should feel under *any* pressure to take any particular stance on issues relating to the identity of indiscernibles.

Ante rem structuralism and indiscernibles

In fact, we think that the version of objects-platonism which is most likely to have a vested interest in the identity of indiscernibles is Shapiro's ante rem structuralism. We introduced this back in §2.4, and we have mentioned it several times since, but we will recap the general idea. Shapiro believes that mathematics is about ante-structures (he calls them 'structures', but they are not structures in the sense of Definition 1.2). These are supposed to be abstract objects, consisting of *places*, with certain intra-structural relations holding between them.[28]

To explain why Shapiro's ante rem structuralist might find the identity of indiscernibles interesting, note the following. According to Shapiro, the *essence* of a place in an ante-structure is its intrastructural relations to the (other) positions in the ante-structure.[29] Clarifying this, Shapiro writes:

> The number 2 [...] has lots of properties [.... But] every property that 2 enjoys comes in virtue of its being that place in the natural number structure. This is because that is what 2 is.[30]

Shapiro seems to be suggesting that there are properties which are *essential* to 2, and that these essential properties (at least partially) ground its *accidental* properties. But, if we accept that places in structures have certain essential properties, then we can formulate new versions of the identity of indiscernibles:

Identity of Essential-Indiscernibles (scheme). *If a and b are places in the same structure, and they are indiscernible in the signature appropriate to the essential properties of that structure, then $a = b$.*

We say that this is a scheme, since there will be different versions of this principle, corresponding to the different grades of indiscernibility. So, we could replace 'indiscernible' in this scheme with 'one-indiscernible$^-$', or 'two-symmetrical', or anything else. But the idea is that, in restricting our attention to *essential* properties, we will rule out 'cheap' routes to discernibility.[31]

Shapiro, however, rejects *all* of these principles, and for a very simple reason. Shapiro believes that there are *cardinal ante-structures*. According to Shapiro, these ante-structures consist *only* of places, with *no* essential (intra-structural) properties

[28] Shapiro (1997: 73–4).

[29] See Shapiro (1997: 5–6, 72, 2006a: 114–21).

[30] Shapiro (2006a: 121).

[31] See e.g. Shapiro (2006a: 140).

or relations. The signature appropriate to the such structures is, then, the empty signature; and so the places in a cardinal ante-structure are indiscernibles according to *any* grade of indiscernibility (short of *bare identity*).[32] As such, Shapiro rejects every version of the principle of the identity of essential-indiscernibles, and allows 'that two distinct objects can [...] have *all* of their essential properties in common'.[33]

There are clear parallels between Shapiro's stance, here, and Leitgeb and Ladyman's. Leitgeb and Ladyman also argued against (any form of) the identity of indiscernibles, by considering structures with, crudely, nothing much going on in them. However, Leitgeb and Ladyman claimed to appeal to little more than a face-value reading of mathematical practice.[34] Shapiro, however, explicitly puts his point in metaphysically loaded terms, speaking of *essential* properties. So our earlier reservations concerning Leitgeb and Ladyman's argument need not apply directly to Shapiro's approach.

Still, we *do* have a reservation concerning Shapiro's argument: we are unclear which properties should count as *essential*. Consider Shapiro's *cardinal-three ante-structure*. Whatever else we want to say, we must surely be allowed to make the innocuous mathematical claim: *there are exactly six symmetries on the cardinal-three ante-structure: the trivial symmetry, three symmetries which simply swap two places and leave the third place undisturbed, and two symmetries which act non-trivially on all three places*. Here, then, we are committed to thinking that there are three *different* functions which swap different pairs of places. Moreover, it is unclear (to us) why the existence of these functions should not count as an *essential* feature of the cardinal-three ante-structure. But, if the existence of these functions is an *essential* feature of the cardinal-three ante-structure, then the places in that structure will be (to some extent) discerned by their essential properties. For example: if the three symmetries h_1, h_2 and h_3 are implemented via unique two-place predicates R_{h_1}, R_{h_2} and R_{h_3}, which are thought of as capturing essential relations, then every place becomes *one-discernible*$^-$ from every other, for exactly one element satisfies $\exists x R_{h_1}(v, x) \wedge \exists x R_{h_2}(v, x)$. In short: unless more can be said concerning 'essential' properties, then even Shapiro's ante rem structuralist yields a *cheap* version of the identity of indiscernibles, in terms of essential properties.

This concludes our main philosophical commentary on indiscernibles. To cut a long story short: within mathematics, we see no obvious reasons to get very stressed about whether or not there are any indiscernibles.

---

[32] Shapiro (2006a: 131–2, 2006b: 167).

[33] Shapiro (2006a: 140); see also Shapiro (2006b: 170–1). This sets him against Burgess (1999: 287–8), Hellman (2001: 193), and Keränen (2001, 2006), who suggest that Shapiro is committed to some non-trivial version of the identity of indiscernibles. For further discussion, see MacBride (2006b).

[34] As something like an inference to the best explanation; see Leitgeb and Ladyman (2008: 389).

15.4 Two-indiscernibles in infinitary logics

In the final two sections of this chapter, we will present two different generalisations of indiscernibility. Both sections can safely be omitted on a first reading. However, they both round out the technical aspects of issues associated with indiscernibles, and also provide results which will be useful in Chapters 16–17.

In this section, we discuss analogues of indiscernibility in *infinitary* logics, with and without identity. We start by sketching the syntax and semantics for infinitary logics, and then show that the grades of indiscernibility for infinitary logics are subsumed by the grades of indiscernibility which we have already discussed.

Syntax and semantics

In ordinary (finitary) first-order logic, given a signature $\mathscr{L}$, every first-order $\mathscr{L}$-formula is of finite length; it contains only finitely many symbols. This feature tracks the intuitive idea of a sentence rather well (cf. §1.6): in principle, we can generate grammatical English sentences of arbitrary finite length, just by repeated conjunction, but natural language seems not to license *infinitely* long sentences. Pure mathematical logic need not, though, be similarly constrained, and nothing prevents us from defining logics with infinitary sentences.

We start by tweaking the syntax of our languages. Instead of regarding conjunction and disjunction as two-place sentential connectives, we regard them as operations on sets of sentences. So, when Φ is a set of formulas (perhaps meeting some further constraint), $\bigwedge \Phi$ will be the (possibly infinitary) conjunction of all the formulas in φ. Equally, instead of having our two quantifiers bind individual variables, we have them bind *sets* of variables. So, when V is a set of variables (perhaps meeting some further constraints) and φ is a formula which does not already bind any of the variables in V, we will have a formula $\forall \mathrm{V}\varphi$. We can vary the constraints on Φ and V to define different infinitary logics. So, where κ and λ are infinite cardinals, $L^{+}_{\kappa\lambda}$ is the infinitary logic which, intuitively, allows up-to-κ-length conjunction and up-to-λ-length quantification and the identity sign '=', and $L^{-}_{\kappa\lambda}$ is the identity-free version of the same infinitary logic. Here is the precise definition.

Definition 15.8: *For any signature* $\mathscr{L}$*, the following—and nothing else—are the* $L_{\kappa\lambda}(\mathscr{L})^{+}$-formulas*:*

- $t_1 = t_2$*, for any* $\mathscr{L}$*-terms* t_1 *and* t_2 *in the sense of Definition 1.3*
- $R(t_1, \ldots, t_n)$*, for any* $\mathscr{L}$*-terms* $t_1, \ldots, t_n$ *in the sense of Definition 1.3 and any n-place relation symbol* $R \in \mathscr{L}$
- $\bigwedge \Phi$ *and* $\bigvee \Phi$*, for any set of* $L_{\kappa\lambda}(\mathscr{L})$*-formulas* Φ *of size* $< \kappa$
- $\exists \mathrm{V}\varphi$ *and* $\forall \mathrm{V}\varphi$*, for any set of variables* V *of size* $< \lambda$*, and any* $L_{\kappa\lambda}(\mathscr{L})$*-formula* φ *such that if* $\exists \mathrm{X}$ *or* $\forall \mathrm{X}$ *occurs in* φ *then* $\mathrm{X} \cap \mathrm{V} = \varnothing$

We say that φ is an $L^{+}_{\infty\lambda}(\mathscr{L})$-formula iff φ is an $L^{+}_{\kappa\lambda}(\mathscr{L})$-formula for some infinite cardinal κ. We say that φ is an $L^{+}_{\infty\infty}(\mathscr{L})$-formula iff φ is an $L^{+}_{\infty\lambda}(\mathscr{L})$-formula for some infinite cardinal λ. We define the $L^{-}_{\kappa\lambda}(\mathscr{L})$-formulas, the $L^{-}_{\infty\lambda}(\mathscr{L})$-formulas, and the $L^{-}_{\infty\infty}(\mathscr{L})$-formulas exactly similarly, except that we prohibit the occurrence of '=' in any formula.

We give a semantics for these infinitary languages just by tweaking the semantics for ordinary first-order logic. On the Tarskian approach, the key clauses are:

$$\mathcal{M}, \sigma \models \neg\varphi \text{ iff } \mathcal{M}, \sigma \not\models \varphi$$
$$\mathcal{M}, \sigma \models \bigwedge \Phi \text{ iff } \mathcal{M}, \sigma \models \varphi \text{ for all } \varphi \in \Phi$$
$$\mathcal{M}, \sigma \models \forall \mathrm{V}\varphi(\mathrm{V}) \text{ iff } \mathcal{M}, \tau \models \varphi(\mathrm{V}) \text{ for every variable-assignment } \tau \text{ which agrees with } \sigma \text{ except perhaps on the values of the members of } \mathrm{V}$$

It is easy to see that $L^{+}_{\omega\omega}$ is essentially first-order logic. After all, $L^{+}_{\omega\omega}$ allows arbitrary finite conjunctions, and we can achieve the same effect via finitely many uses of the two-place connective $\wedge$. Similarly, $L^{+}_{\omega\omega}$ allows a quantifier to bind arbitrary finite numbers of variables, and we can achieve the same thing via finitely many uses of a quantifier which binds a single variable.

Isomorphism and $L^{+}_{\infty\infty}$

These infinitary logics are, though, much richer than ordinary first-order logic. Indeed, $L^{+}_{\infty\infty}$ is powerful enough to pin down a structure's isomorphism type.

Here is the intuitive way in which this is done. First, given a structure, $\mathcal{M}$, we flood it with names, obtaining $\mathcal{M}^{\circ}$, and describe all the basic facts about $\mathcal{M}^{\circ}$ in terms of these names. Second: we form the (often enormous) infinitary conjunction of these facts, and existentially generalise away all the names. Finally: we add a 'totality fact', essentially saying that there are no further objects to consider.

To make this idea precise, we introduce some new notation, for the infinitary notion of an n-type (compare Definition 15.1):

$$\infty\mathrm{tp}^{+}_{\mathcal{M}}(\bar{a}) = \{\varphi \text{ is an } L^{+}_{\infty\infty}(\mathscr{L})\text{-formula} : \mathcal{M} \models \varphi(\bar{a})\}$$

The following Lemma now refines the intuitive idea which we just outlined:

Lemma 15.9: *Let $\mathcal{M}$ and $\mathcal{N}$ be $\mathscr{L}$-structures, and let $\bar{a}$ and $\bar{b}$ be (possibly transfinite) sequences from M and N respectively. The following are equivalent:*

(1) There is an isomorphism $h : \mathcal{M} \longrightarrow \mathcal{N}$ with $h(\bar{a}) = \bar{b}$
(2) $\infty\mathrm{tp}^{+}_{\mathcal{M}}(\bar{a}) = \infty\mathrm{tp}^{+}_{\mathcal{N}}(\bar{b})$

Proof. *(1) $\Rightarrow$ (2).* This is a simple induction on complexity, extending the proof of Theorem 2.3.

(2) ⇒ (1). First: Let $a_1, \ldots, a_\beta, \ldots$ be a complete enumeration of the elements in M, of length κ, such that the *initial* part of this enumeration is just our given $\bar{a}$. (Without loss of generality, we can assume our enumeration involves no repetition.) As in Definition 1.5, let $\mathcal{M}^\circ$ be the signature expansion of $\mathcal{M}$ formed by adding new constant symbols $c_{a_1}, \ldots, c_{a_\beta}, \ldots$. Now let T be the set of all atomic and negated atomic sentences φ in the expanded signature such that $\mathcal{M}^\circ \vDash \varphi$.[35]

Second: Take the infinitary conjunction $\bigwedge T$. Replace each constant c_{a_β} in $\bigwedge T$ with the variable v_β; call this new infinitary formula ψ. Finally: let V be the set of variables $\{v_\beta : a_\beta \text{ is not in the sequence } \bar{a}\}$, and consider the formula:

$$\chi(\bar{v}) := \exists \mathrm{V}\left(\psi \wedge \forall x \bigvee_{\alpha=1}^{\kappa} x = v_\alpha\right)$$

By construction, $\mathcal{M} \vDash \chi(\bar{a})$. Hence, assuming (2), $\mathcal{N} \vDash \chi(\bar{b})$. And, because χ tracks all of the atomic and negated atomic sentences, this characterises $\mathcal{M}$ and $\mathcal{N}$ up to isomorphism. □

We will exploit this Lemma in Chapter 16. But its immediate interest lies in its ability to connect grades of symmetry with certain notions of indiscernibility:

Corollary 15.10: *For any $\mathscr{L}$-structure $\mathcal{M}$, a and b are:*

(1) *one-symmetricals in $\mathcal{M}$ iff* $\infty\mathrm{tp}^+_{\mathcal{M}}(a) = \infty\mathrm{tp}^+_{\mathcal{M}}(b)$
(2) *two-symmetricals in $\mathcal{M}$ iff* $\infty\mathrm{tp}^+_{\mathcal{M}}(a, b) = \infty\mathrm{tp}^+_{\mathcal{M}}(b, a)$

We can gloss this result as follows: the infinitary analogue of indiscernibility$^+$ is symmetricality. So there is no need to complicate Theorem 15.7 by considering grades of $L^+_{\infty\infty}$-indiscernibility; they are subsumed by the grades of symmetry.

Relativeness correspondence and $L^-_{\infty\infty}$

There is an exactly analogous connection between the grades of relativity, and grades of indiscernibility for infinitary languages *without* identity. However, rather than proving these results directly, we invoke some machinery from identity-free model theory; specifically, the idea of quotienting a structure by Leibniz-indiscernibility$^-$:[36]

---

35 The set of formulas T is sometimes called the *complete atomic diagram*. The proof of this direction of the Lemma is thus closely related to the *Diagram Lemma* (cf. Hodges 1993: Lemma 1.4.2 p.17; Marker 2002: Lemma 2.3.3 p.44).

36 See e.g. Monk (1976: Exercises 29.33–34) and Casanovas et al. (1996: Definition 2.3–2.4).

Definition 15.11: *Let $\mathcal{M}$ be any $\mathscr{L}$-structure. Then $\mathcal{M}/\sim$ is the $\mathscr{L}$-structure obtained by quotienting $\mathcal{M}$ by Leibniz-indiscernibility$^-$. So its elements are*

$$[a]_{\mathcal{M}/\sim} = \{b \in M : a \text{ and } b \text{ are Leibniz-indiscernibles}^- \text{ in } \mathcal{M}\}$$

When no confusion can arise, we dispense with the subscript, talking of $[a]$ rather than $[a]_{\mathcal{M}/\sim}$. The $\mathscr{L}$-symbols are interpreted on the domain $\{[a] : a \in M\}$ as follows, for all $\mathscr{L}$-constant symbols c, all n-place $\mathscr{L}$-predicates R, all n-place $\mathscr{L}$-function symbols f, and all $\bar{a}$ from M^n:

$$c^{\mathcal{M}/\sim} = [c^{\mathcal{M}}]$$
$$R^{\mathcal{M}/\sim} = \left\{\overline{[a]} \in [M]^n : \bar{a} \in R^{\mathcal{M}}\right\}$$
$$f^{\mathcal{M}/\sim}\left(\overline{[a]}\right) = [f^{\mathcal{M}}(\bar{a})]$$

A key feature of quotient-structures is that they preserve satisfaction of identity-free formulas. More precisely, when we define:

$$\infty\text{tp}^-_{\mathcal{M}}(\bar{a}) = \{\varphi \text{ is an } L^-_{\infty\infty}(\mathscr{L})\text{-formula} : \mathcal{M} \vDash \varphi(\bar{a})\}$$

a simple induction on complexity yields:

Lemma 15.12: *Let $\mathcal{M}$ be an $\mathscr{L}$-structure. For any sequence $\bar{a}$ of elements from M (possibly transfinite), $\infty\text{tp}^-_{\mathcal{M}}(\bar{a}) = \infty\text{tp}^-_{\mathcal{M}/\sim}\left(\overline{[a]}\right)$.*

The second interesting idea from identity-free model-theory employs these quotient-structures, and reinforces the idea—mentioned in §15.1—that relativeness correspondences are the identity-free analogues of isomorphisms:[37]

Lemma 15.13: *Let $\mathcal{M}$ and $\mathcal{N}$ be $\mathscr{L}$-structures, and let $\bar{a}$ and $\bar{b}$ be (possibly transfinite) sequences from M and N respectively. The following are equivalent:*

(1) *There is an isomorphism $\hbar : \mathcal{M}/\sim \longrightarrow \mathcal{N}/\sim$ with $h\left(\overline{[a]}\right) = \overline{[b]}$*
(2) *There is a relativeness correspondence H from $\mathcal{M}$ to $\mathcal{N}$ such that $a_1Hb_1, \ldots, a_\gamma IIb_\gamma, \ldots$*

By combining these two big ideas, we can very easily 'lift' Lemma 15.9 and Corollary 15.10 into their identity-free analogues:

Lemma 15.14: *Let $\mathcal{M}$ and $\mathcal{N}$ be $\mathscr{L}$-structures, and let $\bar{a}$ and $\bar{b}$ be (possibly transfinite) sequences from M and N respectively. The following are equivalent:*

[37] For this result, see any of Casanovas et al. (1996: Proposition 2.6), Bonnay and Engström (2013: §4), and Button (2017: §4).

(1) *There is a relativeness correspondence H from $\mathcal{M}$ to $\mathcal{N}$ such that $a_1 H b_1, \ldots,$ $a_\gamma H b_\gamma, \ldots$*
(2) $\infty\text{tp}^-_{\mathcal{M}}(\bar{a}) = \infty\text{tp}^-_{\mathcal{M}}(\bar{b})$

Proof. (1) ⇒ (2). Given (1), there is an isomorphism $h : \mathcal{M}/\!\sim \longrightarrow \mathcal{N}/\!\sim$ with $\hbar\left(\overline{[a]}\right) = \overline{[b]}$ by Lemma 15.13. By Lemma 15.9, $\infty\text{tp}^+_{\mathcal{M}/\sim}\left(\overline{[a]}\right) = \infty\text{tp}^+_{\mathcal{N}/\sim}\left(\overline{[b]}\right)$, so that $\infty\text{tp}^-_{\mathcal{M}/\sim}\left(\overline{[a]}\right) = \infty\text{tp}^-_{\mathcal{N}/\sim}\left(\overline{[b]}\right)$ and so $\infty\text{tp}^-_{\mathcal{M}}(\bar{a}) = \infty\text{tp}^-_{\mathcal{N}}(\bar{b})$ by Lemma 15.12.

(2) ⇒ (1). Given (2), we have $\infty\text{tp}^-_{\mathcal{M}/\sim}\left(\overline{[a]}\right) = \infty\text{tp}^-_{\mathcal{N}/\sim}\left(\overline{[b]}\right)$ by Lemma 15.12. Since $\mathscr{L}$ is set-sized, there is an $L^-_{\infty\infty}(\mathscr{L})$-formula which states that x and y are Leibniz-indiscernibles$^-$, given by:

$$\bigwedge\{\varphi(x,x) \leftrightarrow \varphi(x,y) : \varphi \text{ is an } L^-_{\omega\omega}(\mathscr{L})\text{-formula}\}$$

This $L^-_{\infty\infty}(\mathscr{L})$-formula defines *identity* in $\mathcal{M}/\!\sim$ and $\mathcal{N}/\!\sim$. So $\infty\text{tp}^+_{\mathcal{M}/\sim}\left(\overline{[a]}\right) = \infty\text{tp}^+_{\mathcal{N}/\sim}\left(\overline{[b]}\right)$. By Lemma 15.9, there is an isomorphism $h : \mathcal{M}/\!\sim \longrightarrow \mathcal{N}/\!\sim$ such that $\hbar\left(\overline{[a]}\right) = \overline{[b]}$. The result now follows by Lemma 15.13. □

Corollary 15.15: *For any $\mathscr{L}$-structure $\mathcal{M}$, a and b are:*
(1) *one-relatives in $\mathcal{M}$ iff $\infty\text{tp}^-_{\mathcal{M}}(a) = \infty\text{tp}^-_{\mathcal{M}}(b)$*
(2) *two-relatives in $\mathcal{M}$ iff $\infty\text{tp}^-_{\mathcal{M}}(a,b) = \infty\text{tp}^-_{\mathcal{M}}(b,a)$*

Another way to state Corollary 15.15 is as follows: the infinitary analogue of indiscernibility$^-$ is relativity. Again, there is no need to complicate Theorem 15.7 with grades of $L^-_{\infty\infty}$-indiscernibility.

15.5 *n*-indiscernibles, order, and stability

We now consider a second generalisation of indiscernibles. The notions of indiscernibility from Definition 15.2 only involved pairwise comparisons of elements. Sometimes, we may want to go further. To see why, let $\mathcal{G}$ be this graph:

```
1 — 2
|   |
4 — 3
```

Clearly 1 and 2 are two-symmetricals, as are 1 and 3, and likewise 2 and 3. But the formalisation of 'x has an edge to y and y has an edge to z' is satisfied by $(1,2,3)$ but not by $(2,1,3)$, so that $\text{tp}^+_{\mathcal{G}}(1,2,3) \neq \text{tp}^+_{\mathcal{G}}(2,1,3)$. Hence *pairwise* comparisons of elements in the set $\{1,2,3\}$ do not reflect their *trio-wise* discernibility.

n-indiscernibles and ω-indiscernibles

To deal with this, we offer the following definition:

Definition 15.16: *Let $\mathcal{M}$ be an $\mathcal{L}$-structure and let $n \geq 1$. We say that a subset $X \subseteq M$ with at least n-elements is* n-indiscernible$^+$ *iff for any two n-element sequences of distinct elements $\bar{a} = (a_1, \ldots, a_n)$ and $\bar{b} = (b_1, \ldots, b_n)$ from X, we have $\text{tp}^+_{\mathcal{M}}(\bar{a}) = \text{tp}^+_{\mathcal{M}}(\bar{b})$. When $X \subseteq M$ has infinitely many elements, we say that X is* ω-indiscernible$^+$ *iff X is n-indiscernible$^+$ for each $n \geq 1$.*

The adjective 'distinct', in this definition, has the meaning that $a_i \neq a_j$ and $b_i \neq b_j$ for all $1 \leq i < j \leq n$. In the case $n = 2$, this usage clearly agrees with the definition of two-indiscernibles$^+$ from Definition 15.2, and so the notion of n-indiscernible$^+$ can be seen as a natural generalisation of two-indiscernibles$^+$. In our example $\mathcal{G}$, from above, the set $\{1, 2\}$ is two-indiscernible$^+$, but the set $\{1, 2, 3\}$ is not three-indiscernible$^+$. Clearly, Definition 15.16 may be modified in the obvious way to generate notions of n-indiscernibles$^-$, n-indiscernibles^s, and even n-symmetricals and n-relatives, via infinitary logics and the results of §15.4.

We now want to explain how linear orders interact with indiscernibility, in this extended sense.

Orders and indiscernibles

Here is a simple observation: *If $\mathcal{M}$ defines a linear order $<$, then no distinct elements a, b are two-indiscernibles$^+$.* (The axioms for a linear order were set out in §1.12.) To see why, suppose for reductio that $\text{tp}^+_{\mathcal{M}}(a, b) = \text{tp}^+_{\mathcal{M}}(b, a)$ for some $a \neq b$. Now, either $a < b$ or $b < a$: but if $a < b$, then $v_1 < v_2$ is in $\text{tp}^+_{\mathcal{M}}(a, b) = \text{tp}^+_{\mathcal{M}}(b, a)$, so that $b < a$, a contradiction; and a similar contradiction ensues if $b < a$. Generalising this thought, a linear order prevents there from being any n-indiscernible$^+$ set for $n \geq 2$. (However, a linear order is *compatible* with various notions of one-discernibility. For example, in the rationals as a linear order, any distinct elements $a < b$ are one-symmetrical since $x \mapsto x + (b - a)$ is a symmetry sending a to b.)

In fact, linear orders are essentially the *only* obstacle to indiscernibility. To see this connection, we liberalise the notion of a linear order:

Definition 15.17: *We say that a structure $\mathcal{M}$ has an order iff for some $n \geq 1$, there is a 2n-place formula $\varphi(\bar{x}, \bar{y})$ and an infinite sequence of n-tuples $\bar{a}_1, \bar{a}_2, \ldots$ from M such that $\mathcal{M} \vDash \varphi(\bar{a}_i, \bar{a}_j)$ iff $i < j$. We say that a theory T* has an order *iff there is a model $\mathcal{M}$ of T which has an order.*

Another way to phrase the idea that $\mathcal{M}$ has an order, is to say that there is a definable relation $<$ on $M^n \times M^n$, and an infinite set $L \subseteq M^n$ which need *not* be definable, such that $(L, <)$ is a linear order. So 'having an order' is more liberal than 'defining an infinite linear order', since it does not require that the domain of the order be definable.[38] The following theorem then states that having an order is the *only* obstacle to having a model with ω-indiscernibles$^+$:

Theorem 15.18: *Let T be a complete first-order theory with infinite models. If T does not have an order, then there is a model $\mathcal{M}$ of T which contains a set of ω-indiscernibles$^+$.*

The proof of this theorem employs a good deal of machinery which we have not developed here, and so we omit it.[39] But we can contextualise Theorem 15.18, by noting a fundamental, equivalent characterisation of not having an order:[40]

Theorem 15.19: *Let T be a complete first-order theory with infinite models. Then T does not have an order iff T is stable.*

The notion of *stability* mentioned here was introduced in Definition 14.11, and it formalises the idea that a theory has few types. Putting these two theorems together, then, we find that having few types implies the existence of models with ω-indiscernibles$^+$. This is quite intuitive: the types are the different sets of formulas that can be consistently satisfied by an object, and if there are very few of these then it will be harder to discern objects from one another.

Here is another useful and immediate consequence of Theorem 15.19: any theory which defines an infinite linear order is not stable. So, the complete theories of the rationals or the reals, in any signature including $<$, are not stable.

Indeed, stability is the exception, rather than the rule, amongst complete theories. So one might well ask whether there is anything similar to ω-indiscernibles$^+$ in non-stable contexts. To our knowledge, the closest idea is the following:

Definition 15.20: *Let $(L, <)$ be an infinite linear order and let $X = \{a_i \in M : i \in L\}$ be a set of elements of a model $\mathcal{M}$ indexed by L. We say that X is* order-indiscernible$^+$ *iff $\text{tp}^+_{\mathcal{M}}(a_{i_1}, ..., a_{i_n}) = \text{tp}^+_{\mathcal{M}}(a_{j_1}, ..., a_{j_n})$ for all $n \geq 1$ and all increasing sequences $i_1 < ... < i_n$ and $j_1 < ... < j_n$ from L.*

---

[38] As mentioned at the outset of this chapter, usually when considering indiscernibility one restricts attention to formulas without parameters. It is worth then noting that Definition 15.17 is equivalent to a version of it where parameters are allowed in the formula. For, suppose that T has an order in this extended sense, where we display the parameters $\overline{p}$ as $\varphi(\overline{x}, \overline{y}, \overline{p})$ and where the infinite sequence is written as $\overline{a}_i$. Then define $\psi(\overline{x}, \overline{u}, \overline{y}, \overline{v})$ to be $\varphi(\overline{x}, \overline{y}, \overline{u})$ and define $\overline{b}_i = \overline{a}_i\overline{p}$. Then $\psi(\overline{x}, \overline{u}, \overline{y}, \overline{v})$ and $\overline{b}_i$ are witnesses to T having an order in the parameter-free sense.

[39] See Marker (2002: Theorem 5.2.13 p. 184).

[40] For a proof, see Pillay (1983: Theorem 2.15 p.22).

A complicated argument using infinitary combinatorics shows that, for *any* theory T with infinite models, and any infinite linear order $(L, <)$, there is a model of T containing order-indiscernibles$^+$ associated with $(L, <)$.[41] So one can find order-indiscernibles$^+$ in models of PA, ZFC, the theory of the reals, etc.[42]

Sources of stability

We end this chapter by mentioning some important examples of stable theories.

The complex numbers in the field signature $\{0, 1, +, \times\}$ provide an important example of a stable theory, and hence an important example of a theory with ω-indiscernibles$^+$.[43] Indeed, there are very natural examples of ω-indiscernibles$^+$ here. Say that a complex number a is *algebraic* over a set B of complex numbers if $p(a) = 0$ for some polynomial $p(x)$ with coefficients from B or the rationals; and further say that a set A is *independent* if a is not algebraic over $A \setminus \{a\}$ for any a from A. Independent sets which are maximal—in that they cannot be further extended and remain independent—turn out to be ω-indiscernibles$^+$, because any permutation of them extends to a symmetry on the complex field.[44]

Another source of stable theories, and hence ω-indiscernibles$^+$, comes from the simple observation that stability is preserved downwards under interpretability (where interpretability was defined in §5.5). More precisely, T and T^* are complete theories with infinite models, and T^* is stable and interprets T, then T is also stable.[45] So, by the above, the complete theory of any structure definable in the complex field is stable. This includes the general linear group (the set of $n \times n$ matrices over the complex numbers with non-zero determinant) and the special linear group (those such matrices with determinant 1).

Other examples of stable theories generalise the complex field, in that while the complex field is an algebraically closed structure (every non-trivial polynomial has a root), these other stable theories are closed in some other way: for example, differentially closed fields of characteristic zero are stable, as are separably closed fields.[46] But it is worth noting that the stability of a theory is sometimes very difficult to determine. For example, it was only recently shown that the complete theory of a free

---

[41] See Marker (2002: 179) and Pillay (1983: 6–7).

[42] Furthermore, it turns out that: a complete theory with infinite models is stable iff all order-indiscernibles$^+$ are ω-indiscernible$^+$. For the left-to-right direction, see Pillay (1983: 87). The right-to-left direction follows from Theorem 15.19 and the result cited in the previous footnote. Indeed, one normally proves Theorem 15.18 by showing that order-indiscernibles$^+$ are ω-indiscernibles$^+$ in a stable theory.

[43] Marker (2002: Corollary 4.1.18 p. 124). It is hard to show that this theory has *no* order; but it is easy to show that it has no linear order which respects the usual rules concerning how the order interacts with zero, one, addition, and multiplication. If such an order $<$ respected those rules, then: if $i > 0$ then $-1 = i^2 > 0$, a contradiction; and if $i < 0$ then $0 < -i$ and thus $0 < (-i)^2 = -1$, a contradiction again.

[44] See Hungerford (1980: 312).

[45] See Hodges (1993: 307).

[46] See Marker et al. (2006: 49, 142).

group with more than two generators is stable, despite the fact that mathematicians have been aware of such groups since the advent of group theory.[47] By contrast, it has been known since the 1970s that all abelian groups are stable.[48]

15.A Charting the grades of discernibility

In this appendix, we sketch a proof of Theorem 15.7. Most of this is proved elsewhere;[49] our further observations will also give a sense of the full proof.

Proof sketch of Theorem 15.7. Two-symmetricals are two-indiscernibles^{s}*, and one-symmetricals are one-indiscernibles*^{s}. Just by Theorem 2.3.

Two-indiscernibles^{s} *are two-indiscernibles*$^{+}$*, and one-indiscernibles*^{s} *are one-indiscernibles*$^{+}$. This is because every first-order formula is a second-order formula. These implications would continue to hold if '=' were omitted from the second-order vocabulary, since in the full semantics for second-order logic we could define $x = y$ via $\forall X\,(X(x) \leftrightarrow X(y))$.

Two-indiscernibles^{+} *need not be one-indiscernibles*^{s}. Consider $\mathcal{B}$ from §15.1, the disjoint union of a complete countable graph with a complete continuum-sized graph. Let a be from the countable graph and b be from the continuum-sized graph. Now a and b are one-discernibles^{s}, since full second-order logic allows us to formulate 'v has edges to *only countably* many elements'. But a and b are two-indiscernibles^{+}. To see this, use Theorem 7.2(1) to build a countable Skolem hull, $\mathcal{H}$, of $\mathcal{B}$ such that $\{a, b\} \subseteq H$. Clearly $\mathcal{H}$ consists of two (disjoint) complete countable graphs; so there is a symmetry sending a to b on $\mathcal{H}$, so that $\text{tp}^{+}_{\mathcal{H}}(a) = \text{tp}^{+}_{\mathcal{H}}(b)$ by Theorem 2.3. Then $\text{tp}^{+}_{\mathcal{B}}(a) = \text{tp}^{+}_{\mathcal{H}}(a) = \text{tp}^{+}_{\mathcal{H}}(b) = \text{tp}^{+}_{\mathcal{B}}(b)$ as $\mathcal{H} \preceq \mathcal{B}$.

Two-indiscernibles^{s} *need not be one-relatives.* Let $\mathscr{L}$ contain one-place predicates P_n for all $n < \omega$, and a single two-place predicate R. Define a structure C as follows:

$$C = \mathbb{N}$$
$$P_n^{C} = \{2n+1, 2n+2\}, \text{for all } 0 < n < \omega$$
$$R^{C} = \{(1, 2n+1) : 0 < n < \omega\} \cup \{(0, 2n) : 0 < n < \omega\}$$

We can represent C more perspicuously as follows:

47 See Sela (2013).

48 This follows from the stability of modules. See Hodges (1993: 660).

49 Button (2017: §3).

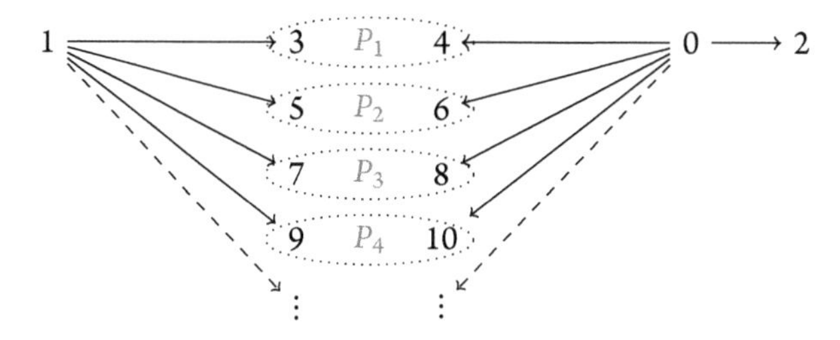

We first show that 0 and 1 are not one-relatives. For reductio, suppose there is a relativity H with $0H1$. Then since H's domain is the whole of C, we have $2Hy$ for some $y \in C$. Since $0H1$, $2Hy$ and $R^C(0,2)$, we have $R^C(1,y)$. So $P_n^C(y)$, for some $n \geq 1$. Now since $2Hy$ and $P_n^C(y)$, we have $P_n^C(2)$, a contradiction, as required.

We now show that 0 and 1 are two-indiscernibles[s]. Let $\varphi(x,y)$ be any second-order $\mathscr{L}$-formula with free variables displayed, and let C_φ be C's reduct to the signature consisting of just the *finitely* many $\mathscr{L}$-symbols appearing in φ. For each of the finitely many one-place predicates P_i appearing in φ, both 0 and 1 are connected via R to exactly one entity picked out by P_i; and both are connected to a countable infinity of entities picked out by no one-place predicate occurring in φ. So 0 and 1 are two-symmetricals in C_φ, so that $C_\varphi \vDash \varphi(0,1) \leftrightarrow \varphi(1,0)$. Since φ was arbitrary, $C \vDash \varphi(0,1) \leftrightarrow \varphi(1,0)$. □

16
Quantifiers

We typically regard the quantifiers $\exists$ and $\forall$ as *logical* expressions. But there are straightforward semantic means for defining all sorts of new quantifiers. This might make us ask: Which of these are relevantly similar to our initial paradigms, $\exists$ and $\forall$, to count as *logical*? More generally: Can we explicate the notion of *logic*, in a way which yields a precise, intuitively satisfying, and fruitful criterion for distinguishing between 'logical' and 'non-logical' quantifiers?

These questions have generated a substantial literature. Our answers to them are somewhat pluralistic. There are many 'intuitive' ideas associated with 'logic'. The relations between these 'intuitive' ideas are rather unclear. By insisting on a very particular explication of each intuitive idea, and in a particular combination, one may be able to draw sharp bounds for logic. But we doubt that any particular combination of explications is forced upon us. To show this, we start by introducing the semantic framework for generalised quantifiers (§16.1). Having elaborated on what we mean by raising the question of logicality (§16.2), we introduce a popular thesis about logicality. It was provided Sher, building on work by Tarski, Mostwoski, and Lindström; and we call it *Tarski–Sher* (§16.3). We spend the remainder of the chapter considering and rebutting attempts to vindicate Tarski–Sher.

16.1 Generalised quantifiers

So far in this book, we have almost entirely restricted our attention to two quantifiers: the universal quantifier, $\forall$, and the existential quantifier, $\exists$. In fact, there have only been two exceptions to this: in §7.10, we considered the finitary quantifier 'there are finitely many...', symbolised $\exists_{<\aleph_0}$, and Härtig's two-place quantifier, 'there are exactly as many φs as ψs'. But natural language teams with quantifier phrases, such as:[1]

there are exactly five φs
two-thirds of things are φ
there are more φs than ψs
the majority of φs are ψs
fewer φs than ψs are ξs

[1] For many, many more examples, see Sher (1991: ch.1) and Peters and Westerståhl (2006: ch.4).

In this chapter, we explore formal languages which have formalisations of such quantifiers, and more besides.[2]

Where s is an expression, recall that we generally write $s^{\mathcal{M}}$ for the interpretation of the expression in the structure $\mathcal{M}$. We will do the same with quantifiers, writing $\mathsf{Q}^{\mathcal{M}}$ for the interpretation of the quantifier Q in the structure $\mathcal{M}$.

In particular, suppose that we stipulate that the interpretation of $\forall$ in a structure $\mathcal{M}$ is always given by $\forall^{\mathcal{M}} = \{M\}$. That is to say: the interpretation of the universal quantifier on a structure is always the singleton of the underlying domain. Where we also define $\varphi(\bar{x})^{\mathcal{M}}$ by $\{\bar{a} \in M^n : \mathcal{M} \vDash \varphi(\bar{a})\}$,[3] we can stipulate:

$$\mathcal{M} \vDash \forall x \varphi(x) \text{ iff } \varphi(x)^{\mathcal{M}} \in \forall^{\mathcal{M}}$$

Given our definition of $\forall^{\mathcal{M}}$, the right-hand-side holds iff $\varphi(x, \bar{a})^{\mathcal{M}} = M$, i.e. iff every entity in the domain satisfies $\varphi(x, \bar{a})$. This provides precisely the usual semantics for universal quantification, in a slightly different way. We can do the same for existential quantification. We stipulate that the interpretation of $\exists$ is always given by $\exists^{\mathcal{M}} = \wp(M) \setminus \varnothing$. Then this clause yields the usual semantics for existentials:

$$\mathcal{M} \vDash \exists x \varphi(x) \text{ iff } \varphi(x)^{\mathcal{M}} \in \exists^{\mathcal{M}}$$

But we need not stop at one-place quantifiers. Consider Härtig's two-place quantifier, H, which we use to formalise the claim 'there are exactly as many φs as ψs'. We can stipulate that $\mathsf{H}^{\mathcal{M}} = \{(X, Y) \in \wp(M) \times \wp(M) : |X| = |Y|\}$, and obtain an appropriate semantics with this clause:

$$\mathcal{M} \vDash \mathsf{H}xy\,(\varphi(x), \psi(y)) \text{ iff } \left(\varphi(x)^{\mathcal{M}}, \psi(y)^{\mathcal{M}}\right) \in \mathsf{H}^{\mathcal{M}}$$

This is the start of a rich framework. For example, we can consider a quantifier expression, Q, which operates on an arbitrary but finite number of one-place predicates. Its extension will be some subset of $\wp(M)^n$, and its semantic clause will be:

$$\mathcal{M} \vDash \mathsf{Q}x_1 \ldots x_n \left(\varphi_1(x_1), \ldots, \varphi_n(x_n)\right) \text{ iff } \left(\varphi_1(x_1)^{\mathcal{M}}, \ldots, \varphi_n(x_n)^{\mathcal{M}}\right) \in \mathsf{Q}^{\mathcal{M}}$$

We can go further still, considering quantifiers which operate on arbitrarily many predicates of arbitrary position. The extension of such a quantifier will be some subset of $\wp(M^{i_1}) \times \ldots \times \wp(M^{i_n})$, and its semantic clause will be:

$$\mathcal{M} \vDash \mathsf{Q}\bar{x}_1 \ldots \bar{x}_n \left(\varphi_1(\bar{x}_1), \ldots, \varphi_n(\bar{x}_n)\right) \text{ iff } \left(\varphi_1(\bar{x}_1)^{\mathcal{M}}, \ldots, \varphi_n(\bar{x}_n)^{\mathcal{M}}\right) \in \mathsf{Q}^{\mathcal{M}}$$

We now have an extremely general framework for considering quantifiers.

---

[2] The original generalisation is due to Mostowski (1957) and Lindström (1966). For more detail see Peters and Westerståhl (2006: §§2.2–2.4).

[3] Formulas mentioned in this section are allowed to include parameters $\bar{b}$ from M. (Compare §1.13, and note that this contrasts with Chapter 15, where by default we *banned* parameters).

16.2 Clarifying the question of logicality

We could happily deploy (some fragment of) this framework to handle the quantifier-expressions that arise in natural language. For instance: first-order logic gives no good way to formalise 'there are as many φs as ψs', even though that is a perfectly reasonable thing to say. To handle this, we could simply augment our logical vocabulary, to include H.

This augmentation would dramatically increase the logic's expressive power. We saw in §7.10 that we can offer a categorical theory of arithmetic if we add H to first-order logic. As an immediate corollary, the resulting logic is not compact. In short: using generalised quantifiers with these semantics has substantial *metalogical* implications (cf. §7.6).

Consequently, the model-theoretical sceptic of Chapter 9 will insist that we are *wrong* to say that using H allows us to provide a categorical theory of arithmetic. She will point out that the categoricity of the axiomatisation requires that H has its 'standard' semantics. And she will then challenge us to explain how we do this, when we could instead consider a *deviant* interpretation of H by considering a 'deviant' notion of cardinality. That is, she will treat the semantic theory governing H as *just more theory*.

We do not want to re-open the wounds of Part B here. So, in what follows, we will set these sceptical challenges entirely to one side; not because they are irrelevant, but because we have nothing to add to our discussion in Part B. So, when we consider the 'logicality' of a quantifier like H in this chapter, we will not consider any 'sceptical' challenges concerning our ability to articulate H's semantics.

The question of 'logicality' that we have in mind can also be refined, by revisiting a point raised in Chapter 13. There, we emphasised two different ways to think about the meaning of (logical) expressions: *semanticist* approaches, which claim that the meaning of an expression is given by its semantic conditions, and *inferentialist* approaches, which claim that the meaning of an expression is given by its inference rules. The framework of generalised quantifiers is *purely* semantic. So the question we are exploring is *purely* directed at semanticists.

In sum: this chapter addresses semanticists who (rightly or wrongly) are not much worried by sceptical concerns, but who want to draw the bounds of logic. Our aim is to critique the best-developed attempt at drawing sharp bounds.

16.3 Tarski and Sher

In a lecture given in 1966 (published as Tarski 1986), Tarski proposed a criterion of logicality. Roughly, his idea was as follows. Let M be a class of basic objects.[4]

[4] Following Tarski, we use the word 'class' to stand, neutrally, for sets or types.

Consider the hierarchy of classes one might construct over M. Now Tarski asked: *Which of these classes are logical?* And he answered:

Tarski's Thesis. *C is logical relative to M iff C is permutation-invariant on M, where C is permutation-invariant on M iff $\hbar(C) = C$ for any bijection $h : M \longrightarrow M$*

Since the interpretation of a quantifier on a structure $\mathcal{M}$ is always some class constructed from M, Tarski's Thesis also suggests a simple answer to our question, of which quantifiers are logical: a quantifier is logical iff every interpretation of the quantifier is permutation-invariant. Tweaking this idea slightly, we obtain:[5]

Tarski–Sher. *Q is logical iff Q is bijection-invariant, where Q is bijection-invariant iff $\hbar(\mathsf{Q}^{\mathcal{M}}) = \mathsf{Q}^{\mathcal{N}}$ for any bijection $h : M \longrightarrow N$.*

For any structures $\mathcal{M}$ and $\mathcal{N}$ with a bijection h, it is easy to see that $\hbar(\forall^{\mathcal{M}}) = \forall^{\mathcal{N}}$, that $\hbar(\exists^{\mathcal{M}}) = \exists^{\mathcal{N}}$ and that $\hbar(\mathsf{H}^{\mathcal{M}}) = \mathsf{H}^{\mathcal{N}}$. So these quantifiers are all logical, according to Tarski–Sher. Indeed, any quantifier which can be defined solely in terms of cardinality will qualify as logical on Tarski–Sher, just because cardinals themselves are defined in terms of bijections.

Tarski–Sher supplies us with a simple, perfectly general answer to the question of which quantifiers are logical. Indeed, Bonnay, who is critical of Tarski–Sher, nevertheless suggests that it 'might be considered as the received view regarding the semantic characterisation of logical constants.'[6] In what follows, we focus on Tarski–Sher, and show just how hard it is to establish it.

16.4 Tarski and Klein's Erlangen Programme

Tarski situated his Thesis against the background of Klein's *Erlangen Programme*.[7] Tarski emphasised two features of this Programme. First: different branches of geometry can be characterised by the fact that they restrict their attention to transformations which are *invariant* in certain ways. Second: as the branch of geometry becomes more abstract, so the transformations they entertain hold fewer features invariant.

To illustrate, the transformations of Euclidean geometry all preserve *ratios between distances*. Such maps will always send a triangle to a *congruent* triangle. The

[5] McGee (1996: 575, example slightly adapted) offers a nice motivation for the tweak. Let the *wombat-quantifier*, W, behave exactly like existential quantification in domains containing wombats, but like universal quantification in wombat-free domains. So $\mathsf{W}^{\mathcal{M}} = \wp(M) \setminus \{\varnothing\}$ if some wombat is in M, and $\mathsf{W}^{\mathcal{M}} = \{M\}$ otherwise. This is permutation-invariant, but not bijection-invariant; but we probably should not count wombat-quantification as *logical*, since it invokes extra-logical considerations, viz., wombats.

[6] Bonnay (2014: 56).

[7] Tarski (1986); see also Mautner (1946), Sher (1991: 61–5, 2008: 301–2, 305–7), and Bonnay (2008: 33–4, 2014: 56, 59).

transformations of affine geometry all preserve *colinearity*. Such maps will always send a triangle to a triangle, but may sacrifice congruence. Last, the transformations of topology preserve continuity. Such maps will always send a triangle to a closed figure, but it need not even be a polygon. Observing this, Tarski asks us to consider what he called the 'extreme case'. According to Tarski, this would be a discipline which entertains *any* transformation (i.e. permutation) on an underlying domain. And this thought led him to Tarski's Thesis.

Unfortunately, Tarski's own lecture provides us with no real clue as to why observations about a *geometric* programme should tell us anything about *logic*. Fortunately, MacFarlane offers a charitable suggestion to plug this gap on Tarski's behalf. Seen through the lens of the Erlangen Programme, bijection-invariance is 'the end point of a chain of progressively more abstract notions defined by their invariance under progressively wider groups of transformations of the domain.'[8] The idea is that we begin with an intuitive notion of logic as *topic-neutral*, and that the analogy with the increasingly abstract branches of geometry shows us that the best way to explicate topic-neutrality is via the 'extreme case', i.e. permutation-invariance.

This line of reasoning is not, though, wholly compelling. To see why, consider three pointed questions.

First, *why go so far?* That is, why should we think that *any* permutation should be relevant to judgements of logicality?

Second, *why stop there?* That is, why not entertain functions other than bijections, or perhaps even relations which are not functional? Given the mention of the Erlangen Programme, it is worth noting that topologists do not *just* consider permutations on a space, or bijections between spaces; they consider all kinds of maps between spaces.

Third, and most fundamentally, *why think that a limiting case of geometry should coincide with logic?* Even if we regard the Erlangen Programme as a triumph for geometry, it is doubtful that a similar programme—characterising the various increasingly abstract sub-disciplines of a discipline in terms of increasingly relaxed transformations—could be carried out in many *other* areas of inquiry. After all, on the face of it, many areas of inquiry having nothing to do with transformations at all. But then it is unclear why anyone who thinks of logic as 'topic neutral', or 'maximally general', should expect logic itself to have much to do with transformations.

We must, then, look for a defence of Tarski–Sher which does not depend upon analogies with the Erlangen Programme. Moreover, we should learn a lesson from the third question which we just posed. A successful defence of Tarski–Sher must connect bijection-invariance with an intuitive idea of what logic *is*. In the remainder of this chapter, we will consider three attempts to do this. We will consider the idea that logic is *non-discriminatory* (§16.5), that logic must meet certain *closure* con-

---

[8] MacFarlane (2000: 175, 2015: §5); cf. also Bonnay (2014: 33).

ditions (§16.6) and that logic must be kept separate from *mathematics* (§16.8). We will argue that none of these attempts is successful.

16.5 The Principle of Non-Discrimination

In defending Tarski–Sher, Sher outlines a conception of logic as both *formal* and *necessary*.[9] This characterisation certainly fits well with a certain 'intuitive' concept of logic. Moreover, it provides a characterisation of logic which does not leave it completely mysterious why logic should be of interest to us. It is, then, an excellent starting point.

Formality and non-discrimination

That said, the idea of *formality* certainly needs development. To see why, consider the following three schematic inference-patterns:

$$\varphi \text{ and } \psi \therefore \varphi$$

$$x \text{ knows that } \varphi \therefore \varphi$$

$$x \text{ is a kitten} \therefore x \text{ is a cat}$$

All three inference-patterns seem to be necessarily truth-preserving. And all three inference-patterns could be codified as *formal rules* for some particular system of reasoning. Since Sher wants to draw the bounds of logic in such a way that only the first of these counts as logical, she must say more about the notion of formality.

Her idea is to characterise *formality* as *non-discrimination*.[10] This idea has deep roots. Mostowski writes that logic 'should not allow us to distinguish between different elements of [the domain]'.[11] Picking up on this, Sher states that logic 'should not distinguish the identity of particular individuals in the universe of a given model'.[12] McGee, in a later defence of Tarski–Sher, claims that 'any consideration which discriminates among individuals lies beyond the reach of logic, whose concerns are entirely general'.[13] Peters and Westerståhl express a similar thought, writing that 'in logic only structure counts, not individual objects, sets, or relations.'[14] And all of this connects with other highly general (but pre-formal) ways to describe what is 'characteristic' about logic, such as the claim that logic is extremely general, or abstract, or topic-neutral, or contentless. In short, we arrive at an intuitive idea:

---

[9] Sher (1991: 40ff). Interestingly, she takes these ideas from Tarski's *earlier* work; cf. Sher (1991: 63).

[10] Cf. MacFarlane's (2000: ch.3) discussion of three different notions of 'formality'.

[11] Mostowski (1957: 13).

[12] Sher (1991: 34, 43, 53); see also Sher (2001: 247–8, 2008: 305–8).

[13] McGee (1996: 567).

[14] Peters and Westerståhl (2006: 95); though they ultimately defend a stricter thesis than Tarski–Sher.

Principle of Non-Discrimination. *Logical expressions are non-discriminatory.*

We agree that the idea is appealing. To unpack it, though, we must explain what it means to call an *expression* discriminatory, for that is not abundantly clear.

Fortunately, we (sometimes) have some relatively clear intuitions on what it means to say that a *class* is discriminatory. Let M be a class of basic objects. Intuitively, $\varnothing$ is non-discriminatory relative to M, since it treats every object equally (by omission). Equally, M itself is non-discriminatory relative to M, since it treats every object equally (by inclusion). But these are intuitively the *only* non-discriminatory subclasses, relative to M. For if N is a subclass of M which is neither empty nor identical to M, then N rules *out* some entities and rules *in* others. And of course, $\varnothing$ and M are the only subclasses deemed logical by Tarski's Thesis.

Expressions can take classes as their interpretations. We therefore recommend understanding the discriminatoriness of expressions in terms of the discriminatoriness of their interpretations. Specifically, we suggest: *an expression is discriminatory iff there is some interpretation of the expression which (intuitively) discriminates between the entities in the domain over which the expression has been interpreted.* So understood, the Principle of Non-Discrimination becomes:

Principle of Non-Discrimination (updated). *No interpretation of a logical expression (intuitively) discriminates between any entities.*

This is how we understand the Principle in what follows. And, so understood, the Principle of Non-Discrimination delivers the intuitively correct verdict in a paradigm case, of one-place predicates.

Consider the natural language one-place predicate '... is a cat'. We might reasonably say that this predicate should always take as its extension (on a given domain) the set of cats. But, if the domain contains both a cat and a wombat, then this interpretation is discriminatory: it includes some things and omits others. As such, the natural language predicate '... is cat' counts as discriminatory, and hence non-logical, according to the Principle of Non-Discrimination. And this is exactly what we should hope for: intuitively, the expression '... is a cat' is not purely logical.

Similar reasoning will convince us that any primitive one-place predicate, in any formal language, will count as discriminatory. After all, it can be interpreted as a non-empty, non-universal subset of a structure's domain, and so is non-logical by the Principle of Non-Discrimination.

Indeed, the only predicates which might qualify as non-discriminatory are predicates like '$x = x$', whose interpretation on a domain is always the domain itself, and '$x \neq x$', whose interpretation on a domain is always $\varnothing$. And it is reasonably plausible to think that these two expressions are, indeed, *logical* expressions.

All of this lends some support to Tarski–Sher. After all, primitive one-place predicates are not bijection-invariant; but both '$x = x$' and '$x \neq x$' *are* bijection-invariant. So Tarski–Sher seems to get the right verdicts here.

Simple rivals to Tarski–Sher

We cannot, though, conclude that Tarski–Sher is *correct.* After all, we formulated the Principle of Non-Discrimination in terms of some intuitive notion of non-discrimination, and it is an open question whether the appropriate notion of non-discrimination should be bijection-invariance (as Tarski–Sher insists), or something else. And, crucially, *many* different notions of non-discrimination will exactly deliver the same verdict concerning one-place predicates.

To see this, we need only revisit our grades of discernibility from Chapter 15. Let (M, U) denote a structure with domain M, whose signature is just a single one-place predicate which is assigned the extension U.[15] Then consider this principle:

U is non-discriminatory on M iff there are no discernibles in (M, U).

The principle is schematic, in that we can replace 'discernibles' with 'Leibniz-discernibles$^-$', or 'one-discernibles$^-$', or anything in between (see §15.1 and Theorem 15.7). But in fact, plugging in any grade of discernibility will yield (extensionally) the same criterion. For if either $U = M$ or $U = \varnothing$, then no elements in $\mathcal{M}$ are pairwise discerned by any of our grades of discernibility from Chapter 15 (except for identity itself). However, if U is non-empty but non-universal, then there will be elements a and b such that a and b are one-discernibles$^-$ in $\mathcal{M}$; and equally a and b will be Leibniz-discernibles$^-$ in $\mathcal{M}$.

In short: *every* grade of discernibility that we considered in Chapter 15 yields a notion of non-discrimination which coincides with Tarski's Thesis when it comes to the logicality of a subclass of a domain.

Now, at the risk of repetition: unlike Tarski, and like Sher, our ultimate interest is with the logicality of *expressions,* rather than classes. However, as explained above, the clearest way to determine whether an *expression* is discriminatory is to consider whether any of its possible *interpretations* is discriminatory. So, the (schematic) suggestion, that the discriminatoriness of classes should be thought of in terms of indiscernibles, generates a scheme of rivals to Tarski–Sher:

s is logical iff for all $\mathcal{M}$ there are no indiscernibles in $(M, s^{\mathcal{M}})$.

And, for any grade of indiscernibility, these rivals agree with Tarski–Sher, provided $s^{\mathcal{M}} \subseteq M$ for all $\mathcal{M}$.

At the risk of further repetition, we do not want to to *endorse* any of these rivals to Tarski–Sher. Our point is just that considering one-place *predicates* does not lend

[15] Compare the notation from Lemma 9.2 of §9.A.

much support to Tarski–Sher. If we want to consider a *test case* for Tarski–Sher, then we must consider more complicated expressions.

One-place quantifiers and pentagons

The next simplest expressions for us to consider are interpretations of one-place *quantifiers*. In the framework of generalised quantifiers outlined in §16.1, these are subsets of $\wp(M)$. But at this level of complexity, we think that no one should have very firm thoughts about what, intuitively, should count as (non-)discriminatory.

We show this by considering some toy examples. All of the toy examples concern quantifiers on the domain $\mathbb{F} = \{1, 2, 3, 4, 5\}$. To specify the examples, we use a simple, graphical notation, depicting a subset $X \subseteq \wp(\mathbb{F})$ with a graph such that a has an edge to b in our graph iff $\{a, b\} \in X$. So, for example, the graph:

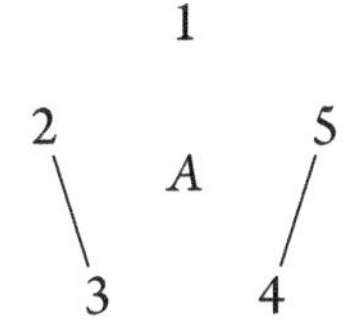

depicts the class $A = \{\{2, 3\}, \{4, 5\}\} \subseteq \wp(\mathbb{F})$.

Intuitively, A is a rather *discriminatory* class. Indeed, A singles out 1, in that, *uniquely*, if $1 \in X \subseteq \mathbb{F}$, then $X \notin A$. Moreover, A is obviously not permutation-invariant. To see this, consider the graph $(\mathbb{F}, A)$—i.e. the graph with domain $\mathbb{F}$ and edges given by A, just as depicted—and observe that when $h : \mathbb{F} \longrightarrow \mathbb{F}$ is a permutation, if $h(1) = 2$ then h is not a symmetry on $(\mathbb{F}, A)$. So, A is a discriminatory class, in Tarski's sense. And, since A is not permutation-invariant, no quantifier which takes A as its interpretation (in some model with domain $\mathbb{F}$) will count as logical according to Tarski–Sher. This all seems fairly plausible.

Again, though, this tells us very little. After all, if $1 \neq x$, then 1 and x are one-discernibles$^-$ in the graph $(\mathbb{F}, A)$. So, 1 is uniquely discriminated within that graph, using even the very weakest of our grades of discernibility from Chapter 15. Otherwise put, and as in the case of subsets of the domain: every plausible notion of discrimination will deliver the same verdict about the class A, and so about expressions which can take A as their interpretation.

So here is a trickier case:

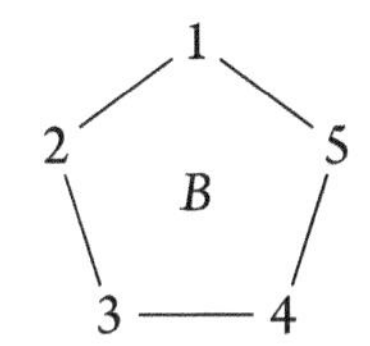

This class, B, is not permutation-invariant. And one way to see this is to consider a permutation $h : \mathbb{F} \longrightarrow \mathbb{F}$ such that $h(1) = 2$ but $h(3) = 3$. This is not a symmetry on the graph $(\mathbb{F}, B)$. More generally, invoking our grades of discernibility from Chapter 15, we have the following: in this graph, if $a \neq b$, then a and b are not Leibniz-relatives (see Definition 15.6). So there is a good sense in which B is discriminatory: distinct entities need not even by Leibniz-relatives.

Equally, though, there are several senses in which B is *non*-discriminatory. Put naïvely: B arranges the points of $\mathbb{F}$ as a *pentagon*, and hence treats every point equally. (Compare: King Arthur designed his Round Table to treat all the knights equally.) Indeed, there are *nine* non-trivial symmetries on the graph $(\mathbb{F}, B)$; and a and b are two-symmetricals in $(\mathbb{F}, B)$ for any $a, b \in \mathbb{F}$ (see Definition 15.4). So there are good senses in which B is *non*-discriminatory.

The real question is: is B discriminatory in a way that matters to *logicality*? Defenders of Tarski–Sher will have to insist that it is. But they cannot do this, just by appealing to '*the* intuitive notion of non-discrimination'. There are simply too *many* intuitive notions of non-discrimination floating around.

To this end, we will now consider three replies that Sher might offer, and explain why they are all inadequate.

'Any discrimination is bad'

Recall that the elements of the graph $(\mathbb{F}, B)$ are not Leibniz-relatives. Consequently, one might say that B is *too* discriminatory to count as logical, precisely because there is *some* grade of discernibility according to which B allows us to discriminate pairwise between the elements of $\mathbb{F}$.

The idea might be correct; but a fan of Tarski–Sher cannot believe it. To see why, consider a third example:

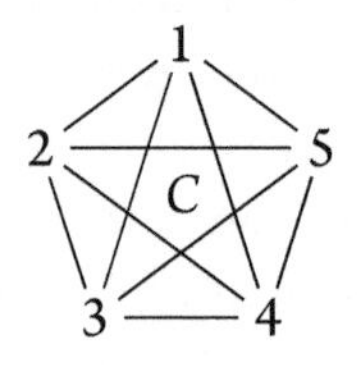

Unsurprisingly, C is permutation-invariant. So according to Tarski's Thesis, C is a logical class. Moreover, Tarski–Sher allows quantifiers to count as logical which have C as their interpretation. For example, the numerically-definite quantifier 'there are exactly two...', i.e. $\exists_{=2}$ with $\exists^{\mathcal{M}}_{=2} = \{X \subseteq M : |X| = 2\}$, counts as logical according to Tarski–Sher, and if $\mathcal{M}$'s domain is $\mathbb{F}$ then $\exists^{\mathcal{M}}_{=2} = C$. Moreover, C certainly treats the points of $\mathbb{F}$ at least as 'equally' as does B, for $(\mathbb{F}, C)$ is the *complete* graph on $\mathbb{F}$.

But even C is *somewhat* discriminatory. In the (multi)graph $(\mathbb{F}, C)$, the node 1 has an edge to 2, but not to *itself*. Otherwise put, $\{1, 2\} \in C$ but $\{1, 1\} \notin C$. So: if $a \neq b$, then a and b are Leibniz-discernibles$^-$ in our (multi)graph. Otherwise put: there is *a* sense in which C is discriminatory, *in spite* of its acceptability to both Tarski and Sher.

'Which pentagon are we considering?'

A second line of thought, from a fan of Tarski–Sher, would be that B somehow *presupposes* access to the 'identities of particular individuals' in the domain of $\mathbb{F}$ in a way which renders it discriminatory.[16] To bring out the idea, consider this subset of $\mathbb{F}$:

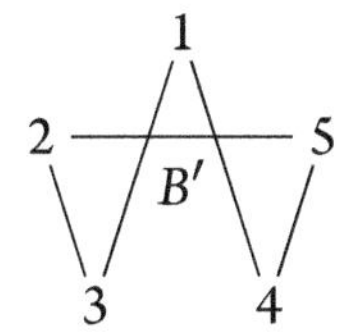

Squint for a moment, and you will see that B' is also a pentagon, just like B.[17] So, considered as a graph, $(\mathbb{F}, B)$ is isomorphic to but distinct from $(\mathbb{F}, B')$. And this raises an interesting question, familiar from §15.2: How can we tell which graph we are considering? Equally: How can we tell whether we are considering B or B'? Of course, if the elements of the domain, $\mathbb{F}$, are given to us *as* the numbers $1, 2, 3, 4, 5$, as it were, then there will be no difficulty. Since we know which number is which, we will know which class is which. But if we lack access to 'the identities of the numbers', as it were, then the isomorphism between B and B' suggests that we will be unable to single out either one of them (see §15.2). And this might suggest that the very *distinctness* of B from B' somehow presupposes, or provides access to, 'the identities of the numbers'. Finally, we might observe that nothing similar holds for C, since in that case there are no distinct, isomorphic structures with the same domain. So: perhaps *this* is why B but not C presupposes 'the identities of the numbers'.

Unfortunately, this line of thought merely rephrases our earlier considerations in more opaque language. The imagery of 'the identities of the numbers' is hard to get

[16] This is close to the phraseology which Sher uses to explicate the relevant notion of non-discrimination; see the quotes in §16.5. Sher (1991: 43) also explains that, by 'the identity of particular individuals', she 'mean[s] the features that make an object what it is, the properties that single it out'. We do not mean to put this argument in Sher's mouth; we simply want to show that these ideas lend no real support to Tarski–Sher.

[17] If precision is needed concerning the relevant notion of a pentagon, here it is. Suppose $|M| = n > 2$; then we can say that any $X \subseteq \wp(M)$ is an n-gon provided: (i) $|X| = n$; (ii) every member of X has exactly two members; (iii) every member of M occurs in exactly two members of X.

to grips with—hence our need for the repeated phrase 'as it were'—but we can try to engage with it via of the discussion of 'singling out' from §15.2. So: suppose that Dappy is (somehow) systematically confused about the 'identities of the numbers'. Dappy consistently mistakes 1 for (what he calls) '2', mistakes 2 for '3', 3 for '4', 4 for '5' and 5 for '1'. Despite Dappy's ignorance of the 'identities of the numbers', his attempt to describe B would be no less successful than someone who was not confused about the 'identities of numbers'. But of course, considering Dappy's particular confusion is just another way to spell out a particular symmetry on $(\mathbb{F}, B)$. Indeed: since there are nine non-trivial symmetries on $(\mathbb{F}, B)$, there are exactly nine non-trivial ways in which Dappy might be confused about the 'identities of the numbers', and yet still succeed in spelling out B rather than B'. In sum, talking about the 'identities' of the numbers provides no further grip on why bijection-invariance should be the *important* notion of discrimination.

'Special features'

So far, we have no clear reason for thinking that B is an *unacceptably* discriminatory class; that is, why it is not a (possible) extension of a logical quantifier. At this stage, a defender of Tarski–Sher might agree that, regarded just as a *class*, there is nothing too objectionable about it. However, she might insist that the test we are using to determine whether an expression is discriminatory—just check whether any of its possible interpretations is discriminatory—is simply too crude.

Here is the how the thought would run. Suppose that some expression Q takes the interpretation B on a structure $\mathcal{F}$ whose domain is $\mathbb{F}$, i.e. that $Q^{\mathcal{F}} = B$. Presumably, Q could also be interpreted as a different pentagon, $Q^{\mathcal{G}}$, on some other structure $\mathcal{G}$, whose domain was some five-element set $\mathbb{G} \neq \mathbb{F}$. Then, trivially, there is a bijection $h : \mathbb{F} \longrightarrow \mathbb{G}$ with $\bar{h}(Q^{\mathcal{F}}) \neq Q^{\mathcal{G}}$. So Q is not bijection-invariant, and hence it is not logical according to Tarski–Sher. But, according to the present line of thought, this is exactly the *right* verdict. For, according to this line of thought, the fact that Q is disrupted by a bijection indicates that it invokes 'some special feature shared by the members of the first domain', i.e. $\mathbb{F}$, 'and lacked by the members of the second domain', i.e. $\mathbb{G}$.[18]

[18] These phrases are quotes from McGee, and they have inspired this argument. However, the quotes are almost entirely out of context, and we owe it to McGee to set them *in* context.

McGee (1996: 567) thinks it is 'clear enough' that permutation-invariance is a necessary condition on the logicality of classes. He then aims to lift this necessary constraint, on logical classes, over to the constraint that logical expressions must be bijection-invariant. To do this, he writes (1996: 576): 'A property which, while invariant under all permutation[s] of a given domain, is disrupted when we move, via a bijection, to a different domain must depend on some special feature shared by the members of the first domain and lacked by the members of the second domain. It is not the sort of purely structural properties that pure logic studies.'

We have been challenging the first step in McGee's reasoning: the case of B highlights that it is *not* 'clear' that permutation-invariance is a necessary condition on the logicality of classes. So we can agree

The problem with this line of thought is entirely at the last stage. If 'special feature' is simply stipulated to mean 'not-bijection-invariant', then the argument is obviously circular. But if it is not stipulated to have this meaning, then we just do not understand how the argument is supposed to work. Suppose we think—consistently with the above—that any pentagon on a five-element domain is sufficiently non-discriminatory to count as a logical class. Then, for any bijection $h : \mathbb{F} \longrightarrow \mathbb{G}$, the elements of $\mathbb{G}$ will be arranged as a pentagon by $\hbar(Q^{\mathcal{F}}) = \hbar(B)$. In which case, a bijection will always take us from a logical class to a logical class, and never to anything which looks very 'special'.

16.6 The Principle of Closure

We have shown that the Principle of Non-Discrimination does not suffice to establish Tarski–Sher. So, if Tarski–Sher is to be defended, we must look for some new ideas concerning what logic is, beyond the idea that it is non-discriminatory. In this section, we introduce a second idea: that logical expressions should be closed under certain minimal operations.

The idea of closure

Consider the quantifier expressions 'there at least three...' and 'there are no more than fifteen...'. What we might call the *conjunction* of these quantifier expressions is also a quantifier expression, i.e. 'there are at least three and no more than fifteen...'. Moreover, all of these are bijection invariant; and indeed we have that $\exists^{\mathcal{M}}_{\geq 3} \cap \exists^{\mathcal{M}}_{\leq 15} = \exists^{\mathcal{M}}_{3 \leq n \leq 15}$. This suggests a simple thought: the conjunction of two logical expressions should itself be a logical expression. And the natural generalisation of this thought is as follows:

Principle of Closure. *The logical expressions should be closed under combination.*

We will subject this Principle to scrutiny in what follows. But first, we should show how it suffices to rule out some of the cases which we have been finding tricky.

Suppose we decided to treat B and B' from the previous section as logical classes. Then, by the Principle of Closure, $B \cap B' = A$ should *also* be a logical class. But A was precisely our initial example of a class which is *too* discriminatory to count as logical. So, by appeal to the Principles of Closure and Non-Discrimination *together*, we cannot treat both B and B' as logical. But if we said that B is logical and B' is not (or vice versa), then we will again violate the Principle of Non-Discrimination, since

---

with McGee's (1996: 576) essentially conditional claim, that *if* a property is permutation-invariant but disrupted by a bijection, *then* it depends 'on some special feature shared by the members of the first domain'. However, the antecedent fails in the case of B.

B and B' differ *only* with respect to their treatment of particular individuals. So, we finally get to say what fans of Tarski–Sher must say: neither B nor B' is logical.

The Principle of Closure deals well with some of the problem cases from the last section. Moreover, the Principle of Closure seems compatible with both Tarski's Thesis and Tarski–Sher. For taking the negation, conjunction, or disjunction of bijection-invariant expressions always results in a bijection-invariant expression. The Principle of Closure therefore seems like a useful weapon for Tarski and Sher.

That said, any defence of Tarski–Sher via the Principle of Closure will take us away from Sher's own thinking. In particular, the Principle of Closure seems not to follow from Sher's idea that logic is non-discriminatory. At best, perhaps, the Principle of Non-Discrimination and the Principle of Closure have a common source, perhaps in the idea of logic as *formal*.

Whatever the source of either principle, though, the obvious question now is whether the Principles of Non-Discrimination and Closure *together* yield Tarski–Sher. And here, again, our answer will be *No*. For there are accounts of logicality which obey both Principles but which are 'weaker' than Tarski–Sher.

Bonnay's framework for generalised invariance

To explain this, we must first expand our framework. Here we follow Bonnay, who shows how to generate potential candidates for logicality of quantifiers from various invariance relations.[19]

Let $\sim$ be any relation which may hold between structures. To begin with an easy case, let Q be a quantifier whose interpretation on a structure $\mathcal{M}$ is $\mathrm{Q}^{\mathcal{M}} \subseteq \wp(M)$. Using the same notation for (M, U) as in §16.5, we say:

Q *is* $\sim$*-invariant iff: if* $(M, U) \sim (N, V)$*, then* $U \in \mathrm{Q}^{\mathcal{M}}$ *iff* $V \in \mathrm{Q}^{\mathcal{N}}$

The more general case is for a quantifier whose interpretation is $\mathrm{Q}^{\mathcal{M}} \subseteq \wp(M^{i_1}) \times \ldots \wp(M^{i_n})$. In that case, writing $(M, \overline{U})$ for $(M, U_1, \ldots, U_n)$, we say:

Q *is* $\sim$-invariant *iff: if* $(M, \overline{U}) \sim (N, \overline{V})$*, then* $\overline{U} \in \mathrm{Q}^{\mathcal{M}}$ *iff* $\overline{V} \in \mathrm{Q}^{\mathcal{N}}$

The following Lemma both illustrates the framework, and also shows that we can use this framework to (re)formulate Tarski–Sher in terms of *isomorphism-invariance*:

Lemma 16.1: Q *is isomorphism-invariant (in the sense just described) iff* Q *is bijection-invariant (in the sense of §16.3).*

[19] Bonnay (2008).

Proof. Left-to-right. Let Q be isomorphism-invariant. For any bijection $h : M \longrightarrow N$, we have $(M, \overline{U}) \cong (N, \hbar(\overline{U}))$ by the Push-Through Construction, so that $\overline{U} \in \mathrm{Q}^{\mathcal{M}}$ iff $\hbar(\overline{U}) \in \mathrm{Q}^{\mathcal{N}}$. Hence $\hbar(\mathrm{Q}^{\mathcal{M}}) = \mathrm{Q}^{\mathcal{N}}$.

Right-to-left. Let Q be bijection-invariant, in the sense of §16.3. Let h be an isomorphism witnessing $(M, \overline{U}) \cong (N, \overline{V})$. So $\hbar(U_i) = V_i$ for each $1 \leq i \leq n$ and so $\hbar(\overline{U}) = \overline{V}$. Now $\overline{U} \in \mathrm{Q}^{\mathcal{M}}$ iff $\overline{V} = \hbar(\overline{U}) \in \hbar(\mathrm{Q}^{\mathcal{M}}) = \mathrm{Q}^{\mathcal{N}}$, since Q is bijection-invariant. □

In what follows, we will plug different relations between structures into the scheme. This will yield a variety of notions of invariance, and hence a variety of (potential) criteria for logicality. But we will see that several of these meet reasonable versions of the Principle of Closure.

Families of invariance relations and definability

To show this, though, we must make the Principle of Closure more precise. Our first attempt at doing so will be to consider the idea that the logical expressions should be closed under *definability*. The guiding thought was expressed by McGee: 'intuitively, anything definable from logical connectives is again a logical connective.'[20] This was repeated by Bonnay: 'Operators which are definable in a purely logical manner are logical.'[21] However, we need to know what *definability* amounts to in this context.

To explain the idea, we start with a simple example. The interpretation of the quantifier $\exists_{=2}$ is given by $\exists_{=2}^{\mathcal{M}} = \{X \subseteq M : |X| = 2\}$. And it makes good sense to say that $\exists_{=2}$ is definable in first-order logic, thanks to the obvious logical equivalence between $\exists_{=2} x F(x)$ and

$$\exists x_1 \exists x_2 (x_1 \neq x_2 \wedge F(x_1) \wedge F(x_2) \wedge \forall y (F(y) \leftrightarrow (y = x_1 \vee y = x_2)))$$

Now, $\exists_{=2}$ was an easy example, because it operates on a single, one-place predicate. But we can easily generalise the idea. Let Q be a quantifier which can operate on an α-length sequence of predicates $R_1, \ldots, R_\beta, \ldots$. Where L is a logic, we say that the $L(\overline{R})$-formulas are exactly those formulas whose syntax and semantics is given by the logic L, in the signature whose predicates are just among $\overline{R}$. Finally, we say that Q is *L-definable* iff (for suitable $\overline{R}$) there is some $L(\overline{R})$-formula $\chi(\overline{R})$ such that $\overline{U} \in \mathrm{Q}^{\mathcal{M}}$ iff $(M, \overline{U}) \vDash \chi(\overline{U})$ for any structure $\mathcal{M}$.

We will now survey some logics which satisfy the Principle of Closure. Recall from §15.4 that $L^{+}_{\kappa\lambda}$ is the logic which, intuitively, allows for κ-sized conjunctions

[20] McGee (1996: 571).

[21] Bonnay (2008: 50).

and λ-sized quantification, and which has '=' as a primitive, and $L^{-}_{\kappa\lambda}$ is the identity-free version of that logic. In particular, $L^{+}_{\omega\omega}$ is effectively standard first-order logic, and $L^{-}_{\omega\omega}$ is effectively first-order logic without identity.

For both of these finitary logics, there is a standard, crude, measure of complexity, known as quantifier rank:[22]

$$\begin{aligned} \mathrm{qr}(\varphi) &= 0 \text{ if } \varphi \text{ is atomic} \\ \mathrm{qr}(\neg\varphi) &= \mathrm{qr}(\varphi) \\ \mathrm{qr}(\bigwedge \Phi) &= \sup\{\mathrm{qr}(\varphi) : \varphi \in \Phi\} \\ \mathrm{qr}(\exists x\varphi) &= \mathrm{qr}(\varphi) + 1 \end{aligned}$$

We write $\mathcal{M} \equiv^{+}_{n} \mathcal{N}$ to indicate that $\mathcal{M} \models \varphi$ iff $\mathcal{N} \models \varphi$ for every $L^{+}_{\omega\omega}$-sentence φ with $\mathrm{qr}(\varphi) \leq n$; and similarly we write $\mathcal{M} \equiv^{-}_{n} \mathcal{N}$ to indicate that $\mathcal{M} \models \varphi$ iff $\mathcal{N} \models \varphi$ for every $L^{-}_{\omega\omega}$-sentence φ with $\mathrm{qr}(\varphi) \leq n$. And now, following Feferman, we can now characterise first-order logic—with or without identity—in terms of invariance:[23]

Theorem 16.2: *Let Q operate on a finite sequence of predicates:*

(1) *Q is $L^{+}_{\omega\omega}$-definable iff Q is $\equiv^{+}_{n}$-invariant for some natural number n.*
(2) *Q is $L^{-}_{\omega\omega}$-definable iff Q is $\equiv^{-}_{n}$-invariant for some natural number n.*

Proof. Throughout, let $\overline{R}$ be the finite sequence of predicates on which Q operates. We will prove (1), but (2) is exactly similar.

Left-to-right. Let $\chi(\overline{R})$ define Q, i.e. $\overline{U} \in \mathrm{Q}^{\mathcal{M}}$ iff $(M, \overline{U}) \models \chi(\overline{U})$ for any structure $\mathcal{M}$. Let $n = \mathrm{qr}(\chi(\overline{R}))$. Now if $(M, \overline{U}) \equiv^{+}_{n} (N, \overline{V})$, then $(M, \overline{U}) \models \chi(\overline{U})$ iff $(N, \overline{V}) \models \chi(\overline{V})$, and hence $\overline{U} \in \mathrm{Q}^{\mathcal{M}}$ iff $\overline{V} \in \mathrm{Q}^{\mathcal{N}}$; so Q is $\equiv^{+}_{n}$-invariant.

Right-to-left. Fix n, and assume that Q is $\equiv^{+}_{n}$-invariant. Up to logical equivalence, there are only finitely many $L^{+}_{\omega\omega}(\overline{R})$-formulas with quantifier-rank $\leq n$. So take an enumeration of them $\varphi_1, \ldots, \varphi_m$ and define, for each $\mathcal{M}$ and $\overline{U}$:

$$\chi^{M}_{\overline{U}} = \bigwedge\{\varphi_i : (M, \overline{U}) \models \varphi_i(\overline{U}), \text{and } 1 \leq i \leq m\}$$

Now consider the set of $L^{+}_{\omega\omega}(\overline{R})$-formulas:

$$\Phi = \{\chi^{M}_{\overline{U}} : \overline{U} \in \mathrm{Q}^{\mathcal{M}}, \text{for some } \mathcal{M} \text{ and some } \overline{U}\}$$

It is easy to see that $|\Phi| \leq 2^m$, so that $\bigvee \Phi$ is an $L^{+}_{\omega\omega}(\overline{R})$-formula. We claim that $\bigvee \Phi$ defines Q. If $\overline{U} \in \mathrm{Q}^{\mathcal{M}}$, then $\chi^{M}_{\overline{U}} \in \Phi$ and so $(M, \overline{U}) \models \bigvee \Phi(\overline{U})$. Conversely, if $(M, \overline{U}) \models \bigvee \Phi(\overline{U})$, then $(M, \overline{U}) \models \chi^{N}_{\overline{V}}$ for some $\chi^{N}_{\overline{V}} \in \Phi$, i.e. for some $\mathcal{N}$ and $\overline{V}$ with

[22] In the third clause, the expression on the right-hand side denotes the supremum of a set of ordinals, that is, the least upper bound of the set of ordinals.

[23] Feferman (2010: Theorem 4.4). As Feferman notes, the proof employs a technique due to Fraïssé (1954b).

$\overline{V} \in Q^{\mathcal{N}}$. Since $\chi^{N}_{\overline{U}}$ pins down the truth or falsity of every formula with quantifier-rank $\leq n$, we have $(M, \overline{U}) \equiv^{+}_{n} (N, \overline{V})$. So since Q is $\equiv^{+}_{n}$-invariant, we have that $\overline{U} \in Q^{\mathcal{M}}$. □

The key to the proof is that every $L^{+}_{\omega\omega}(\overline{R})$-formula has some finite quantifier-rank, and that the number of $L^{+}_{\omega\omega}(\overline{R})$-formulas with a bounded quantifier-rank is well behaved. This basic thought can now be carried over from finitary to infinitary logics. For any $L^{+}_{\infty\infty}$-formula φ, there is some least κ such that φ is an $L^{+}_{\kappa\kappa}$-formula; and, for each κ and fixed $\overline{R}$, there is some cardinal μ such that there are exactly μ-many $L^{+}_{\kappa\kappa}(\overline{R})$-sentences up to logical equivalence.[24] Similarly, for any $L^{+}_{\infty\omega}(\overline{R})$-sentence φ, there is some (smallest) κ such that φ is an $L^{+}_{\kappa\omega}(\overline{R})$-sentence; and, for each κ, there is some cardinal μ such that there are exactly μ-many $L^{+}_{\kappa\omega}(\overline{R})$-sentences up to logical equivalence. Where $\mathcal{M}$ and $\mathcal{N}$ are $\overline{R}$-structures we now write $\mathcal{M} \equiv^{+}_{\kappa\lambda} \mathcal{N}$ to indicate: $\mathcal{M} \vDash \varphi$ iff $\mathcal{N} \vDash \varphi$ for all $L^{+}_{\kappa\lambda}(\overline{R})$-sentences φ, and similarly for $\mathcal{M} \equiv^{-}_{\kappa\lambda} \mathcal{N}$. Then a tiny tweak to Theorem 16.2 yields:

Theorem 16.3: *Let Q be any quantifier:*

(1) *Q is $L^{+}_{\infty\infty}$-definable iff Q is $\equiv^{+}_{\kappa\kappa}$-invariant for some cardinal κ*

(2) *Q is $L^{-}_{\infty\infty}$-definable iff Q is $\equiv^{-}_{\kappa\kappa}$-invariant for some cardinal κ*

Moreover, for any cardinal $\lambda \geq \omega$:

(3) *Q is $L^{+}_{\infty\lambda}$-definable iff Q is $\equiv^{+}_{\kappa\lambda}$-invariant for some cardinal $\kappa \geq \lambda$.*

(4) *Q is $L^{-}_{\infty\lambda}$-definable iff Q is $\equiv^{-}_{\kappa\lambda}$-invariant for some cardinal $\kappa \geq \lambda$*

Proof sketch. We will sketch case (3), leaving the rest to the reader.

Left-to-right. If Q is $L^{+}_{\infty\lambda}$-definable, then it is defined by some $L^{+}_{\kappa\lambda}(\overline{R})$-formula $\chi(\overline{R})$ with $\kappa \geq \lambda$, i.e. $\overline{U} \in Q^{\mathcal{M}}$ iff $(M, \overline{U}) \vDash \chi(\overline{U})$ for any structure $\mathcal{M}$. Now if $(M, \overline{U}) \equiv^{+}_{\kappa\lambda} (N, \overline{U})$ then $\overline{U} \in Q^{\mathcal{M}}$ iff $\overline{V} \in Q^{\mathcal{N}}$, as in Theorem 16.2.

Right-to-left. Assume Q is $\equiv^{+}_{\kappa\lambda}$-invariant. Let $\varphi_1, \ldots, \varphi_a, \ldots$ exhaustively list the $L^{+}_{\kappa\lambda}(\overline{R})$-formulas, and define:

$$\chi^{M}_{\overline{U}} = \bigwedge\{\varphi_a : (M, \overline{U}) \vDash \varphi_a(\overline{U})\}$$

Now consider the set of $L^{+}_{\kappa\lambda}(\overline{R})$ formulas:

$$\Phi = \{\chi^{M}_{\overline{U}} : \overline{U} \in Q^{\mathcal{M}}, \text{ for some } \mathcal{M} \text{ and some } \overline{U}\}$$

Since this is a set, $\bigvee \Phi$ is an $L^{+}_{\infty\lambda}(\overline{R})$-formula. And it defines Q, exactly as in Theorem 16.2. □

There are, then, *proper-class-many* logics L which come equipped with a family of relations, such that a quantifier is L-definable iff it is invariant under some relation

[24] This holds since the $L^{+}_{\kappa\kappa}(\overline{R})$-sentences form a set (Dickmann 1975: 64–5).

in this family. Abusing notation slightly, these are the proper-class-many logics $L_{\kappa\lambda}$, with or without identity, for $\kappa = \infty$ or $\kappa = \omega$, and $\kappa \geq \lambda \geq \omega$.

The philosophical upshot is that we have proper-class many logics which can plausibly claim to satisfy both the Principle of Closure and the Principle of Non-Discrimination. After all, Theorems 16.2–16.3 *precisely* guarantee that our logics satisfy Closure. And as for Non-Discrimination: $\equiv^+_n$-invariance is increasingly non-discriminatory as n increases; and $\equiv^-_{\kappa\lambda}$ is increasingly non-discriminatory as κ increases; so it seems quite reasonable to think that the *limit* of such an increasing process is 'appropriately' non-discriminatory.

It is not that we want to endorse any one of these proper-class-many logics as *the* correct logic. Our point is simply this: if we want to draw precise bounds to logic, then we must invoke further intuitions.

Single invariance relations and weak-definability

It is worth emphasising that our Theorems 16.2–16.3 invoke *families* of similarity relations. Crucially, we characterised $L^+_{\omega\omega}$ using countably many relations, $\equiv^+_n$ for each natural number n. Worse, we characterised $L^+_{\infty\infty}$ using proper-class many relations, $\equiv^+_{\kappa\kappa}$ for each infinite cardinal κ. Significantly, we have not found a *single* relation, $\sim$, with the following property: Q is $\sim$-invariant iff Q is definable using a formula whose only quantifiers are themselves $\sim$-invariant.

We call a relation which satisfies this property a *fixed-point-invariant* relation. Finding such a relation would certainly be mathematically elegant. However, it is unclear that finding a fixed-point-invariant relation should make us believe that we have found the (once and for all) bounds of logic. Even if we accept both the Principle of Non-Discrimination and the Principle of Closure, there is no *obvious* reason to think that either Principle should force us to search for just *one* relation, rather than for a family of them (as in Theorems 16.2–16.3).[25]

More immediately, though, we know of no interesting fixed-point-invariant relations, and it is not obvious (to us) that any exist. But we should pause here.

Earlier, we said that Q is L-definable iff there is some $L(\overline{R})$-formula $\chi(\overline{R})$ such that $\overline{U} \in Q^{\mathcal{M}}$ iff $\mathcal{M} \vDash \chi(\overline{U})$ for any structure $\mathcal{M}$. That notion is quite demanding, so here is a weaker notion. Say that Q is *weakly-L-definable* iff: for any cardinal κ, there is some $L(\overline{R})$-formula $\chi_\kappa(\overline{R})$ such that $\overline{U} \in Q^{\mathcal{M}}$ iff $\mathcal{M} \vDash \chi_\kappa(\overline{U})$ for any structure $\mathcal{M}$ of cardinality κ.

The following Theorem shows that weak-definability *does* align with certain single invariance-relations. (The first three results are due, respectively, to McGee, Bonnay and Engström, and Barwise;[26] the definitions of potential-isomorphism

[25] This is essentially Feferman's (2010: 13) point.

[26] McGee (1996), Bonnay and Engström (2013: Theorem 11.1), and Barwise (1973). Barwise's result is cited in this connection by Bonnay (2008: 62).

and partial-relativity are relegated to a footnote):[27]

Theorem 16.4: *Let Q be any quantifier:*

(1) *Q is weakly-$L^{+}_{\infty\infty}$-definable iff Q is isomorphism-invariant*
(2) *Q is weakly-$L^{-}_{\infty\infty}$-definable iff Q is relativeness-correspondence-invariant*
(3) *Q is weakly-$L^{+}_{\infty\omega}$-definable iff Q is potential-isomorphism-invariant*
(4) *Q is weakly-$L^{-}_{\infty\omega}$-definable iff Q is partial-relativity-invariant*

Proof sketches. In all four cases, the difficult direction is left-to-right, and the proof strategy is the same. First, given our interest in *weak*-definability, we can fix some cardinal, κ, and let $(M, \overline{U})$ have $|M| = \kappa$. We then find a sentence $\sigma^{M}_{\overline{U}}$ such that any structure satisfying $\sigma^{M}_{\overline{U}}$ is isomorphic (respectively: relative, potentially-isomorphic, or partially-relative) to $(M, \overline{U})$. We then follow the strategy of Theorems 16.2–16.3, replacing $\chi^{M}_{\overline{U}}$ with $\sigma^{M}_{\overline{U}}$. So it just remains to show how to obtain an appropriate $\sigma^{M}_{\overline{U}}$.

(1). By Lemma 15.9, there is an $L^{+}_{\infty\infty}$-sentence satisfied by precisely the structures which are isomorphic to $(M, \overline{U})$.

(2). Exactly as in case (1), but invoking Lemma 15.14.

(3). The Scott sentence of $(M, \overline{U})$ is an $L^{+}_{\infty\omega}$-sentence which is satisfied precisely by those structures which satisfy exactly the same $L^{+}_{\infty\omega}$-sentences as $(M, \overline{U})$.[28] By Karp's Theorem, these are precisely the structures which are potentially-isomorphic to $(M, \overline{U})$.[29]

(4). Let $\rho^{M}_{\overline{U}}$ be the Scott sentence of the quotiented structure $(\mathcal{M}, \overline{U})/\!\sim$, defined in Definition 15.11. Turn this it into an $L^{-}_{\infty\omega}$-sentence $\sigma^{M}_{\overline{U}}$, by replacing each instance of '=' with an $L^{-}_{\infty\omega}$-formula abbreviating Leibniz-indiscernibility^{-} (see the proof of Lemma 15.14). By Lemmas 15.12 and 15.14, $\sigma^{M}_{\overline{U}}$ is satisfied precisely by those structures which satisfy exactly the same $L^{-}_{\infty\omega}$-sentences as $(M, \overline{U})$. By a result of Casanovas et al., these are precisely the structures which are partial-relatives of $(M, \overline{U})$.[30] □

Now, by Lemma 15.9, structures are isomorphic iff they satisfy exactly the same $L^{+}_{\infty\infty}$-sentences. So, were it not for the restriction to *weak*-definability, case (1) of

[27] Let $\mathcal{M}$ and $\mathcal{N}$ be two structures. A function g is a *partial isomorphism* from $\mathcal{M}$ to $\mathcal{N}$ iff there are substructures $\mathcal{A} \subseteq \mathcal{M}$ and $\mathcal{B} \subseteq \mathcal{N}$ such that g is an isomorphism $\mathcal{A} \longrightarrow \mathcal{B}$. A *potential isomorphism* I between two structures $\mathcal{M}$ and $\mathcal{N}$ is a non-empty set of partial isomorphisms from $\mathcal{M}$ to $\mathcal{N}$ such that, for any $g \in I$:

- for any $a \in M$, there is an $h \in I$ with $g \subseteq h$ and $a \in \text{domain}(h)$; and
- for any $b \in N$, there is an $h \in I$ with $g \subseteq h$ and $b \in \text{range}(h)$

Partial relativity is then the obvious 'relativeness correspondence' version of potential isomorphism; see Casanovas et al. (1996: Definitions 4.1, 4.7).

[28] For details, see e.g. Marker (2002: Exercise 2.5.33(b)).

[29] Karp (1965) and Marker (2002: Exercise 2.5.34(c)).

[30] Casanovas et al. (1996: 521).

Theorem 16.4 would provide us with a fixed-point-invariant relation. Exactly similar points hold for cases (2)–(4).

Sadly, though, all four cases essentially depend upon the restriction to *weak*-definability. As Feferman observes, to overcome the restriction to weak-definability, we would have to 'disjoin' all of the formulas χ_κ which (weakly-)define Q for κ-sized structures. Since there are proper-class-many cardinals, the ensuing 'disjunction' would therefore have to be proper-class-sized. As such, it would have to go 'well beyond $L_{\infty\infty}$ as ordinarily conceived'.[31]

We agree with Feferman's concern, and would push the point further. *If* we can even make sense of a proper-class-sized *sentence*—i.e. a single entity consisting of proper-class-many symbols—then we should equally well be able to make sense of proper-class-sized *domains*. But now the notion of weak-definability should be strengthened to accommodate proper-class-sized structures. In particular, to weakly-L-define a quantifier, we should *also* require that there be an $L(\overline{R})$-formula $\chi_\Omega(\overline{R})$ such that $\overline{U} \in Q^{\mathcal{M}}$ iff $\mathcal{M} \vDash \chi_\Omega(\overline{U})$ for any proper-class-sized structure $\mathcal{M}$. And now the proof-strategy of Theorem 16.4 breaks down at an earlier stage. To illustrate the problem, we focus on case (1). Given a proper-class-sized structure, $\mathcal{M}$, we may be able to find a proper-class-sized sentence $\sigma^{\mathcal{M}}_{\overline{U}}$ which pins down proper-class-sized structures up to isomorphism. But to find the appropriate formula $\chi_\Omega(\overline{R})$, we would now need to disjoin all the combinatorially possible σs for all proper-class-sized structures (see the strategy of Theorems 16.2–16.3). And there may be two-to-the-power-proper-class-many such sentences. If that is even intelligible, it is strictly more than proper-class-many. Cantor's Theorem keeps chasing us; we will get nowhere near a fixed-point-invariant relation by this route.

Similar considerations show that Tarski–Sher counts *more* as logical than does $L_{\infty\infty}$. Intuitively, there are two-to-the-power-proper-class-many functions from the class of cardinals to the set $\{0, 1\}$. So, there are two-to-the-power-proper-class-many bijection-invariant quantifiers which behave like existential quantification on domains of *certain* cardinalities (those mapped to 0), and like universal quantification on domains of certain *other* cardinalities (those mapped to 1). But since there are only (!) proper-class-many $L^+_{\infty\infty}$-sentences, *most* bijection-invariant quantifiers are not $L^+_{\infty\infty}$-definable.

The situation, then, is as follows. There are *many* logics which satisfy certain versions of Non-Discrimination and Closure. These are the proper-class-many logics $L_{\kappa\lambda}$ with or without identity (for $\kappa = \infty$ or $\kappa = \omega$, and $\kappa \geq \lambda \geq \omega$); and the four logics we obtain by treating isomorphism-, relativeness-correspondence-, potential-isomorphism-, and partial-relativity-invariance as necessary and sufficient for logicality. The proper-class-many logics of the form $L_{\kappa\lambda}$ certainly *look* like logics, but they cannot obviously be characterised by a single relation. The last four can be

[31] Feferman (2010: 8).

characterised by a single relation, but do not obviously *look* like logics, in that they countenance notions more *logical quantifiers* than *sets*. Without invoking further intuitions about the nature of logic, it just is not clear how to choose any *one* of these as 'the bounds of logic'.

16.7 McGee's squeezing argument

At this point, we turn to McGee's argument for Tarski–Sher. This invokes both the Principles of Non-Discrimination and Closure, and might *hope* to trade on no further principles. We formulate McGee's Squeezing Argument as follows:[32]

(a) If Q is logical, then Q is bijection-invariant
(b) If Q is weakly-$L^{+}_{\infty\infty}$-definable, then Q is logical.
(c) Q is bijection-invariant iff Q is weakly-$L^{+}_{\infty\infty}$-definable
So: Q is logical iff Q is bijection-invariant.

This argument is distinctive, in that McGee only begins with the idea that Non-Discrimination is *necessary* for logicality, rather than both necessary and sufficient. Moreover, the argument is valid, and (c) is just part of Theorem 16.4. So we must focus on (a) and (c).

McGee's own attempt to establish (a) involves very little more than a few remarks to the effect that logic should be *non-discriminatory*. Now we saw in §16.5 that this will not work: (a) cannot be established just by appealing to the Principle of Non-Discrimination. However, we might be able to establish (a) by combining Non-Discrimination with the Principle of Closure. After all, we saw in §16.6 that these two Principles together entail that our pentagons B and B' should not be treated as logical. Moreover, although we saw in §16.6 that various different notions of invariance respect both Principles, all of these notions *entail* bijection-invariance. (For example, anything which is $\equiv^{+}_{n}$-invariant is bijection-invariant, just by Lemma 16.1 and the fact that if $\mathcal{M} \cong \mathcal{N}$ then $\mathcal{M} \equiv^{+}_{n} \mathcal{N}$.) So, if McGee is happy to invoke the Principle of Closure, then he may be able to secure (a).

In fact, the Principle of Closure is a key component of McGee's defence of premise (b). His defence comes in two parts:

(b1) 'the primitive connectives of $L^{+}_{\infty\infty}$ are all intuitively clearly logical connectives'
(b2) 'intuitively, anything definable from logical connectives is a logical connective'.[33]

Clearly (b2) is just a version of the Principle of Closure. However, (b1) is a separate idea, eminently contestable. We, indeed, do not find it 'intuitively clear' that the

[32] We call it this, since it has the same shape as Kreisel's (1967: 152–7) squeezing argument.

[33] Both quotes from McGee (1996: 571); we have inserted the numbering and the superscript '+'.

primitive connectives of $L^{+}_{\infty\infty}$ are 'logical'. After all, these are strongly infinitary operations.

But even if we waive this point, there is a further problem with McGee's argument. For reasons we saw §16.6, (b1) and (b2) are jointly *insufficient* to establish (b). Instead, we must replace (b2) with the stronger claim that anything *weakly*-definable from logical connectives is a logical connective. But the status of this last claim is even less obvious. Bonnay attempts to justify the Principle of Closure by stating that, if we have *defined* an operation using only logical vocabulary, then he simply could 'not see how a non-logical element could creep in [among] the logical elements of the definition and make the defined operator non logical.'[34] However, if we have only *weakly*-defined an operation, then perhaps a non-logical element creeps at exactly the moment when we effectively try to entertain a *class-length* sentence and thereby embrace the idea that there are more logical quantifiers than there are sets. Bluntly: it is not at all obvious that $L^{+}_{\infty\infty}$-*indefinable* quantifiers must count as logical.

As such, we must set aside McGee's argument for Tarski–Sher. And so we still have proper-class-many candidate logics to contend with.

16.8 Mathematical content

At this point, the most usual way to attempt to rule out some of these candidate logics is to maintain that some of them provide us with too much *mathematical* content to count as logic. For example, Feferman alleges that Tarski–Sher 'assimilates logic to mathematics, more specifically to set theory'.[35] For three reasons, we are not sure that this is a fair objection.

Policing boundaries

The key new idea here is that it is important to police some boundary between logic and mathematics. This idea does not seem to have anything to do with the idea that logic should be formal or necessary, or that logic should obey Non-Discrimination or Closure. It seems, to us, to be an entirely *additional* idea. As such, it is hardly surprising that Sher is totally comfortable in saying that '[t]he bounds of logic, on my view, are the bounds of mathematical reasoning', given that her aim was to explicate the idea of logic as formal and necessary.[36] It is equally unsurprising, though, that other authors—such as Feferman and Bonnay—feel strongly the other way.[37]

---

[34] Bonnay (2008: 50).
[35] Feferman (1999: 37, 2010: 8).
[36] Sher (1991: xii); cf. also Tarski (1986: §4).
[37] Feferman (1999: 37–9) and Bonnay (2008: 35–8).

Outsiders to this debate, though, might well wonder why it *matters* how we carve up the territory between logic and mathematics. Granted, there is a (slightly porous) *institutional* boundary, arising from the way in which mathematicians and logicians tend to label themselves and each other. But it is not obvious that we need to assign any deep *philosophical* significance to that boundary.

Non-logical assumptions

We should also take care in articulating the accusation that Tarski–Sher 'assimilates logic to mathematics'. To make the point, consider the simple claim that $2 + 2 = 4$. This has what we might call a 'logical analogue', namely this truth of logic:

$$[\exists_{=2}xF(x) \wedge \exists_{=2}xG(x) \wedge \neg\exists x(F(x) \wedge G(x))] \rightarrow \exists_{=4}x(F(x) \vee G(x))$$

But now suppose that—somehow—there were no structures with cardinality more than 4. Then the 'logical analogue' of the *false* arithmetical claim that $2 + 2 = 5$, i.e.:

$$[\exists_{=2}xF(x) \wedge \exists_{=2}xG(x) \wedge \neg\exists x(F(x) \wedge G(x))] \rightarrow \exists_{=5}x(F(x) \vee G(x))$$

would be true in all structures, and hence would be a truth of logic. Contraposing: to make it false, we need there to be sufficiently large finite structures. Moreover, since Tarski–Sher treats arbitrarily large cardinality quantifiers as logical, the 'logical analogues' of elementary facts from transfinite cardinal arithmetic require that there are *arbitrarily large* structures.

We do not want to *contest* the claim that there are arbitrarily large structures. Our aim is only to draw attention to its *status*. If the claim that there are arbitrarily large structures is not itself a truth of *logic*, then Tarski–Sher does *not* assimilate mathematics to logic. Rather, it assimilate mathematics to logic *plus the extra-logical claim* that there are arbitrarily large cardinals.[38]

The indeterminacy of logic?

The greatest resistance to assimilating mathematics to logic comes from combining the idea that some mathematical claims are genuinely indeterminate, with the idea that anything expressed using only logical language is (determinately) either a theorem or not. As Feferman notes, these two ideas can come into conflict. Indeed, Feferman thinks that the continuum hypothesis, CH, should be left indeterminate, but worries that 'we can express [CH...] as logically determinate' given Tarski–Sher.[39] We reconstruct Feferman's complaint as follows. Consider these sentences:

[38] Cf. the status of the Axiom of Infinity for Russell's logicism, as discussed by Potter (2000: 150–2).

[39] Feferman (1999: 38).

$$\mathrm{CH}^{\mathrm{S}} := \exists_{=\aleph_1} x F(x) \leftrightarrow \exists_{=2^{\aleph_0}} x F(x)$$
$$\mathrm{NCH}^{\mathrm{S}} := \neg(\exists_{=\aleph_1} x F(x) \wedge \exists_{=2^{\aleph_0}} x F(x))$$

According to Tarski–Sher, CH^{S} is a *logical* truth iff CH is true, and $\mathrm{NCH}^{\mathrm{S}}$ is a *logical* truth iff CH is false. So: it seems that logical facts will now determine the truth or falsity of the Continuum Hypothesis.[40]

There is, though, a lacuna in this argument. For suppose we think that CH is indeterminate. Then nothing, prima facie, would prevent us from thinking that it is *similarly indeterminate* whether CH^{S} or $\mathrm{NCH}^{\mathrm{S}}$ is a logical truth.[41]

This is consonant with Hamkins's reaction to second-order logic, as discussed in §8.4. Hamkins seems to have no problem with employing second-order logic *per se*; his complaint is against those who think that there is such a thing as *full* second-order logic, since he believes that there are many equally admissible interpretations of the powerset operator. Similarly here: someone who thinks that there are many equally good interpretations of set theory—some of which obey CH, and some of which do not—does not need to *reject* Tarski–Sher. Instead, she might simply insist that there is no *single* standard semantics for the (cardinality-)quantifiers, such as $\exists_{=\aleph_1}$ and $\exists_{=2^{\aleph_0}}$, even though (she insists) they count as *logical*.

In saying all of this, we are not intending to endorse it. We do not claim to have a completely firm grasp on what the 'indeterminacy' of logic would amount to. Equally, we do not claim to have a completely firm grasp on what it means to say, in the present context, that CH is indeterminate. We just want to suggest that these two ideas may well go hand in hand.

16.9 Explications and pluralism

In this chapter, we have focussed on the intuitive ideas that logic should satisfy some Principle of Non-Discrimination and some Principle of Closure. We have also considered whether there should be some intuitive separation between logic and mathematics. Such intuitive guides, we think, fail to establish *sharp* bounds for logic.

This is not, though, to say that Tarski–Sher is wrong. It is simply to say that, if Tarski–Sher is to be established, it must be established by other means. One way to do that would be to suggest that it is simply the best *explication* of logic. That is, it is the best purely formal notion which is both useful and a reasonable descendant of some pre-formal, intuitive notion of logic.

There is no doubt that Tarski–Sher has many merits. It is simple to state. It draws the bounds of logic precisely, and in a way which clearly satisfies the Principles of Non-Discrimination and of Closure. Granted, it leaves the boundary between

[40] Cf. Shapiro (1991: 105) on second-order logic.

[41] This possibility was also suggested by Sher in conversation.

mathematics and logic blurrier than expected; but, as discussed in §16.8, it is unclear how much this matters. If logic is to be explicated in semanticist terms at all (see §16.2), then Tarski–Sher is a great explication.

Nevertheless, it is unclear that any *unique* explication will be the best explication for all possible circumstances. In this chapter, we have argued that proper-class many logics satisfy Non-Discrimination and Closure. Any of these formal logics could serve as an explication of the intuitive notion of *logic*, depending on what one *wanted* from logic; and what one wants might vary from circumstance to circumstance. In certain circumstances, we might insist that the logic under discussion should be associated with a finitary deductive system, whereupon $L_{\omega\omega}$ suggests itself; in other circumstances, we might insist that every logical operation should be definable by some (infinitary) formula, in which case we will stop short of bijection-invariance and reach only for $L_{\infty\infty}$; and so on.

This naturally leads to a *pluralism* about the bounds of logic. But the pluralism in question is quite mundane. Consider the following inference pattern:

there are exactly $\aleph_{14}$ φs
there are exactly $\aleph_{14}$ ψs
So: there are exactly $\aleph_{14}$ things which are either φ or ψ

We can all agree that there is an *entailment* here. The scope for pluralism is really only over whether the entailment holds 'as a matter of *logic*' or 'for other reasons'. Indeed, the scope of pluralism is really very narrow; for certainly the entailment holds 'thanks to considerations which can be formulated in bijection-invariant terms'. And at this point, we may well wonder whether it is worth bothering to ask whether that entailment also merits the honorific, that it is *logical*.

17

Classification and uncountable categoricity

A cursory glance at contemporary model theory indicates an intense focus on *classification*. 'Classification' is literally the term which Shelah used to describe his work, concerning the number of models of a given complete theory in a given cardinality. And Zilber's ambitious Trichotomy Conjecture predicted that the phenomenon of *uncountable categoricity*—the situation where there is only one model of a given uncountable cardinality—can be classified into exactly three basic kinds. This chapter considers both Shelah's programme (§17.2) and uncountable categoricity (§17.3).

Classification is, though, a near-ubiquitous phenomena, not only in model theory, but in other areas of mathematics. Despite this, it has been largely neglected by philosophers of mathematics. So, we begin the chapter by presenting a general framework for conceiving of classification within mathematics. This will allow us to discuss the extent to which contemporary developments in model theory can be seen as instances of classificatory activity, so conceived.

17.1 The nature of classification

If one asks a mathematician what she is working on, often the answer will be that she is trying to prove or refute a given conjecture. But, equally as often, the answer will be that she is seeking to classify a certain kind of mathematical object.

Obviously these two activities—proving and classifying—are complementary in several ways. Most immediately, a successful classification programme will eventually result in the articulation and proof of a theorem. Equally, the resolution of a long-standing conjecture might be rendered feasible by means of the classification of some related phenomena. And proofs *themselves* are often classified in proof theory and complexity theory.

These connections between proving and classifying raise two basic questions.

The first basic question is this: *What are the means and aims of classification?* This question can be made vivid by thinking about how we answer the corresponding question about proof. We are likely to say that *one* characteristic aim of proof is the extension of knowledge, and this aim is effected via formal deductions from known axioms.[1] Even this brief answer suggests that understanding the activity of classification will involve two things: (a) identifying the aims of classification, and (b) identifying the various mechanisms by which this aim is typically achieved. Given what we have said about proof, we might anticipate that answers to (b) are accessible by inspection of mathematical texts, in a way in which answers to (a) need not be.

The second basic question is this: *Can classification be reduced to proof?* For example, one might wonder whether, for every classification programme, there is a specific theorem such that the classificatory programme is successful iff the theorem is successfully proven from accepted axioms.

The aim of this section is to suggest an answer to the first basic question, which then naturally recommends an answer to the second.

Paradigmatic examples and the general mechanism

We begin by setting down three paradigmatic examples of classification, from three different branches of mathematics: algebra, probability, and topology.

Algebra. In 1910, Steinitz classified the uncountable algebraically closed fields, such as the complex numbers. Steinitz's result says that two uncountable algebraically closed fields are isomorphic iff they have the same characteristic and the same cardinality.[2]

Topology. A second well-known example is the classification of compact connected surfaces.[3] This states: any compact connected surface is homeomorphic to the sphere, a connected sum of n-tori, or a connected sum of n-projective planes, and no two distinct surfaces on this list are homeomorphic to one another. This result can also be stated as follows: two compact connected surfaces are homeomorphic iff they have the same Euler characteristic and either both are orientable or both are non-orientable. Both of the invariants—Euler characteristic and orientability—can be represented as an integer.[4]

Probability. A final celebrated example of classification is Ornstein's classification of isomorphism of Bernoulli shifts. Suppose you have an n-sided die, with faces $1, \ldots, n$, and suppose you roll it once per minute, with no first roll, so that the sequences $\bar{x} = (\ldots, x_{-2}, x_{-1}, x_0, x_1, x_2, x_3, \ldots)$ correspond to individual histories of

---

[1] Obviously there are other aims too, like 'being explanatory'.

[2] Some fields are such that one can have $1 + \ldots + 1 = 0$. For these fields, the smallest number of times 1 may be summed with itself to produce 0 is called the *characteristic* of the field. Fields that do not exhibit this property are said to be of *characteristic zero*. Steinitz's result follows from considerations regarding so-called transcendence bases which feature in almost every introductory algebra textbook, e.g. Lang (2002: §VIII.1) and Hungerford (1980: §VI.1, esp. Theorem 1.12, p.317).

[3] For a discussion of the history, from Möbius onwards, see e.g. Gallier and Xu (2013: 151–7).

[4] This result is mentioned in many introductory topology texts, e.g. Kinsey (1993: 79, 107) and Lawson (2003: 120). For an accessible proof, see Munkres (2000: ch.12).

die-rolling. Let Ω denote the set of all such sequences. Since we assumed that there is no first roll, there is a natural operation of 'fast-forwarding' on this space Ω given by moving the ith entry in a sequence to the $i+1^{\text{th}}$ slot.[5] Where $p_1, \ldots, p_n$ are positive real numbers which sum to one, define a corresponding probability μ by saying that there is probability $p_{j_1} \cdot p_{j_2} \cdot \ldots \cdot p_{j_k}$ of landing j_1 on roll t_1, and landing j_2 on roll $t_2, \ldots, j_k$ on roll t_k, for distinct rolls. The pair (Ω, μ) is called a *Bernoulli shift*, where the word 'shift' refers to the fast-forwarding operation.[6] Where (Ω, μ) is a Bernoulli shift, its *entropy* is given by $-\sum_{i=1}^{n} p_i \log p_i$. In 1970, Ornstein showed that any two Bernoulli shifts (Ω_1, μ_1) and (Ω_2, μ_2)—which may concern dice with different numbers of sides—are metrically isomorphic iff they have the same entropy.[7] To say that they are 'metrically isomorphic' is to say that there is a bijection $\Omega_1 \longrightarrow \Omega_2$ which preserves probabilities and respects fast-forwarding almost everywhere and whose domain and range need only be measure one sets.[8]

These three examples suggests the following general mechanism of classification. The initial data are given by a class C of mathematical objects and an equivalence relation E on C induced by a certain type of bijection between the objects. The classification is then effected by identifying two further pieces of data: a class *Inv* of invariants, and an assignment of invariants in *Inv* to objects in C that respects equivalence. Writing the assignment as $\iota : C \longrightarrow Inv$, the requirement is that $E(X, Y)$ iff $\iota(X) = \iota(Y)$, as X, Y ranges over the classified objects in C.[9] So, in our examples:

	C	E	*Inv*
Algebra	uncountable algebraically closed fields	isomorphism	characteristic and cardinality
Topology	compact connected surfaces	homeomorphism	Euler characteristic and orientability
Probability	Bernoulli shift	metric isomorphism	entropy

Several modifications of this general framework are possible. First, one might consider notions of partial classification in which just the forward direction of the

[5] For instance, if $n = 3$, then we map $(\ldots, 3, 2, 1, 3, 2, 1, \ldots)$ to $(\ldots, 1, 3, 2, 1, 3, 2, \ldots)$.

[6] In contexts where one is considering a wider class of operations, one might rather use 'Bernoulli shift' to refer to the triple formed by adding the fast-fowarding operation to the pair (Ω, μ).

[7] Ornstein (1970), Petersen (1983: 281), and Rudolph (1990: §7).

[8] Petersen (1983: 4) and Rudolph (1990: 7).

[9] We have seen this kind of general framework set out in Rosendal (2011: 1252) and Gowers (2008: 51). We learnt of the Bernoulli shift example from Rosendal and the surfaces example from Gowers.

biconditional '$E(X, Y)$ iff $\iota(X) = \iota(Y)$' is available. Second, one might liberalise $E(X, Y)$ so that the equivalence relation need not be given by a bijection between X and Y.[10] More radically, one might allow that $E(X, Y)$ is not an equivalence relation at all, but rather a metric-like similarity relation which expresses that X, Y are close to one another in some sense.[11] So far as we can tell, everything we say in what follows is compatible with any of these modifications.

Calculable mechanisms

The general mechanism described above is a good start for understanding classification programmes. However, it is excessively permissive. To illustrate the point, let C be any class of objects C, with any equivalence relation E on them; put a well-order $\lhd$ on C by appealing to the Axiom of Choice and let $Inv \subseteq C$ consist of those elements of C which are the $\lhd$-least elements of their E-equivalence class; finally, let $\iota : C \longrightarrow Inv$ send each element to the unique element of C with which it is E-equivalent. This will satisfy the minimum conditions stated above, but it does no useful classificatory work. Indeed, if such uninteresting appeals to Choice sufficed, then all classification problems would be immediately and trivially resolved.

This problem arises because we have not yet imposed any constraints on the nature of the invariants and their relations to the original class of objects. Indeed, the issue here is similar to what happens by (mistakenly) regarding an *arbitrary* deduction from entirely *arbitrary* axioms as sufficient for engaging in serious mathematical proof. Not only would this prevent you from accurately describing the activity of proof in mathematics; it would also blind you to the *aims* of proof.

To deal with this, we must ask: *What distinguishes the invariants and assignments used in classification in mathematics from arbitrary invariants and assignments?* To begin answering this, consider Ornstein's classification of isomorphism of Bernoulli shifts. The invariant here is entropy, which is given by $-\sum_{i=1}^{n} p_i \log p_i$. Evidently, this is an easily calculated function of the tuple $(p_1, \ldots, p_n)$, and this tuple is itself prominent in the canonical presentation of the system (Ω, μ).

This observation leads directly to the following thesis concerning how we should view mathematical classification. The invariants Inv and the function ι used in classifications in mathematics are such that:

(a) ascertaining the particular invariant assigned to an object is easily calculable from a canonical presentation of that object, i.e. $\iota(X)$ is calculable from a canonical presentation of X; and,

---

[10] One does just this in the theory of Borel equivalence relations; see e.g. Gao (2009). A representative example is when X, Y are sequences of natural numbers and we define: $E_0(X, Y)$ iff there is some point after which X and Y agree.

[11] This is what happens in Gowers' (2000) notion of 'rough classification'.

(b) the comparison of invariants can likewise be easily effected, i.e. it is easy to determine whether $\iota(X) = \iota(Y)$.

This thesis resonates well with our other paradigmatic examples. In the example of compact connected surfaces, we think about the surface as 'triangulated', i.e. as broken up into a finite number of triangles, lines, and points, from which the Euler characteristic (for example) may be calculated. In the example of algebraically closed fields, we conceive of the algebraically closed field as a set-sized structure which possesses a cardinality which may be easily ascertained.[12]

But the strongest evidence for our thesis comes from the fact that mathematicians routinely talk about classification in patently computational terms, even in areas far removed from mathematical logic and the theory of computation. For instance, here is the start of a recent research monograph in differential topology:

> A classification of manifolds up to diffeomorphism requires the construction of a complete set of algebraic invariants such that: [¶] (i) the invariants of a manifold are computable, [¶] (ii) two manifolds are diffeomorphic if and only if they have the same invariants, [...].[13]

Similarly, in speaking of classifications, Gowers writes that 'as often as possible one should actually be able to establish when $\iota(X)$ is different from $\iota(Y)$. There is not much use in having a fine invariant if it is impossible to calculate'.[14]

It is worth noting that our thesis presupposes that canonical presentations are readily available to us (somehow). This is no surprise: the thesis would be fairly ineffectual otherwise, since proceeding by way of the canonical presentations might be just as difficult as enumerating all of the equivalence classes.

Now, someone might worry that a 'canonical presentation' can end up misidentifying the 'topic' of the relevant mathematical enquiry. For instance, in the topological case, one might have thought one was studying the *surfaces themselves*, and not their triangulations. Relatedly, one might worry that what counts as a 'canonical presentation' is historically contingent: a contemporary 'canonical presentation' of a surface might not have counted as 'canonical' in previous eras.

These are not, though, serious objections *against* the thesis. After all, exactly the same issues pervade our ordinary ways of talking about proofs. In developing innovative proof techniques, one often appeals to new resources, and this can generate a concern that the topic has been changed. For instance, Bolzano used the completeness of the real line to establish the intermediate value theorem, where previous mathematicians had sought to use considerations more closely related to the

[12] This is in contrast to working with an all-encompassing 'universal domain', as is the default in some treatments of algebraically closed fields; see Weil (1946: 242ff).

[13] Ranicki (2002: 1). But for some examples in topology where the invariants are not computable, see Poonen (2014: §7 pp.223ff).

[14] Gowers (2008: 54), with variables changed to match preceding text.

geometry of curves themselves.[15] Likewise, students nowadays reason about products as sets of ordered pairs, or as objects of a certain category, whereas previous eras might have rather talked about shapes of different dimensions. Phenomena like these generate deeply interesting philosophical questions, such as *are proofs which do not introduce new concepts better?*, and *how should we think about theory change in mathematics?*[16] But, presumably everyone accepts that this phenomena is present in the activity of proving. It should not be surprising that the same is true of the activity of classifying.

Now, the thesis is a proposal for how to think about the *activity* of classification within mathematics. But it also naturally suggests a conception of the *aim* of classification: classification is valuable because it leaves us better placed to calculate whether objects X and Y are (dis)similar, in that we are better positioned to calculate whether $E(X, Y)$. So, on this picture, the aim of the classification of compact surfaces was to leave us better placed to tell whether two surfaces are (dis)similar, by calculating and comparing their Euler characteristic and orientability.

Of course, this tells us nothing about *why* we might value the ability to determine whether various objects objects are (dis)similar. But this is just as it should be: the answer to that general question will vary from case to case. Our reasons for valuing the capacity to discern similar from dissimilar surfaces may be very different from our reasons for valuing the capacity to discern similar from dissimilar Bernoulli shifts. However, all of our examples presuppose that one task for mathematics is to provide a taxonomy of the most frequently encountered mathematical structures.

This view of classification also helps to explain some initially puzzling remarks about the kind of *completeness* which classifications sometimes give us. For instance, Steinitz motivates his classification of algebraically closed fields as follows:

> Our program in this work is to obtain an overview of all possible fields and to ascertain their relations to one another with regard to their main features.[17]

We need to understand Steinitz's idea of *obtaining an overview* of all possible fields. At its most basic, we need to say why the truism 'all fields are isomorphic to the reals, or to some other field' fails to provide an overview in the relevant sense. Our thesis suggests the following reading. The hope is that calculating invariants will provide an easy way to test for isomorphism of fields, where for each invariant there is also a simple example of a field with that invariant. More generally, classifications yield the relevant type of *completeness*, when it is possible both to describe all the invariants and provide examples for each invariant.

---

[15] See Lützen (2003: 174–5).

[16] Arana and Detlefsen (2011) and S. R. Smith (2015) provide really interesting work on these questions.

[17] Steinitz (1910: 167).

Conversely, this suggests a way in which classification can be *unsuccessful*: namely, when it turns out that identifying and individuating the proposed invariants is just as hard as discerning the similarity of the classified objects in the first place. In short: successful classification must employ invariants that are somehow 'simpler' than the objects to be classified. And it is notable that in our paradigmatic examples of classification, all the invariants were finite sequences of natural numbers, integers, or real numbers. It would, of course, be lovely to have some greater understanding of what makes something fit to be an invariant; that is, to have a deeper grasp of the relevant notion of 'simplicity'. But, returning once again to the parallel with proof, this question may well be just as hard as the question of what makes something fit to be an axiom.

The relationship to theorem-proving

We began this section by raising a basic question, namely: *What are the means and aims of the activity of classification?* Briefly stated, our answer is: *To provide 'easily calculable' E and Inv, for certain objects C, and explicit examples of each member of E.* We now show how our answer to this first question suggests an answer to the second basic question which we raised at the start of this section, namely: *Can classification be reduced to proof?* Having specified all the components of the mechanism—the equivalence relation, E, the invariants Inv, and the mapping ι—there is a clear theorem whose proof is necessary for completing the classification. However, providing that theorem is not *sufficient* for success, since the mapping must be 'easily calculable', and one ought to be able to find explicit members of each equivalence class.

Moreover, the invariants are rarely given at the outset of the enquiry. Instead, a classification problem begins with the objects-to-be-classified, C, and the similarity relation, E, and the task is to find the appropriate invariants. This is one good reason to resist offering a one-one 'reduction' of classification problems to specific theorems-to-be-proved. But there is also a second good reason. Whereas the proof of a theorem from accepted axioms is ultimately an all-or-nothing affair, the success of a classification programme is a matter of degree. After all, one can debate the degree to which something is 'easily calculable', and one can debate the degree to which an element of each equivalence class has been explicitly described.

17.2 Shelah on classification

In the previous section, we considered the nature of classification in mathematics in general. We now focus on classification in model theory. In this section, we describe Shelah's work on classification, and suggest a way to conceptualise it using the rubric from the previous section.

Shelah and Morley's conjecture

Shelah's classification programme culminated in the resolution of the Morley conjecture. Morley had conjectured that the number of non-isomorphic models of a complete theory, of a given uncountable cardinality, does not decrease as the cardinality increases.[18] Let us write $I(T,\kappa)$ for the number of non-isomorphic models of T whose underlying domain has cardinality κ. So, $I(T,\kappa)$ is the number of isomorphism types of models of T, where we restrict our attention to models of size κ. In this terminology, Morley had conjectured that if $\kappa \leq \lambda$ are both uncountable, then $I(T,\kappa) \leq I(T,\lambda)$. In the early 1980s, Shelah proved Morley's conjecture when T is countable; and the proof is contained in his *Classification Theory and the Number of Nonisomorphic Models*.[19]

The machinery which Shelah used to derived the Morley conjecture also led to a result which he called the *Main Gap Theorem*, and stated this way in 1985:[20]

Theorem 17.1 (Main Gap Theorem): *Let T be a complete theory in a countable language. Then one and only one of the following happens:*

(1) $I(T,\kappa) = 2^{\kappa}$, *for all uncountable* κ.
(2) $I(T,\aleph_{\gamma}) \leq \beth_{\omega_1}(\max(|\gamma|,\omega))$, *and T has a structure theory with countable depth.*

In the next subsection, we will discuss the philosophical significance which Shelah ascribed to the Main Gap Theorem. First, we must explain the terminology introduced in condition (2).

The expression $\aleph_{\gamma}$ denotes the γth infinite cardinal (cf. §1.B). The cardinal $\beth_{\alpha}(\kappa)$ is a relativisation of the usual beth function, enumerating the cardinality of successive iterations of the power-set operator starting at κ. To be more exact, we offer the following recursive definition:

$$\beth_0(\kappa) = \kappa, \quad \beth_{\alpha+1}(\kappa) = 2^{\beth_{\alpha}(\kappa)}, \quad \beth_{\alpha}(\kappa) = \sup_{\beta<\alpha}\beth_{\beta}(\kappa), \text{if } \alpha \text{ limit} \qquad (beth)$$

However, the notion of 'having a structure theory'—which we should read as synonymous with 'being classifiable'—will take require rather more explanation.

The key idea behind Shelah's notion of 'having a structure theory', or 'being classifiable', is that the relevant invariants are 'cardinal-like invariants'. To motivate this

[18] The conjecture is problem 19 in H. Friedman (1975: 116), where it is attributed to Morley. As far as we know, Morley never himself published this conjecture.

[19] Shelah (1990); the first edition was Shelah 1978.

[20] Shelah (1985: 228); see also Shelah (1990: 620), Harrington and Makkai (1985: 140), and Baldwin (1988: 3).

idea, we start with Steinitz's example of an algebraically closed field of a fixed characteristic, discussed in §17.1. To classify these, we merely need a single cardinal number, providing the size of the field's underlying domain. But suppose we want a theory which states that there are multiple disjoint algebraically closed fields of fixed characteristic. To describe the models of this 'disjoint theory', we need to say how many fields of each cardinality we have. This is naturally represented by a function from cardinals less than or equal to the size of the domain, to these same cardinals.

We can indefinitely iterate this idea of partitions, and with each step of the iteration the invariants become slightly more complex. To make this precise, we define the set of *cardinal-like invariants* $Inv_\alpha(\kappa)$ *of depth* α, by recursion on α. Intuitively, α records the length of the iterative process, while κ records that these invariants are reserved for models whose underlying domain has cardinality κ. The recursive definition proceeds in three steps:

- $Inv_0(\kappa)$ is the set of all cardinals $\lambda \leq \kappa$, which for ease we write just as the set $\{\lambda : \lambda \leq \kappa\}$
- $Inv_{\alpha+1}(\kappa)$ is the set of sequences of length less than or equal to the cardinality of the continuum, with each element of the sequence being a function $f : Inv_\alpha(\kappa) \longrightarrow \{\lambda : \lambda \leq \kappa\}$
- $Inv_\alpha(\kappa)$ is $\bigcup_{\beta<\alpha} Inv_\beta(\lambda)$, when α is a limit.

Finally, we define Inv_α as the union of $Inv_\alpha(\lambda)$ as λ ranges over all infinite cardinals.

This notion of cardinal-like invariance is the key component to Shelah's explication of 'having a structure theory' or 'being classifiable'.[21] In particular, Shelah stipulates that T has a structure theory of depth α if there is a function ι from the set of models of T to Inv_α such that:

(a) if $\mathcal{M}$ has size κ then $\iota(\mathcal{M})$ is in $Inv_\alpha(\kappa)$, and
(b) if $\mathcal{M}, \mathcal{N}$ are two models of T, then $\mathcal{M}$ is isomorphic to $\mathcal{N}$ iff $\iota(\mathcal{M}) = \iota(\mathcal{N})$.

Finally, we say that T *has a structure theory* if there is an α such that T has a structure theory of depth α. This is the definition of 'having a structure theory' which occurs in condition (2) of the Main Gap Theorem 17.1. In terms of the intuitive picture of the 'iterated partitions', the constraint in condition (2) is that the iteration need only proceed a countable number of times.

Given this, Shehah's restriction to 'the cardinality of the continuum', in the definition of $Inv_{\alpha+1}$, might seem *ad hoc*. Elsewhere, though, Shelah relaxes this clause, allowing the continuum to be replaced with any fixed infinite cardinality.[22] The relaxed version captures the motivating idea, while the more specific version is the one which operates in the Main Gap Theorem 17.1.

---

[21] Shelah (1985: 228); cf. Shelah (1987b: §1.4 p.155, 2009b: §2.9 p.25) and Baldwin (1987b: 5).

[22] Shelah (1987b: §1.4 p.155); Shelah uses χ for the parameter which replaces the cardinality of the continuum. So, the successor step reads: $Inv_{\alpha+1}(\kappa)$ is the set of sequences of length $\leq \chi$, with each element of the sequence being a function $f : Inv_\alpha(\kappa) \longrightarrow \{\lambda : \lambda \leq \kappa\}$.

Shelah's classification programme

Shelah had a deep interest in the boundary between *classifiable* and *non-classifiable* theories. He believed that this could be explicated by his Main Gap Theorem 17.1. In particular, he suggested the following:

(a) theories are classifiable iff they do not have 'too many models';

(b) when theories are classifiable, we can characterise each of their models up to isomorphism by invariants which are 'cardinal-like'.

Straightforwardly, claim (a) corresponds with condition (1) of the Main Gap Theorem 17.1, and claim (b) corresponds with condition (2) of the Theorem.

Moreover, we can explain Shelah's classification programme in terms of the general framework from §17.1. The objects to be classified, C, are models of a certain complete theory in a countable signature. The equivalence relation, E, is isomorphism. And Shelah's invariants, Inv, are the 'cardinal-like' invariants, Inv_a, described above. Then clause (a) purports to establish a limit on when the theories are too 'wild' for us to hope for a successful classification programme, and clause (b) aims to deliver a classification in 'tame' cases.

However, there is a wrinkle in the carpet. In §17.1, we argued for a connection between classification and calculability. Shelah's cardinal-like invariants do not, though, seem to have anything to do with calculability. So, as it stands, Shelah's programme does not exactly fit our rubric.

There is, though, much more to say here. In the next subsection, we shall argue that clause (a) of Shelah's programme is rather unmotivated. And we shall then argue that a *better*-motivated condition actually fits the rubric of §17.1 rather well.

Unclassifiability as having too many models

Shelah does not say much by way of support for his explication of the idea that 'having a structure theory' amounts to classifiability, in any intuitive sense. The single consideration which he repeats in three places is this proposition:[23]

Proposition 17.2: *Given GCH, if there is infinite κ such that $I(T,\lambda) = 2^\lambda$ for all $\lambda \geq \kappa$, then T does not have a structure theory.*

The proof of this proposition is elementary, but for ease of readability we defer its proof until §17.A. The statement of the proposition is related to the following idea about the nature of classification: *if a theory has too many models, then it is not classifiable.* Proposition 17.2 is then supposed to lend further credence to the extensional correctness of Shelah's analysis of 'having a structure theory' or 'being classifiable'.

[23] Shelah (1985: 228); cf. Shelah (1987b: §1.6 p.155, 2009b: 25–6 immediately below Corollary 2.12). GCH is the generalised continuum hypothesis (defined in footnote 36 of chapter 8).

The intuitive idea here is essentially point (a), above: *unclassifiability is having too many models*. We think, though, that this idea is rather unmotivated.

We begin by noting Hodges' challenge to (a). In particular, Hodges notes that group theorists have provided an apparently successful classification of 'totally projective abelian p-groups', *despite* the fact that there are 2^{λ} of them in any uncountable cardinality λ.[24] However, there is some indirect evidence that group-theorists' views on this matter are less than univocal. For instance, Problem 51 in the 1973 version of Fuchs' *Infinite Abelian Groups* was to 'Characterize the separable p-groups by invariants'.[25] In 1974, Shelah showed that for regular uncountable λ there are 2^{λ} non-isomorphic separable p-groups of cardinality λ.[26] Shelah wrote of this that 'the proof indicates to me that separable p-groups cannot be characterized by any reasonable set of invariants. (This answers Problem 51 of Fuchs [...])'.[27] In the later editions of Fuchs' book, Problem 51 no longer appears; in its place, special cases of Shelah's result are given.[28]

So the situation is this: we apparently classified the totally projective abelian p-groups, despite there being many of them; but then Shelah proved that there were also maximally many separable p-groups, and it appears as though group-theorists have inferred from this that these groups are not classifiable. In short: our two anecdotal examples from group theory seem to point us in different directions, and we cannot hope to infer too much from them concerning (a). Indeed, if we were serious in the project of trying to determine whether mathematical practice (be it in group theory, or elsewhere) counted for or against (a), we would need to carry out a proper survey.

Lacking such a survey, we will instead consider a *general* argument for (a). The argument is due to Hrushovski, and is mentioned by Hodges.[29] Hrushovski's argument begins from the reasonable supposition that there is no sense in which *all models* can be classified. Now, if we are inclined to restrict attention to countable signatures anyway, then we may as well view all structures under consideration as structures in a maximally generous signature with countably many constant symbols and countably many relation and function symbols of all numbers of places. And for a given infinite cardinality κ, there are exactly 2^{κ}-many non-isomorphic models in that signature. By appeal to the premise that there is no reasonable sense in which one can usefully classify all models of a given infinite cardinality, one then concludes that the same fate befalls any theory which has just as many models.

This argument is not very robust though. For instance, if the continuum hypoth-

---

[24] Hodges (1987: 231, 221).

[25] Fuchs (1973: 55).

[26] Shelah (1974b: Theorem 1.2 pp.245–6).

[27] Shelah (1974b: 244).

[28] Fuchs (2015: 332–3)

[29] Hodges (1987: 232).

esis holds, then there are exactly $2^{\aleph_0}$ countable well-orders (up to isomorphism). But it does not seem that anyone has *ever* thought that the classification of the countable well-orders was ever an open question. Indeed, one when learns very elementary set theory, one learns various methods for determining whether two well-orders are isomorphic, or if rather one is isomorphic to a proper initial segment of the other. In short: the countable well-orders are extremely well-behaved, despite the fact that they can have 'too many' models. And this sinks the idea that unclassifiability, in general, amounts to having too many models.

Consequently, we must abandon principle (a). Shelah's programme will need to be thought of in slightly different terms.

Unclassifiability and definability

We just considered Hrushovski's attempt to connect unclassifiability with having too many models, in general. But perhaps a restricted version of this argument can be rescued, by describing a principled difference between countable well-orders, on the one hand, and countable structures in a countable signature, on the other, which explains why the former are 'tame' and the latter are too 'wild' to classify.

With this in mind, we turn to recent work on generalised descriptive set theory. The idea is to view models of uncountable cardinality κ as points in a topological space.[30] Each model in a countable signature whose underlying domain has cardinality κ can be naturally coded as a function from κ to κ, and the underlying domain of the topological space is the set of all such functions.[31] There is a measure of complexity on subsets of this space whereby: open sets are least complex; the Borel sets (those obtained from the opens through complementation and κ-sized unions) are more complex; and the analytic sets (those formed from projection over closed sets) are yet more complex. This measure of complexity can be extended naturally to the product spaces, so that it makes sense to ask after the complexity of relations between structures.

Väänänen is one of the first to have studied model theory from this perspective, and he writes: 'It turns out that stability theory and the topological approach proposed here give similar suggestions as to what is complicated and what is not'.[32] This has recently been confirmed in a startling way by results of Friedman, Hyttinen, and Kulikov. They show that for certain infinite cardinals κ,[33] the relation of

---

[30] See again §14.A for a review of topological spaces, including definitions and motivating examples.

[31] The open sets in this space are sets of the form $\{f : (\forall \beta < \alpha) f(\beta) = s(\beta)\}$ where $s : \alpha \longrightarrow \kappa$ is a function for some $\alpha < \kappa$. Further, it turns out that one must restrict attention to cardinals κ such that $\kappa^{<\kappa} = \kappa$, where '$\kappa^{<\kappa}$' refers to the cardinality of all the functions $f : \alpha \longrightarrow \kappa$ for some $\alpha < \kappa$. Recall that if GCH holds then $\kappa^{<\kappa} = \kappa$ for all and only regular κ (cf. Jech 2003: 55).

[32] Väänänen (2008: 117).

[33] In particular, for regular limit cardinals $> 2^{\aleph_0}$. Regular cardinals are defined in Definition 8.3. κ is a *limit cardinal* iff $\kappa > \omega$ and κ is not the least cardinal larger than some other cardinal.

isomorphism between κ-sized models of T is Borel *iff* T falls on the 'has a structure theory' side of Shelah's Main Gap Theorem 17.1.[34] Moreover, the proof of this theorem involves carefully noting how elements of Shelah's own proofs correspond to ideas stemming from this topological measure of complexity.

All of this provides a natural response to our 'countable well-order' counterexample to Hrushovki's general argument. Given two countable well-orders, the relation of isomorphism between them is virtually a Borel condition on this topological measure of complexity.[35] However, by the result described in the previous paragraph, if a theory T does *not* fall on the 'having a structure side' of Shelah's Main Gap Theorem 17.1, then there are many uncountable cardinals κ in which the relation of isomorphism between κ-sized models of T is *more* complex than being Borel. So there is, after all, a good sense in which the countable well-orders look more 'tame' than such 'wild' theories.

Ultimately, though, approaching classifiability in this way suggests that we should *abandon* (a). After all, the explication of classifiability is no longer in terms of 'having few models', but instead in terms of 'definability' or 'higher-order computational' resources.

Moreover, approaching classifiability in terms of Borel conditions resolves the wrinkle which emerged earlier. As we explained above, Shelah's objects-to-be-classified, C, are models of a certain complete theory in a countable signature; the equivalence relation, E, is isomorphism; and the invariants, Inv, were cardinality-like notions which had little to do with calculability. But once we approach classifiability in terms of Borel conditions, we find that *having a structure* in Shelah's sense coincides with isomorphism being a Borel condition (in the generalised sense). Hence in the case where C consists of the models of a classifiable theory in Shelah's sense, the equivalence relation E is itself calculable.

We are now left only with a minor dissimilarity between classifiability, in Shelah's sense, and the paradigmatic cases from §17.1. In our paradigmatic cases, to determine whether $\iota(X) = \iota(Y)$, one first computed $\iota(X)$ from X and $\iota(Y)$ from Y, and then compared the results of these two computations. Here, when the theory is classifiable in Shelah's sense, it turns out that the relation $E(X, Y)$ is itself computable in this generalised sense. But this of course is completely compatible with subsequent identifications of invariants which may be easily calculated from

[34] S.-D. Friedman et al. (2014: Theorem 63 p.55). As in footnote 31, above, the restriction that $\kappa^{<\kappa} = \kappa$ is still in force.

[35] This is because two well-orders are not isomorphic iff one is isomorphic to a proper initial segment of the other. Hence, in the case of well-orders, both being isomorphic and being non-isomorphic are analytic conditions. And Souslin's Theorem says that conditions which are both analytic and co-analytic are Borel. Of course being a countable well-order is a co-analytic condition in the first place. Hence the qualifier 'virtually': if you *already* know that you are dealing with two well-orders, then the relation of isomorphism is analytic and co-analytic. We do not think that this difference between 'Borel' and 'analytic and co-analytic on a co-analytic set' matters much for the philosophical point we are making here.

the presentations of the objects, in the case where the theory is classifiable or 'has a structure theory'.[36]

17.3 Uncountable categoricity

Outside of Shelah's work, another primary classificatory programme in recent model theory has been Zilber's attempt to classify uncountably categorical theories. By the Löwenheim–Skolem Theorem 7.2, no first-order theory with infinite models is categorical. But, as Zilber retrospectively put the point, this obvious unavailability simply 'entailed a rethinking of the concept of categoricity'.[37]

Rethinking categoricity

Before we explain the rethought version of categoricity, we should say more to motivate this change of perspective. In §7.2, we suggested two reasons for treating categoricity a desirable property of theories: first, categorical theories pin down mathematical structure (in the intuitive sense); second, via a supervaluational semantics, one can believe that every sentence in the theory's language has a determinate truth value. These are *not* the ambitions which drive contemporary work on categoricity in mathematical logic. The aim, rather, is to use the 'rethought' notion of categoricity as part of a *classificatory* project, which we will explain in terms of the general framework from §17.1.

The 'rethought' notion of categoricity gets its impetus from restricting attention to structures of a *given* infinite cardinality. For a fixed infinite cardinality, there are many natural examples of theories that have exactly one model up to isomorphism of this cardinality. For example, Cantor proved that the complete first-order theory of the rationals as a linear order has exactly one countable model up to isomorphism.[38] However, this theory has many non-isomorphic models of higher cardinalities: the real numbers as a linear order satisfy it, as do the real numbers minus a single given real number, but these two models cannot be isomorphic. As another example, take the complete theory of the integers with just the successor. Models of this theory consist of one or more copies of the integers, with no relations 'between' any of these copies. So this theory has, up to isomorphism, countably many models of countable cardinality, corresponding simply to the number of copies of

[36] That said, it is not obvious that the invariants will necessarily be things like natural or real numbers. For instance, even in the ordinary setting of Borel equivalence relations, only the most well-understood and well-behaved classifications have such invariants. See the notion of 'smoothness' in Gao (2009: 128ff). And if the invariants are more complicated objects, this might make them less easily calucable, and this might make producing examples of each invariant more difficult.

[37] Zilber (1993: 1).

[38] See Marker (2002: 48) and Hodges (1993: 100).

the integers in the model. However, it has exactly one model in any uncountable cardinality, because the only way that a union of copies of the integers will get to an uncountable cardinality is if there are uncountably many such copies.

Given such examples, Łoś asked whether there were theories with exactly one model up to isomorphism in one uncountable cardinality, but not in others.[39] Morley showed that this could not happen:[40]

Theorem 17.3 (Morley): *Let T be a complete consistent theory in a countable signature. T has exactly one model up to isomorphism of* some *uncountable cardinality iff T has exactly one model up to isomorphism of* every *uncountable cardinality; i.e., $I(T,\kappa) = 1$ for some uncountable κ iff $I(T,\kappa) = 1$ for all uncountable κ.*

Such theories which satisfy either side of the biconditional in Morley's Theorem are called *uncountably categorical*. Uncountable categoricity is the 'rethought' notion of categoricity mentioned above, and Zilber's programme was to classify such theories. In this section, we describe the rudiments of this classification programme.

For simplicity, in the remainder of this chapter, we assume that all theories in question are complete and in a countable signature. The restriction to completeness is justified, because any theory in a countable language which satisfies either side of the biconditional in Morley's Theorem 17.3 will be complete, by the Löwenheim–Skolem Theorem 7.2. The restriction to theories in countable signature is simply because less is known about the uncountable case. While Shelah proved the analogue of Morley's Theorem 17.3 for complete theories in uncountable languages in 1974,[41] a great many complications arise in treating the uncountable case, and the work in the tradition of Zilber, which is our focus, is always done in the context of countable languages.

Pregeometries and dimension

The proof of Morley's Theorem 17.3 has been refined since Morley's own work. These refinements have revealed a connection between uncountably categoricity and a geometrical notion of independence and dimension.

The notion of a *pregeometry* is sufficient to provide us with an abstract, algebraic, axiomatic treatment of certain basic ideas related to dimension:[42]

Definition 17.4: *Let $\mathbb{G}$ be a set and* $\mathrm{cl} : \wp(\mathbb{G}) \longrightarrow \wp(\mathbb{G})$ *be a function. Then* $(\mathbb{G}, \mathrm{cl})$ *is a* pre-geometry *iff it satisfies the following four axioms:*

---

39 Łoś (1954: 62).

40 Morley (1965a).

41 Shelah (1974a).

42 See Marker (2002: 289), Hodges (1993: 170–1), Buechler (1996: 52), and Baldwin (2014: §4.2).

(1) $A \subseteq \mathrm{cl}(A)$ *and* $\mathrm{cl}(\mathrm{cl}(A)) = \mathrm{cl}(A)$
(2) *If* $A \subseteq B$ *then* $\mathrm{cl}(A) \subseteq \mathrm{cl}(B)$
(3) *If* $a \in \mathrm{cl}(A \cup \{b\}) \setminus \mathrm{cl}(A)$ *then* $b \in \mathrm{cl}(A \cup \{a\})$
(4) *If* $a \in \mathrm{cl}(A)$ *then there is a finite* $A_0 \subseteq A$ *such that* $a \in \mathrm{cl}(A_0)$

Where $(\mathbb{G}, \mathrm{cl})$ *is any pregeometry:*

(1) *A set* $B \subseteq \mathbb{G}$ *is* independent *iff* $c \notin \mathrm{cl}(B \setminus \{c\})$ *for all* $c \in B$
(2) *A set* $A \subseteq \mathbb{G}$ *is* closed *iff* $A = \mathrm{cl}(A)$
(3) *A subset B of a closed set A is a* basis *of A iff B is independent and* $\mathrm{cl}(B) = A$
(4) *The* dimension *of a closed set A is the cardinality of any basis for A*

These definitions straightforwardly generalise the notion of dimension that we encounter when we deal with Euclidean space. In more detail: the elements of n-dimensional Euclidean space, $\mathbb{R}^n$, are vectors $\bar{v}$, each of whose entries are real numbers. Along with the operation of pointwise addition $\bar{v} + \bar{u} = (v_1 + u_1, \ldots, v_n + u_n)$, we have the operation of scalar multiplication $c \cdot \bar{u} = (c \cdot u_1, \ldots, c \cdot u_n)$, for any $c \in \mathbb{R}$. Given a subset $A \subseteq \mathbb{R}^n$, its *linear span* is $\mathrm{span}(A) = \{c_1 \cdot \bar{a}_1 + \ldots + c_k \cdot \bar{a}_k : c_i \in \mathbb{R}, \bar{a}_i \in A\}$. If we now define $\mathrm{cl}(A) = \mathrm{span}(A)$, then it is easy to check that have a pregeometry. Indeed, the ensuing definitions of a *closed* set, a *basis*, and a *dimension* are exactly the standard ones, such as one encounters in elementary linear algebra. To illustrate, suppose we are working in $\mathbb{R}^3$, and let $A = \{(0,0,1)\}$, $B = \{(0,0,1),(0,1,0)\}$ and $C = \{(0,0,1),(0,1,0),(1,0,0)\}$; then we have $\dim(\mathrm{span}(A)) = 1$, $\dim(\mathrm{span}(B)) = 2$, and $\dim(\mathrm{span}(C)) = 3$.

For model-theoretic purposes, an important example of pregeometry comes from the idea of a strongly minimal set. This uses the notion of an *elementary extension*, from Definition 4.3:

Definition 17.5: *Let* $\mathbb{G} = \{\bar{a} \in M^n : \mathcal{M} \models \varphi(\bar{a})\}$ *be any infinite definable subset in the structure* $\mathcal{M}$. *Then* $\mathbb{G}$ *is* strongly minimal *iff, for every elementary extension* $\mathcal{N}$ *of* $\mathcal{M}$, *the set* $\mathbb{G}(\mathcal{N}) = \{\bar{a} \in N^n : \mathcal{N} \models \varphi(\bar{a})\}$ *has only finite or cofinite definable subsets.*[43]

Given a strongly minimal set $\mathbb{G}$, defined by a formula with parameters from some finite A_0, we define a closure operation by $\mathrm{cl}(A) = \mathrm{acl}(A \cup A_0) \cap \mathbb{G}$. Here, $\mathrm{acl}(B)$ is the model-theoretic 'algebraic closure' of B: the set of elements $c \in M$ such that there is a formula $\psi(x)$ with parameters from B such that $\mathcal{M} \models \psi(c)$ and only finitely many elements of M satisfy $\psi(x)$ in $\mathcal{M}$. So defined, $(\mathbb{G}, \mathrm{cl})$ is a pregeometry.[44]

This can seem rather technical at first. However, it describes some very classical situations and examples. Here are three which are particularly important:[45]

[43] I.e. if Y is an $\mathcal{N}$-definable subset of $\mathbb{G}(\mathcal{N})$, then either Y is finite or $\mathbb{G}(\mathcal{N}) \setminus Y$ is finite.
[44] For more, see Marker (2002: 208, 290), Hodges (1993: 134, 164, 171), and Buechler (1996: 15, 51–3).
[45] See Marker (2002: 291), Hodges (1993: 164, 167), and Buechler (1996: 51–2).

(a) *The integers under successor*, i.e. the structure $(\mathbb{Z}, S)$. Here, $\mathbb{G} = \mathbb{Z}$ is strongly minimal, and $\mathrm{acl}(A)$ is the set of points which are 'finitely far away' from some element of A.
(b) *The rationals as a vector space*, i.e. the structure $(\mathbb{Q}^n, 0, +)$, augmented with linear maps f_p for each $p \in \mathbb{Q}$ such that $f_p(\bar{a}) = p \cdot \bar{a}$. Here, $\mathbb{G} = \mathbb{Q}^n$ is strongly minimal and $\mathrm{acl}(A) = \mathrm{span}(A)$, in the Euclidean sense of 'linear span' described above.
(c) *The complex field*, i.e. the structure $(\mathbb{C}, +, \times)$. Here, $\mathbb{G} = \mathbb{C}$ is strongly minimal and $\mathrm{acl}(A)$ is the smallest subfield of $\mathbb{C}$ containing A such that every non-zero polynomial with coefficients in the field has a root in the field.

With all of these definitions in place, Morley's Theorem is a direct consequence of the following result (which is typically regarded as implicit in the proof of the Baldwin–Lachlan Theorem):[46]

Theorem 17.6: *Suppose that T has only one model up to isomorphism for some uncountable cardinality. Then T has a countable model $\mathcal{M}$ with a strongly minimal set $\mathbb{G}$ such that all of the following hold:*

(1) *For any model $\mathcal{N}$ of T there is an elementary embedding from $\mathcal{M}$ to $\mathcal{N}$*
(2) *Any model $\mathcal{N}$ of T of cardinality $\lambda > \omega$ satisfies* $\dim(\mathbb{G}(\mathcal{N})) = \lambda$
(3) *Any models $\mathcal{N}, \mathcal{O}$ of T with* $\dim(\mathbb{G}(\mathcal{N})) = \dim(\mathbb{G}(\mathcal{O}))$ *are isomorphic*

The notion of dimension in the statement of Theorem 17.6 is given via the pregeometry $(\mathbb{G}, \mathrm{cl})$ where $\mathrm{cl}(A) = \mathrm{acl}(A \cup A_0) \cap \mathbb{G}$ and A_0 is the finite set of parameters used to define $\mathbb{G}$. The statement of Theorem 17.6 retains the convention, flagged above, that all theories are complete first-order theories in countable signatures.

The idea arising from these refinements of the proof of Morley's Theorem 17.3 can be summarised as follows: uncountable categoricity requires the presence of geometrical resources like *dimension*. Zilber expresses this idea as follows:

> [...] the main logical problem after answering the question of J. Łoś was *what properties of $\mathcal{M}$ make it κ-categorical for uncountable κ?* [¶] The answer is now reasonably clear: *the key factor is that we can measure definable sets by a rank-function (dimension) and the whole construction is highly homogeneous.*[47]

We have just explained the notion of a 'dimension' to which Zilber is referring, and we see it occurring explicitly in conditions (2) and (3) of Theorem 17.6. The

[46] Baldwin and Lachlan (1971), Marker (2002: 213–4), and Buechler (1996: 68). The Baldwin–Lachlan Theorem states that T is uncountably categorical iff both (i) T has no Vaughtian pairs (see Marker 2002: 151; Buechler 1996: 58) and (ii) T is ω-stable (see Definition 14.11). This should not be confused with a related theorem of Baldwin–Lachlan which states: if T is a (complete) uncountably categorical theory (in a countable language), then $I(T, \aleph_0)$ is either 1 or $\aleph_0$ (see Marker 2002: 215; Buechler 1996: 92).

[47] Zilber (2010: 200).

notion of 'homogeneity' which Zilber mentions is related to indiscernibility, as discussed in Chapter 15. In particular, if a structure $\mathcal{M}$ is strongly minimal and $\mathrm{tp}^{\mathcal{M}}(\bar{a}) = \mathrm{tp}^{\mathcal{M}}(\bar{b})$, then there is an automorphism of the structure which sends $\bar{a}$ to $\bar{b}$.[48] In terms of the various grades of discernibility defined in §15.1, this entails that, in such structures, the n-indiscernibles$^+$ are precisely the n-symmetricals. Now, models of an uncountably categorical theory need not themselves be strongly minimal, but the import of Theorem 17.6 is that the models of this theory are 'controlled' by a strongly minimal set in the countable model, and this strongly minimal set can be viewed as a model in its own right, with whatever structure is definable from the countable model.[49] Hence, an important idea behind Theorem 17.6 is that the strongly minimal structure itself has high levels of indiscernibility; and this together with the dimension function is what accounts for uncountable categoricity.

The Trichotomy Conjecture and extreme classification

Zilber is well-known for advancing an ambitious research programme for classifying uncountably categorical theories. By the previous discussion, this can be reduced to classifying strongly minimal structures, that is, structures that are themselves strongly minimal. However, distinct strongly minimal structures can have different signatures, and so one needs some signature-insensitive notion of equivalence. The requisite notion of equivalence is *mutual interpretability*, as introduced in §5.3.

Zilber's classification was formulated in the following conjecture:[50]

The Trichotomy Conjecture. *Every strongly minimal structure is either trivial, or is mutually interpretable with a vector space over a division ring, or is mutually interpretable with an algebraically closed field.*

Here, 'trivial' is supposed to capture the sense in which an infinite structure in a signature containing just constant symbols is strongly minimal. In such a structure, it is easy to see that the algebraic closure of the union of two sets is equal to the union of respective closures, and this is one of the equivalent formulations of the relevant notion of 'triviality.' Hence the conjecture expresses the idea that all strongly minimal structures are similar to one of the three examples from the previous section, namely: (a) the integers under successor; (b) the rationals as a vector space; and (c) the complex field. The motivation for the Trichotomy Conjecture is, therefore,

[48] See Marker (2002: 133) and Zilber (2010: 189). A structure $\mathcal{M}$ is said to be *strongly minimal* if its underlying domain M is itself strongly minimal within $\mathcal{M}$, in the sense of Definition 17.5 (cf. Hodges 1993: 164; Marker 2002: 78).

[49] In particular, the structure on the strongly minimal set is given by taking it to have an n-ary relation symbol corresponding to every subset of it which is definable in the original structure.

[50] See Zilber (1984b: 362, 2010: 201). Division rings differ from fields only in that their multiplication operation does not need to be commutative. See Hungerford (1980: 116).

that it would indicate that all of the examples of strongly minimal sets are both (essentially) already known and well-understood. The conjecture therefore expresses 'a belief in a strong logical predetermination of basic mathematical structures.'[51]

In 1993, Hrushovski found a family of counterexamples to the Trichotomy Conjecture as just formulated.[52] To date, though, these are the only known counterexamples to the Conjecture. Hence the classification of strongly minimal sets is still very much open. Indeed, in terms of the framework of §17.1, at this point even the *invariants* have not yet been isolated. The classificatory problem is only cast in terms of the to-be-classified objects C, which here are the strongly minimal structures, and the relevant equivalence relation E, which here is mutual interpretability.

Some special cases of the Trichotomy Conjecture have been established by Hrushovski and Zilber. These cases strengthen the hypothesis of a strongly minimal set to that of a so-called Zariski geometry, and show that certain of these are mutually interpretable with an algebraically closed field.[53] These Zariski geometries then correspond to the most interesting of the three cases mentioned the Trichotomy Conjecture, since the model theory of the complex field can be viewed as a part of algebraic geometry. This led Hrushovski to say of these special proven cases that 'this was originally conceived as a foundational result, showing that algebraic geometry is sui generis.'[54] Hodges cites the work of Hrushovski as part of the motivation for viewing model theory as 'algebraic geometry minus fields'.[55]

As a final illustration of the successes of this classificatory programme, consider this remark by Macintyre:

> A useful contribution of post-Morley model theory is to explain these extreme classifications in terms of a geometrical independence theory [...] From these explanations one does understand why there are so few extreme classifications in algebra, and one understands some absolutely new things, for example that there are no such extreme classifications in ordered algebra.[56]

To unpack Macintyre's claim, note that Theorem 17.6 tells us that uncountable categoricity involves a geometric-like notion of dimension. However, the availability of such a notion is the exception rather than the rule, in contemporary algebra writ large. Hence, we see why there are 'so few extreme classifications in algebra'. (Moreover, if any suitable modification of the Trichotomy Conjecture is true, then the scope of uncountable categoricity will be even more tightly circumscribed.)

---

[51] Zilber (2010: 201).
[52] See Hrushovski (1993) and Ziegler (2013).
[53] See Hrushovski and Zilber (1996: 2) and Zilber (2010: ch.4).
[54] Hrushovski (1998: 288).
[55] Hodges (1997a: vii).
[56] Macintyre (2003: 199).

Macintyre's point that 'there are no such extreme classifications in ordered algebra' then relates to another result which follows from considerations about Theorem 17.6: uncountably categorical theories are *stable,* in the sense of Definition 14.11.[57] From the equivalent characterisation of stability in terms of not having an order, Theorem 15.19, it follows that no structure which defines a linear order is uncountably categorical. Hence, while there are uncountably categorical theories amongst theories familiar from linear algebra and algebraic geometry, as soon as one goes to a setting in mathematics where there is a linear order, uncountable categoricity is simply unavailable.

17.4 Conclusions

In this final chapter we have engaged with two important episodes in post-Morley model theory: Shelah's classification programme and Zilber's Trichotomy Conjecture. The latter is borne of an attempt to gain a better understanding of the possibilities for uncountably categorical theories. But the motivation for the study of uncountable categoricity is rather different than what prompted the discussion of categoricity in Chapters 7–8 and 10–11. Instead of being concerned with pinning down an isomorphism type or establishing determinacy of truth-value, we have suggested that work on uncountable categoricity should be understood in terms of the initial stages of a classification programme.

However, saying this requires some prior understanding of what classification programmes are. Hence, at the outset of this chapter we laid down a general framework for understanding the nature of classification in mathematics. Briefly stated: in order to discern structures from one another, mathematicians seek out invariants that both respect the relevant similarity relation and are easily calculable from canonical presentations of the structures.

17.A Proof of Proposition 17.2

Here is the proof of Proposition 17.2 from §17.2. In fact, as mentioned in footnote 22 (immediately prior to Proposition), this holds for the more general notion of a parameterised cardinal-like invariant, χ, so long as χ is infinite.

Proposition (17.2): *Given* GCH, *if there is infinite κ such that $I(T,\lambda) = 2^{\lambda}$ for all $\lambda \geq \kappa$, then T does not have a structure theory.*

Proof. First, we show by induction on α that we have the following

[57] Indeed, as remarked in footnote 46, they are ω-stable.

$$\text{If } \gamma \geq \chi, \text{ then } \left|Inv_\alpha(\aleph_\gamma)\right| \leq \beth_\alpha(|\gamma|) \tag{17.1}$$

In this, $\aleph_\gamma$ denotes the γ-th infinite cardinal (cf. §1.B), and $\beth_\alpha(|\gamma|)$ is defined in (*beth*) from §17.2. For $\alpha = 0$, one has that $Inv_0(\aleph_\gamma) = \{\lambda : \lambda \leq \aleph_\gamma\}$, which has cardinality $|\gamma|$. And similarly $\beth_0(|\gamma|) = |\gamma|$. For $\alpha + 1$, note that if X is an infinite set, then the set of sequences of length $\leq \chi$ with values in X is of the same cardinality as the set of sequences of length χ with values in X; for, we could identify a function $f : \delta \longrightarrow X$ where $\delta < \chi$ with a function $\bar{f} : \chi \longrightarrow (X \cup \{x_0\})$ by setting $\bar{f}(\beta) = f(\beta)$ when $\beta < \delta$ and $\bar{f}(\beta) = x_0$ for all $\beta \geq \delta$ with $\beta < \chi$, where x_0 is a set not in X. Hence, $\left|Inv_{\alpha+1}(\aleph_\gamma)\right|$ is $\leq$ the cardinality of the set of functions from χ to $\{f : Inv_\alpha(\aleph_\gamma) \longrightarrow \{\lambda : \lambda \leq \aleph_\gamma\}\}$. By the induction hypothesis, $\left|Inv_\alpha(\aleph_\gamma)\right| \leq \beth_\alpha(|\gamma|)$. Thus, $\left|Inv_{\alpha+1}(\aleph_\gamma)\right|$ is $\leq$ the cardinality of the set of functions from χ to $\{f : \beth_\alpha(|\gamma|) \longrightarrow |\gamma|\}$. Since $\gamma \geq \chi$, one has $\chi \leq \beth_\alpha(|\gamma|)$. Then one has that $\left|Inv_{\alpha+1}(\aleph_\gamma)\right| \leq |\gamma|^{\chi \cdot \beth_\alpha(|\gamma|)} \leq |\gamma|^{\beth_\alpha(|\gamma|)} = 2^{\beth_\alpha(|\gamma|)} = \beth_{\alpha+1}(|\gamma|)$. For α a limit, one has that $\left|Inv_\alpha(\aleph_\gamma)\right| \leq \sup_{\beta<\alpha} \left|Inv_\beta(\aleph_\gamma)\right| \leq \sup_{\beta<\alpha} \beth_\beta(|\gamma|) = \beth_\alpha(|\gamma|)$. This finishes the inductive argument for (17.1).

We now show that for all infinite α there are unboundedly many $\gamma \geq \alpha$ such that

$$\beth_\alpha(|\gamma|) < \aleph_{\gamma+1} \tag{17.2}$$

Now GCH implies that if $|\gamma| = \aleph_\beta$ then $\beth_\alpha(|\gamma|) = \beth_\alpha(\aleph_\beta) = \aleph_{\beta+\alpha}$. Hence, for (17.2), it suffices to find unboundedly many γ with $|\gamma| = \aleph_\beta$ and $\beta + \alpha \leq \gamma$. To obtain this, let $|\alpha| = \aleph_\theta$ and let $\theta' > \theta$ and set $\beta = \aleph_{\theta'+1}$. Then $\theta' + 2 \leq \aleph_{\theta'} + 2 < \aleph_{\theta'+1}$. Thus we have $|\beta| = \beta = \aleph_{\theta'+1} < \aleph_{\theta'+2} < \aleph_{\aleph_{\theta'+1}} = \aleph_\beta$. Hence there are unboundedly many β with $|\alpha| < |\beta| < \aleph_\beta$. Then $\beta + \alpha$ has cardinality $|\beta|$ and thus $\beta + \alpha < \aleph_\beta$. Setting $\gamma = \aleph_\beta$ we obtain unboundedly many γ with $|\gamma| = \aleph_\beta$ and $\beta + \alpha \leq \gamma$.

To finish the argument, suppose there is infinite κ with $I(T, \lambda) = 2^\lambda$ for all $\lambda \geq \kappa$. We must show that T does not have a structure theory. For reductio, suppose T has a structure theory of depth α. By (17.2), choose γ with $\aleph_\gamma > \kappa$ and $\gamma \geq \chi$ and $\beth_\alpha(|\gamma|) < \aleph_{\gamma+1}$. Then by the definition of having a structure theory of depth α, $I(T, \aleph_\gamma) \leq \left|Inv_\alpha(\aleph_\gamma)\right|$. Then by (17.1) one has $I(T, \aleph_\gamma) \leq \beth_\alpha(|\gamma|) < \aleph_{\gamma+1}$. But this contradicts that $I(T, \aleph_\gamma) = 2^{\aleph_\gamma} = \aleph_{\gamma+1}$. □

D

Historical appendix

Introduction to Hodges' essay

This book ends with Hodges' essay, 'A short history of model theory'. While the previous parts of the book were organised around various philosophical topics and issues, Hodges' essay is organised chronologically. In tracing the development of model theory, from its origins to the present, it covers the history of many of the results stated and proved earlier in the book (and much more besides). For example, we stated and proved the Löwenheim–Skolem Theorem 7.2 in Chapter 7, and discussed its philosophical implications throughout Part B; Hodges discusses its history in §18.4.

But Hodges does more than simply describe model theory's history: he contextualises its practitioners' motivations. This, too, is philosophically informative, and we shall mention two examples of this. First: in Chapter 1 we surveyed different approaches to the semantics for first-order logic. Elements of §18.3 of Hodges' essay complement this, by looking more closely at the principal architects of these approaches—Tarski and Robinson—and setting their motivations in their original historical context. Second: in Chapters 7–8, 10–11, and 17 we considered the notion of categoricity, as it connected to considerations of determinacy of reference and truth-value and the nature of classification. Similarly, the history of categoricity in §18.2 and §18.7 constitute two book-ends of Hodges history, starting with Veblen in the early twentieth century and ending with Shelah in the late twentieth century.

Hodges' essay is intended to be readable in isolation from the rest of this book. However, for the benefit of those who have read other parts of this book, we (Sean and Tim) have inserted some cross-references to relevant portions of Parts A–C. These occur as additions of the form '[*See such-and-such.*] '.

18
Wilfrid Hodges
A short history of model theory

18.1 'A new branch of metamathematics'

In 1954, Alfred Tarski wrote:

> Within the last years a new branch of metamathematics has been developing. It is called the *theory of models* and can be regarded as a part of the semantics of formalized theories. The problems studied in the theory of models concern mutual relations between sentences of formalized theories and mathematical systems in which these sentences hold.[1]

In these words Tarski defined and named a new branch of mathematics, which today we know as *mathematical model theory*, or simply as *model theory*. The present essay will trace some of the main themes in the history of mathematical model theory, roughly up to the beginning of the twenty-first century. (What would non-mathematical model theory be? One example—there are others—is the 'model-theoretic syntax' developed by the linguists Pullum and Scholz;[2] it has historic links with mathematical model theory.)

Although Tarski named the new subject, he certainly didn't own it. Already before 1954 Anatoliĭ Mal'tsev and Abraham Robinson had published results that became as characteristic of the subject as any of Tarski's own contributions to it; we will come to their work below. Tarski's main role—apart from collecting a stellar group of young researchers around him in Berkeley and giving them problems to work on—had been to take up some earlier questions from the heuristic fringes of mathematics, and show how to give them mathematical precision.

Tarski refers to 'mathematical systems'. He means what we now usually call *structures*—they have a domain of elements, and a collection of relations, functions, and distinguished elements defined in this domain and named by specified relation symbols, function symbols and individual constants. Structures in this sense are an invention of the second half of the nineteenth century—for example David Hilbert handled them freely in his *Grundlagen der Geometrie*.[3] A system is a collection of things brought together in an orderly way. For Hilbert and his German predecessors, it seems that the things brought together were the elements of the structure. Thus, Richard Dedekind used the name 'System' both for structures and for

---

[1] Tarski (1954: 572). [2] Pullum and Scholz (2001). [3] Hilbert (1899).

sets—apparently he thought of a structure as a set that comes with added features.[4] Heinrich Weber and Hilbert spoke of 'Systeme von Dingen' [systems of things], to distinguish from axiom systems. On the other hand George Boole,[5] adapting the language of George Peacock,[6] had spoken of a 'system of interpretation'; for Boole, the things brought together were the operations as interpretations of symbols, for example + and × as function symbols and 0 as individual constant. Logicians regarded the interpretation of symbols by relations etc. of the structure as central, so one finds structures being referred to as 'interpretations' well into the twentieth century.

In model theory the name 'system' persisted until it was replaced by 'structure' in the late 1950s, it seems under the influence of Robinson and Bourbaki.[7]

Tarski speaks of 'sentences'. Mostly these were taken as concatenated strings of formal symbols. But already in the 1930s, Kurt Gödel was handling languages of uncountable cardinality, with arbitrary objects as symbols, using any suitable functions to replace concatenation of symbols.[8] Mal'tsev did likewise.[9] By the 1950s it was taken for granted that a 'sentence' could be a purely set-theoretic object.

Tarski also refers to the notion of a sentence 'holding in' a structure. The notion of a statement 'holding in' some contexts and not others is not a particularly mathematical one; for example a legal journal of 1900 speaks of 'contravening the rule held in the above cases'. It was one of a number of idioms that mathematicians had used to express what we now mean by saying that a structure is a *model of*, or *satisfies*, a formal sentence. Alessandro Padoa spoke of a structure 'verifying' axioms.[10] The word 'satisfy' in this context may be due to Edward V. Huntington;[11] Huntington was a member of the group of American mathematicians around Eliakim H. Moore and Oswald Veblen who, in the early twentieth century, made a systematic study of axiomatically defined classes of structures.[12] We can trace back the use of the word 'model' itself to the seventeenth century geometers who spoke of gypsum or paper 'models' of geometrical axioms. The term 'model' for abstract structures appeared during the 1920s in writings of the Hilbert school.[13]

18.2 Replacing the old metamathematics

One feature of the early work on models of axioms was the looseness of some of the formulations. Three examples follow. In each of them an informal method was in use around 1900, then Tarski attempted a non-model-theoretic formalisation in

[4] Dirichlet and Dedekind (1871) and Dedekind (1872). [5] Boole (1847: 3). [6] Peacock (1833). [7] A. Robinson (1952) and Bourbaki (1951). [8] Gödel (1932). [9] Mal'tsev (1936). [10] Padoa (1900). [11] For example in Huntington (1902). [12] Scanlan (2003). [13] von Neumann (1925) and Fraenkel (1928: 342). R. Muller (2009) gives historical information on the use of the word 'model' in model theory and elsewhere.

the 1930s, and finally in the 1950s a model-theoretic formalisation was given which is now widely regarded as canonical.

Categoricity

Veblen introduced the notion of categoricity:

[...] a system of axioms is categorical if it is sufficient for the complete *determination* of a class of objects or elements.[14]

to which he added a brief informal explanation of isomorphisms. [*See* §7.2 *footnote* 3.] Veblen's word 'sufficient' harks back to Huntington's paper,[15] where a set of postulates (i.e. axioms) is said to be 'sufficient' if 'there is essentially *only one*' structure that satisfies the postulates. In 1935, Tarski attempted to tidy up the notion of categoricity as follows.[16] First he assumed that the system of axioms is finite, so that its conjunction can be written as a single formula of an appropriate higher-order logic

$$a(x, y, z, \dots)$$

where the variables 'x' etc. represent the non-logical notions in the axioms (for example 'point', 'line'). Then he wrote

$$R\frac{(x', y', z', \dots)}{(x'', y'', z'', \dots)}$$

for the formal statement that R is a permutation of the universe of individuals, which takes x' to x'', y' to y'' etc. Finally he defined the axiom system $a(x, y, z, \dots)$ to be categorical if the higher-order statement

$$\forall x' \forall y' \forall z' \dots \forall x'' \forall y'' \forall z'' \dots \left(a(x', y', z', \dots) \wedge a(x'', y'', z'', \dots) \to \exists R R\frac{(x', y', z', \dots)}{(x'', y'', z'', \dots)} \right)$$

is 'logically provable'. Note that at this date, Tarski's notion of 'categorical' made no use of the notion of an axiom 'holding in' a structure. In short, it was not model-theoretic. Nor was it objective, since the notion of 'logically provable' in higher-order logic depends on what axioms you accept for this logic.

By the early 1950s, all the definitions were in place to allow the definition that a theory T is categorical if and only if T has exactly one model up to isomorphism. [*See* §7.2.] But by the 1950s the preferred logical language had become first-order logic, and the Upward Löwenheim–Skolem Theorem implied that no first-order theory with infinite models is categorical. Accordingly Vaught defined a theory T

---

[14] Veblen (1904: 347). [15] Huntington (1902). [16] Tarski (1935a).

to be λ-*categorical* (for a cardinal λ) if T has, up to isomorphism, exactly one model of cardinality λ.[17] [*See* §17.3.] (The cardinality of a structure is that of its domain of elements.) We will see below how this became one of the most fertile definitions in model theory.

Padoa's method

Padoa proposed a criterion for showing that in the context of an axiomatic theory T, no formal definition of a notion A in terms of notions $B_1, \ldots, B_n$ can be deduced from the axioms T.[18] The criterion was that there exist two interpretations of T which agree in how they interpret $B_1, \ldots, B_n$ but disagree in the interpretation of A. Padoa sketched proofs of the necessity and sufficiency of this criterion. But today it is obvious that he couldn't hope to prove necessity without saying more about how he understood 'deducible from T'; and in fact his proof of necessity is just a blurred repetition of his proof of sufficiency. Today 'Padoa's method' is generally taken to consist of a model-theoretic criterion for a syntactic notion. But Tarski's reformulation of Padoa's proposal removed all model-theoretic notions and translated Padoa's proposal into pure syntax.[19]

Padoa's method had a bumpy ride into the new context of model theory. In 1953, Evert Beth proved that Padoa's claim was true at least for first-order logic.[20] Beth took Padoa's criterion model-theoretically. But since at this date there was no clear model-theoretic route from the absence of a definition to the truth of the criterion, Beth translated the criterion into proof theory along Tarski's lines but within first-order logic, and then used his own adaptation of Gentzen's cut-free proofs to build the required models. Tarski, through his student Solomon Feferman,[21] responded that, since Beth's Theorem was proof-theoretic, it would be best to play down the model-theoretic form of the criterion, which was only incidental to the main result. Soon afterwards another member of the Berkeley group, William Craig, reworked Beth's use of cut-free proofs, and thereby discovered the Craig Interpolation Theorem.[22] Almost at once it came to notice that Abraham Robinson in Toronto had already proved a model-theoretic result equivalent to the Interpolation Theorem, using purely model-theoretic methods.[23] From this point onwards it was accepted that model theory and proof theory could each feed useful information to the other. In particular, Feferman proved a number of model-theoretic results by giving proof-theoretic demonstrations of a range of interpolation theorems.[24]

---

[17] Vaught (1954). [18] Padoa (1900). [19] Tarski (1935a). [20] Beth (1953). [21] van Ulsen (2000: 138). [22] Craig (1957a,b). [23] A. Robinson (1956a). [24] For example in Feferman (1974).

Proofs of logical independence

Padoa (1900) related his proposal to another heuristic that was already in use. Namely, we can show that a formal axiom ψ doesn't follow from formal axioms $\varphi_1, \ldots, \varphi_n$ by exhibiting an interpretation of the symbols in these axioms, which makes $\varphi_1, \ldots, \varphi_n$ hold but ψ fail to hold. This method had been used by Felix Klein and Eugenio Beltrami to show that Euclid's parallel postulate doesn't follow from his other axioms. In the years around 1900, Giuseppe Peano, Hilbert, and Huntington all applied the method.[25]

Gottlob Frege took umbrage at Hilbert's use of this method. One assumption that Hilbert made was that the non-logical symbols in the axioms are ambiguous in the sense that they can be interpreted in different ways in different structures, even within the same mathematical discourse. Frege commented:

> In der Tat, wenn es sich darum handelte, sich und andere zu täuschen, so gäbe es kein besseres Mittle dazu, als vieldeutige Zeichen. [Indeed, if it were a matter of deceiving oneself and others, there would be no better means than ambiguous signs.][26]

Frege's comments were not all negative. He went on to sketch a way in which Hilbert's arguments could be brought into a formal deductive system, by replacing the 'ambiguous signs' by higher-order variables and then proving formal statements that quantified universally over these variables, very much as in Tarski's later work of the 1930s.[27]

In this case it will be best to jump straight to the 1950s to see how Frege's concerns were answered within model theory. A paper of Tarski and Vaught indicates how to write within pure set theory a recursive definition of the relation:[28]

$$\text{Sentence } \varphi \text{ is true in structure } \mathcal{M}. \tag{1}$$

Standard methods allow this recursive definition to be reduced to a set-theoretic formula $\theta(\mathcal{M}, \varphi)$. The independence notion mentioned by Padoa above can then be formalised in pure set theory as

$$\exists \mathcal{M}(\theta(\mathcal{M}, \varphi_1) \wedge \ldots \wedge \theta(\mathcal{M}, \varphi_n) \wedge \neg\theta(\mathcal{M}, \psi)).$$

Hilbert's independence proofs in his *Grundlagen der Geometrie* can be read as proving set-theoretic sentences of this form,[29] and it then becomes a standard but tedious exercise to translate Hilbert's proofs into purely set-theoretic arguments. In these resulting arguments there is no mention of the meanings of symbols, since 'meaning' is not a set-theoretic notion. Thus Frege's complaint about ambiguous signs is met. (Tarski and Vaught use first-order logic, and some of Hilbert's formulations were not first-order; but set-theoretic formulas corresponding to θ can be

[25] Peano (1891), Hilbert (1899), and Huntington (1902). [26] Frege (1906: 307). [27] Frege (1906) and Tarski (1935a). [28] Tarski and Vaught (1958). [29] Hilbert (1899).

found for any other reasonable logic.) Frege had other objections, for example to Hilbert's use of the word 'axiom'. But 'axiom' is not a set-theoretic notion either, so this and all similar objections lose their purchase.

Now we can go back to the 1930s to see where the formula $\theta(\mathcal{M}, \varphi)$ came from. In 1933 Tarski published a paper in which he considered any formalised theory T satisfying certain conditions;[30] one of the conditions was that the symbols of T have fixed and known meanings, in such a way that every sentence of T is either true or false. This included the case where $\mathcal{M}$ is a fixed structure and φ is a formal sentence whose non-logical symbols are interpreted as in $\mathcal{M}$. He showed how to construct a metamathematical formula θ', using only higher order logic, syntax and symbols expressing the notions expressible by symbols of T, such that $\theta'(\varphi)$ is true if and only if φ is a true sentence of T. [*See §§1.3, 12.4, 12.A.*]

Tarski's famous 'Concept of Truth' paper is a translation of the expanded German version of his 1933 paper.[31] None of these versions of the paper should be counted as model-theoretic; in fact neither the word 'model' nor any equivalent expression occurs in any of them. But Tarski wanted to show that his truth definition could be used to give a precise and rigorously defined meaning to the relation (1) with $\mathcal{M}$ variable. Here he ran up against the problem that had vexed Frege. Namely, how do we deal with the notion of giving a meaning in $\mathcal{M}$ to a symbol in φ which might already have another meaning? Tarski came to suppy an answer remarkably close to Frege's.[32] Namely, he replaces the non-logical symbols in φ by variables $\bar{x}$, and then uses his truth definition to express that $\mathcal{M}$ satisfies the resulting formula $\varphi_0(\bar{x})$. From the later point of view of model theory, this procedure carries irrelevant clutter. But it can be converted into a formula $\theta(\mathcal{M}, \varphi)$ expressing (1), in set theory or some suitable higher order logic.

Tarski in the 1950s had a clean mathematical definition of (1), but he still tended to avoid the use of any notation such as $\mathrm{Mod}(T)$ for the class of models of the theory T.[33] If one also writes K for the set of sentences true in all the structures of the class K, then there are certain fundamental facts that we expect to see set down, for example

$$T \subseteq \mathrm{Th}(K) \text{ iff } \mathrm{Mod}(T) \supseteq K$$

But this group of facts are found in Abraham Robinson's doctoral thesis of 1949,[34] not in Tarski's model-theoretic papers.

[30] Tarski (1933). [31] Tarski (1983: Paper VIII). [32] Tarski (1936, 1994); followed by Tarski's student Andrzej Mostowski in his (1948). [33] For T a single sentence this notion does appear briefly in Definition 14(ii) on p.710 of his 1952. [34] A. Robinson (1951: 36–7).

18.3 Definable relations in one structure

The method of quantifier elimination

In his 'Concept of truth' paper, Tarski presents several examples of truth definitions for different kinds of language. He describes one of them as 'purely accidental'.[35] In this example he considers what today we would call the structure $\mathcal{M}$ of all subsets of a given set a, with relation $\subseteq$; he discusses what can be said about $\mathcal{M}$ using the corresponding first-order language L. (This may be an anachronism; one could also describe his example as the structure consisting of the set a with no relations, and a corresponding monadic second-order language.) Tarski works out an explicit definition of the relation 'φ is true in $\mathcal{M}$', where φ ranges over the sentences of L.

This truth definition might be accidental, but Tarski's decision to mention it was not. Leopold Löwenheim had already studied the same example within the context of the Peirce–Schröder calculus of relatives, and he had proved a very suggestive result.[36] In modern terms, Löwenheim had shown that there is a set of 'basic' formulas of the language L with the property that every formula φ of L can be reduced to a Boolean combination ψ of basic formulas which is equivalent to φ in the sense that exactly the same assignments to variables satisfy it in $\mathcal{M}$. Thoralf Skolem and Heinrich Behmann had reworked Löwenheim's argument so as to replace the calculus of relatives by more modern logical languages.[37] In 1927, Cooper H. Langford applied the same ideas to dense or discrete linear orderings.[38]

Tarski realised that not only the arguments of Löwenheim and Skolem, but also the heuristics behind them, provided a general method for analysing structures. This method became known as the *method of quantifier elimination*. In his Warsaw seminar, starting in 1927, Tarski and his students applied it to a wide range of interesting structures. An important example was the ordered abelian group of integers—not the natural numbers—with symbols for 0, 1, + and <.[39] Another was the ordered field of real numbers.[40] In both these cases the method yielded (a) a small and easily described set of basic formulas, (b) a description of all the relations definable in the structure by first-order formulas, (c) an axiomatisation of the set of all first-order sentences true in the structure, and (d) an algorithm for testing the truth of any sentence in the structure. (Here (b) comes at once from (a). For (c), one would write down any axioms needed to reduce all formulas to Boolean combinations of basic formulas, and all axioms needed to determine the truth or falsehood of basic sentences. Then (d) follows since the procedure for reducing to basic formulas is effective.)

In principle the method of quantifier elimination tells us, for any structure M, what are the sets and relations on the domain of $\mathcal{M}$ that are definable by formulas

[35] Tarski (1933: §3). [36] Löwenheim (1915: §4). [37] Skolem (1919: §4) and Behmann (1922).
[38] Langford (1926/27a,b). [39] Presburger (1930) and supplement. [40] Tarski (1931).

of the first-order language appropriate for $\mathcal{M}$. In practice we may lack the skill or the information needed to carry the method to a conclusion. But thanks to earlier work using this method, model theorists in the 1950s had at their disposal a large amount of information about the first-order definable relations in various important mathematical structures. This certainly helped to make the definable relations of a structure one of the fundamental tools of model theory. (In 1910, Hermann Weyl had introduced the class of first-order definable relations of a relational structure, but without using a formal language.)[41]

In some cases, but not all, the method showed that every definable relation in the structure is defined by a quantifier-free formula. Joseph Shoenfield, in his textbook,[42] said that a theory T *admits elimination of quantifiers* if every formula of the language of T is equivalent, provably in T, to a quantifier-free formula. He gave a model-theoretic sufficient condition for a first-order theory to admit elimination of quantifiers, and showed that some of the results of the method of quantifier elimination could be recovered easily by using this condition. Soon afterwards, necessary and sufficient model-theoretic conditions for admitting quantifier elimination were found.[43]

For most model theorists, these new methods won hands down against the sometimes heavy syntactic calculations that were needed for the method of quantifier elimination. Tarski dissented. As late as 1978 he was defending the method of quantifier elimination against modern methods

> [...] which often prove more efficient. [...] It seems to us that the elimination of quantifiers, whenever it is applicable to a theory, provides us with direct and clear insight into both the syntactical structure and the semantical contents of that theory—indeed, a more direct and clearer insight than the modern more powerful methods to which we referred above.[44]

The method of quantifier elimination works on just one structure at a time. It involves no comparison of structures. For example Tarski applied it to the ordered field of reals, and discovered among other things that the sets of reals definable in this field by first-order formulas are precisely the unions of finitely many sets, each of which is either a singleton or an open interval with endpoints either in the field or $\pm\infty$. Ordered structures with this property are said to be *o-minimal*, following Anand Pillay and Charles Steinhorn.[45] [*See §4.10, Definition 4.19.*] Tarski also found a set T of sentences which axiomatises the field, in the sense that a first-order sentence is true in the field if and only if it is provable from T. It was realised some time later that T is precisely the set of axioms defining real-closed fields. From the calculations in the quantifier elimination, it then followed at once that every real-closed field is o-minimal. So Tarski proved a theorem about a class of structures, but the

41 Weyl (1910). 42 Shoenfield (1967: 83). 43 For example Feferman (1968: 81–2).
44 Doner et al. (1978: 1–2). 45 Pillay and Steinhorn (1984).

theorem was proved by a procedure that applied separately to each structure in the class. There was never any direct comparison of structures.

In fact Tarski's quantifier elimination for the reals had much wider ramifications even than this. Lou van den Dries had pointed out in 1984 that the o-minimality of the field of real numbers already gives strong information about definable relations of higher arity—in particular, it allows one to recover the cell decomposition of semialgebraic sets in real geometry.[46] Julia Knight, Pillay, and Steinhorn generalised this cell decomposition to all o-minimal structures, and showed that any structure elementarily equivalent to an o-minimal structure is also o-minimal.[47] O-minimal structures became one of the most productive tools for applications of model theory, thanks largely to the insightful enthusiasm of van den Dries and some deep applications by Alex Wilkie.[48]

In 1959, Feferman and Vaught published a paper in which they study a structure $\mathcal{M}$ of the following form.[49] An indexed family $(\mathcal{N}_i : i \in I)$ of structures is given, and $\mathcal{M}$ is the Cartesian product. [*See §13.c for notation.*] They apply the method of quantifier elimination to $\mathcal{M}$, but with a twist: instead of showing that each formula $\varphi(\bar{x})$ is equivalent to a Boolean combination of basic formulas, they find for each formula $\varphi(\bar{x})$ a formula Φ in the language of the powerset Boolean algebra $\wp(I)$, and formulas $\theta_1(\bar{x}), \ldots, \theta_n(\bar{x})$ such that, writing $X_k(\bar{a})$ for the set of indices $i \in I$ such that the projection of $\bar{a}$ to $\mathcal{N}_i$ satisfies $\theta_i(\bar{x})$ in $\mathcal{N}_i$, the statement

$$\bar{a} \text{ satisfies } \varphi(\bar{x}) \text{ in } \mathcal{M}$$

holds if and only if

$$(X_1(\bar{a}), \ldots, X_n(\bar{a})) \text{ satisfies } \Phi \text{ in } \wp(I).$$

Having got this far, they were able to prove analogous theorems for various other constructions besides Cartesian product. (The list has been expanded since.)[50] The mind boggles at how these results could ever have been discovered. In fact we know the history, and an important ancestor of the results is work of Mostowski,[51] applying a form of quantifier elimination to show, for example, that the set of sentences true in an initial ordinal with the operation of natural addition of ordinals is a decidable set.

Before we leave the topic of quantifier elimination, we should note a quantifier elimination given by Angus Macintyre for p-adic number fields in a suitable first-order language.[52] Macintyre's reduction of the definable sets to Boolean combinations

[46] van den Dries (1984). [47] J. F. Knight et al. (1986). [48] van den Dries (1998) and, for example, Wilkie (1996). [49] Feferman and Vaught (1959). [50] Makowsky (2004). [51] Mostowski (1952). [52] Macintyre (1976).

of basic sets was exactly what Jan Denef needed in order to evaluate certain p-adic integrals.[53] (This marriage of quantifier elimination and integration was soon extended to other cases.) One of Macintyre's concerns throughout his career has been to use first-order logic in order to bring mathematical notions into tractable forms. A more recent example is his reduction of a significant part of William Fulton's scheme-theoretic *Intersection theory* to first-order form, by careful rearrangement of the material.[54] Macintyre's paper illustrates how much useful work in areas related to model theory can be done by concentration and intelligence, with only minimal recourse to model-theoretic devices. (He uses some ultraproducts, but little else.)

The definition of satisfaction

Tarski's truth definition of the 1930s gave, for each structure $\mathcal{M}$ and logic $\mathcal{L}$, a formula $\theta(x)$ of some appropriate form of higher-order logic such that

$$\theta(\varphi) \text{ iff } \varphi \text{ is a sentence of } \mathcal{L} \text{ that is true in } M.$$

[*See* §12.A.] The revised form in his later paper with Vaught gave a formula $\theta(x, y)$ of set theory such that for every structure $\mathcal{M}$ and first-order sentence with symbols appropriate for $\mathcal{M}$,[55]

$$\theta(\mathcal{M}, \varphi) \leftrightarrow \mathcal{M} \text{ is a model of } \varphi. \quad (2)$$

[*See* §1.3.] Both truth definitions used induction on the complexity of formulas, and as a result of this the revised form actually gave a set-theoretic definition of the relation

$$\text{The sequence } \bar{a} \text{ of elements of } \mathcal{M} \text{ satisfies the formula } \varphi.$$

The earlier definition was given for a single fixed structure; the later allowed the structure to vary, but also involved no comparison of structures.

It is rather rare for model theorists to give arguments that refer to the existence of set-theoretic formulas defining truth or satisfaction in structures.[56] On the other hand the recursive clauses of Tarski's truth definition are used constantly, often without explicit mention. For example $\exists x\varphi(x)$ is true in $\mathcal{M}$ if and only if some element of $\mathcal{M}$ satisfies $\varphi(x)$.

Already in 1949 Abraham Robinson gave a recursive definition of a formula θ as in (2), but without invoking the notion of elements satisfying a formula.[57] He was able to do this by adding an assumption that every element of a structure is associated with an individual constant. [*See* §1.5, *Definition 1.5*.] This association could

[53] Denef 1984. [54] Fulton (1984) and Macintyre (2000b). [55] Tarski and Vaught (1958).
[56] Such arguments do occur in what Barwise (1972) called 'soft model theory', which deduces model-theoretic theorems from the fact that the formula defining satisfaction is set-theoretically absolute. [*See* §9.A.] [57] A. Robinson (1951: 19–21).

be 'possibly only in passing': if a structure $\mathcal{M}$ has elements with no corresponding individual constant, then new individual constants can be added for purposes of the truth definition. The assumption proved to be a valuable device for mathematical purposes, because it led directly to Robinson's notion of the *diagram* of a structure. The diagram D of $\mathcal{M}$ is the set of all atomic or negated atomic sentences true in $\mathcal{M}$, in a language where every element has a corresponding individual constant. [*See §15.4, footnote 35.*] Then $\mathcal{M}$ is embeddable in $\mathcal{N}$ if and only if $\mathcal{N}$ is a model of D.[58] Likewise, we can take the *complete diagram* of $\mathcal{M}$ to be the set of all first-order sentences true in $\mathcal{M}$ with constants for all elements; then $\mathcal{M}$ is elementarily embeddable in $\mathcal{N}$ if and only if $\mathcal{N}$ is a model of the complete diagram of $\mathcal{M}$. These devices became valuable tools of the paradigm shift which Robinson initiated, to make mappings between structures a central notion of model theory; see §18.5 below.

Robinson's truth definition was serendipity. His original reason for assuming the individual constants was that he learned his logic not from Tarski but from Rudolf Carnap, and Carnap had assumed that each element of a 'state-description'—his nearest counterpart of a structure—was named by a constant.[59] Carnap's involvement in this area was almost as old as Tarski's. In 1932, Gödel wrote to Carnap that he was intending to publish "eine Definition für 'wahr'";[60] this was in the context of arithmetic, where every element is named by a constant term. Gödel never published it, and we can only guess how it would have gone.[61]

Following Mal'tsev,[62] many authors have found it convenient to use the notion of the *signature* of a structure or a language, which is the set of relation, function and individual constant symbols of the language. [*See §1.1, Definition 1.1.*] An earlier notion playing a similar role was the *similarity type*, following McKinsey and Tarski: 'Two algebras [...] are called *similar* if the number of operations is the same in both algebras and if the corresponding operations [...] are operations with the same number of terms.'[63]

18.4 Building a structure

The method of quantifier elimination serves to analyse structures that we already have. But model theory relies also on methods for building new structures with specified properties.

In his paper of 1915 on the calculus of relatives, Löwenheim showed that every sentence of first-order logic, if it has a model, has a model with at most countably many

[58] Cf. A. Robinson (1956b: 24). [59] See for example Carnap's definition of 'holds in a state-description', Carnap (1947: 9). [60] Gödel (2003: 346–7). [61] See Feferman (1998). [62] Mal'tsev (1962). [63] McKinsey and Tarski (1944: 190).

elements. His proof has several interesting features, including his introduction of function symbols to reduce the satisfiability of a sentence:[64]

$$\forall x \exists y \varphi(x, y)$$

to the satisfiability of the sentence

$$\forall x \varphi(x, F(x)).$$

Thus it seems that Löwenheim invented Skolem functions, if we forgive him his bizarre explanation of the passage from the first sentence to the second. Löwenheim's starting assumption is that a given sentence φ is 'satisfied' in some domain; this means the same as saying that some structure is a model of φ, but Löwenheim never mentions the structure, which is another reason why his proof is hard to follow.[65]

Skolem tidied up Löwenheim's argument and strengthened the result.[66] He showed, using a coherent account of Skolem functions, that if T is a countable first-order theory with a model $\mathcal{M}$, then T has a model $\mathcal{N}$ with at most countably elements. (In fact he allowed countable conjunctions and disjunctions in the sentences of T too, and infinite quantifier strings.) The proof shows that $\mathcal{N}$ can be taken as an elementary substructure of $\mathcal{M}$, but at this date Skolem lacked even the notion of substructure. [*See §4.1 Definition 4.3, §3.8 Definition 3.7.*] Because values have to be chosen for the Skolem functions, and the starting structure need not allow these values to be defined explicitly (for example it may have too many automorphisms), Skolem had to assume the axiom of choice.

Skolem's argument was adapted and generalised in many ways. For example if κ is an infinite cardinal, $\mathcal{L}$ is a signature of cardinality at most κ, and $\mathcal{M}$ is a structure of signature $\mathcal{L}$ containing a set of elements X of cardinality at most κ, then $\mathcal{M}$ has an elementary substructure of cardinality at most κ containing all the elements of X. This is for first-order logic, but most logics allow analogous results. Theorems of this type came to be called *Downward Löwenheim–Skolem Theorems.* [*See §7.3, Theorem 7.2(1).*] Takeuti is said to have joked that Downward must be a very clever person to have so many theorems.

Later, Skolem made an adjustment of his argument which was fateful for model theory.[67] Starting from a structure $\mathcal{M}$, he built a new structure $\mathcal{N}$; but the elements of $\mathcal{N}$ were not elements of $\mathcal{M}$, they were all the ordinals below an ordinal α. (He chose $\alpha = \omega$, so that the elements of $\mathcal{N}$ were natural numbers.) The construction of $\mathcal{N}$ was inductive, with infinitely many steps. At each step a choice was made that ensured that certain elements would satisfy a certain formula. (For example

64 Löwenheim (1915: ¶4 in the proof of Theorem 2).
65 See the analysis in Badesa (2004).
66 Skolem (1920).
67 Skolem (1922).

if the formula was $\exists xR(x, y)$ and n was a given natural number, then it might be specified that $R(m, n)$ holds, where m is the first natural number not so far used; the well-ordering of α made this choice well-defined.) Some combinatorics was invoked to ensure that by the end of the construction $\mathcal{N}$ would have all the required properties.[68]

Using this scheme, Skolem showed, without using the axiom of choice, that if T is a countable first-order theory and T has a model, then T has a model with at most countably many elements.[69] It was on this basis that he stated *Skolem's Paradox*: if Zermelo–Fraenkel set theory is consistent then it has a countable model, so that 'There are uncountable cardinals' is satisfied in a countable domain. [*See §8.2.*]

The scheme allows many variations: a larger ordinal can be used, different starting assumptions can be fed in, different combinatorics can be invoked. The earliest variation came in the 1930 doctoral thesis of Gödel.[70] Gödel started not with Skolem's assumption that the theory T has a model, but with the assumption that no contradiction can be deduced from T within a standard proof calculus. In this way Gödel proved *completeness* for first-order logic: if no contradiction can be deduced from the countable first-order theory T, then T has a model. [*See §4.A, Theorem 4.24.*] Using the fact that proofs are finite, he pointed out the consequence that a countable first-order theory has a model if and only if every finite subset of it has a model; this is the *Compactness Theorem* for countable first-order logic. [*See §4.1, Theorem 4.1.*] In fact we can prove the Compactness Theorem without mentioning formal deductions, by moving back halfway to Skolem's construction; instead of assuming, as Skolem did, that T has a model, we assume that every finite subset of T has a model. (This device is not in Gödel's paper, but later it became common knowledge.)

To prove Completeness for uncountable theories in first-order logic, the same scheme works but with an uncountable cardinal in place of ω, and more careful combinatorics to justify the induction. This was done first by Mal'tsev (1936), and later but independently by Leon Henkin and by Abraham Robinson.[71] Probably the version most commonly used today is Henkin's neat second attempt, as filtered through Gisbert Hasenjaeger.[72] Henkin's method prepares the theory before the inductive construction begins. The preparation includes expanding T to a maximal syntactically consistent set—which in general requires the axiom of choice. Again we can convert the proof to a proof of the Compactness Theorem for first-order theories of any cardinality, by the same device as in the previous paragraph. [*See §§4.A–4.B.*]

[68] In Skolem (1922) a finite set of alternative choices were made at each step, creating a tree of choices; then a form of König's tree lemma was invoked to ensure that at least one branch of the tree is infinite and hence meets the requirements. [69] Skolem (1922). [70] Gödel (1931). There is some doubt how far Gödel was aware of Skolem (1922); see van Atten and Kennedy 2009. [71] Henkin (1949) from his PhD thesis of 1947; A. Robinson (1951) from his PhD thesis of 1949. [72] Hasenjaeger (1953).

Another variation of Skolem's scheme is omitting types. The *type* of a tuple $\bar{a}$ of elements in a structure $\mathcal{N}$ is the set $\Phi(\bar{x})$ of all formulas $\varphi(\bar{x})$ such that $\bar{a}$ satisfies $\varphi(\bar{x})$ in $\mathcal{N}$; $\mathcal{N}$ is said to *realise* the types of its tuples of elements. If X is a set of elements of N and the formulas $\varphi(\bar{x})$ are allowed to contain constants for the elements of X, we say that $\Phi(\bar{x})$ is a *type over X*. [*See §14.1.*]

The type of a tuple $\bar{a}$ of elements of $\mathcal{N}$ is an infinite set of formulas. This allows the possibility that the type of $\bar{a}$ is not yet determined at any finite step in the construction of $\mathcal{N}$; so if $\Phi(\bar{x})$ is a particular set of formulas, we have enough opportunities in the construction to ensure that the type of $\bar{a}$ in $\mathcal{N}$ is not $\Phi(\bar{x})$. If $\mathcal{N}$ is countable then there are countably many tuples of elements, and we can interweave the requirements so as to ensure that each of countably many sets $\Phi(\bar{x})$ is *omitted* in $\mathcal{N}$, in the sense that no tuple in $\mathcal{N}$ has $\Phi(\bar{x})$ as its type. (This presupposes that the sets $\Phi(\bar{x})$ are *non-principal*, i.e. not determined by a finite part of themselves.) Each set omitted can be 'over' a finite number of elements of $\mathcal{N}$.

In 1959 Vaught gave the classic omitting types theorem for countable models of complete first-order theories.[73] This theorem allows one to omit countably many types at once; Vaught attributes this feature to Andrzej Ehrenfeucht. The paper also contains *Vaught's Conjecture* as a question: 'Can it be proved, without the use of the continuum hypothesis, that there exists a complete theory having exactly $\aleph_1$ non-isomorphic denumerable models?' (The Conjecture is that there is no such theory. Some special cases of the Conjecture have been proved; at the time of writing it is still unresolved whether a counterexample has been given.)

There were several close variants of omitting types. The Henkin–Orey theorem was one that appeared before Vaught's paper, while Robinson's finite forcing and Grilliot's theorem on constructing families of models with few types in common were two that came later.[74] Martin Ziegler made finite forcing more palatable by recasting it in terms of Banach–Mazur games;[75] the same recasting works for all versions of omitting types.

Finite forcing builds existentially closed models; these were introduced into model theory by Michael Rabin and Per Lindström.[76] During the 1970s Oleg Belegradek, Ziegler, Saharon Shelah and others put a good deal of energy into constructing existentially closed groups, after Macintyre had shown that they have remarkable definability properties.[77]

Skolem's scheme also allows the use of set-theoretic prediction principles. These are set-theoretic statements, some provable in Zermelo–Fraenkel set theory and some true in the constructible universe or merely consistent, which tell us that certain things are guaranteed to happen a large number of times (for example on a

---

[73] Vaught (1961). [74] Orey (1956), Barwise and A. Robinson (1970), and Grilliot (1972). [75] Ziegler (1980). [76] Rabin (1964) and Lindström (1964). [77] Macintyre (1972).

stationary subset of an uncountable cardinal). Such principles were first pointed out by Ronald Jensen;[78] Shelah added Jensen's principles and some of his own to the arsenal of model-theoretic techniques.[79] In this work, the boundaries between set theory, model theory, and abelian group theory become very thin.

The Compactness Theorem can often allow us to build structures without having to go through the combinatorics needed to prove the Compactness Theorem itself. For example, given the Compactness Theorem, it is easy to prove that if λ is an infinite cardinal, $\mathscr{L}$ is a signature of cardinality at most λ, L is a first-order language of signature σ, and $\mathcal{M}$ is an infinite structure of signature $\mathscr{L}$ with fewer than λ elements, then $\mathcal{M}$ has an elementary extension of cardinality λ. One takes the complete diagram of $\mathcal{M}$, adds λ new individual constants together with inequations to express that the new constants stand for distinct elements, and then notes that every finite subset of the resulting theory has a model by interpreting the finitely many new constants in $\mathcal{M}$. This result became known as the *Upward Löwenheim–Skolem–Tarski Theorem*—though Tarski's name was generally dropped. [*See §7.3, Theorem 7.2(2).*] The irony was that it was Skolem,[80] not Tarski, who for anti-platonist reasons refused to accept that the theorem was true (though he allowed that it might be deducible within some formal set theories).

The Upward Löwenheim–Skolem Theorem above was first stated by Tarski and Vaught, though the proof above by Compactness was essentially as in Mal'tsev's proof of a weaker result.[81] Tarski had claimed in 1934 that in 1927/8 he had proved that every consistent first-order theory with no finite model has a model with uncountably many elements.[82]

Combinatorics could be added to Compactness to get further results. Ehrenfeucht and Mostowski showed, using Compactness and Ramsey's Theorem, that if T is a complete first-order theory with infinite models and $(X, <)$ is a linearly ordered set, then T has a model $\mathcal{M}$ whose domain includes X, and for each finite n, any two strictly increasing n-tuples from X satisfy the same formulas in $\mathcal{M}$.[83] Thus $(X, <)$ is what later came to be called an *indiscernible sequence* in $\mathcal{M}$. [*See §15.5, Definition 15.20.*] If $\mathcal{M}$ is the closure of X under Skolem functions (as we can always arrange), $\mathcal{M}$ is said to be an *Ehrenfeucht–Mostowski model* of T.

Ehrenfeucht–Mostowski models have tightly controlled properties. For example they realise few types (see their use in §18.7 below). By choosing $(X, <)$ and $(X', <')$ sufficiently different, we can often ensure that the Ehrenfeucht–Mostowski models constructed over these two ordered sets are not isomorphic; this is the basic idea underlying many of Shelah's constructions of large families of nonisomorphic

[78] Jensen (1972). [79] See for example the use of Shelah's 'black box' to construct abelian groups with interesting properties, in Corner and Göbel (1985). [80] Skolem (1955). [81] Tarski and Vaught (1958) and Mal'tsev (1936). [82] The claim is in a note added by the editors to the end of Skolem (1934). Vaught (1954: 160) reports the few facts that are known about this early proof by Tarski. [83] Ehrenfeucht and Mostowski (1956).

models (again see §18.7). One can also construct Ehrenfeucht–Mostowski models of infinitary theories, using various theorems of the Erdős–Rado partition calculus in place of Ramsey's Theorem. As a byproduct we get a versatile way of building *two-cardinal models*, i.e. models of first-order theories in which some definable parts have one infinite cardinality and others have another infinite cardinality, as Michael Morley showed.[84] (Vaught had obtained two-cardinal results earlier by other methods.)

In his doctoral dissertation of 1966 Jack Silver, building on work of Haim Gaifman and Frederick Rowbottom, showed that if the set-theoretic universe contains a measurable cardinal (or even an Erdős cardinal), then the constructible universe forms an Ehrenfeucht–Mostowski model whose indiscernibles are a class of ordinals which includes all uncountable cardinals. Silver's dissertation was published as 'Some applications of model theory in set theory';[85] but the Silver indiscernibles rapidly took on a life of their own as one of the fundamental notions of large cardinal theory.

Other proofs of the Compactness Theorem were found later. Among the most elegant, one was found by Edward Frayne, Anne Morel, and Dana Scott using ultraproducts (on which see §18.6 below),[86] after Tarski had noticed that reduced products can be used to prove Compactness for sets of Horn sentences. [*See §13.c, Corollary 13.23.*] A quirky but extremely neat proof of the Compactness Theorem was found later by Itai Ben-Yaacov, using a fragment of first-order logic called positive logic.[87]

There is another general procedure for building structures; it goes by the name of *interpretation*. [*See Chapter 5.*] We illustrate with the familiar construction of the field $\mathbb{Q}$ of rational numbers from the ring $\mathbb{Z}$ of integers. Suppose $\mathbb{Z}$ is given. We select a definable relation on $\mathbb{Z}$, namely the set of all ordered pairs (m, n) with $n \neq 0$; a formula $\varphi_{\text{dom}}(x, y)$ defines this relation in $\mathbb{Z}$. We define an equivalence relation on these pairs: $(m, n) \sim (m', n')$ if and only if $mn' = m'n$; a formula $\varphi_{\sim}(x, y, x', y')$ defines this relation. The elements of the structure $\mathbb{Q}$ will be the equivalence classes of $\sim$. We define the operation $\times$ on the equivalence classes, by defining it on representatives:

$$(m, n) \times (m', n') = (m'', n'') \text{ iff } mm'n'' = m''nn'.$$

Again this is definable in $\mathbb{Z}$ by a formula $\varphi_{\times}(x, y, x', y', x'', y'')$. Likewise with $+$ and $-$, and $^{-1}$ too if we find a suitable conventional value for 0^{-1}. The outcome is that the instructions for building $\mathbb{Q}$ from $\mathbb{Z}$ are coded up as a bundle Γ of formulas in the language of $\mathbb{Z}$, indexed by the operations of $\mathbb{Q}$ together with formulas defining the

[84] Morley (1965b). [85] Silver (1971). [86] Frayne et al. (1962/1963). [87] Ben-Yaacov (2003).

equivalence classes that form the elements of $\mathbb{Q}$. We can summarise the situation by writing $\mathbb{Q} = \Gamma(\mathbb{Z})$. The bundle Γ is the *interpretation.*

Note that if R is any other integral domain then we can form $\Gamma(R)$ with the same Γ; it will be the field of fractions of R. Note also that if ψ is any sentence in the first-order language of $\mathbb{Q}$, then via Γ there is a sentence ψ^Γ such that ψ^Γ holds in R if and only if ψ holds in $\Gamma(R)$. If ψ^Γ can be effectively calculated from ψ, and the set of sentences true in R is recursive, then it is decidable whether or not ψ holds in $\Gamma(R)$.

Mostowski, Tarski, Mostowski, et al., Mal'tsev, and Ershov gave definitions of the notion of interpretation.[88] To construct the domain of the new structure, Mostowski and Tarski used single elements; Mal'tsev used ordered triples of elements, and Ershov introduced a definable equivalence relation on n-tuples. In the 1970s model theorists became interested in the question what structures are interpretable in a given structure, and Ershov's notion of interpretation was generally the one they used. Shelah described how one might think of the elements of structures interpretable in a structure $\mathcal{M}$ as *imaginary elements* of $\mathcal{M}$.[89]

Hilbert and Bernays noticed that if the theory T, suitably encoded as a set of natural numbers, is definable in the structure $\mathbb{N}$ of natural numbers, then Gödel's completeness proof can be carried out within first-order arithmetic, and the effect is that the built model $\mathcal{N}$ of T has the form $\Gamma(\mathbb{N})$ for an interpretation Γ defined in terms of T.[90] They also put a bound on the arithmetical complexity of the relations of $\mathcal{N}$. This suggestive result points in a number of directions; we mention two.

One direction is to consider structures that are encoded in the natural numbers in such a way that all their relations and functions are recursive. Model theory with the structures taken to be of this form is called *recursive model theory*. Mal'tsev took some early steps in this direction.[91] The textbook of Sergei Goncharov and Ershov could cite nearly 400 references.[92]

Another direction is to exploit the idea of doing model theory within arithmetic, for example constructing models of arithmetic within arithmetic. Ideas akin to this allowed Jeff Paris and Leo Harrington to find, for the first time, a naturally occurring theorem of arithmetic that is provable in set theory but independent of the first-order Peano axioms.[93]

18.5 Maps between structures

During the period 1930–50, mathematicians generally had begun to take a closer interest in the maps between structures. This was the period that saw the invention

[88] Mostowski (1948: 270), Tarski, Mostowski, et al. (1953: 20ff), Mal'tsev (1960a), and Ershov (1974). [89] Shelah (1978: chIII, §6). [90] Hilbert and Bernays (1939). [91] Mal'tsev (1960b). [92] Goncharov and Ershov (1999); see also Ershov et al. (1998). [93] Paris and Harrington (1977).

of category theory. The trend naturally made its way into model theory.

Garrett Birkhoff published his famous characterisation of the classes of models of sets of identities in 1935.[94] Birkhoff's paper uses a number of straightforward model-theoretic facts about mappings, for example that universally quantified equations are preserved under taking homomorphic images; Edward Marczewski extended this fact to all positive first-order sentences and asked for a converse.[95] Tarski reported that his own work on formulas preserved in substructures (the Łoś–Tarski Theorem) was done in 1949–50.[96]

In §18.6 we will examine how these new ideas played out in model theory. In the present section we will see how maps between structures came to play a deeper role in model theory, not just as possible topics but as essential tools of the subject. One can trace this development to two model theorists, Abraham Robinson and Roland Fraïssé. I begin with Robinson.

Abraham Robinson

In his PhD thesis, Robinson considered two algebraically closed fields $\mathcal{M}$ and $\mathcal{N}$ of the same characteristic.[97] By juggling upwards and downwards Löwenheim–Skolem arguments, he found algebraically closed fields $\mathcal{M}^*$ and $\mathcal{N}^*$ which both have transcendence degree ω, such that the same first-order sentences hold in $\mathcal{M}^*$ and $\mathcal{M}$ (so that $\mathcal{M}$ and $\mathcal{M}^*$ have the same characteristic), and the same holds for $\mathcal{N}$ and $\mathcal{N}^*$. Then he quoted Steinitz's Theorem, that two algebraically closed fields of the same characteristic and the same transcendence degree are isomorphic. From this he deduced that the same first-order sentences hold in $\mathcal{M}^*$ and $\mathcal{N}^*$, and hence also in $\mathcal{M}$ and $\mathcal{N}$. So the first-order theory of algebraically closed fields of a given characteristic is a complete theory—it settles all questions in the language.

There were two major novelties here. First, Robinson used a known algebraic fact about maps between structures (Steinitz's Theorem) in order to deduce a model-theoretic conclusion. Second, he used complete diagrams so as to construct elementary embeddings. At this date the use of elementary embeddings was only implicit. [*See §4.1, Definitions 4.3–4.4.*] Tarski defined elementary extensions in 1952/3 (though at that date he called them arithmetical extensions) and published them some years later.[98] Conspicuously, Tarski failed even then to define elementary embeddings; 'elementary imbeddings' [sic] appeared in a paper first published in 1961.[99]

Between Robinson's doing this work and publishing it, Tarski published the completeness of the theory of algebraically closed fields of a given characteristic, which he had discovered by the method of quantifier elimination. So the method of quantifier elimination gave Robinson's result, together with other results that didn't

---

[94] Birkhoff (1935). [95] Marczewski (1951). [96] Tarski (1954). [97] A. Robinson (1951: 59–60). [98] Tarski and Vaught (1958). [99] Kochen (1961).

obviously yield to Robinson's new methods. The next few years saw Robinson working hard to extend his methods to capture Tarski's results and more besides. To this work we owe the notions of model completeness, model companion, differentially closed field, an amalgamation criterion for quantifier elimination, model-theoretic forcing, and Robinson's joint consistency theorem that gave the Craig Interpolation Theorem.

Vaught was one of the first model theorists to exploit the new methods. For example he pointed out, using essentially Robinson's argument, that any countable theory that is λ-categorical for some infinite λ and has no finite models must be complete; this is *Vaught's Test.*[100] [*See §3.B, Proposition 3.10.*] Robinson wrote appreciatively of Vaught's Test, noting that his own argument could be simplified by taking λ uncountable.[101]

Roland Fraïssé

In 1953/4 Fraïssé published two papers in which he pointed out that certain countable structures are in a sense determined by the families of finite structures embeddable in them.[102] Taking the ordered set of rational numbers as a paradigm, he made two important observations.

(a) We can characterise those classes of finite structures which are of the form

all finite structures embeddable in $\mathcal{M}$

for some countable structure $\mathcal{M}$. (Following Fraïssé I shall call these *γ-classes*—it is not a standard name.)

(b) A γ-class has the amalgamation property if and only if $\mathcal{M}$ can be chosen to be homogeneous, and in this case $\mathcal{M}$ is determined up to isomorphism by the γ-class. (A class **K** has the *amalgamation property* if for all embeddings $e_1 : A \longrightarrow B_1$ and $e_2 : A \longrightarrow B_2$ within **K** there are embeddings $f_1 : B_1 \longrightarrow C$ and $f_2 : B_2 \longrightarrow C$, also within **K**, such that $f_1 \circ e_1 = f_2 \circ e_2$. A is *homogeneous* if every isomorphism between finite substructures of $\mathcal{A}$ extends to an automorphism of $\mathcal{A}$.)

By observation (a), Fraïssé introduced into model theory a kind of Galois theory of structures: it invited one to think of a structure as built up by a pattern of amalgamated extensions of smaller structures. This idea became important in stability theory.

By observation (b), Fraïssé introduced the amalgamation property into model theory (though the name came later). Also he provided a way of building countable structures by assembling a suitable γ-class of finite structures; intuitively, one keeps extending in all possible ways, amalgamating the resulting extensions as one goes.

---

[100] Vaught (1954). [101] A. Robinson (1956b: 11). [102] Fraïssé (1953, 1954a).

His version of the idea was modest, but it continues to be widely used as a source of ω-categorical structures. Ehud Hrushovski used a version of it to construct his 'new strongly minimal set'.[103] [*See §17.3.*]

In 1956 and 1960 Bjarni Jónsson, who had reviewed Fraïssé's 1953-paper, published two papers removing the limitation to finite and countable structures in Fraïssé's construction of homogeneous structures.[104] The price he had to pay was that the generalised continuum hypothesis was needed at some cardinals. Morley realised almost at once that, thanks to the Compactness Theorem, Jónsson's assumptions on the γ-class are verified if one considers the class of all 'small' subsets of models of a complete theory T and replaces embeddings by partial elementary maps—i.e. elementary maps defined on a subset of a model.[105] One feature of the resulting structures $\mathcal{M}$, at least under suitable conditions on the cardinals involved, was that if X was a set of elements of $\mathcal{M}$, of smaller cardinality than M itself, then every type of T over X would be realised in $\mathcal{M}$. This property of $\mathcal{M}$ was called *saturation* (generalising Vaught's notion of a saturated countable structure).[106]

The Morley–Vaught theory tells us that under suitable set-theoretic assumptions, every structure has a saturated elementary extension. These set-theoretic assumptions were always a stumbling block, and so weak forms of saturation were devised that served the same purposes without special assumptions. For example every structure has an elementary extension that is special.[107] Every countable structure has a recursively saturated elementary extension.[108] For every structure $\mathcal{M}$ and cardinal κ, $\mathcal{M}$ has an elementary extension that is κ-saturated, meaning that every type over fewer than κ elements is realised.

Saunders Mac Lane reports that when his student Morley first brought him the material that led to Morley and Vaught 1962, '[...] I said, in effect: "Mike, applications of the compactness theorem are a dime a dozen. Go do something better."'[109] Mac Lane adds that Morley's Theorem (see §18.7 below) was the fruit of this advice. [*See §17.3, Theorem 17.3.*]

In the 1970s there was some debate about how best to handle the Morley–Vaught γ-class. Gerald Sacks proposed one should think of it as a category with partial elementary maps as morphisms.[110] Shelah went straight to a very large saturated model C (but we never ask exactly how large); in his picture the γ-class is simply the class of all small subsets of the domain of C, and the partial elementary maps are the restrictions of automorphisms of C.[111] Shelah's view prevailed. The structure C was known as the *big model* or (following John Baldwin) the *monster model*. Studying

[103] Hrushovski (1992, 1993). [104] Jónsson (1956, 1960). [105] Vaught had come to similar conclusions independently. They published this in Morley and Vaught 1962. Morley and Vaught used a trick from Skolem 1920, adding relation symbols so that partial elementary maps become embeddings. [106] Vaught (1961). [107] Chang and Keisler (1990: 217). [108] Barwise and Schlipf (1976). [109] Mac Lane (1989). [110] Sacks (1972). [111] Shelah (1978: chI §1).

models of a complete first-order T, one could go to a monster model and restrict oneself to subsets of the domain of this model, and elementary maps between them.

In practice the monster model came to be used in a way that reflected Robinson's approach with complete diagrams. Morley and Vaught speak of Jerome Keisler's '"one element at a time" property'.[112] Keisler himself compared his procedure with the element-at-a-time methods used by Cantor and Hausdorff to build up isomorphisms between densely ordered sets.[113] Briefly, the idea was to define a partial elementary map by starting with a well-ordered listing of elements, say $(a_i : i < \kappa)$, and constructing a corresponding listing $(c_i : i < \kappa)$ by induction on i, so that each c_i realises the same type over $(c_j : j < i)$ as a_i realises over $(a_j : j < i)$. Then the mapping $a_i \mapsto c_i$ is elementary. An initial segment of $(c_i : i < \kappa)$ might be given by the problem in hand, and then κ-saturation was invoked to find the remaining elements. Amalgamations would be built up one element at a time: for example given $Y \supset X$ and an element b, one would amalgamate Y and $X \cup \{b\}$ over X, and speak of extending the type of b over X to a type over Y.

Around 1970 category theory was developing fast. People noted that by going with Shelah rather than with Sacks, the model-theoretic community had opted for the analogue of André Weil's 'universal domain',[114] rather than the more recent category-theoretic language of Grothendieck. But other model theorists kept the category connection alive. Michael Makkai and colleagues did some groundwork,[115] but the categorical approach never came to centre stage. Perhaps model theorists enjoy handling elements and dislike morphisms between theories. Nevertheless we can point to one useful outcome: Daniel Lascar visited Makkai and discussed with him the category of elementary embeddings between models of a complete theory. Lascar's enquiries threw up the idea of *Lascar strong type*,[116] which plays a significant role in the study of simple theories and elsewhere.

In the 1970s Saharon Shelah was looking for suitable abstract settings for work in stability theory for infinitary languages. He called one such setting *abstract elementary classes*.[117] An abstract elementary class is a class of structures of some given signature, together with a relation $\prec$ between structures, satisfying certain axioms. The axioms include a variant of Jońsson's axiom of unions of chains; they don't include joint embedding or amalgamation, though these two axioms are added for many applications. Shelah restored the amalgamation viewpoint with a vengeance:[118] to construct structures of cardinality ω_n from countable pieces, he formed n-dimensional amalgams. Shelah carries a remarkable amount of model

---

[112] Morley and Vaught (1962). [113] Keisler (1961: footnote on Theorem 2.2), his doctoral dissertation. See Cantor (1895) and Hausdorff (1908). [114] Weil (1946: ch.IX §1). [115] Makkai and Paré (1989). [116] Lascar (1982). [117] Shelah (1987a) and Grossberg (2002). [118] Shelah (1983).

theory over into the setting of abstract elementary classes, considering that the axioms make no reference to any language—in fact the blurb of his 2009a includes the remark that 'Abstract elementary classes provide one way out of the cul de sac of the model theory of infinitary languages which arose from over-concentration on syntactic criteria.' This is partly explained by Shelah's Presentation Theorem, which states that every abstract elementary class can be got by taking the class of all models of some given first-order theory which omit certain types, and then forming reducts to a smaller signature.

Abstract elementary classes turned out to be a suitable setting for various analogues of first-order model theory. For example Zilber, discussing his 'analytic Zariski geometries', used a notion of stability got by considering these geometries within a suitable abstract elementary class.[119] Also work of Hrushovski, Pillay, and Ben-Yaacov led to the notion of a *compact abstract theory*, or *cat* for short, which forms a setting for the model theory of Banach spaces or of Hilbert spaces.[120] The motivations behind cats and abstract elementary classes are different, but there are links.[121]

In 1964 Jan Mycielski noticed that Kaplansky's notion of an algebraically compact abelian group (today more often called a pure-injective abelian group) has a purely model-theoretic characterisation that is a close analogue of saturation.[122] With colleagues in Wrocław, Mycielski developed this observation into a theory of *atomic compact structures*, which was useful on the borderline between model theory and universal algebra.

Since atomic compact structures have a large amount of symmetry, they tend to have neat algebraic structural descriptions too; in fact this was the reason for Kaplansky's interest in them. To some extent the same holds for saturated structures, and even for κ-saturated structures when κ is large enough. For example in 1970 Paul Eklof and Edward Fischer (and independently Gabriel Sabbagh) noted that every ω_1-saturated abelian group is algebraically compact, and so one can read off the results of Wanda Szmielew's quantifier elimination for abelian groups rather easily from Kaplansky's structure theory.[123] Likewise, Ershov used ω_1-saturated Boolean algebras to recover Tarski's quantifier elimination results for Boolean algebras.[124] Clean methods of this kind quickly became standard practice.

18.6 Equivalence and preservation

Tarski tells us that by 1930 he had defined the relation of *elementary equivalence*, in modern symbols: $\mathcal{M} \equiv \mathcal{N}$ if the same first-order sentences are true in $\mathcal{M}$ as in

---

[119] Zilber (2010: 137). [120] Ben-Yaacov (2003). [121] See Baldwin (2009: 36), and his references there. [122] Mycielski (1964). [123] Eklof and Fischer (1972) and Szmielew (1955). [124] Ershov (1964).

$\mathcal{N}$.[125] [*See* §2.4, *Definition* 2.4.] But it was only in 1950 that he claimed to have a mathematical (as opposed to metamathematical) definition of this notion.[126] His definition went by cylindrifications and made no reference to sentences or formulas being satisfied in structures. In 1946 he had asked for 'a theory of [elementary] equivalence of algebras as deep as the notions of isomorphism, etc. now in use'.[127]

Model theorists evidently found Tarski's cylindrical definition of $\equiv$ unappealing, and soon two other 'mathematical' characterisations of the notion appeared.

Ultraproducts

In 1955 Jerzy Łoś described a construction based on Cartesian products $\mathcal{M} = \text{Prod}_{i \in I}\mathcal{N}_i$ of structures of some fixed signature $\mathcal{L}$.[128] An ultrafilter D on I (i.e. a maximal filter on the powerset $\wp(I)$) is given. [*See* §13.2, *Definition* 13.3.] Each relation symbol R of $\mathcal{L}$ is defined to hold of a tuple $\bar{a}$ of elements of $\mathcal{M}$ if and only if the set

$$\{i \in I : R\bar{x} \text{ is satisfied in } \mathcal{N}_i \text{ by the projection of } \bar{a} \text{ at } \mathcal{N}_i\}$$

is in the ultrafilter D; and corresponding clauses hold for function and constant symbols. Equality is read this way too, so that any two elements of the product are identified if and only if the set of indices where they agree is in D. [*See* §13.C.] The resulting structure is called an *ultraproduct* of the $\mathcal{N}_i$, or an *ultrapower* if the $\mathcal{N}_i$ are all equal. Łoś showed that if $\varphi(\bar{x})$ is a first-order formula of signature $\mathcal{L}$, and $\bar{a}$ a tuple of elements of the product, then $\bar{a}$ satisfies $\varphi(\bar{x})$ in the ultraproduct if and only if the set of indices i at which the projection of $\bar{a}$ satisfies $\varphi(\bar{x})$ in N_i is a set in the ultrafilter; this is *Łoś's Theorem*. [*See* §13.C, *Theorem* 13.22.] Łoś's Theorem was new, but it came to light that ultraproducts or their close relatives had been used earlier by Skolem, Hewitt, and Arrow.[129] Skolem's application was model-theoretic, to build a structure elementarily equivalent to the natural numbers with + and × but not isomorphic to them.

We remarked in §18.4 above that ultraproducts give a fast and efficient proof of the Compactness Theorem. It can be done in several ways. For example let T be a nonempty first-order theory such that every finite subset of T has a model. Let I be the set of finite subsets of T, and for each $i \in I$ let $\mathcal{N}_i$ be a model of i. For each sentence $\varphi \in T$ let X_φ be the set of finite subsets of T that contain φ. Then all intersections of finitely many sets X_φ are nonempty, so there is an ultrafilter D on I containing each X_φ. It follows at once by Łoś' Theorem that the resulting ultraproduct is a model of T. [*See* §13.C, *Corollary* 13.23.]

By suitable choice of index set and ultrafilter one can ensure that ultraproducts are κ-saturated, for any required κ. Keisler exploited this fact to show, with the

[125] Tarski (1935b: Appendix). [126] Tarski (1952: 712). [127] Tarski (2000: 27). [128] Łoś (1955b).
[129] Skolem (1931), Hewitt (1948), and Arrow (1950).

help of the generalised continuum hypothesis, that two structures are elementarily equivalent if and only if they have isomorphic ultrapowers.[130] Ten years later Shelah proved the same theorem without assuming the generalised continuum hypothesis, and hence gave a 'purely mathematical' characterisation of elementary equivalence.[131] Kochen gave another characterisation of elementary equivalence, using direct limits of ultrapowers.[132]

Thus it turned out that ultraproducts were useful largely because of their high saturation. Since there are other ways of getting highly saturated models of a theory, this made ultraproducts one of the less essential tools of model theory—though some model theorists keep them on hand as a concrete and transparent construction. There are also a few important theorems for which ultraproducts give the only known reasonable proofs; one is Keisler's theorem that uncountably categorical theories fail to have the finite cover property.[133] But they never achieved in model theory the central role that they came to play in set theory, thanks to Scott.[134]

Back-and-forth equivalence

Fraïssé found another way of characterising elementary equivalence without mentioning formulas. He described a hierarchy of interrelated families of partial isomorphisms between structures.[135] In terms of this hierarchy he gave necessary and sufficient conditions for two relational structures to agree in all prenex first-order sentences with at most n alternations of quantifier, where n is any natural number. So $\mathcal{M} \equiv \mathcal{N}$ if $\mathcal{M}$ and $\mathcal{N}$ agree in this sense for all finite n. Fraïssé's paper was unfortunately hard to read, and his ideas became known through a paper of Ehrenfeucht who recast them in terms of games.[136] Soon afterwards they were rediscovered by the Kazakh mathematician Asan Taimanov.[137]

In Ehrenfeucht's version, two players play a game to compare two structures $\mathcal{M}$ and $\mathcal{N}$. The players alternate; in each step, the first player chooses an element of one structure and the second player then chooses an element of the other structure. The second player loses as soon as the elements chosen from one structure satisfy a quantifier-free formula not satisfied by the corresponding elements from the other structure. (Mention of formulas here is easily eliminated.) This is the *Ehrenfeucht-Fraïssé back-and-forth game* on the two structures. For a first-order language with finitely many relation and individual constant symbols and no function symbols, one could show that $\mathcal{M}$ and $\mathcal{N}$ agree in all sentences of quantifier rank at most k if and only if the second player has a strategy that keeps her alive for at least k steps. [*See §16.6.*] Hence $\mathcal{M}$ is elementarily equivalent to $\mathcal{N}$ if and only if for each finite k, the second player can guarantee not to lose in the first k steps.

---

[130] Keisler (1961). [131] Shelah (1971a). [132] Kochen (1961). [133] Keisler (1967). [134] Scott (1961). [135] Fraïssé (1956). [136] Ehrenfeucht (1960/1961). [137] Taimanov (1962).

With this equipment it is very easy to show, for example, that if $\mathcal{G}, \mathcal{G}'$ are elementarily equivalent groups and $\mathcal{H}, \mathcal{H}'$ are elementarily equivalent groups, then the product group $\mathcal{G} \times \mathcal{H}$ is elementarily equivalent to $\mathcal{G}' \times \mathcal{H}'$.

The beauty of this idea of Fraïssé and Ehrenfeucht was that nothing tied it to first-order logic. Ehrenfeucht himself used it to prove the equivalence of various ordinal numbers as ordered sets with predicates for + and ×, in a language with a second-order quantifier ranging over finite sets.[138] Carol Karp adapted it to infinitary logics,[139] and it reappeared in Chen Chung Chang's construction of Scott sentences.[140] [*See §16.6 Theorem 16.4.*] Today, theoretical computer scientists know it in a thousand different forms.

We turn to applications of all this machinery. One striking application of elementary equivalence was Abraham Robinson's creation of nonstandard analysis in 1961.[141] [*See Chapter 4.*] He used the Compactness Theorem to form an elementary extension $^*\mathbb{R}$ of the field $\mathbb{R}$ of real numbers (with any further relations attached) containing infinitesimal elements. He noted that if a theorem of real analysis can be written as a first-order sentence φ, then to prove φ it suffices to use the infinitesimals to show that φ is true in $^*\mathbb{R}$ (a typical example of what Robinson called a *transfer argument*).

James Ax and Kochen in 1965/6 used the new model-theoretic methods to find a complete set of axioms for the field of p-adic numbers (uniformly for any prime p).[142] Their approach was completely different from the method of quantifier elimination, and it seems likely that any proof by that method would have been hopelessly unwieldy. Instead they considered saturated valued fields of cardinality ω_1. Using algebraic and number-theoretic arguments, Ax and Kochen were able to show that under certain conditions, any two such fields are isomorphic. They then wrote down these conditions as a first-order theory T. Assuming the generalised continuum hypothesis, any two countable models $\mathcal{M}, \mathcal{N}$ of T have saturated elementary extensions of cardinality ω_1, which are isomorphic, so that $\mathcal{M}$ and $\mathcal{N}$ must be elementarily equivalent. This proves the completeness of T (and hence its decidability since the axioms are effectively enumerable); a similar argument using saturated structures shows that T is model-complete, and one more push shows that the theory admits elimination of quantifiers. There are various tricks that one can use to eliminate the generalised continuum hypothesis.

This work of Ax and Kochen, together with very similar but independent work of Yuri Ershov,[143] marked the beginning of a long line of research in the model theory of fields with extra structure (for example with valuations or automorphisms). But it hit the headlines because it gave a proof of an 'almost everywhere' version of a

---

[138] Ehrenfeucht (1960/1961). [139] Karp (1965). [140] Chang (1968). [141] A. Robinson (1961).
[142] Ax and Kochen (1965a,b, 1966). [143] Ershov (1965).

conjecture of Emil Artin on c_2 fields. Since counterexamples to the full conjecture appeared shortly afterwards, 'almost everywhere' was about as much as one could hope for, short of an explicit list of the exceptions.

A notion different from elementary equivalence, but somewhere in the same ballpark, is as follows. Suppose F is a class of mappings between structures, and $\varphi(\bar{x})$ a formula. We say that F *preserves* $\varphi(\bar{x})$ if the following holds: *whenever* $f : \mathcal{M} \longrightarrow \mathcal{N}$ *is a mapping in* F *and* $\bar{a}$ *is a tuple of elements satisfying* $\varphi(\bar{x})$ *in* $\mathcal{M}$, *then* $f(\bar{a})$ *satisfies* $\varphi(\bar{x})$ *in* $\mathcal{N}$. A *preservation theorem* is a theorem characterising, for some class F of mappings, the class of formulas that are preserved by F. For example the Łoś–Tarski Theorem can be paraphrased as characterising the class of formulas preserved by embeddings between models of a given theory.[144]

Stretching the definition above a little, we say that a formula $\varphi(\bar{x})$ is *preserved in unions of chains* when for every chain $(\mathcal{M}_i : i < \beta)$ of structures with union $\mathcal{M}_\beta$ and every tuple $\bar{a}$ of elements of $\mathcal{M}_0$, if $\bar{a}$ satisfies $\varphi(\bar{x})$ in $\mathcal{M}_i$ for each $i < \beta$ then it also satisfies $\varphi(\bar{x})$ in $\mathcal{M}_\beta$. Chang and Łoś and Suszko showed that a first-order formula $\varphi(\bar{x})$ is preserved in unions of chains if and only if it is logically equivalent to a formula of the form $\forall y_1 \ldots \forall y_m \exists z_1 \ldots \exists z_n \psi$ where ψ has no quantifiers (such formulas are called $\forall_2$ formulas, or Π_2 formulas).[145] In the case where φ is a sentence (no free variables), the main thing to be proved is that if Θ is the set of all $\forall_2$ sentences θ that are consequences of φ, then every model of Θ is elementarily equivalent to the union of a chain of models of φ (and hence is a model of φ). This can be proved by building up a chain whose even-numbered members form an elementary chain of models of Θ, and whose odd-numbered members are models of φ.

The model-building techniques of the previous section were honed on this and many similar problems. The text of Chang and Keisler, first published in 1973, is a compendium of the main achievements of model theory up to that date.[146]

It was natural to ask how far the results of this section could be generalised to other languages; in the 1950s and 1960s this usually meant languages with infinitary features or generalised quantifiers. When someone had introduced a technique for first-order languages, he or she could move on to testing the same technique on stronger and stronger languages. Often a variant of the technique would still work, but set theoretic assumptions and arguments would begin to appear. An observation of William Hanf helped to organise this area: he noted that for any reasonable language L there is a least cardinal κ (which became known as the *Hanf number* of L) such that if a sentence of L has a model of cardinality at least κ then it has arbitrarily large models.[147] A great deal of work and ingenuity went into finding the Hanf numbers of a range of languages.

---

[144] Tarski (1954) and Łoś (1955a). [145] Chang (1959) and Łoś and Suszko (1957). [146] Chang and Keisler (1990). [147] Hanf (1960).

One effect of this trend was that during the period from 1950–70 the centre of gravity of research moved away from first-order languages and towards infinitary languages, bringing a heady dose of set theory into the subject. Allow me two anecdotes. In about 1970 a Polish logician reported that a senior colleague of his had advised him not to publish a textbook on first-order model theory, because the subject was dead. And in 1966 David Park, who had just completed a PhD in first-order model theory with Hartley Rogers at MIT, visited the research group in Oxford and urged us to get out of first-order model theory because it no longer had any interesting questions. (Shortly afterwards he set up in computer science, where he applied back-and-forth methods.)

18.7 Categoricity and classification theory

In 1959, Lars Svenonius showed that among countable structures, the models of ω-categorical theories are precisely those structures whose automorphism group has finitely many orbits of n-element sets, for each finite n.[148] Permutation groups with this property are said to be *oligomorphic*.[149] Other model theorists gave other characterisations of ω-categoricity.[150]

Łoś asked: If T is a complete theory in a countable first-order language, and T is λ-categorical for some uncountable λ, then is T also λ-categorical for every uncountable λ?[151] [*See* §*17.3.*] With hindsight we can see that this was an extraordinarily fortunate question to have asked in 1955, for two main reasons. The first was that at just this date the tools for starting to answer the question were becoming available. If T is λ-categorical and $\mathcal{M}$, $\mathcal{N}$ are models of T of cardinality λ which are respectively highly saturated and Ehrenfeucht–Mostowski, then $\mathcal{M}$ and $\mathcal{N}$ are isomorphic and we deduce that models of T of cardinality λ have very few types to realise. This is strong information. Thus Łoś's question 'stimulated quite a bit of the work concerning models of arbitrary complete theories'.[152]

Second, Łoś's question was unusual in that it called for a description of *all* the uncountable models of a theory. The answer would involve finding a *structure theorem* to explain how any model of the theory is put together. This pointed in a very different direction from Tarski's 'mutual relations between sentences of formalised theories and mathematical systems in which these sentences hold'.[153] One mark of the change of focus was that expressions like 'uncountably categorical' (i.e. λ-categorical for all uncountable λ) and 'totally categorical' (i.e. λ-categorical for all infinite λ), which originally applied to theories, came to be used chiefly for *models* of those theories. For example Walter Baur wrote of '$\aleph_0$-categorical modules'.[154]

148 Svenonius (1959). 149 Cf. Cameron (1990). 150 Notably Ryll-Nardzewski (1959). 151 Łoś (1954). 152 Vaught (1963). 153 Tarski (1954). 154 Baur (1975).

In 1965, Michael Morley answered Łoś's question in the affirmative; this is *Morley's Theorem.*[155] [*See §17.3, Theorem 17.3.*] Amid all the literature of model theory, Morley's paper stands out for its clarity, its elegance and its richness in original ideas. Morley's central innovation was *Morley rank,* which assigns an ordinal rank to each definable relation in any model of a theory T which is λ-categorical for some uncountable λ. (In Morley's presentation the rank was assigned to complete types, but later workers generally used the induced rank on formulas or definable relations.) Morley gave the name *totally transcendental* to theories that assign a Morley rank to all definable relations in their models; the terminology came from transcendental extensions in field theory. Morley conjectured that the Morley rank of any uncountably categorical structure (i.e. the Morley rank of the formula $x = x$) is always finite; this was proved soon afterwards by Baldwin and Zilber independently.[156] (As a special case, the Morley rank of an algebraic set over an algebraically closed field is equal to its Krull dimension and hence is finite.)

Baldwin and Lachlan reworked and strengthened Morley's results.[157] [*See §17.3, Theorem 17.6.*] Building on the unpublished dissertation of William Marsh,[158] they showed that each model of an uncountably categorical theory carries a definable *strongly minimal set* with an abstract dependence relation that defines a dimension for the model. Once the strongly minimal set is given, the rest of the model is assembled around it in a way that is unique up to isomorphism. They also showed that the number of countable models of such a theory, up to isomorphism, is either 1 or ω. [*See §17.2.*]

A few young researchers set to work to extend Morley's result to uncountable first-order languages. One of them was Frederick Rowbottom, who introduced the name 'λ-stable' for theories with at most λ types over sets of λ elements;[159] hence the name *stability theory* for this general area.

In 1969, Saharon Shelah began to publish in stability theory.[160] With his formidable theorem-proving skill, he reshaped the subject almost from the start (and some other model theorists fled from the field rather than compete with him). By 1971 he had proved the uncountable analogue of Morley's Theorem.[161] But more important, he had formulated a plan of action for classifying complete theories.

Ehrenfeucht had already noticed that a theory which defines an infinite linear ordering on n-tuples of elements must have a large number of non-isomorphic models of the same cardinality.[162] Shelah saw this result as marking a division between 'good' theories that have few models of the same cardinality, and 'bad' theories that have many. Shelah's strategy was to hunt for possible bad features that a theory might have (like defining an infinite linear ordering), until the list was so comprehensive that a theory without any of these features is pinned down to the point

[155] Morley (1965a). [156] Baldwin (1973) and Zilber (1974). [157] Baldwin and Lachlan (1971).
[158] Marsh (1966). [159] Rowbottom (1964). [160] Shelah (1969). [161] Shelah (1974a).
[162] Ehrenfeucht (1960/1961).

where we can list all of its models in a structure theorem. [*See* §17.2.] As Shelah once explained it in conversation, the outcome should be to show that whenever **K** is the class of all models of a complete first-order theory, 'if **K** is good, it is very very good, but if **K** is bad it is horrid'. Shelah coined the word *nonstructure* for the horrid case, and he suggested several definitions of nonstructure.[163] In one definition, a nonstructure theorem finds a family of 2^λ models of cardinality λ, none of which is elementarily embeddable in any other. In another definition, a nonstructure theorem finds two nonisomorphic models of cardinality λ that are indistinguishable by strong infinitary languages.

Pursuing this planned dichotomy, Shelah wrote some dozens of papers and one large and famously difficult book.[164] Shelah also wrote a number of papers on analogous dichotomies for infinitary theories or abstract classes of structures.[165] His own name for this area of research was *classification theory*. The name applies at two levels: first-order theories classify structures, and Shelah's theory classifies first-order theories.

Shelah himself sometimes suggested that his main interests lay on the nonstructure side:

> I was attracted to mathematics by its generality, its ability to give information where apparently total chaos prevails, rather than by its ability to give much concrete and exact information where we a priori know a great deal.[166]

We should be careful not to deduce too much from this. Shelah's own work on the 'good' side vastly expanded the range of the new tools introduced by Morley. Also it gradually came to light, again mainly through Shelah's own work, that there is not just one dichotomy between good and bad theories; there are many good/bad dichotomies, and they partition the world of complete first-order theories in a complicated pattern. Generally speaking, each dichotomy is defined by the fact that models of theories on the bad side of it have some combinatorial property.[167]

It seemed at first that a minimal requirement for any good structure theory was that the theory should be *stable*, i.e. λ-stable for some cardinal λ. For stable theories, Shelah introduced a notion of relative dependence called *forking*, which reduced to linear or algebraic dependence in classical structures. In terms of forking he defined a class of types which he called *regular*, which carry a dependence relation that gives a cardinal dimension to the set of elements realising them. By the late 1980s it was becoming clear that much of the resulting machinery still worked in theories that were not stable. For example forking still behaved well in a larger class of theories that Shelah had introduced under the name *simple*.[168]

---

[163] Shelah (1985). [164] Shelah (1978); the second edition in 1990 reports the successful completion of the programme for countable first-order theories in 1982. [165] E.g. Shelah (1978). [166] Shelah (1987b: 154). [167] At the time of writing, Gabriel Conant has a web page with a map of the main dichotomies: http://www.forkinganddividing.com. [168] Shelah (1980) and Kim (1998).

In stable theories any complete type is in a certain sense 'definable' by first-order formulas.[169] Shelah showed that the definition can always be taken over a *canonical base* which is a family of 'imaginary' elements of the model. A special case of his construction is André Weil's field of definition of a variety,[170] except that here the field of definition consists of ordinary elements, not imaginary ones. Bruno Poizat explained this in 1985 by showing that algebraically closed fields have *elimination of imaginaries*, in the sense that their genuine elements can stand in for their imaginary ones.[171]

Stable groups turned out to have an unexpectedly large amount of structure, much of which carried over to modules (which are always stable). Poizat created a rich theory of stable groups by generalising ideas from Baldwin and Jan Saxl, Zilber, and Cherlin and Shelah.[172] Poizat's framework allows one to rely on intuitions from algebraic geometry in handling stable groups; for example their behaviour is strongly influenced by their generic elements.

One response to the work of Morley and Shelah was to ask what their classifications meant in concrete mathematical situations. The result was a series of papers determining what structures in various natural classes were categorical, totally transcendental and so forth. The first nontrivial paper of this kind was by Joseph Rosenstein on ω-categorical linear orderings.[173] But certainly the most influential was a paper of Macintyre, where he showed that an infinite field is totally transcendental if and only if it is algebraically closed.[174]

Cherlin and Shelah showed that every superstable skew field is an algebraically closed field.[175] In the course of this and related work, both Zilber and Cherlin independently noticed that a group definable in an uncountably categorical structure has many of the typical features of an algebraic group; in Russia the group theorists Vladimir Remeslennikov and Alexandre Borovik were having similar thoughts. Cherlin conjectured that every totally transcendental simple group is up to isomorphism an algebraic group over an algebraically closed field.[176] This became known as *Cherlin's Conjecture*. It was an invitation to model theorists to blend their techniques with those of the classification of finite simple groups. In 2008 Tuna Altınel, Borovik, and Cherlin published a report on the substantial results achieved.[177]

In the preface to that work, the authors wisely comment:

> [...] much of the history of pure model theory, which underwent a revolution beginning in the late sixties, and even (or perhaps, particularly) for those who lived through much of the latter, is not easy to reconstruct in a balanced way.[178]

---

[169] Shelah (1971b), and independently Lachlan (1972). [170] Weil (1946: 68). [171] Poizat (1985: §16e). [172] Poizat (1985); cf. Baldwin and Saxl (1976), Zilber (1977), and Cherlin and Shelah (1980). [173] Rosenstein (1969). [174] Macintyre (1971). [175] Cherlin and Shelah (1980). [176] Cherlin (1979). [177] Altınel et al. (2008). [178] Altınel et al. (2008: xvii).

This is a warning to readers of the three sections below. These parts of model theory are still on the move. I have recorded events and discoveries as I learned of them at the time, but future historians will be much better placed to distinguish the chassis from the bumper stickers.

18.8 Geometric model theory

Geometric model theory classifies structures in terms of their combinatorial geometries and the groups and fields that are interpretable in the structures. The roots of this theory go back to work of Lachlan, Cherlin, and above all Zilber in stability theory in the 1970s, and for this reason the theory is also known as *geometric stability theory*.[179] But by the early 1990s it emerged that the same ideas sometimes worked well in structures that were by no means stable.

An abstract dependence relation gives rise to a combinatorial geometry—in what follows I say just 'geometry'. In this geometry certain sets of points are closed, i.e. they contain all points dependent on them. Zilber classified geometries into three classes:[180] (a) *trivial* or *degenerate*, where all sets of points are closed; (b) non-trivial locally modular, which are not trivial but if a finite number of points are fixed (i.e. made dependent on the empty set), then the resulting lattice is modular—for brevity this case is often called *modular*; (c) the remainder, known briefly as *non-modular*. Classical examples are: for (a), the dependence relation where an element is dependent only on sets containing it; for (b), linear dependence in a vector space; for (c), algebraic dependence in an algebraically closed field.

This classification made its way into model theory rather indirectly. Zilber was working on a proof that no complete totally categorical theory is finitely axiomatisable. (His first announcement of his proof of this result in 1980 was flawed by a writing-up error which was later repaired.)[181] In work on ω-categorical stable theories, Lachlan had introduced a combinatorial structure which he called a *pseudoplane*.[182] A key step in Zilber's argument was to show that no totally categorical structure contains a definable pseudoplane. From this he deduced that the geometry of the strongly minimal set must be either trivial or modular, and his main result followed in turn from this. Cherlin, on reading Zilber's 1980-paper and seeing the error, went to the classification of finite simple groups and proved directly that the strongly minimal set must be either trivial or modular.[183] This result has a purely group-theoretic formulation. In fact several people discovered it independently, and it became known as the *Cherlin–Mills–Zilber Theorem* in honour of three of them. Zilber's proof, which avoids the error mentioned above, reaches the result without the classification of finite simple groups.

---

[179] As in the title of Pillay (1996). [180] Zilber (1981). [181] Zilber (1980), and then Zilber (1993). [182] Lachlan (1973/74). [183] Cherlin, Harrington, et al. (1985).

Zilber also called attention to the following combinatorial configuration:[184]

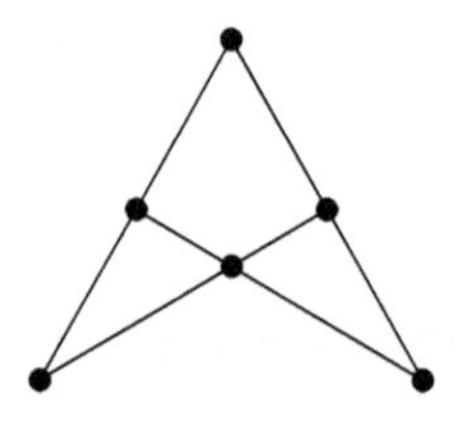

which occurs in modular strongly minimal sets. (The blobs are points of the geometry. All points are pairwise independent. A line between three points means they form a dependent set.) He showed how to construct a group from the configuration; but since this was in the middle of an argument by reductio ad absurdum and quite strong assumptions were in force, it was less than the definitive result. Hrushovski looked closer and showed, using Zilber's configuration, that every modular regular type has an infinite group interpretable in it (in a generalised sense).[185]

When Baldwin and Lachlan had shown that every uncountably categorical structure consists of a strongly minimal set D and other elements attached around it, they found they needed to say something about the way these other elements are attached.[186] Because of categoricity, something in the theory has to prevent the set of attached elements being larger than D. The simplest guess would be that each attached element has to satisfy an algebraic formula (i.e. one satisfied by only finitely many elements) with parameters in D. Baldwin and Lachlan finished their paper with a complicated example to show that this need not hold. Later Baldwin realised that an easy example was already to hand: a direct sum G of countably many cyclic groups of order p^2 for a prime p. The socle (the set of elements of order at most p) is strongly minimal, in fact a vector space over the p-element field. An element a of order p^2 is described by saying what pa is; but if b is any element of the socle then some automorphism of G fixes the socle pointwise and takes a to $a + b$. In fact the orbit of a over the socle is parametrised by elements of the socle. This parametrisation keeps the orbit from having cardinality greater than that of the socle.

Zilber realised that this was a common pattern in uncountably categorical structures.[187] Each such structure is a finite tower; at the bottom is a strongly minimal set, and as we go up the tower, the orbit of an element over the preceding level in the tower is always parametrised by some group interpretable in that preceding level. He called these groups *binding groups*. There are some cohomological

[184] Zilber (1984a: Lemma 3.3). [185] Hrushovski (1987). [186] Baldwin and Lachlan (1971).
[187] Zilber (1993).

constraints, which allowed Ahlbrandt and Ziegler to begin cataloguing the possibilities.[188] Cherlin and Hrushovski, drawing on these ideas of Zilber and work of Lachlan, proved deep classification results on families of finite structures.[189]

In the light of Zilber's work on uncountable categoricity and its extension by Cherlin, Harrington, and Lachlan,[190] model theorists looked to see what other structures might have modular geometries. One particularly influential result was proved independently by Hrushovski and Pillay, and published jointly:[191] a group $\mathcal{G}$ is modular (i.e. has only modular or trivial geometries) if and only if for each finite n, all definable subsets of G^n are Boolean combinations of cosets of definable subgroups.

We saw that Zilber first applied his trichotomy of geometries by showing that in the particular structures he was considering, the non-modular case never occurred. Zilber now proposed to apply the same trichotomy to another question, namely whether every simple group interpretable in an uncountably categorical structure must be an algebraic group over an algebraically closed field. (Cf. Cherlin's Conjecture above.) Algebraically closed fields themselves have non-modular geometry; at the 1984 International Congress Zilber conjectured the converse, viz. that any uncountably categorical structure with non-modular geometry must be—up to interpretability both ways—an algebraically closed field.[192] This was known as *Zilber's Conjecture*. [*See* §17.3.]

A word about Zilber's motivation may be in order. Macintyre said in 1988 that 'Purely logical classification[s] give only the most superficial general information' (and attributed the point to Georg Kreisel).[193] Zilber was convinced that the opposite must be true: if classical mathematics rightly recognises certain structures as 'good', then it should be possible to say in purely model-theoretic terms what makes these structures good. In fact Zilber in conversation quoted Macintyre (1971) as an example of how a purely model-theoretic condition (total transcendence) can be a criterion for an algebraic property (algebraic closure). Zilber was also convinced that being a model of an uncountably categorical countable first-order theory is an extremely strong property with rich mathematical consequences, among them strong homogeneity and the existence of a definable dimension.

In 1988 Hrushovski refuted Zilber's Conjecture using an ingenious variant of Fraïssé's construction from §18.4 above.[194] But for both Zilber and Hrushovski this meant only that the right condition hadn't yet been found. Since it seemed to be particularly hard to recover the Zariski topology from purely model-theoretic data, a possible next step was to axiomatise the Zariski topology. This is not straightforward: it has to be done in all finite dimensions simultaneously, since the closed sets in dimension n don't determine those in dimension $n + 1$. But Hrushovski de-

[188] Ahlbrandt and Ziegler (1991). [189] Cherlin and Hrushovski (2003). [190] Cherlin, Harrington, et al. (1985). [191] Hrushovski and Pillay (1987). [192] Zilber (1984b). [193] Macintyre (1989). [194] Hrushovski (1993).

scribed a set of axioms, and Zilber and Hrushovski found that, by putting together what they knew, they could prove that Zilber's Conjecture holds for models of the axioms.[195]

Hrushovski proved, for the first time, the geometric Mordell–Lang Conjecture in all characteristics.[196] Key ingredients of his argument were the results on the Zariski topology and on weakly normal groups, and earlier results on the stability of separably closed fields and differentiably closed fields. Hrushovski went on to apply a similar treatment to the Manin–Mumford Conjecture.[197] This case was a little different: the structures in question were unstable. But Hrushovski showed that they inherited enough stability from a surrounding algebraically closed field; and in any case they were 'simple' in Shelah's classification.

18.9 Other languages

In 1885 Charles Peirce, fresh from inventing quantifiers, mentioned that the universal and the existential quantifier are not the only examples.[198] He gave the example of the quantifier 'For two-thirds of all x'. Unfortunately, nobody picked up Peirce's idea, until Mostowksi called attention to the quantifiers 'For at least $\aleph_\alpha$ x'.[199] Mostowski's paper was timely, because it was useful to have in the 1960s a variety of extensions of first-order logic for testing out new constructions. [*See Chapter 16.*]

Lindström 1969 was another timely paper, in which he gave model-theoretic necessary and sufficient conditions for a logic to have the same expressive power as first-order logic. His result suggested that it might be possible to fit the various logics studied during the previous decade into some higher organisation of logics, within a *generalised* (or *abstract*) *model theory*. Alas, the facts weren't there to support such a theory. The 1970s saw some valiant efforts in this direction, and by the mid-1980s a large amount was known about many different logics extending first-order logic.[200] But the most quotable outcome was that very few logics apart from first-order logic satisfy a Craig Interpolation Theorem.

The mathematical logicians within computer science shrugged their shoulders and asked what is the interest of a logic in which it is impossible to express everyday notions like connectedness, even on finite structures. Thus, for example, Yuri Gurevich:

> The question arises how good is first-order logic in handling finite structures. It was not designed to deal exclusively with finite structures. [...] One would like to enrich first-order logic so that the enriched logic fits better the case of finite structures.[201]

[195] Hrushovski and Zilber (1996). For details see Zilber (2010: Appendix B.2). [196] Hrushovski (1996). [197] Hrushovski (2001). [198] Peirce (1885). [199] Mostowski (1957). [200] See Barwise and Feferman (1985). [201] Gurevich (1984).

One solution was first-order logic with a fixed-point operator added, as proposed by Ashok Chandra and David Harel.[202] The model theory of this logic and its relatives were studied mostly by computer scientists, but this seems to be purely an accident of history; these languages would have been good to have available in the 1960s.

Also of interest to computer scientists were languages with only a finite number of variables. Michael Mortimer launched the topic by showing that in a signature with no function symbols, any consistent first-order sentence using at most two variables has a finite model.[203] Barwise and, independently, Neil Immerman showed how to modify Ehrenfeucht-Fraïssé games to languages with at most n variables;[204] Immerman called the result *pebble games*.

Tarski, in his truth definitions, had taken universal and existential quantification to be dual to each other. This was at variance with a tradition running from Aristotelian logic up to modern formal linguistics, according to which existential quantifiers should be read as disguised Skolem functions.[205] Henkin and Jaakko Hintikka (with the collaboration of Gabriel Sandu) brought this tradition into model theory.[206] These authors noted that by suppressing some of the arguments of the Skolem functions we can increase the expressive power of the language, so as to express independence of one variable from another (as in Henkin's *branching quantifiers*).[207] They also noted that the Skolem functions can be read as strategies for the player $\exists$ in a semantic game between players $\forall$ and $\exists$ that can be used to give a truth definition for sentences; the suppressed arguments correspond to places where the information available to $\exists$ in the game is imperfect.

For model theory a difficulty was that the Skolem function approach to existential quantifiers made it impossible to give a sensible interpretation of subformulas within the scope of an existential quantifier. This problem was resolved by Hodges,[208] who replaced the notion 'tuple $\bar{a}$ satisfies $\varphi(\bar{x})$ in $\mathcal{M}$' by the notion 'the set $\bar{a}, \bar{b}, \ldots$ of tuples satisfies $\varphi(\bar{x})$ in $\mathcal{M}$', thus introducing what Väänänen later called *team semantics*. Peter van Emde Boas noticed that some of the conditions in Hodges' truth definition were identical to conditions appearing in the study of database dependencies. This point was taken up by Väänänen and his colleagues in Helsinki, to develop a model theory of teams, with hopes of using it to bring logic to bear on questions in database theory, statistics, and quantum theory.[209]

---

[202] Chandra and Harel (1980). [203] Mortimer (1975). [204] Barwise (1977) and Immerman (1982). [205] See the references in Hodges (2015). [206] Henkin (1961) and Hintikka (1996). [207] See also Blass and Gurevich (1986). [208] Hodges (1997b). [209] Abramsky et al. (2016).

18.10 Model theory within mathematics

In their addresses to the 1950 International Congress of Mathematicians at Harvard and MIT, both Abraham Robinson and Tarski expressed the hope that the new subject of model theory—for which neither of them had a name yet—would contribute to algebra and beyond:

> [...] contemporary symbolic logic can produce useful tools—though by no means omnipotent ones—for the development of actual mathematics, more particularly for the development of algebra and, it would appear, algebraic geometry.[210]

> [Model theory has applications] which may be of general interest to mathematicians and especially to algebraists; in some of these applications the notions of [model theory] itself are not involved at all.[211]

Compare these remarks with the comment of Ludwig Faddeev, an observer on the sidelines, in the closing ceremony of the 2002 International Congress of Mathematicians at Beijing:

> Take for instance the sections of logic, number theory and algebra. The general underlining mathematical structures as well as language, used by speakers, were essentially identical.[212]

Job done, one might well say!

Model theory had grown fast. Already the Omega Group bibliography of model theory in 1987 ran to 617 pages.[213] By the mid-1980s there were too many dialects of mathematical model theory for anybody to be expert in more than a fraction. For example, very few model theorists could claim to understand both the work of Zilber and Hrushovski at the edge of algebraic geometry, and the studies by Immerman, Dawar, and other theoretical computer scientists on definable classes of finite structures.

Right from the beginning, model theorists found themselves engaging with other areas of mathematics. In the period from 1950 –70 most of these interactions were with set theory, not with algebra or number theory. From around 1970 there was less interaction with axiomatic set theory. But recent years have seen an increasing amount of discussion between model theory and descriptive set theory. For example when it was realised that Fraïssé's construction in §18.5 above had already been applied by Pavel Urysohn to finite metric spaces with rational distances, the way was open to apply ideas of topological dynamics to Fraïssé-type constructions, as in Alexander Kechris, Vladimir Pestov, and Stevo Todorcevic.[214]

[210] A. Robinson (1952: 694). [211] Tarski (1952: 717). [212] Li (2002: 35). [213] G. H. Müller et al. (1987). [214] Urysohn (1927) and Kechris et al. (2005).

The title of Kechris et al.'s paper mentions Ramsey's Theorem, a reminder that this combinatorial theorem was used in the 1950s to construct Ehrenfeucht–Mostowski models. Links between model theory and combinatorics never ceased, as witness the paper of Maryanthe Malliaris and Shelah relating Szemerédi's Regularity Lemma to the structure theory of graphs stable in the model-theoretic sense.[215] Shelah's book had a twenty-page Appendix of 'the combinatorial theorems needed in the book'.[216]

We have described above some of the interactions between model theory, algebraic geometry and number theory, mostly before the year 2000. More recent years have seen dramatic advances in this area, resting on the earlier work. One example is the *Pila–Wilkie Theorem*, which applies o-minimality in order to bound the numbers of rational points in various sets definable in the real numbers,[217] building on earlier work of Wilkie with o-minimal structures.

Another advance, also closely tied to earlier notions, is the work of Zlil Sela which gives a positive answer to Tarski's question whether all nonabelian finitely generated free groups are elementarily equivalent.[218] The proof ran through several papers and involved building an analogue of diophantine geometry for such groups.[219]

These examples can serve as an indication that future historians of model theory will have plenty of high quality material to write about.

18.11 Notes

Several model theory texts give more detailed historical information about particular theorems; for example Chang and Keisler 1990, Hodges 1993, and Pillay 1996. Dawson 1993 and Lascar 1998 both overlap the present essay. There are surveys on the model-theoretic work of Skolem by Hao Wang (Skolem 1970, 17–52) and on that of Tarski in Vaught 1986.

18.12 Acknowledgments

This essay was originally written in the 1990s at the invitation of Dirk van Dalen for a projected volume on the history of mathematical logic, which sadly never appeared. It was good that Tim Button and Sean Walsh offered a congenial home for it. For their volume I made some corrections and brought the perspective more up to date, but otherwise the content is unaltered.

---

[215] Malliaris and Shelah (2014). [216] Shelah (1978). [217] Pila and Wilkie (2006). [218] Sela (2006). [219] See also Kharlampovich and Myasnikov (2006) for an alternative approach.

Everybody I ever encountered in model theory should be thanked for their implicit contributions. But I tried to keep a note of those people who helped with specific points in it, and the list is as follows: Zofia Adamowicz, Bektur Baizhanov, John Baldwin, Oleg Belegradek, Tim Button, Greg Cherlin, John W. Dawson, John Doner, Yuri Ershov, Solomon Feferman, Ivor Grattan-Guinness, Marcel Guillaume, Angus Macintyre, Dugald Macpherson, Maria Panteki, Anand Pillay, Gabriel Sabbagh, Hourya Sinaceur, Jouko Väänänen, Robert Vaught, Jan Woleński, Carol Wood, Boris Zilber, Jan Zygmunt. Very probably other people have slipped through the net—my apologies to them.

Bibliography

Abramsky, S. et al., eds. (2016). *Dependence Logic: Theory and Applications*. Basel: Birkhäuser.

Adams, R. M. (1974). 'Theories of actuality'. *Noûs* 8, pp.211–31.

Addison, J. et al., eds. (1965). *The Theory of Models*. Amsterdam: North-Holland.

Ahlbrandt, G. and M. Ziegler (1991). 'What's so special about $(\mathbf{Z}/4\mathbf{Z})^{\omega}$?' *Archive for Mathematical Logic* 31.2, pp.115–32.

Ainsworth, P. M. (2009). 'Newman's objection'. *British Journal for the Philosophy of Science* 60.1, pp.135–71.

Altınel, T. et al. (2008). *Simple Groups of Finite Morley Rank*. Providence, RI: American Mathematical Society.

Arana, A. and M. Detlefsen (2011). 'Purity of methods'. *Philosopher's Imprint* 11.2, pp.1–20.

Arrow, K. J. (1950). 'A difficulty in the concept of social welfare'. *The Journal of Political Economy*, pp.328–46.

Artin, E. (1957). *Geometric Algebra*. New York: Interscience.

Ash, C. and J. Knight (2000). *Computable Structures and the Hyperarithmetical Hierarchy*. Amsterdam: North-Holland.

Atten, M. van and J. Kennedy (2009). 'Gödel's logic'. In: *Logic from Russell to Church*. Ed. by D. M. Gabbay and J. Woods. Amsterdam: North-Holland, pp.449–509.

Awodey, S. (2010). *Category Theory*. Oxford: Oxford University Press.

Awodey, S. and E. H. Reck (2002). 'Completeness and categoricity. I. Nineteenth-century axiomatics to twentieth-century metalogic'. *History and Philosophy of Logic* 23.1, pp.1–30.

Ax, J. and S. Kochen (1965a). 'Diophantine problems over local fields I'. *American Journal of Mathematics* 87, pp.605–30.

— (1965b). 'Diophantine problems over local fields II'. *American Journal of Mathematics* 87, pp.631–48.

— (1966). 'Diophantine problems over local fields III'. *Annals of Mathematics* 83, pp.437–56.

Bader, R. M. (2011). 'Supervenience and infinitary property-forming operations'. *Philosophical Studies* 160.3, pp.415–23.

Badesa, C. (2004). *The Birth of Model Theory: Löwenheim's Theorem in the Frame of the Theory of Relatives*. Princeton: Princeton University Press.

Balaguer, M. (1998a). 'Non-uniqueness as a non-problem'. *Philosophia Mathematica* 6.1, pp.63–84.

— (1998b). *Platonism and Anti-Platonism in Mathematics*. Oxford: Oxford University Press.

Baldwin, J. T. and A. H. Lachlan (1971). 'On strongly minimal sets'. *The Journal of Symbolic Logic* 36, pp.79–96.

Baldwin, J. T. and J. Saxl (1976). 'Logical stability in group theory'. *Journal of the Australian Mathematical Society* 21.3, pp.267–76.

Baldwin, J. T. (1973). 'a_T is finite for $\aleph_1$-categorical T'. *Transactions of the American Mathematical Society* 181, pp.37–51.

— ed. (1987a). *Classification Theory*. Berlin: Springer.

— (1987b). 'Classification theory: 1985'. In: Baldwin 1987a, pp.1–23.

— (1988). *Fundamentals of Stability Theory*. Berlin: Springer.

— (2009). *Categoricity*. Providence, RI: American Mathematical Society.

— (2014). 'Completeness and categoricity (in power): Formalization without foundationalism'. *Bulletin of Symbolic Logic* 20.1, pp.39–79.

Barton, N. (2016). 'Multiversism and concepts of set: How much relativism is acceptable?' In: *Objectivity, Realism, and Proof*. Switzerland: Springer, pp.189–209.

Barwise, J. (1972). 'Absolute logics and $L_{\infty\omega}$'. *Annals of Pure and Applied Logic* 4, pp.309–40.

— (1973). 'Back and forth through infinitary logic'. In: *Studies in Model Theory*. Ed. by M. Morley. Buffalo: Mathematical Association of America, pp.5–34.

— (1977). 'On Moschovakis closure ordinals'. *The Journal of Symbolic Logic* 42.2, pp.292–6.

Barwise, J. and S. Feferman (1985). *Model-Theoretic Logics*. New York: Springer.

Barwise, J. and A. Robinson (1970). 'Completing theories by forcing'. *Annals of Pure and Applied Logic* 2.2, pp.119–42.

Barwise, J. and J. Schlipf (1976). 'An introduction to recursively saturated and resplendent models'. *The Journal of Symbolic Logic* 41.2, pp.531–6.

Baur, W. (1975). '$\aleph_0$-categorical modules'. *The Journal of Symbolic Logic* 40, pp.213–20.

Bays, T. (2001). 'On Putnam and his models'. *The Journal of Philosophy* 98.7, pp.331–50.

— (2007). 'More on Putnam's models: A reply to L. Belloti'. *Erkenntnis* 67.1, pp.119–35.

— (2008). 'Two arguments against realism'. *The Philosophical Quarterly* 58.231, pp.193–213.

— (2014). 'Skolem's paradox'. *The Stanford Encyclopedia of Philosophy* Spring 2014 Edition. Ed. by E. N. Zalta.

Behmann, H. (1922). 'Beiträge zur Algebra der Logik, insbesondere zum Entscheidungsproblem'. *Mathematische Annalen* 86.3-4, pp.163–229.

Bell, J. L. (2005). *Set Theory: Boolean-Valued Models and Independence Proofs*. 3rd ed. Oxford: Clarendon Press.

Bellotti, L. (2005). 'Putnam and constructibility'. *Erkenntnis* 62.3, pp.395–409.

Benacerraf, P. (1960). 'Logicism, some considerations'. PhD Thesis, Princeton University.

— (1965). 'What numbers could not be'. *The Philosophical Review* 74.1, pp.47–73.

— (1973). 'Mathematical truth'. *The Journal of Philosophy* 70, pp.661–80.

— (1984). 'Comments on Maddy and Tymoczko'. *PSA: Proceedings of the Biennial Meeting of the Philosophy of Science Association* 2, pp.476–85.

— (1985). 'Skolem and the skeptic'. *Proceedings of the Aristotelian Society, Supplementary Volume* 59, pp.85–115.

— (1996). 'What mathematical truth could not be – I'. In: *Benacerraf and His Critics*. Ed. by A. Morton and S. P. Stich. Blackwell: Blackwell, pp.9–59.

Ben-Yaacov, I. (2003). 'Positive model theory and compact abstract theories'. *Journal of Mathematical Logic* 3.1, pp.85–118.

Beth, E. W. (1953). 'On Padoa's method in the theory of definition'. *Nederlandse Akademie van Wetenschappen. Proceedings. Series A. Indagationes Mathematicae* 15, pp.330–9.

Bezboruah, A. and J. C. Shepherdson (1976). 'Gödel's second incompleteness theorem for Q'. *The Journal of Symbolic Logic* 41.2, pp.503–12.

Birkhoff, G. (1935). 'On the structure of abstract algebras'. *Proceedings of the Cambridge Philosophical Society* 31.4, pp.433–54.

Black, M. (1952). 'The identity of indiscernibles'. *Mind* 61.242, pp.153–64.

Blanchette, P. A. (2001). 'Logical consequence'. In: *The Blackwell Guide to Philosophical Logic*. Malden: Blackwell, pp.115–35.

Blass, A. and Y. Gurevich (1986). 'Henkin quantifiers and complete problems'. *Annals of Pure and Applied Logic* 32.1, pp.1–16.

Bloom, S. L. and R. Suszko (1971). 'Semantics for the sentential calculus with identity'. *Studia Logica* 28, pp.77–82.

— (1972). 'Investigations into the sentential calculus with identity'. *Notre Dame Journal of Formal Logic* 13, pp.289–308.

Bonnay, D. (2008). 'Logicality and invariance'. *The Bulletin of Symbolic Logic* 14.1, pp.29–68.

— (2014). 'Logical constants, or how to use invariance in order to complete the explication of logical consequence'. *Philosophy Compass* 9.1, pp.54–65.

Bonnay, D. and F. Engström (2013). 'Invariance and definability, with and without equality'. URL: arXiv:1308.1565.

Boole, G. (1847). *The Mathematical Analysis of Logic. Being an Essay Towards a Calculus of Deductive Reasoning*. Cambridge: Macmillan.

Boolos, G. (1971). 'The iterative conception of set'. *The Journal of Philosophy* 68, pp.215–32.

— (1975). 'On second-order logic'. *The Journal of Philosophy* 72.16, pp.509–27.

— (1984). 'To be is to be the value of a variable (or some values of some variables)'. *The Journal of Philosophy* 81.8, pp.430–9.

Boolos, G. and R. Jeffrey (1974). *Computability and Logic*. Cambridge: Cambridge University Press.

Bos, H. J. M. (1974). 'Differentials, higher-order differentials and the derivative in the Leibnizian calculus'. *Archive for History of Exact Sciences* 14, pp.1–90.

Bourbaki, N. (1951). *Algèbre, ch. I*. Paris.

— (1954). *Théorie des ensembles, chs. I et II*. Paris.

Bouvère, K. de (1965a). 'Logical synonymity'. *Indagationes Mathematicae* 27, pp.622–9.

— (1965b). 'Synonymous theories'. In: Addison et al. 1965, pp.402–6.

Brandom, R. (1996). 'The significance of complex numbers for Frege's philosophy of mathematics'. *Proceedings of the Aristotelian Society* 96, pp.293–315.

Bricker, P. (1983). 'Worlds and propositions: The structure and ontology of logical space'. PhD thesis. Princeton University.

Buechler, S. (1996). *Essential Stability Theory*. Berlin: Springer.

Bueno, O. and Ø. Linnebo, eds. (2009). *New Waves in the Philosophy of Mathematics*. New York: Palgrave.

Burgess, J. P. (1985). 'Review of Reinhardt 1974 and others'. *The Journal of Symbolic Logic* 50.2, pp.544–7.

— (1999). 'Book review: Stewart Shapiro. *Philosophy of Mathematics: Structure and Ontology*'. *Notre Dame Journal of Formal Logic* 40.2, pp.283–91.

Buss, S. R. et al. (2001). *Bulletin of Symbolic Logic* 7.2, pp.169–96.

Button, T. (2006). 'Realistic structuralism's identity crisis: A hybrid solution'. *Analysis* 66.3, pp.216–22.

— (2010). 'Dadism: Restrictivism as militant quietism'. *Proceedings of the Aristotelian Society* 110.3, pp.387–98.

— (2011). 'The metamathematics of Putnam's model-theoretic arguments'. *Erkenntnis* 74.3, pp.321–49.

— (2013). *The Limits of Realism*. Oxford: Oxford University Press.

— (2014). 'The weight of truth: Lessons for minimalists from Russell's Gray's Elegy argument'. *Proceedings of the Aristotelian Society* 114.3, pp.261–89.

— (2016a). 'Brains in vats and model theory'. In: *The Brain in a Vat*. Ed. by S. Goldberg. Cambridge: Cambridge University Press, pp.131–54.

— (2016b). 'Knot and tonk: Nasty connectives on many-valued truth-tables for classical sentential logic'. *Analysis* 76.1, pp.7–19.

— (2017). 'Grades of discrimination: Identity, symmetry, and relativity'. *Notre Dame Journal of Formal Logic* 58.4, pp.527–53.

Button, T. and P. Smith (2012). 'The philosophical significance of Tennenbaum's theorem'. *Philosophia Mathematica* 20.1, pp.114–21.

Button, T. and S. Walsh (2016). 'Structure and categoricity: Determinacy of reference and truth value in the philosophy of mathematics'. *Philosophia Mathematica* 24.3, pp.283–307.

Butzer, P. L. and H. Berens (1967). *Semi-Groups of Operators and Approximation*. Springer.

Caleiro, C., W. Carnielli, et al. (2005). 'Two's company: "The humbug of many logical values"'. In: *Logica Universalis*. Ed. by J.-Y. Beziau. Basel: Birkhäuser, pp.169–89.

Caleiro, C. and J. Marcos (2009). 'Classic-like analytic tableaux for finite-valued logics'. *Lecture Notes in Computer Science* 5514, pp.268–80.

Caleiro, C., J. Marcos, and M. Volpe (2014). 'Bivalent semantics, generalized compositionality and analytic classic-like tableaux for finite-valued logics'. URL: arXiv:1408.3775.

Cameron, P. J. (1990). *Oligomorphic Permutation Groups*. Cambridge: Cambridge University Press.

Cantor, G. (1895). 'Beiträge zur Begründung der transfiniten Mengenlehre'. *Mathematische Annalen* 46.4, pp.481–512.

Carnap, R. (1943). *Formalization of Logic*. Cambridge MA: Harvard University Press.

— (1947). *Meaning and Necessity. A Study in Semantics and Modal Logic*. Chicago: The University of Chicago Press.

— (1958). 'Beobachtungssprache und theoretische Sprache'. *Dialectica* 12.3–4. Translated in Carnap 1975, pp.236–48.

— (1959). 'Theoretical concepts in science'. *Studies in History and Philosophy of Science* 31.1. Ed. by S. Psillos. Lecture delivered at American Philosophical Association, Pacific Division, at Santa Barbara, California, on 29 December 1959. Edited by S. Psillos. Published in 2000.

— (1963). 'Replies and expositions. Carl G. Hempel on scientific theories'. In: *The Philosophy of Rudolf Carnap*. Ed. by P. A. Schilpp. La Salle: Open Court, pp.958–66.

— (1966). *Philosophical Foundations of Physics: An Introduction to the Philosophy of Science*. Ed. by M. Gardner. New York: Basic Books.

— (1975). 'Observation language and theoretical language'. In: *Rudolf Carnap, Logical Empiricist*. Ed. by J. Hintikka. Dordrecht: Riedel, pp.75–85.

Casanovas, E. et al. (1996). 'On elementary equivalence for equality-free logic'. *Notre Dame Journal of Formal Logic* 37.3, pp.506–22.

Cassirer, E. (1998–2009). *Gesammelte Werke*. 26 volumes. Edited by Birgit Recki. Hamburg: Meiner.

— (1910). *Substanzbegriff und Funktionsbegriff*. Reprinted in Cassirer 1998–2009 vol. 6. Page references to translation by W. C. Swabey as *Substance and Function, and Einstein's Theory of Relativity* (1923), Chicago: Open Court. Berlin.

Caulton, A. and J. Butterfield (2012). 'On kinds of indiscernibility in logic and metaphysics'. *British Journal for the Philosophy of Science* 63, pp.27–84.

Chandra, A. K. and D. Harel (1980). 'Structure and complexity of relational queries'. In: *21st Annual Symposium on Foundations of Computer Science (Syracuse, NY, 1980)*. New York: IEEE, pp.333–47.

Chang, C. C. (1959). 'On unions of chains of models'. *Proceedings of the American Mathematical Society* 10, pp.120–7.

— (1968). 'Some remarks on the model theory of infinitary languages'. In: *The Syntax and Semantics of Infinitary Languages*. Ed. by J. Barwise. Springer, pp.36–63.

Chang, C. C. and H. J. Keisler (1990). *Model Theory*. 3rd ed. Amsterdam: North-Holland.

Cherlin, G. (1979). 'Groups of small Morley rank'. *Annals of Mathematical Logic* 17.1-2, pp.1–28.

Cherlin, G., L. Harrington, et al. (1985). '$\aleph_0$-categorical, $\aleph_0$-stable structures'. *Annals of Pure and Applied Logic* 28.2, pp.103–35.

Cherlin, G. and E. Hrushovski (2003). *Finite Structures with few Types*. Princeton, NJ: Princeton University Press.

Cherlin, G. and S. Shelah (1980). 'Superstable fields and groups'. *Annals of Mathematical Logic* 18.3, pp.227–70.

Chihara, C. S. (2004). *A Structural Account of Mathematics*. Oxford: Clarendon Press.

Church, A. (1944). 'Review of Carnap 1943'. *The Philosophical Review* 53.5, pp.493–8.

— (1953). 'Non-normal truth-tables for the propositional calculus'. *Boletín de la Sociedad Matemática Mexicana* 10.1–2.

— (1956). *Introduction to Mathematical Logic. Vol. 1*. Princeton, New Jersey: Princeton University Press.

Clarke-Doane, J. (2016). 'What is the Benacerraf problem?' In: *Truth, Objects, Infinity: New Perspectives on the Philosophy of Paul Benacerraf*. Ed. by F. Pataut. Springer, pp.17–43.

Cohen, P. J. (1963). 'The independence of the continuum hypothesis'. *Proceedings of the National Academy of Sciences of the United States of America* 50.6, pp.1143–8.

— (1964). 'The independence of the continuum hypothesis, II'. *Proceedings of the National Academy of Sciences of the United States of America* 51.1, pp.105–10.

Coppelberg, S. (1989). *Handbook of Boolean Algebras*. Vol. 1. Amsterdam: North Holland.

Corcoran, J. (1980a). 'Categoricity'. *History and Philosophy of Logic* 1, pp.187–207.

— (1980b). 'On definitional equivalence and related topics'. *History and Philosophy of Logic* 1, pp.231–4.

— (1981). 'From categoricity to completeness'. *History and Philosophy of Logic* 2, pp.113–19.

Corcoran, J. et al. (1974). 'String theory'. *The Journal of Symbolic Logic* 39.4, pp.625–37.

Corner, A. L. S. and R. Göbel (1985). 'Prescribing endomorphism algebras, a unified treatment'. *Proceedings of the London Mathematical Society* 50.3, pp.447–79.

Craig, W. (1953). 'On axiomatizability within a system'. *Journal of Symbolic Logic* 18, pp.30–2.

— (1957a). 'Linear reasoning. A new form of the Herbrand–Gentzen theorem'. *The Journal of Symbolic Logic* 22, pp.250–68.

— (1957b). 'Three uses of the Herbrand–Gentzen theorem in relating model theory and proof theory'. *The Journal of Symbolic Logic* 22, pp.269–85.

Cruse, P. (2005). 'Ramsey sentences, structural realism and trivial realization'. *Studies in History and Philosophy of Science* 36, pp.557–76.

Curry, H. B. (1933). 'Apparent variables from the standpoint of combinatory logic'. *Annals of Mathematics* 34.3, pp.381–404.

Dasgupta, A. (2014). *Set Theory: With an introduction to real point sets*. New York: Birkhäuser/Springer.

Davidson, D. (1965). 'Theories of meaning and learnable languages'. In: *Proceedings of the International Congress for Logic, Methodology, and Philosophy of Science*. Ed. by Y. Bar-Hillel. Amsterdam: North-Holland, pp.3–17.

— (1990). 'The structure and content of truth'. *The Journal of Philosophy* 87.6, pp.279–328.

Dawson Jr., J. W. (1993). 'The compactness of first-order logic: From Gödel to Lindström'. *History and Philosophy of Logic* 14.1, pp.15–37.

De Clercq, R. (2012). 'On some putative graph-theoretic counterexamples to the principle of the identity of indiscernibles'. *Synthese* 187, pp.661–72.

Dean, W. (2002). 'Models and recursivity'. Pittsburgh Graduate Philosophy Conference. URL: citeseerx.ist.psu.edu/%20viewdoc/summary?doi=10.1.1.136.446.

— (2014). 'Models and computability'. *Philosophia Mathematica* 22.2, pp.143–66.

— (2015). 'Arithmetical reflection and the provability of soundness'. *Philosophia Mathematica* 23.1, pp.31–64.

Dean, W. and S. Walsh (2017). 'The prehistory of the subsystems of second-order arithmetic'. *The Review of Symbolic Logic* 10.2, pp.357–96.

Dedekind, R. (1930–32). *Gesammelte mathematische Werke*. Braunschweig: Vieweg.

— (1872). *Stetigkeit und irrationale Zahlen*. Braunschweig: Vieweg.

— (1888). *Was sind und was sollen die Zahlen?* Braunschweig: Vieweg.

Dedekind, R. and H. Weber (1882). 'Theorie der algebraischen Funktionen einer Veränderlichen'. *Journal für die reine und angewandte Mathematik* 92, pp.181–290.

Demopoulos, W. (2003). 'On the rational reconstruction of our theoretical knowledge'. *British Journal for the Philosophy of Science* 54, pp.371–403.

— (2011). 'Three views of theoretical knowledge'. *British Journal for the Philosophy of Science* 62, pp.177–205.

Demopoulos, W. and M. Friedman (1985). 'Bertrand Russell's *The Analysis of Matter*: Its historical context and contemporary interest'. *Philosophy of Science* 52, pp.621–39.

Denef, J. (1984). 'The rationality of the Poincaré series associated to the *pf*-adic points on a variety'. *Inventiones Mathematicae* 77.1, pp.1–23.

Detlefsen, M. (1986). *Hilbert's Program*. Dordrecht: Kluwer.

— (1996). 'Philosophy of mathematics in the twentieth century'. In: *Philosophy of Science, Logic and Mathematics in the Twentieth Century*. Ed. by S. G. Shanker. London: Routledge, pp.50–123.

— (2014). 'Duality, epistemic efficiency, and consistency'. In: *Formalism and Beyond: On the Nature of Mathematical Discourse*. Ed. by G. Link. Berlin: de Gruyter, pp.1–24.

Dickmann, M. A. (1975). *Large Infinitary Languages*. Amsterdam: North Holland.

Dieudonnè, J. (1985). *History of Algebraic Geometry*. Belmont: Wadsworth.

Dirichlet, P. G. L. and R. Dedekind (1871). *Vorlesungen über Zahlentheorie*. 2nd ed. Braunschweig: Vieweg.

Dizadji-Bahmani, F. et al. (2010). 'Who's afraid of Nagelian reduction?' *Erkenntnis* 73.3, pp.393–412.

Doner, J. E. et al. (1978). 'The elementary theory of well-ordering—A metamathematical study'. In: *Logic Colloquium '77*. Vol. 96. Amsterdam: North-Holland, pp.1–54.

Duke, G. (2012). *Dummett on Abstract Objects*. New York: Palgrave.

Dummett, M. (1956). 'Nominalism'. *The Philosophical Review* 65.4. Reprinted in Dummett 1978., pp.491–505.

— (1959). 'Truth'. *Proceedings of the Aristotelian Society* 59, pp.141–62.

— (1963). 'The philosophical significance of Gödel's theorem'. In: Dummett 1978, pp.186–201.

— (1978). *Truth and Other Enigmas*. London: Duckworth.

— (1981). *Frege: Philosophy of Language*. 2nd ed. New York: Harper & Row.

— (1996). 'Frege and the paradox of analysis'. In: *Frege and Other Philosophers*. Oxford: Oxford University Press, pp.17–52.

Ehrenfeucht, A. (1960/1961). 'An application of games to the completeness problem for formalized theories'. *Fundamenta Mathematicae* 49, pp.129–41.

Ehrenfeucht, A. and A. Mostowski (1956). 'Models of axiomatic theories admitting automorphisms'. *Fundamenta Mathematicae* 43, pp.50–68.

Eklof, P. C. and E. R. Fischer (1972). 'The elementary theory of abelian groups'. *Annals of Pure and Applied Logic* 4, pp.115–71.

Enayat, A. et al. (2011). 'ω-models of finite set theory'. In: Kennedy and Kossak 2011. La Jolla, CA, pp.43–65.

Enderton, H. B. (2001). *A Mathematical Introduction to Logic*. 2nd ed. Burlington: Harcourt.

Engler, A. J. and A. Prestel (2005). *Valued Fields*. Berlin: Springer.

Epple, M. (2003). 'The end of the science of quantity: Foundations of analysis 1860–1910'. In: *A History of Analysis*. Ed. by H. N. Jahnke. Vol. 24. Providence, RI: American Mathematical Society.

Ershov, Y. L. (1964). 'Decidability of the elementary theory of relatively complemented lattices and of the theory of filters'. *Algebra i Logika* 3.3, pp.17–38.

— (1965). 'On elementary theories of local fields'. *Algebra i Logika* 4.2, pp.5–30.

— (1974). 'Theories of nonabelian varieties of groups'. In: Henkin 1974, pp.255–64.

Ershov, Y. L. et al., eds. (1998). *Handbook of Recursive Mathematics*. Vol. 138. Amsterdam: North-Holland.

Euler, L. (1755). *Foundations of Differential Calculus*. Translated by J. D. Blanton (2009). New York: Springer.

— (1780). 'On the infinity of infinities of orders of the infinitely large and infinitely small'. Translated by J. Bell (2009). Item E507 in the Eneström index. URL: arXiv: 0905.2254v2.

Evans, G. (1978). 'Can there be vague objects?' *Analysis* 38.4, p.208.

Ewald, W. (1996). *From Kant to Hilbert: A Source Book in the Foundations of Mathematics*. New York: Oxford University Press.

Feferman, S. and R. L. Vaught (1959). 'The first order properties of products of algebraic systems'. *Fundamenta Mathematicae* 47, pp.57–103.

Feferman, S. (1960/1961). 'Arithmetization of metamathematics in a general setting'. *Fundamenta Mathematicae* 49, pp.35–92.

— (1968). 'Lectures on proof theory'. In: *Proceedings of the Summer School in Logic*. Berlin: Springer, pp.1–107.

— (1974). 'Applications of many-sorted interpolation theorems'. In: Henkin 1974, pp.205–23.

— (1991). 'Reflecting on incompleteness'. *Journal of Symbolic Logic* 56.1, pp.1–49.

— (1998). 'Kurt Gödel: Conviction and caution'. In: *In the Light of Logic*. New York: Oxford University Press, pp.150–64.

— (1999). 'Logic, logics and logicicism'. *Notre Dame Journal of Formal Logic* 40.1, pp.31–54.

— (2010). 'Set-theoretical invariance criteria for logicality'. *Notre Dame Journal of Formal Logic* 51.1, pp.3–20.

Feferman, S., H. M. Friedman, et al. (2000). 'Does mathematics need new axioms?' *The Bulletin of Symbolic Logic* 6.4, pp.401–46.

Field, H. (1975). 'Conventionalism and instrumentalism in philosophy of science'. *Noûs* 9.4, pp.375–405.

— (1980). *Science without Numbers: A Defence of Nominalism*. Princeton: Princeton University Press.
— (1989). *Realism, Mathematics and Modality*. New York: Blackwell.
— (1994). 'Are our logical and mathematical concepts highly indeterminate?' *Midwest Studies in Philosophy* 19.
— (2001). *Truth and the Absence of Fact*. Oxford: Clarendon Press.
Fine, K. (2003). 'The role of variables'. *The Journal of Philosophy* 100.12, pp.605–31.
— (2007). *Semantic Relationism*. Oxford: Blackwell.
Fletcher, P. (1989). 'Nonstandard set theory'. *The Journal of Symbolic Logic* 54.3, pp.1000–8.
Florio, S. and Ø. Linnebo (2016). 'On the innocence and determinacy of plural quantification'. *Noûs* 50.3, pp.565–83.
Fraenkel, A. (1928). *Einleitung in die Mengenlehre*. Berlin: Springer.
Fraïssé, R. (1953). 'Sur certaines relations qui généralisent l'ordre des nombres rationnels'. *Comptes Rendus de l'Académie des Sciences de Paris* 237, pp.540–2.
— (1954a). 'Sur l'extension aux relations de quelques propriétés des ordres'. *Annales Scientifiques de l'École Normale Supérieure* 71, pp.363–88.
— (1954b). 'Sur quelques classifications des systemes de relations'. *Publications scientifiques de l'Université d'Alger. Série A, Sciences mathématiques* 1.1, pp.35–182.
— (1956). 'Application des γ-opérateurs au calcul logique du premier échelon'. *Zeitschrift für Mathematische Logik und Grundlagen der Mathematik* 2, pp.76–92.
Frayne, T. et al. (1962/1963). 'Reduced direct products'. *Fundamenta Mathematicae* 51, pp.195–228.
Frege, G. (1893). *Grundgesetze der Arithmetik: Begriffsschriftlich abgeleitet*. Vol. 1. Jena: H. Pohle.
— (1906). 'Über die Grundlagen der Geometrie'. *Jahresbericht der Deutschen Mathematiker-Vereinigung* 15, pp.293–309, 377–403, 423–30.
— (1980). *The Foundations of Arithmetic: A Logico-Mathematical Enquiry into the Concept of Number*. second. Evanston: Northwestern University Press.
Friedman, S.-D. et al. (2014). 'Generalized descriptive set theory and classification theory'. *Memoirs of the American Mathematical Society* 230.1081.
Friedman, H. (1975). 'One hundred and two problems in mathematical logic'. *The Journal of Symbolic Logic* 40, pp.113–29.
Fuchs, L. (1973). *Infinite Abelian Groups. Vol. II*. London: Academic Press.
— (2015). *Abelian Groups*. Cham: Springer.
Fulton, W. (1984). *Intersection Theory*. Vol. 2. Berlin: Springer.
Galebach, J. E. (2016). 'Visualization and o-minimality'. Unpublished.
Gallier, J. and D. Xu (2013). *A Guide to the Classification Theorem for Compact Surfaces*. Berlin: Springer.
Gamut, L. (1991). *Logic, Language, and Meaning. Volume 1: Introduction to Logic*. Chicago: University of Chicago Press.
Gao, S. (2009). *Invariant Descriptive Set Theory*. Boca Raton: CRC Press.
Gauss, C. F. (1863–1929). *Werke*. Hidesheim: Olms.

Geach, P. T. (1962). *Reference and Generality: An examination of some Medieval and Modern Theories*. Ithaca: Cornell University Press.

George, A. (1985). 'Skolem and the Löwenheim–Skolem theorem: A case study of the philosophical significance of mathematical results'. *History and Philosophy of Logic* 6.1, pp.75–89.

Giaquinto, M. (2015). 'The epistemology of visual thinking in mathematics'. In: *The Stanford Encyclopedia of Philosophy*. Ed. by E. N. Zalta. Winter 2015.

Givant, S. and P. Halmos (2009). *Introduction to Boolean algebras*. New York: Springer.

Glanzberg, M. (2001). 'Supervenience and infinitary logic'. *Noûs* 35.3, pp.419–39.

Gödel, K. (1931). 'Über formal unentscheidbare Sätze der *Principia Mathematica* und verwandter Systeme I'. *Monatshefte für Mathematik und Physik* 38. Translated in Gödel 1986, pp.173–98.

— (1932). 'Eine Eigenschaft der Realisierungen des Aussagenkalküls'. *Ergebnisse eines mathematischen Kolloquiums* 3. Reprinted with translation in Gödel 1986, pp.20–1.

— (1986). *Collected Works. Vol. I. Publications 1929–1936*. Ed. by S. Feferman et al. New York: Clarendon Press.

— (2003). *Collected Works. Vol. IV. Correspondence A–G*. New York: Clarendon Press.

Goldblatt, R. (1998). *Lectures on the Hyperreals: An Introduction to Nonstandard Analysis*. Springer.

Goncharov, S. S. and Y. L. Ershov (1999). *Konstruktivnye Modeli*. New York: Plenum.

Gowers, W. T. (2000). 'Rough structure and classification'. *Geometric and Functional Analysis* Special Volume, Part I, pp.79–117.

— (2008). 'Introduction'. In: *The Princeton Companion to Mathematics*. Ed. by W. T. Gowers. Princeton, NJ: Princeton University Press, pp.1–72.

Grilliot, T. J. (1972). 'Omitting types: Application to recursion theory'. *The Journal of Symbolic Logic* 37, pp.81–9.

Grishin, V. N. (2011). 'Boolean-valued model'. *Encyclopedia of Mathematics*. URL: www.encyclopediaofmath.org/index.php?title=Boolean-valued_model.

Grossberg, R. (2002). 'Classification theory for abstract elementary classes'. In: *Logic and Algebra*. Providence, RI: American Mathematical Society, pp.165–204.

Grothendieck, A. (1997). 'Esquisse d'un programme'. In: *Geometric Galois Actions*. Cambridge: Cambridge University Press, pp.5–48, 243–83.

Grzegorczyk, A. (1962). 'On the concept of categoricity'. *Studia Logica* 13, pp.39–66.

Gurevich, Y. (1984). 'Toward logic tailored for computational complexity'. In: *Computation and Proof Theory (Aachen, 1983)*. Springer, Berlin, pp.175–216.

Haffner, E. (forthcoming). 'Strategical use(s) of arithmetic in Richard Dedekind and Heinrich Weber's Theorie der algebraischen Funktionen einer Veränderlichen'. *Historia Mathematica*.

Hájek, P. and P. Pudlák (1998). *Metamathematics of First-Order Arithmetic*. Berlin: Springer.

Halbach, V. (2011). *Axiomatic Theories of Truth*. Cambridge: Cambridge University Press.

Halbach, V. and L. Horsten (2005). 'Computational structuralism'. *Philosophia Mathematica* 13, pp.174–86.

Hale, B. (1987). *Abstract Objects*. Oxford: Basil Blackwell.
Hale, B. and C. Wright (1997). 'Putnam's model-theoretic argument against metaphysical realism'. In: *A Companion to the Philosophy of Language*. Ed. by B. Hale and C. Wright. Malden: Blackwell, pp.427–57.
— (2003). 'Responses to commentators'. *Philosophical Books* 44.3, pp.245–63.
Hamkins, J. D. (2012). 'The set-theoretic multiverse'. *Review of Symbolic Logic* 5.3, pp.416–49.
Hamkins, J. D. and R. Yang (forthcoming). 'Satisfaction is not absolute'. *Review of Symbolic Logic*, pp.1–34. URL: arXiv:1312.0670.
Hanf, W. P. (1960). 'Models of languages with infinitely long expressions'. In: *Abstracts of contributed papers from the First Logic, Methodology and Philosophy of Science Congress*. Palo Alto: Stanford University Press, p.24.
Harrington, L. and M. Makkai (1985). 'An exposition of Shelah's "main gap": Counting uncountable models of ω-stable and superstable theories'. *Notre Dame Journal of Formal Logic* 26.2, pp.139–77.
Hartshorne, R. (1967). *Foundations of Projective Geometry*. Benjamin.
— (2000). *Geometry: Euclid and Beyond*. New York: Springer.
Hasenjaeger, G. (1953). 'Eine Bemerkung zu Henkin's Beweis für die Vollständigkeit des Prädikatenkalküls der ersten Stufe'. *The Journal of Symbolic Logic* 18, pp.42–8.
Hausdorff, F. (1908). 'Grundzüge einer Theorie der geordneten Mengen'. *Mathematische Annalen* 65.4, pp.435–505.
Hawley, K. (2009). 'Identity and indiscernibility'. *Mind* 118.1, pp.101–19.
Heijenoort, J. van (1967). *From Frege to Gödel: A Source Book in Mathematical Logic, 1879–1931*. Cambridge: Harvard University Press.
Hellman, G. (1989). *Mathematics without Numbers: Towards a modal-structural interpretation*. Oxford: Oxford University Press.
— (1996). 'Structuralism without structures'. *Philosophia Mathematica* 4, pp.100–23.
— (2001). 'Three varieties of mathematical structuralism'. *Philosophia Mathematica* 9.3, pp.184–211.
— (2005). 'Structuralism'. In: Shapiro 2005a, pp.536–62.
Henkin, L. (1949). 'The completeness of the first-order functional calculus'. *The Journal of Symbolic Logic* 14.3, pp.159–66.
— (1961). 'Some remarks on infinitely long formulas'. In: *Infinitistic Methods*. Oxford: Pergamon Press, pp.167–83.
— ed. (1974). *Proceedings of the Tarski Symposium*. Providence, RI: American Mathematical Society.
Hewitt, E. (1948). 'Rings of real-valued continuous functions. I'. *Transactions of the American Mathematical Society* 64, pp.45–99.
Hilbert, D. (1899). *Grundlagen der Geometrie*. Citations to 14th edition, with a supplement by Paul Bernays, published in La Salle: Open Court. Teubner.
— (1925). 'Über das Unendliche'. *Mathematische Annalen* 95, pp.161–90.
Hilbert, D. and P. Bernays (1934). *Grundlagen der Mathematik. I*. Berlin: Springer.
— (1939). *Grundlagen der Mathematik. II*. Berlin: Springer.

Hintikka, J. (1996). *The Principles of Mathematics Revisited*. Cambridge: Cambridge University Press.

Hjortland, O. T. (2014). 'Speech acts, categoricity, and the meanings of logical connectives'. *Notre Dame Journal of Formal Logic* 55.4, pp.445–67.

Hodes, H. (1984). 'Logicism and the ontological commitments of arithmetic'. *The Journal of Philosophy* 81.3, pp.123–49.

— (1990). 'Where do the natural numbers come from?' *Synthese* 84.3, pp.347–407.

— (1991). 'Where do sets come from?' *The Journal of Symbolic Logic* 56.1, pp.150–75.

Hodesdon, K. (forthcoming). 'Structuralism and semantic glue'. In: *Proceedings to Philosophy in an Age of Science*. Ed. by A. Berger. Oxford: Oxford University Press.

Hodges, W. (1987). 'What is a structure theory?' *The Bulletin of the London Mathematical Society* 19.3, pp.209–37.

— (1993). *Model Theory*. Cambridge: Cambridge University Press.

— (1997a). *A Shorter Model Theory*. Cambridge: Cambridge University Press.

— (1997b). 'Compositional semantics for a language of imperfect information'. *Logic Journal of the IGPL. Interest Group in Pure and Applied Logics* 5.4, pp.539–63.

— (2015). 'Notes on the history of scope'. In: *Logic Without Borders: Essays on Set Theory, Model Theory, Philosophical Logic and Philosophy of Mathematics*. Berlin: de Gruyter, pp.215–40.

Horsten, L. (2012). 'Vom Zählen zu den Zahlen: On the relation between computation and arithmetical structuralism'. *Philosophia Mathematica* 20.3, pp.275–88.

Hrbáček, K. (1978). 'Axiomatic foundations for nonstandard analysis'. *Fundamenta Mathematicae* 98.1, pp.1–19.

— (2006). 'Nonstandard objects in set theory'. In: *Nonstandard methods and applications in mathematics*. La Jolla: Association for Symbolic Logic, pp.80–120.

Hrbáček, K. and T. Jech (1999). *Introduction to Set Theory*. 3rd ed. New York: Dekker.

Hrushovski, E. and A. Pillay (1987). 'Weakly normal groups'. In: Paris Logic Group 1987, pp.233–44.

Hrushovski, E. (1987). 'Locally modular regular types'. In: Baldwin 1987a, pp.132–64.

— (1992). 'Strongly minimal expansions of algebraically closed fields'. *Israel Journal of Mathematics* 79.2–3, pp.129–51.

— (1993). 'A new strongly minimal set'. *Annals of Pure and Applied Logic* 62.2, pp.147–66.

— (1996). 'The Mordell–Lang conjecture for function fields'. *Journal of the American Mathematical Society* 9.3, pp.667–90.

— (1998). 'Geometric model theory'. In: *Proceedings of the International Congress of Mathematicians, Vol. I (Berlin, 1998)*, pp.281–302.

— (2001). 'The Manin–Mumford conjecture and the model theory of difference fields'. *Annals of Pure and Applied Logic* 112, pp.43–115.

Hrushovski, E. and B. Zilber (1996). 'Zariski geometries'. *Journal of the American Mathematical Society* 9.1, pp.1–56.

Humberstone, L. (2011). *The Connectives*. Cambridge: MIT Press.

Hungerford, T. W. (1980). *Algebra*. New York: Springer.

Huntington, E. V. (1902). 'A complete set of postulates for the theory of absolute continuous magnitude'. *Transactions of the American Mathematical Society* 3.2, pp.264–79.

Hurd, A. E. and P. A. Loeb (1985). *An Introduction to Nonstandard Real Analysis*. Orlando: Academic Press.

Ikegami, D. and J. Väänänen (2015). 'Boolean-valued second-order logic'. *Notre Dame Journal of Formal Logic* 56.1, pp.167–90.

Immerman, N. (1982). 'Upper and lower bounds for first order expressibility'. *Journal of Computer and System Sciences* 25.1, pp.76–98.

Incurvati, L. (2010). 'Set theory: Its justification, logic, and extent'. PhD thesis. University of Cambridge.

— (2016). 'Can the cumulative hierarchy be categorically characterized?' *Logique et Analyse* 59.236, pp.367–87.

Incurvati, L. and P. Smith (2010). 'Rejection and valuations'. *Analysis* 70.1, pp.3–10.

Isaacson, D. (1994). 'Mathematical intuition and objectivity'. In: *Mathematics and Mind*. Ed. by A. George. New York: Oxford University Press, pp.118–40.

— (2011). 'The reality of mathematics and the case of set theory'. In: *Truth, Reference, and Realism*. Ed. by Z. Novák and A. Simonyi. Budapest: Central European University Press, pp.1–75.

Jech, T. (1973). *The Axiom of Choice*. Amsterdam: North-Holland.

— (2003). *Set Theory*. 3rd ed. Berlin: Springer.

Jensen, R. B. (1972). 'The fine structure of the constructible hierarchy'. *Annals of Mathematical Logic* 4, pp.229–308.

— (1995). 'Inner models and large cardinals'. *Bulletin of Symbolic Logic* 1.4, pp.393–407.

Jesseph, D. (1998). 'Leibniz on the foundations of the calculus: The question of the reality of infinitesimal magnitudes'. *Perspectives on Science* 6.1-2, pp.6–40.

— (2008). 'Truth in fiction: Origins and consequences of Leibniz's doctrine of infinitesimal magnitudes'. In: *Infinitesimal Differences: Controversies between Leibniz and his Contemporaries*. Ed. by U. Goldenbaum and D. Jesseph. de Gruyter, pp.215–33.

Johnstone, P. T. (1982). *Stone Spaces*. Cambridge: Cambridge University Press.

Jónsson, B. (1956). 'Universal relational systems'. *Mathematica Scandinavica* 4, pp.193–208.

— (1960). 'Homogeneous universal relational systems'. *Mathematica Scandinavica* 8, pp.137–42.

Just, W. and M. Weese (1997). *Discovering Modern Set Theory. II*. Providence, RI: American Mathematical Society.

Kanamori, A. (2003). *The Higher Infinite*. 2nd ed. Berlin: Springer.

Kant, I. (1787). *Kritik der reinen Vernunft*. Trans. by N. Kemp Smith. Revised (2003) edition, from Palgrave Macmillan.

Kaplan, D. (1986). 'Opacity'. In: *The Philosophy of W. V. Quine*. Ed. by L. E. Hahn and P. A. Schilpp. Chicago: Open Court.

Karp, C. R. (1965). 'Finite-quantifier equivalence'. In: Addison et al. 1965, pp.407–12.

Kaye, R. (1991). *Models of Peano Arithmetic*. Oxford: Clarendon Press.

— (2007). *The Mathematics of Logic: A guide to completeness theorems and their applications*. Cambridge: Cambridge University Press.

Kaye, R. (2011). 'Tennenbaum's theorem for models of arithmetic'. In: Kennedy and Kossak 2011, pp.66–79.

Kaye, R. and T. L. Wong (2007). 'On interpretations of arithmetic and set theory'. *Notre Dame Journal of Formal Logic* 48.4, pp.497–510.

Kechris, A. S. et al. (2005). 'Fraïssé limits, Ramsey theory, and topological dynamics of automorphism groups'. *Geometric and Functional Analysis* 15.1, pp.106–89.

Kechris, A. S. (1995). *Classical Descriptive Set Theory*. New York: Springer.

Keisler, H. J. (1961). 'Ultraproducts and elementary classes'. *Indagationes Mathematicae* 23, pp.477–95.

— (1967). 'Ultraproducts which are not saturated'. *The Journal of Symbolic Logic* 32, pp.23–46.

Kennedy, J. (2014). *Interpreting Gödel: Critical Essays*. Cambridge University Press.

Kennedy, J. and R. Kossak, eds. (2011). *Set Theory, Arithmetic, and Foundations of Mathematics*. Cambridge: Association for Symbolic Logic.

Keränen, J. (2001). 'The identity problem for realist structuralism'. *Philosophia Mathematica* 9.3, pp.308–30.

— (2006). 'The identity problem for realist structuralism II: A reply to Shapiro'. In: MacBride 2006a, pp.146–63.

Ketland, J. (2004). 'Empirical adequacy and ramsification'. *British Journal for the Philosophy of Science*, pp.287–300.

— (2006). 'Structuralism and the identity of indiscernibles'. *Analysis* 66.4, pp.303–15.

— (2009). 'Empirical adequacy and ramsification, II'. In: *Reduction, Abstraction, Analysis*. Ed. by A. Hieke and H. Leitgeb. Heusenstamm: Ontos, pp.29–45.

— (2011). 'Identity and indiscernibility'. *Review of Symbolic Logic* 4, pp.171–85.

Kharlampovich, O. and A. Myasnikov (2006). 'Elementary theory of free non-abelian groups'. *Journal of Algebra* 302.2, pp.451–552.

Kim, B. (1998). 'Forking in simple unstable theories'. *Journal of the London Mathematical Society* 57.2, pp.257–67.

Kinsey, L. C. (1993). *Topology of Surfaces*. New York: Springer.

Klenk, V. (1976). 'Intended models and the Löwenheim–Skolem theorem'. *Journal of Philosophical Logic* 5.4, pp.475–89.

Kline, M. (1980). *Mathematics: The Loss of Certainty*. New York: Oxford University Press.

Knight, J. F. et al. (1986). 'Definable sets in ordered structures. II'. *Transactions of the American Mathematical Society* 295.2, pp.593–605.

Kochen, S. (1961). 'Ultraproducts in the theory of models'. *Annals of Mathematics* 74, pp.221–61.

Koellner, P. (2009). 'Truth in mathematics: The question of pluralism'. In: Bueno and Linnebo 2009, pp.80–116.

Krajewski, S. (1974). 'Mutually inconsistent satisfaction classes'. *Bulletin de L'Académie Polonaise des Sciences. Série des sciences math., astr., et physc.* 22.10, pp.983–7.

Kreisel, G. (1967). 'Informal rigour and completeness proofs [with discussion]'. In: Lakatos 1967, pp.138–86.

— (1971). 'Observations on popular discussions of foundations'. In: *Axiomatic Set Theory*. Ed. by D. Scott. Providence, RI: American Mathematical Society, pp.189–98.

Kunen, K. (1980). *Set Theory*. Amsterdam: North-Holland.

Lachlan, A. H. (1973/74). 'Two conjectures regarding the stability of ω-categorical theories'. *Fundamenta Mathematicae* 81.2, pp.133–45.

— (1972). 'A property of stable theories'. *Fundamenta Mathematicae* 77.1, pp.9–20.

Ladyman, J. et al. (2012). 'Identity and discernibility in philosophy and logic'. *The Review of Symbolic Logic* 5.1, pp.162–86.

Lakatos, I., ed. (1967). *Problems in the Philosophy of Mathematics*. Amsterdam: North-Holland.

Lang, S. (2002). *Algebra*. 3rd ed. New York: Springer.

Langford, C. H. (1926/27a). 'Some theorems on deducibility'. *Annals of Mathematics* 28.1-4, pp.16–40.

— (1926/27b). 'Theorems on deducibility (second paper)'. *Annals of Mathematics* 28.1-4, pp.459–71.

Lascar, D. (1982). 'On the category of models of a complete theory'. *The Journal of Symbolic Logic* 47.2, pp.249–66.

— (1998). 'Perspective historique sur les rapports entre la théorie des modèles et l'algèbre. Un point de vue tendancieux'. *Revue d'Histoire des Mathématiques* 4.2, pp.237–60.

Laskowski, M. C. (2006). 'Descriptive set theory and uncountable model theory'. In: *Logic Colloquium '03*. La Jolla: Association for Symbolic Logic, pp.133–45.

Lavine, S. (1994). *Understanding the Infinite*. Cambridge MA: Harvard University Press.

— (1999). 'Skolem was wrong'. Unpublished. Dated June.

— (2000). 'Quantification and ontology'. *Synthese* 124.1, pp.1–43.

Lawson, T. (2003). *Topology: A Geometric Approach*. Oxford: Oxford University Press.

Leibniz, G. W. (1849–63). *Mathematische Schriften*. Ed. by C. I. Gerhardt. Berlin, von Asher; Halle, von Schmidt.

Leigh, G. E. and C. Nicolai (2013). 'Axiomatic truth, syntax and metatheoretic reasoning'. *Review of Symbolic Logic* 6.4, pp.613–36.

Leitgeb, H. and J. Ladyman (2008). 'Criteria of identity and structuralist ontology'. *Philosphia Mathematica* 16, pp.388–96.

Levin, M. (1997). 'Putnam on reference and constructible sets'. *British Journal for the Philosophy of Science* 48.1, pp.55–67.

Lewis, D. K. (1970). 'How to define theoretical terms'. *The Journal of Philosophy* 67.13, pp.427–46.

— (1984). 'Putnam's paradox'. *Australasian Journal of Philosophy* 62.3, pp.221–36.

— (1986). *On the Plurality of Worlds*. Oxford: Basil Blackwell.

l'Hôpital, G. d. (2015). *L'Hôpital's Analyse des infiniments petits. An Annotated Translation with Source Material by Johann Bernoulli*. Ed. by R. E. Bradley et al. New York. Vol. 50. l'Hôpital's text was first published in 1696. Birkhäuser.

Li, T. (2002). *Proceedings of the International Congress of Mathematicians. Vol. I.* Beijing: Higher Education Press.

Lindström, P. (1964). 'On model-completeness'. *Theoria* 30, pp.183–96.

— (1966). 'First order predicate logic with generalized quantifiers'. *Theoria* 32.3, pp.186–95.

— (1969). 'On extensions of elementary logic'. *Theoria* 35, pp.1–11.

— (2003). *Aspects of Incompleteness*. 2nd ed. Urbana, IL: Association for Symbolic Logic.

Linnebo, Ø. and A. Rayo (2012). 'Hierarchies ontological and ideological'. *Mind* 142.482, pp.269–308.

Łoś, J. (1954). 'On the categoricity in power of elementary deductive systems and some related problems'. *Colloquium Mathematicum* 3, pp.58–62.

— (1955a). 'On the extending of models. I'. *Fundamenta Mathematicae* 42, pp.38–54.

— (1955b). 'Quelques remarques, théorèmes et problèmes sur les classes définissables d'algèbres'. In: *Mathematical interpretation of formal systems*. Amsterdam: North-Holland, pp.98–113.

Łoś, J. and R. Suszko (1957). 'On the extending of models. IV'. *Fundamenta Mathematicae* 44, pp.52–60.

Löwenheim, L. (1915). 'Über Möglichkeiten im Relativkalkül'. *Mathematische Annalen* 76.4. Translated in Heijenoort 1967, pp.447–70.

Lutz, S. (2014). 'The semantics of scientific theories'. In: *Księga pamiątkowa Marianowi Przełęckiemu w darze na 90-lecie urodzin*. Ed. by A. Bro zek and J. Jadacki. Authoritative preprint at philsci-archive.pitt.edu/id/eprint/9630. Norbertinum, Lublin, pp.33–67.

Lützen, J. (2003). 'The foundation of analysis in the 19th century'. In: *A History of Analysis*. Providence, RI: American Mathematical Society, pp.155–95.

Mac Lane, S. (1989). 'The education of Ph.D.s in mathematics'. In: *A Century of Mathematics in America, part III*. Vol. 3. Providence, RI: American Mathematical Society, pp.517–23.

— (1998). *Categories for the Working Mathematician*. 2nd ed. New York: Springer.

MacBride, F. (2005). 'Structuralism reconsidered'. In: Shapiro 2005b, pp.563–89.

— (2006a). 'The Julius Caesar objection: More problematic than ever'. In: *Identity and Modality*. Oxford: Oxford University Press, pp.174–202.

— (2006b). 'What constitutes the numerical diversity of mathematical objects?' *Analysis* 66.1, pp.63–9.

MacFarlane, J. (2000). 'What does it mean to say that logic is formal?' PhD thesis. University of Pittsburgh.

— (2015). 'Logical constants'. *Stanford Encyclopedia of Philosophy* Fall 2015. Ed. by E. N. Zalta.

Macintyre, A. (1971). 'On ω_1-categorical theories of fields'. *Fundamenta Mathematicae* 71.1, pp.1–25.

— (1972). 'On algebraically closed groups'. *Annals of Mathematics* 96, pp.53–97.

— (1976). 'On definable subsets of p-adic fields'. *The Journal of Symbolic Logic* 41.3, pp.605–10.

— (1989). 'Trends in logic'. In: *Logic Colloquium '88*. Ed. by R. Ferro et al. Vol. 127. Amsterdam: North-Holland, pp.365–67.

— ed. (2000a). *Connections between Model Theory and Algebraic and Analytic Geometry.* Naples: Seconda Università di Napoli.
— (2000b). 'Weil cohomology and model theory'. In: Macintyre 2000a. Vol. 6, pp.179–99.
— (2003). 'Model theory: Geometrical and set-theoretic aspects and prospects'. *The Bulletin of Symbolic Logic* 9.2, pp.197–212.
Macintyre, A. and A. J. Wilkie (1996). 'On the decidability of the real exponential field'. In: *Kreiseliana*. Ed. by P. Odifreddi. Wellesley: Peters, pp.441–67.
Maddy, P. (2005). 'Three forms of naturalism'. In: Shapiro 2005b, pp.437–59.
Makkai, M. and R. Paré (1989). *Accessible Categories: The Foundations of Categorical Model Theory*. Vol. 104. Providence, RI: American Mathematical Society.
Makowsky, J. A. (2004). 'Algorithmic uses of the Feferman–Vaught theorem'. *Annals of Pure and Applied Logic* 126.1-3, pp.159–213.
Malinowski, G. (1993). *Many-Valued Logics*. New York: Clarendon Press.
Malliaris, M. and S. Shelah (2014). 'Regularity lemmas for stable graphs'. *Transactions of the American Mathematical Society* 366.3, pp.1551–85.
Mal'tsev, A. I. (1936). 'Untersuchungen aus dem Gebiete der mathematischen Logik'. *Matematicheskie Sbornik* 43.1. Translated in Mal'tsev 1971, pp.323–36.
— (1960a). 'A correspondence between rings and groups'. *Matematicheskie Sbornik* 50.92. Russian. Translated in Mal'tsev 1971, pp.257–66.
— (1960b). 'Constructive algebras I'. *Uspekhi Matematicheskih Nauk* 16.99. Translated in Mal'tsev 1971, pp.3–60.
— (1962). 'Axiomatizable classes of locally free algebras of certain types'. *Sibirskii Matematicheskii Zhurnal* 3. Russian. Translated in Mal'tsev 1971, pp.729–43.
— (1971). *The Metamathematics of Algebraic Systems. Collected Papers: 1936–1967*. Ed. by B. F. Wells. Vol. 66. Amsterdam: North-Holland.
Mancosu, P. et al. (2009). 'The development of mathematical logic from Russell to Tarski, 1900–1935'. In: *The Development of Modern Logic*. Ed. by L. Haaparanta. Oxford: Oxford University Press, pp.318–470.
Manzano, M. (1996). *Extensions of First Order Logic*. Cambridge: Cambridge University Press.
Marczewski, E. (1951). 'Sur les congruences et les propriétés positives d'algèbres abstraites'. *Colloquium Mathematicum* 2, 220–8 (1952).
Marker, D. (2002). *Model Theory: An Introduction*. New York: Springer.
— (2006). 'Introduction to the model theory of fields'. In: *Model Theory of Fields*. 2nd ed. La Jolla: Association for Symbolic Logic, pp.1–37.
Marker, D. et al. (2006). *Model Theory of Fields*. 2nd ed. La Jolla: Association for Symbolic Logic.
Marsh, W. E. (1966). 'On ω_1-categorical and not ω-categorical theories'. Dissertation, Dartmouth College.
Martin, D. A. (1970). 'Review of Quine 1969'. *The Journal of Philosophy* 67.4, pp.111–14.

Martin, D. A. (1976). 'Hilbert's first problem: The continuum hypothesis'. In: *Mathematical Developments Arising from Hilbert Problems*. Providence, R. I.: American Mathematical Society, pp.81–92.

— (2001). 'Multiple universes of sets and indeterminate truth values'. *Topoi* 20.1, pp.5–16.

— (2005). 'Gödel's conceptual realism'. *Bulletin of Symbolic Logic* 11.2, pp.207–24.

— (2015). 'Completeness or incompleteness of basic mathematical concepts'. Unpublished. Dated October 20.

Mates, B. (1965). *Elementary Logic*. New York: Oxford University Press.

Mautner, F. I. (1946). 'An extension of Klein's Erlanger program: Logic as invariant-theory'. *American Journal of Mathematics* 68.3, pp.345–84.

Maxwell, G. (1962). 'The ontological status of theoretical entities'. In: *Scientific Explanation, Space, and Time*. Ed. by H. Feigl and G. Maxwell. Minneapolis: University of Minnesota Press, pp.3–15.

— (1971). 'Structural realism and the meaning of theoretical terms'. In: *Analyses of Theories and Methods of Physics and Psychology*. Ed. by S. Winokur and M. Radner. University of Minnesota Press, pp.181–92.

McCarty, C. (1987). 'Variations on a thesis: Intuitionism and computability'. *Notre Dame Journal of Formal Logic* 28.4, pp.536–80.

McGee, V. (1993). 'A semantic conception of truth?' *Philosophical Topics* 21.2, pp.83–111.

— (1996). 'Logical operations'. *Journal of Philosophical Logic* 25.6, pp.567–80.

— (1997). 'How we learn mathematical language'. *Philosophical Review* 106.1, pp.35–68.

— (2000). '"Everything"'. In: *Between Logic and Intuition*. Ed. by G. Sher and R. Tieszen. Cambridge: Cambridge University Press, pp.54–78.

— (2005). 'A semantic conception of truth? [and afterward]'. In: *Deflationary Truth*. Ed. by B. P. Amour-Garb and J. Beall. Open Court, pp.111–52.

McIntosh, C. (1979). 'Skolem's criticisms of set theory'. *Noûs* 13, pp.313–34.

McKinsey, J. C. C. and A. Tarski (1944). 'The algebra of topology'. *Annals of Mathematics. Second Series* 45, pp.141–91.

Meadows, T. (2013). 'What can a categoricity theorem tell us?' *The Review of Symbolic Logic* 6.3, pp.524–44.

Mendelson, E. (1997). *Introduction to Mathematical Logic*. 4th ed. London: Chapman & Hall.

Menzel, C. (2014). 'Wide sets, ZFCU, and the iterative conception'. *The Journal of Philosophy* 111.2, pp.57–83.

Merrill, G. H. (1980). 'The model-theoretic argument against realism'. *Philosophy of Science* 47.1, pp.69–81.

Monk, J. D. (1969). *Introduction to Set Theory*. New York: McGraw-Hill.

— (1976). *Mathematical Logic*. New York: Springer.

Moore, A. W. (2001). *The Infinite*. 2nd ed. London: Routledge.

— (2011). 'Vats, sets, and tits'. In: *Transcendental Philosophy and Naturalism*. Ed. by P. Sullivan and J. Smith. Oxford: Oxford University Press, pp.42–54.

Moore, G. E. (1939). 'Proof of an external world'. *Proceedings of the British Academy* 25, pp.273–300.

Moore, G. H. (1982). *Zermelo's Axiom of Choice*. New York: Springer.
Morley, M. (1965a). 'Categoricity in power'. *Transactions of the American Mathematical Society* 114, pp.514–38.
— (1965b). 'Omitting classes of elements'. In: *Theory Of Models*, pp.265–73.
Morley, M. and R. L. Vaught (1962). 'Homogeneous universal models'. *Mathematica Scandinavica* 11, pp.37–57.
Mortimer, M. (1975). 'On languages with two variables'. *Zeitschrift für Mathematische Logik und Grundlagen der Mathematik* 21, pp.135–40.
Morton, A. and S. P. Stich (1996). *Benacerraf and His Critics*. Oxford: Blackwell.
Mostowski, A. (1948). *Logika Matematyczna*. Warsaw.
— (1952). 'On direct products of theories'. *The Journal of Symbolic Logic* 17, pp.1–31.
— (1957). 'On a generalization of quantifiers'. *Fundamenta Mathematicæ* 44.1, pp.12–36.
— (1967). 'Recent results in set theory [with discussion]'. In: Lakatos 1967. Ed. by I. Lakatos. Amsterdam: North-Holland, pp.82–108.
— (1969). *Constructible Sets with Applications*. Amsterdam: North-Holland.
Mukhopadhyay, S. N. (2012). *Higher Order Derivatives*. Boca Raton: CRC Press.
Müller, G. H. et al. (1987). *Ω-Bibliography of Mathematical Logic. Vol. III: Model Theory*. Berlin: Springer.
Müller, R. (2009). 'The notion of a model: A historical overview'. In: *Philosophy of Technology and Engineering Sciences*. Ed. by D. M. Gabbay et al. Amsterdam: Elsevier, pp.637–64.
Munkres, J. R. (2000). *Topology*. 2nd ed. Upper Saddle River: Prentice Hall.
Murzi, J. and O. T. Hjortland (2009). 'Inferentialism and the categoricity problem: Reply to Raatikainen'. *Analysis* 69.3, pp.480–8.
Mycielski, J. (1964). 'Some compactifications of general algebras'. *Colloquium Mathematicum* 13, pp.1–9.
Nagel, E. (1961). 'The reduction of theories'. In: *The Structure of Science: Problems in the Logic of Scientific Explanation*. New York: Harcourt, Brace & World, pp.336–97.
Nelson, E. (1977). 'Internal set theory: A new approach to nonstandard analysis'. *Bulletin of the American Mathematical Society* 83.6, pp.1165–98.
Newman, M. H. A. (1928). 'Mr. Russell's causal theory of perception'. *Mind* 5.146, pp.26–43.
Niebergall, K.-G. (2000). 'On the logic of reducibility: Axioms and examples'. *Erkenntnis* 53, pp.27–61.
Okasha, S. (2002). 'Underdetermination, holism and the theory/data distinction'. *The Philosophical Quarterly* 52.208, pp.303–19.
Oliver, A. and T. Smiley (2006). 'What are sets and what are they for?' *Philosophical Perspectives* 20, pp.123–55.
— (2013). *Plural Logic*. Oxford: Oxford University Press.
Orey, S. (1956). 'On ω-consistency and related properties'. *The Journal of Symbolic Logic* 21, pp.246–52.
Ornstein, D. (1970). 'Bernoulli shifts with the same entropy are isomorphic'. *Advances in Mathematics* 4, pp.337–52.

Padoa, A. (1900). 'Essai d'une théorie algébrique des nombres entiers, précédé d'une introduction logique à une théorie déductive quelconque'. In: *Bibliothèque du Congrès international de philosophie*. Vol. 3. Translated in Heijenoort 1967. Paris: Armand Colin, pp.309–65.

Paris Logic Group, ed. (1987). *Logic Colloquium '85 (Orsay, 1985)*. Amsterdam: North-Holland.

Paris, J. and L. Harrington (1977). 'A mathematical incompleteness in Peano arithmetic'. In: *Handbook of Mathematical Logic*. Amsterdam: North-Holland, pp.1133–42.

Parsons, C. (1990a). 'The structuralist view of mathematical objects'. *Synthese* 84.3, pp.303–46.

— (1990b). 'The uniqueness of the natural numbers'. *Iyyun* 39.1, pp.13–44.

— (2008). *Mathematical Thought and Its Objects*. Cambridge: Harvard University Press.

Paseau, A. (2010). 'Proofs of the compactness theorem'. *History and Philosophy of Logic* 31.1, pp.73–98.

Peacock, G. (1833). 'Report on the recent progress and present state of certain branches of analysis'. *British Association for the Advancement of Science* 3, pp.185–352.

Peano, G. (1891). 'Sul concetto di numero'. *Rivista di matematica* 1, pp.87–102, 256–67.

Pedersen, N. J. L. L. and M. Rossberg (2010). 'Open-endedness, schemas and ontological commitment'. *Noûs* 44.2, pp.329–39.

Peirce, C. S. (1885). 'On the algebra of logic: A contribution to the philosophy of notation'. *American Journal of Mathematics* 7.2, pp.180–96.

Peters, S. and D. Westerståhl (2006). *Quantifiers in Logic and Language*. Oxford: Oxford University Press.

Petersen, K. (1983). *Ergodic Theory*. Cambridge: Cambridge University Press.

Pickel, B. and B. Rabern (2017). 'The antinomy of the variable: A Tarskian resolution'. *The Journal of Philosophy* 113.3, pp.137–70.

Pila, J. and A. J. Wilkie (2006). 'The rational points of a definable set'. *Duke Mathematical Journal* 133.3, pp.591–616.

Pillay, A. (1983). *An Introduction to Stability Theory*. New York: Clarendon Press.

— (1996). *Geometric Stability Theory*. Vol. 32. New York: Clarendon Press.

Pillay, A. and C. Steinhorn (1984). 'Definable sets in ordered structures'. *Bulletin of the American Mathematical Society* 11.1, pp.159–62.

— (1986). 'Definable sets in ordered structures. I'. *Transactions of the American Mathematical Society* 295.2, pp.565–92.

Poizat, B. (1985). *Cours de Théorie des Modèles*. Translated as Poizat 2000. Paris: Nur Al-Mantiq Wal-Ma'rifah.

— (2000). *A Course in Model Theory*. New York: Springer.

Poonen, B. (2014). 'Undecidable problems: A sampler'. In: Kennedy 2014, pp.211–41.

Potter, M. (1993). 'Iterative set theory'. *Philosophical Quarterly* 43.171, pp.178–93.

— (2000). *Reason's Nearest Kin: Philosophies of Arithmetic from Kant to Carnap*. Oxford: Oxford University Press.

— (2004). *Set Theory and its Philosophy*. Oxford: Oxford University Press.

Prawitz, D. (1965). *Natural Deduction: A proof-theoretical study*. Dover.

Presburger, M. (1930). 'Über die Vollständigkeit eines gewissen Systems der Arithmetik ganzer Zahlen, in welchem die Addition als einzige Operation hervortritt'. In: *Sprawozdanie z 1. Kongresu matematyków krajów slowianskich. Comptes-rendus du 1. Congrès des mathématiciens des pays slaves, Warszawa, 1929*. Warszawa: Ksiaznica atlas t.n.s.w, pp.92–101.

Price, H. (2003). 'Truth as convenient friction'. *The Journal of Philosophy* 100.4, pp.167–90.

Priest, G. (2005). *Towards Non-Being: The Logic and Metaphysics of Intentionality*. Oxford: Oxford University Press.

Przełęckie, M. (1973). 'A model-theoretic approach to some problems in the semantics of empirical languages'. In: *Logic, Language, and Probability*. Ed. by R. J. Bogdan and I. Niiniluoto. Dordrecht: D. Reidel, pp.285–90.

Psillos, S. (1999). *Scientific Realism: How Science Tracks Truth*. New York and London: Routledge.

Pullum, G. K. and B. C. Scholz (2001). 'On the distinction between model-theoretic and generative-enumerative syntactic frameworks'. In: *International Conference on Logical Aspects of Computational Linguistics*. Springer, pp.17–43.

Putnam, H. (1967). 'Mathematics without foundations'. *The Journal of Philosophy* 64.1, pp.5–22.

— (1975). 'What is mathematical truth?' In: *Mathematics, Matter and Method: Philosophical Papers, Volume I*. Cambridge: Cambridge University Press, pp.60–78.

— (1977). 'Realism and reason'. *Proceedings and Addresses of the American Philosophical Association* 50.6, pp.483–98.

— (1980). 'Models and reality'. *Journal of Symbolic Logic* 45.3, pp.464–82.

— (1981). *Reason, Truth and History*. Cambridge: Cambridge University Press.

— (1982). 'A defense of internal realism'. In: Putnam 1990, pp.30–42.

— (1983). *Realism and Reason*. Cambridge: Cambridge University Press.

— (1987). *The Many Faces of Realism*. La Salle, Illinois: Open Court.

— (1990). *Realism with a Human Face*. Ed. by J. Conant. London: Harvard University Press.

— (1994). 'Comments and replies'. In: *Reading Putnam*. Oxford: Blackwell, pp.242–95.

Quine, W. v. O. (1946). 'Concatenation as a basis for arithmetic'. *The Journal of Symbolic Logic* 11.4, pp.105–14.

— (1951). 'Ontology and ideology'. *Philosophical Studies* 2.1, pp.11–15.

— (1960). *Word and Object*. Cambridge MA: MIT Press.

— (1969). *Set Theory and Its Logic*. Cambridge: Harvard University Press.

— (1976). 'Grades of discriminability'. *The Journal of Philosophy* 73.5, pp.113–16.

— (1981). *Mathematical Logic*. Revised edition. Cambridge MA: Harvard University Press.

Quinon, P. (2010). 'Le modèle attendu de l'arithmétique. L'argument du théorème de Tennenbaum'. PhD thesis. Institute d'historie et philosophie des sciences et des techniques.

Raatikainen, P. (2008). 'On rules of inference and the meanings of logical constants'. *Analysis* 68.4, pp.282–7.

Rabin, M. O. (1964). 'Non-standard models and independence of the induction axiom'. In: *Essays on the foundations of mathematics*. Jerusalem: Magnes Press, pp.287–99.

Ramsey, F. P. (1931). 'Theories'. In: *The Foundations of Mathematics and Other Logical Essays*. Routledge & Kegan Paul, pp.212–36.

Ranicki, A. (2002). *Algebraic and Geometric Surgery*. Oxford: Oxford University Press.

Rautenberg, W. (2010). *A Concise Introduction to Mathematical Logic*. 3rd ed. New York: Springer.

Rayo, A. and G. Uzquiano (1999). 'Toward a theory of second-order consequence'. *Notre Dame Journal of Formal Logic* 40.3, pp.315–25.

Read, S. (1997). 'Completeness and categoricity: Frege, Gödel and model theory'. *History and Philosophy of Logic* 18, pp.79–93.

Reinhardt, W. N. (1974). 'Remarks on reflection principles, large cardinals, and elementary embeddings'. In: *Axiomatic Set Theory*. Ed. by T. Jech. Providence, RI: American Mathematical Society, pp.189–205.

Resnik, M. D. (1981). 'Mathematics as a science of patterns: Ontology and reference'. *Nous* 15.4, pp.529–50.

— (1997). *Mathematics as a Science of Patterns*. Oxford: Clarendon Press.

Rin, B. G. and S. Walsh (2016). 'Realizability semantics for quantified modal logic: Generalizing Flagg's 1985 construction'. *The Review of Symbolic Logic* 9.4, pp.752–809.

Robinson, A. (1951). *On the Metamathematics of Algebra*. Amsterdam: North Holland.

— (1952). 'On the application of symbolic logic to algebra'. In: *Proceedings of the International Congress of Mathematicians, Cambridge, Mass., 1950, vol. 1*. Providence, RI: American Mathematical Society, pp.686–94.

— (1956a). 'A result on consistency and its application to the theory of definition'. *Indagationes Mathematicae* 18, pp.47–58.

— (1956b). *Complete Theories*. Amsterdam: North-Holland.

— (1961). 'Non-standard analysis'. *Indagationes Mathematicae* 23, pp.432–40.

— (1965). 'Formalism 64'. In: *Proceedings of the International Congress for Logic, Methodology and Philosophy of Science*, pp.228–46.

— (1966). *Non-Standard Analysis*. Amsterdam: North-Holland.

— (1967). 'The metaphysics of the calculus [with discussion]'. In: Lakatos 1967, pp.28–46.

— (1968). 'Model theory'. In: *Contemporary Philosophy. A Survey*. Ed. by R. Klibansky. Vol. 1. Firenze: La nuova Italia, pp.61–73.

Robinson, J. (1949). 'Definability and decision problems in arithmetic'. *The Journal of Symbolic Logic* 14, pp.98–114.

Rosen, G. (2003). 'Platonism, semiplatonism and the Caesar problem'. *Philosophical Books* 44.3, pp.229–44.

Rosendal, C. (2011). 'Descriptive classification theory and separable Banach spaces'. *Notices of the American Mathematical Society* 58.9, pp.1251–62.

Rosenstein, J. G. (1969). '$\aleph_0$-categoricity of linear orderings'. *Fundamenta Mathematicae* 64, pp.1–5.

Rosser, B. (1936). 'Extensions of some theorems of Gödel and Church'. *Journal of Symbolic Logic* 1.3, pp.87–91.

Rowbottom, F. (1964). 'The Łoś conjecture for uncountable theories'. *Notices of the American Mathematical Society* 11, p.248.

Rudolph, D. J. (1990). *Fundamentals of Measurable Dynamics*. New York: The Clarendon Press.

Russell, B. (1927). *The Analysis of Matter*. London: Kegan Paul.

Ryll-Nardzewski, C. (1959). 'On the categoricity in power $\leq \aleph_0$'. *Bulletin de l'Académie Polonaise des Sciences. Série des Sciences Mathématiques, Astronomiques et Physiques* 7, pp.545–8.

Sacks, G. E. (1972). *Saturated Model Theory*. Reading, Massachusetts: W A Benjamin.

Scanlan, M. (2003). 'American postulate theorists and Alfred Tarski'. *History and Philosophy of Logic* 24.4, pp.307–25.

Scanlon, T. (2012). 'A proof of the André–Oort conjecture via mathematical logic [after Pila, Wilkie and Zannier]'. *Astérisque* 348, Exp. No. 1037, ix, 299–315.

Schappacher, N. (2010). 'Rewriting points'. In: *Proceedings of the International Congress of Mathematicians. Volume IV*. Hindustan Book Agency, New Delhi, pp.3258–91.

Scott, D. (1960). 'The notion of rank in set theory'. In: *Summaries of the talks presented at the Summer Institute for Symbolic Logic, Cornell University 1957*. 2nd ed. Princeton: Communications Research Division, Institute for Defense Analysis, pp.267–69.

— (1961). 'Measurable cardinals and constructible sets'. *Bulletin de l'Académie Polonaise des Sciences. Série des Sciences Mathématiques, Astronomiques et Physiques* 9, pp.521–24.

— (1974). 'Axiomatizing set theory'. In: *Axiomatic Set Theory*. Ed. by T. J. Jech. Providence, Rhode Island: American Mathematical Society, pp.207–14.

Scowcroft, P. (2012). 'Review of Button 2011'. Mathematical Reviews MR2785345 (2012e:03005).

Sela, Z. (2006). 'Diophantine geometry over groups. VI. The elementary theory of a free group'. *Geometric and Functional Analysis* 16.3, pp.707–30.

— (2013). 'Diophantine geometry over groups VIII: Stability'. *Annals of Mathematics* 177.3, pp.787–868.

Shapiro, S. (1991). *Foundations without Foundationalism: A Case for Second-Order Logic*. New York: The Clarendon Press.

— (1997). *Philosophy of Mathematics: Structure and Ontology*. New York: Oxford University Press.

— (2005a). 'Logical consequence, proof theory, and model theory'. In: Shapiro 2005b, pp.651–70.

— ed. (2005b). *The Oxford Handbook of Mathematics and Logic*. Oxford: Oxford University Press.

— (2006a). 'Structure and identity'. In: MacBride 2006a, pp.109–45.

— (2006b). 'The governance of identity'. In: MacBride 2006a, pp.164–73.

— (2012). 'Higher-order logic or set theory? A false dilemma'. *Philosophia Mathematica* 20, pp.305–23.

Sheffer, H. M. (1926). 'Review of *Principia Mathematica*'. *Isis* 8.1, pp.226–31.

Shelah, S. (1969). 'Stable theories'. *Israel Journal of Mathematics* 7, pp.187–202.
— (1971a). 'Every two elementarily equivalent models have isomorphic ultrapowers'. *Israel Journal of Mathematics* 10, pp.224–33.
— (1971b). 'Stability, the f.c.p., and superstability; model theoretic properties of formulas in first order theory'. *Annals of Mathematical Logic* 3.3, pp.271–362.
— (1974a). 'Categoricity of uncountable theories'. In: Henkin 1974. Ed. by L. Henkin et al., pp.187–203.
— (1974b). 'Infinite abelian groups, Whitehead problem and some constructions'. *Israel Journal of Mathematics* 18, pp.243–56.
— (1978). *Classification Theory and the Number of Nonisomorphic Models*. Amsterdam: North-Holland.
— (1980). 'Simple unstable theories'. *Annals of Mathematical Logic* 19.3, pp.177–203.
— (1983). 'Classification theory for nonelementary classes. I. The number of uncountable models of $\psi \in L_{\omega_1,\omega}$'. *Israel Journal of Mathematics* 46.3, pp.212–40, 241–73.
— (1985). 'Classification of first order theories which have a structure theorem'. *American Mathematical Society. Bulletin* 12.2, pp.227–32.
— (1987a). 'Classification of nonelementary classes. II. Abstract elementary classes'. In: Baldwin 1987a, pp.419–97.
— (1987b). 'Taxonomy of universal and other classes'. In: *Proceedings of the International Congress of Mathematicians, Vol. 1, 2 (Berkeley, California, 1986)*. Providence, RI: American Mathematical Society, pp.154–62.
— (1990). *Classification Theory and the Number of Nonisomorphic Models*. 2nd ed. Amsterdam: North-Holland.
— (2009a). *Classification Theory for Abstract Elementary Classes*. 2nd ed. London: College Publications.
— (2009b). 'Introduction to: Classification theory for abstract elementary class'. Introduction to Shelah 2009a. URL: arxiv.org/abs/0903.3428v1.
Sher, G. (1991). *The Bounds of Logic: A generalized viewpoint*. Cambridge MA: MIT Press.
— (2001). 'The formal-structural view of logical consequence'. *The Philosophical Review* 110.2, pp.241–61.
— (2008). 'Tarski's thesis'. In: *New Essays on Tarski and Philosophy*. Ed. by D. Patterson. Oxford: Oxford University Press, pp.300–39.
Shoenfield, J. R. (1967). *Mathematical logic*. Reading: Addison-Wesley.
Sider, T. (2011). *Writing the Book of the World*. Oxford: Oxford University Press.
Silver, J. H. (1971). 'Some applications of model theory in set theory'. *Annals of Mathematical Logic* 3.1, pp.45–110.
Simmons, H. (2011). *An Introduction to Category Theory*. Cambridge: Cambridge University Press.
Simpson, S. G. (1988). 'Partial realizations of Hilbert's program'. *The Journal of Symbolic Logic* 53.2, pp.349–63.
— (2009). *Subsystems of Second Order Arithmetic*. 2nd ed. Cambridge: Cambridge University Press.

Simpson, S. G. and K. Yokoyama (2012). 'Reverse mathematics and Peano categoricity'. *Annals of Pure and Applied Logic* 164.3, pp.284–93.

Sinaceur, H. (1991). *Corps et modèles: Essai sur l'histoire de l'algèbre réelle*. Paris: Vrin.

— (1994). 'Calculation, order and continuity'. In: *Real Numbers, Generalizations of the Reals, and Theories of Continua*. Ed. by P. Ehrlich. Dordrecht: Kluwer, pp.191–206.

Skolem, T. (1919). 'Untersuchungen über die Axiome des Klassenkalküls und über Produktations- und Summationsprobleme, welche gewisse Klassen von Aussagen betreffen'. *Videnskapsselskapets Skrifter, I. Matem.-naturv. klasse* 3.

— (1920). 'Logisch-kombinatorische Untersuchungen über die Erfüllbarkeit und Beweisbarkeit mathematischen Sätze nebst einem Theoreme über dichte Mengen'. *Videnskapsselskapets Skrifter, I. Matem.-naturv. klasse I* 4. Translated in part in Heijenoort 1967, pp.1–36.

— (1922). 'Einige Bemerkungen zur axiomatischen Begründung der Mengenlehre'. *Proceedings of 5th Scandinavian Mathematical Congress*. Reprinted in Skolem 1970, pp.217–32.

— (1931). 'Über einige Satzfunktionen in der Arithmetik'. *Skrifter utgitt av Det Norske Videnskaps-Akademi i Oslo* 1.7. Reprinted in Skolem 1970, pp.1–28.

— (1934). 'Über die nicht-charakterisierbarkeit der Zahlenreihe mittels endlich oder abzählbar unendlich vieler Aussagen mit ausschliesslich Zahlenvariablen'. *Fundamenta Mathematicae* 23, pp.150–61.

— (1941). 'Sur la porté du théorème Löwenheim–Skolem'. *Les entretiens de Zurich sur les fondements et la méthode des sciences mathématiques*. Reprinted in Skolem 1970, pp.25–52.

— (1955). 'A critical remark on foundational research'. *Kongelige Norske Videnskabsselskabs Forhandlinger, Trondheim* 28.20. Reprinted in Skolem 1970, pp.100–5.

— (1970). *Selected Works in Logic*. Ed. by J. E. Fenstad. Oslo: Universitetsforlaget.

Slaman, T. A. (2008). 'Global properties of the Turing degrees and the Turing jump'. In: *Computational Prospects of Infinity. Part I. Tutorials*. Ed. by C. Chong et al. Hackensack, NJ: World Scientific, pp.83–101.

Smiley, T. (1996). 'Rejection'. *Analysis* 56.1, pp.1–9.

Smith, S. R. (2015). 'Incomplete understanding of concepts: The case of the derivative'. *Mind* 124.496, pp.1163–99.

Stalnaker, R. (1976). 'Possible worlds'. *Noûs* 10, pp.65–75.

— (2012). *Mere Possibilities: Metaphysical Foundations of Modal Semantics*. Princeton: Princeton University Press.

Steinitz, E. (1910). 'Algebraische Theorie der Körper'. *Journal für die reine und angewandte Mathematik* 137, pp.167–309.

Stone, M. (1936). 'The theory of representations of Boolean algebras'. *Transactions of the American Mathematical Society* 40, pp.37–111.

Strawson, P. F. (1959). *Individuals*. London: Methuen.

— (1966). *The Bounds of Sense: An Essay on Kant's Critique of Pure Reason*. London: Methuen.

Suszko, R. (1971). 'Identity connective and modality'. *Studia Logica* 27, pp.7–41.

Suszko, R. (1975a). 'Abolition of the Fregean axiom'. In: *Logic Colloquium (Boston, Mass., 1972–1973)*. Springer, Berlin, pp.169–239.

— (1975b). 'Remarks on Lukasiewicz's three-valued logic'. *Bulletin of the Section of Logic* 4.3, pp.87–90.

— (1977). 'The Fregean axiom and Polish mathematical logic in the 1920s'. *Studia Logica* 36.4, pp.377–80.

Svenonius, L. (1959). '$\aleph_0$-categoricity in first-order predicate calculus'. *Theoria* 25, pp.82–94.

Szmielew, W. (1955). 'Elementary properties of abelian groups'. *Fundamenta Mathematicae* 41, pp.203–71.

Taimanov, A. D. (1962). 'The characteristics of axiomatizable classes of models'. *Algebra i Logika* 1.4, pp.5–31.

Tait, W. M. (1998). 'Zermelo's conception of set theory and reflection principles'. In: *The Philosophy of Mathematics Today*. Ed. by M. Schirn. Oxford: Clarendon Press, pp.469–83.

Takeuti, G. (1987). *Proof Theory*. 2nd ed. Amsterdam: North-Holland.

Tarski, A. (1931). 'Sur les ensembles définissables de nombres réels I.' *Fundamenta Mathematicae* 17. Citations to translation in Tarski 1983, pp.210–39.

— (1933). 'Pojecie prawdy w jezykach nauk dedukcyjnych'. Translated as 'The Concept of Truth in Formalized Languages' in Tarski 1983.

— (1935a). 'Einige methodologische Untersuchungen über die Definierbarkeit der Begriffe'. *Erkenntnis* 5. Translated as 'Some methodological investigations on the definability of concepts' in Tarski 1983, pp.80–100.

— (1935b). 'Grundzüge des Systemenkalküls'. *Fundamenta Mathematicae* 25. Translated as 'Foundations of the calculus of systems' in Tarski 1983 pp.342–83, pp.503–26.

— (1936). 'O pojęciu wynikania logicznego'. *Przegląd Filozoficzny* 39. Translated as 'On the concept of logical consequence' in Tarski 1983, pp.58–68.

— (1948). *A Decision Method for Elementary Algebra and Geometry*. Santa Monica: RAND Corporation.

— (1952). 'Some notions and methods on the borderline of algebra and metamathematics'. In: *Proceedings of the International Congress of Mathematicians, Cambridge, Mass., 1950, vol. 1*. Providence, RI: American Mathematical Society, pp.705–20.

— (1954). 'Contributions to the theory of models. I'. *Indagationes Mathematicae* 57.

— (1967). *The Completeness of Elementary Algebra and Geometry*. Paris: Centre National de la Recherche Scientifique. Institute Blaise Pascal.

— (1983). *Logic, Semantics, and Metamathematics*. Ed. by J. H. Woodger and J. Corcoran. 2nd ed. Indianapolis: Hackett.

— (1986). 'What are logical notions?' *History and Philosophy of Logic* 7.2. Ed. by J. Corcoran, pp.143–54.

— (1994). *Introduction to Logic and to the Methodology of the Deductive Sciences*. 4th ed. New York: Clarendon Press.

— (2000). 'Address at the Princeton university bicentennial conference on problems of mathematics (December 17–19, 1946)'. *Bulletin of Symbolic Logic* 6.1, pp.1–44.

Tarski, A., A. Mostowski, et al. (1953). *Undecidable Theories*. North-Holland.

Tarski, A. and R. L. Vaught (1958). 'Arithmetical extensions of relational systems'. *Compositio Mathematica* 13, pp.81–102.

Tennenbaum, S. (1959). 'Non-Archimedian models for arithmetic'. *Notices of the American Mathematical Society* 6, p.270.

Tiercelin, C. (2013). 'No pragmatism without realism. Review of Huw Price, *Naturalism without mirrors*'. *Metascience* 22, pp.659–65.

Troelstra, A. S. and D. van Dalen (1988). *Constructivism in Mathematics. An Introduction*. Vol. 2. Amsterdam: Elsevier.

Trueman, R. (2012). 'Neutralism within the semantic tradition'. *Thought* 1.3, pp.246–51.

— (2014). 'A dilemma for neo-Fregeanism'. *Philosophia Mathematica* 22.3, pp.361–79.

Tutte, F., ed. (1984). *Graph Theory*. Menlo Park, California: Addison-Wesley.

Tymoczko, T. (1989). 'In defense of Putnam's brains'. *Philosophical Studies* 57.3, pp.281–97.

Urysohn, P. (1927). 'Sur un espace métrique universel'. *Bulletin des Sciences Mathématiques* 51.2, pp.43–64.

Uzquiano, G. (1999). 'Models of second-order Zermelo set theory'. *Bulletin of Symbolic Logic* 5.3, pp.289–302.

Väänänen, J. (2008). 'How complicated can structures be?' *Nieuw Archief voor Wiskunde*. June, pp.117–21.

— (2012a). 'Lindström's theorem'. In: *Universal Logic: An Anthology*. Ed. by J.-Y. Béziau. Springer, pp.231–6.

— (2012b). 'Second order logic or set theory?' *Bulletin of Symbolic Logic* 18.1, pp.91–121.

Väänänen, J. and T. Wang (2015). 'Internal categoricity in arithmetic and set theory'. *Notre Dame Journal of Formal Logic* 56.1, pp.121–34.

van Ulsen, P. (2000). 'E. W. Beth als logicus'. Doctoral Dissertation, Institute for Logic, Language and Computation, Amsterdam. ILLC Dissertation Series DS–2000–04.

van den Dries, L. (1984). 'Remarks on Tarski's problem concerning ($\mathbf{R}$, +, ·, exp)'. In: *Logic colloquium '82*. Amsterdam: North-Holland, pp.97–121.

— (1998). *Tame Topology and O-Minimal Structures*. Cambridge: Cambridge University Press.

— (2014). 'Lectures on the model theory of valued fields'. In: *Model Theory in Algebra, Analysis and Arithmetic*. Heidelberg: Springer, pp.55–157.

Vaught, R. L. (1954). 'Applications to the Löwenheim–Skolem–Tarski theorem to problems of completeness and decidability'. *Indagationes Mathematicae* 16, pp.467–72.

— (1961). 'Denumerable models of complete theories'. In: *Infinitistic Methods (Proc. Sympos. Foundations of Math., Warsaw, 1959)*. Warsaw: Państwowe Wydawnictwo Naukowe, pp.303–21.

— (1963). 'Models of complete theories'. *Bulletin of the American Mathematical Society* 69, pp.299–313.

— (1986). 'Alfred Tarski's work in model theory'. *The Journal of Symbolic Logic* 51.4, pp.869–82.

Veblen, O. (1904). 'A system of axioms for geometry'. *Transactions of the American Mathematical Society* 5.3, pp.343–84.

Veblen, O. and J. W. Young (1965). *Projective Geometry*. New York: Blaisdell.

Velleman, D. J. (1998). 'Review of Levin 1997'. Mathematical Reviews MR1439801 (98c:03015).

Verbrugge, R. (1993). 'Feasible interpretability'. In: *Arithmetic, proof theory, and computational complexity (Prague, 1991)*. Oxford: Oxford University Press, pp.387–428.

Visser, A. (2006). 'Categories of theories and interpretations'. In: *Logic in Tehran*. Ed. by A. Enayat et al. La Jolla: Association for Symbolic Logic, pp.284–341.

— (2012). 'The second incompleteness theorem and bounded interpretations'. *Studia Logica* 100.1-2, pp.399–418.

Visser, A. and H. M. Friedman (2014). 'When bi-interpretability implies synonymy'. *Logic Group Preprint Series* 320, pp.1–19.

von Neumann, J. (1925). 'Eine Axiomatisierung der Mengenlehre'. *Journal für die Reine und Angewandte Mathematik* 154. Translated in Heijenoort 1967, pp.219–40.

Wallace, J. (1979). 'Only in the context of a sentence do words have any meaning'. In: *Contemporary Perspectives in the Philosophy of Language*. Ed. by P. A. French et al. Minneapolis: University of Minnesota Press, pp.305–25.

Walmsley, J. (2002). 'Categoricity and indefinite extensibility'. *Proceedings of the Aristotelian Society* 102, pp.239–57.

Walsh, S. (2012). 'Comparing Hume's principle, Basic Law V and Peano arithmetic'. *Annals of Pure and Applied Logic* 163, pp.1679–709.

— (2014). 'Logicism, interpretability, and knowledge of arithmetic'. *The Review of Symbolic Logic* 7.1, pp.84–119.

Walsh, S. and S. Ebels-Duggan (2015). 'Relative categoricity and abstraction principles'. *The Review of Symbolic Logic* 8.3, pp.572–606.

Wehmeier, K. F. (forthcoming). 'The proper treatment of variables in predicate logic'. *Linguistics and Philosophy*.

Weil, A. (1946). *Foundations of Algebraic Geometry*. New York: American Mathematical Society.

Weir, A. (2010). *Truth Through Proof*. Oxford: Oxford University Press.

Weston, T. (1976). 'Kreisel, the continuum hypothesis and second order set theory'. *Journal of Philosophical Logic* 5.2, pp.281–98.

Weyl, H. (1910). 'Über die Definitionen der mathematischen Grundbegriffe'. *Mathematisch-naturwissenschaftliche Blätter* 7, pp.93–5, 109–13.

Wilkie, A. J. (1996). 'Model completeness results for expansions of the ordered field of real numbers by restricted Pfaffian functions and the exponential function'. *Journal of the American Mathematical Society* 9.4, pp.1051–94.

Williams, J. R. G. (2007). 'Eligibility and inscrutability'. *Philosophical Review* 116.3, pp.361–99.

Winnie, J. A. (1967). 'The implicit definition of theoretical terms'. *British Journal for the Philosophy of Science* 18.3, pp.223–9.

Worrall, J. (2007). 'Miracles and models: Why reports of the death of structural realism may be exaggerated'. *Royal Institute of Philosophy Supplement* 82.61, pp.125–54.

Wright, C. (1983). *Frege's Conception of Numbers as Objects*. Aberdeen: Aberdeen University Press.

— (1985). 'Skolem and the skeptic'. *Proceedings of the Aristotelian Society* 59, pp.117–37.

— (1992). 'On Putnam's proof that we are not brains in a vat'. *Proceedings of the Aristotelian Society* 92, pp.67–94.

Zahar, E. and J. Worrall (2001). 'Appendix IV'. In: *Poincaré's Philosophy: From Conventionalism to Phenomenology*, by Elie Zahar. Chicago and La Salle, Illinois: Open Court, pp.236–51.

Zarach, A. (1982). 'Unions of ZF^--models which are themselves ZF^--models'. In: *Logic Colloquium '80 (Prague, 1980)*. Amsterdam: North-Holland, pp.315–42.

Zermelo, E. (1930). 'Über Grenzzahlen und Mengenbereiche: Neue Untersuchungen über die Grundlagen der Mengenlehre'. *Fundamenta Mathematicæ* 16, pp.29–47.

— (2010). *Ernst Zermelo: gesammelte Werke. Band I*. Ed. by C. G. F. Heinz-Dieter Ebbinghaus and A. Kanamori. Berlin: Springer.

Ziegler, M. (1980). 'Algebraisch abgeschlossene Gruppen'. In: *Word Problems, II*. Amsterdam: North-Holland, pp.449–576.

— (2013). 'An exposition of Hrushovski's new strongly minimal set'. *Annals of Pure and Applied Logic* 164.12, pp.1507–19.

Zilber, B. (1974). 'On the transcendence rank of formulas in an $\aleph_1$-categorical theory'. *Matematicheskie Zametki* 15. Russian, pp.321–29.

— (1977). 'Groups and rings whose theory is categorical'. *Fundamenta Mathematicae* 95.3, pp.173–88.

— (1980). 'Strongly minimal countably categorical theories'. *Sibirskii Matematicheskii Zhurnal* 21.98–112. Russian.

— (1981). 'Totally transcendental structures and combinatorial geometries'. *Doklady Akademii Nauk SSSR* 259.5. Russian, pp.1039–41.

— (1984a). 'Strongly minimal countably categorical theories'. *Sibirskii Matematicheskii Zhurnal*. Russian. Part II, 25 No. 3 (1984) 71–88; Part III, 25 No. 4 (1984) 63–77.

— (1984b). 'The structure of models of uncountably categorical theories'. In: *Proceedings of the International Congress of Mathematicians (Warsaw, 1983)*. Vol. 1. Warsaw: PWN, pp.359–68.

— (1993). *Uncountably Categorical Theories*. Providence, RI: American Mathematical Society.

— (2010). *Zariski Geometries: Geometry from a Logician's Point of View*. Cambridge: Cambridge University Press.

Index

Index of names

Index of symbols and definitions